THE VEGETATION OF WISCONSIN

THE
VEGETATION
OF WISCONSIN

An Ordination of Plant Communities

JOHN T. CURTIS

THE UNIVERSITY OF WISCONSIN PRESS

Published 1959
The University of Wisconsin Press
Box 1379, Madison, Wisconsin 53701
The University of Wisconsin Press, Ltd.
70 Great Russell Street, London

Copyright © 1959, 1971; in Canada, 1959
The Regents of the University of Wisconsin
All rights reserved

Printings 1959, 1971, 1974, 1978

Printed in the United States of America
ISBN 0-299-01940-3; LC 59-5308

Foreword to the second printing

When *The Vegetation of Wisconsin* was first published in 1959, it aroused controversy among the critics. There was unanimous agreement that the subject was treated with thoroughness, the data were assembled and presented with care, and the writing was graceful and clear. The book was praised as the most comprehensive study of the vegetation of a state ever published, and recognized as being of more than local significance. Many reviewers, like R. Elfyn Hughes in *Ecology*, seeing Curtis' work as a "major contribution to the development of plant ecology," expressed the belief that "the application and elaboration of his methods and concepts elsewhere should do much to advance ecological research."

At the same time, however, other scholars voiced doubts about the basic concepts and methodology of the book. The position they took is indicated by the comment of E. W. Tisdale, writing in the *Journal of Wildlife Management*: "The ordination of vegetation stands along an environmental gradient has much to recommend it in theory. Whether we possess sufficient knowledge of either plants or environment to develop such patterns is another matter. In addition, the analytical methods on which the present work is based will raise doubts in the minds of many students of vegetation. . . . Whether this approach is conducive to sound biological results is open to question."

The principles on which *The Vegetation of Wisconsin* was based did not originate with Curtis. He readily acknowledged the work of many scientists, particularly that of H. A. Gleason, who as early as 1916 had presented the theories on which Curtis based much of his own work. The major contribution of *The Vegetation of Wisconsin* to plant ecology was its documentation—the study and analysis of over 1,400 stands, and the synthesis of these data into a convincing exposition of the relationships of the vegetation of a large area. "The general conclusions," wrote Stanley Cain in *Forest Science,* "carry force not only by the reasonableness of the ideas; the weight of data is convincing. Perhaps for the first time since the turn of the century . . . American ecology again has a school."

As might be expected, the publication of this book was the stimulus for a large amount of research. In the twelve years since its appearance, many papers dealing with the concepts and methods presented in *The Vegetation of Wisconsin* have been published. Methodologies have been refined, especially methods of computation of ordinations based on the rapidly improving capabilities of computers. Acceptance of the basic theories has been widespread. Controversy still abounds, which is good, because it generates still more research and discussion of the mechanisms by which vegetation sorts itself, and ultimately leads to a better understanding of that vegetation.

Professor Curtis died shortly after *The Vegetation of Wisconsin* was published, and this book is a summary of his work and thought. It is tragic that the science of ecology has been denied the benefits that the continuing work of this perceptive ecologist might have produced. But by itself, this work has had a lasting influence on the development of modern ecology. It is an ecological classic.

GRANT COTTAM

Madison, Wisconsin
March 1, 1971

Contents

Part 1
BACKGROUND

	Introduction	3
1	Flora	7
2	Environment	25
3	Plant communities and their distribution	49
4	Vegetation study methods	63

Part 2
SOUTHERN FORESTS

5	Southern forests—general	87
6	Southern forests—mesic	103
7	Southern forests—xeric	132
8	Southern forests—lowland	156

Part 3
NORTHERN FORESTS

9	Northern forests—general	171
10	Northern forest—mesic	184

11	Northern forests—xeric	202
12	Northern forests—lowland	221
13	Boreal forest	243

Part 4
GRASSLANDS

| 14 | Prairie | 261 |
| 15 | Sand barrens and bracken-grassland | 308 |

Part 5
SAVANNA AND SHRUB COMMUNITIES

| 16 | Savanna | 325 |
| 17 | Tall shrub communities | 352 |

Part 6
LESSER COMMUNITIES

18	Fen, meadow, and bog	361
19	Aquatic communities	385
20	Beach, dune, and cliff communities	402
21	Weed communities	412

Part 7
THE VEGETATION AS A WHOLE

22	Postglacial history	437
23	The effect of man on the vegetation	456
24	Interrelations of communities	476

APPENDIX

Tables for Chapter 4	515
Table for Chapter 5	519
Figures and tables for Chapter 6	519
Figures and tables for Chapter 7	522
Figures and tables for Chapter 8	528

Figures and tables for Chapter 10 533
Figures and tables for Chapter 11 536
Figures and tables for Chapter 12 541
Figures and tables for Chapter 13 548
Figures and tables for Chapter 14 552
Figures and tables for Chapter 15 562
Figures and tables for Chapter 16 566
Figures and tables for Chapter 17 575
Figures and tables for Chapter 18 578
Figures and tables for Chapter 19 584
Figures and tables for Chapter 20 587
Figure and tables for Chapter 21 590
Figure for Chapter 22 596

Glossary, 599
Bibliography, 604
Species list, 633
Index, 645

Illustrations

LIST OF PLATES

following page 92
1 Residual soil over limestone
2 Varying thickness of loess blanket
3 Colony of Dutchman's breeches *(Dicentra cucullaria)*
4 Spring-beauty *(Claytonia virginica)*
5 Bloodroot *(Sanguinaria canadensis)*
6 Hepatica *(Hepatica acutiloba)*
7 Southern mesic forest in midsummer
8 Southern mesic forest in early spring
9 Circle of basswood trunks
10 Wyalusing Scientific Area

following page 158
11 White and black oak forest
12 Red oak forest
13 Open-grown oak
14 Southern lowland forest in dry spring
15 Flooded southern lowland forest
16 Summer view of forest
17 Hemlock forest
18 Mature white pine—red pine forest

following page 208
19 Young stand of mixed jack pine and red pine
20 Mature stand of red pine

21 Second-growth stand of aspen, white birch, and red oak
22 Bunchberry *(Cornus canadensis)*

following page 234
23 Tamarack *(Larix laricina)* and black spruce *(Picea mariana)*
24 Interior of northern wet-mesic forest
25 Bog mat at Devil's Track Lake (1940)
26 Devil's Track Lake (1957)

following page 246
27 Mature stand of white spruce and balsam fir at Bailey's Harbor
28 Young stand under mature aspens
29 Intermediate stand of spruce and fir at High Lake Scientific Area
30 Young stand under maturing jack pine

following page 276
31 Dry prairie on steep hillside
32 Dry prairie landscape
33 Pasque flower *(Anemone patens)*
34 Compass plant *(Silphium laciniatum)*

ILLUSTRATIONS

following page 292
35 View of mesic prairie with compass plant
36 Big bluestem (*Andropogon gerardi*) along path
37 Moss hummocks
38 Desert pavement with hudsonia (*Hudsonia tomentosa*)
39 View of blowout
40 Exposed roots of grass
41 Hudsonia in stabilized area
42 Bracken-grassland

following page 356
43 Oak opening
44 Bur oak grub
45 Pine barrens
46 Cedar glade
47 Dense cedar glade
48 Creeping juniper (*Juniperus horizontalis*)
49 Southern sedge meadow
50 Hummocks in a grazed meadow

following page 388
51 Northern sedge meadow
52 Open bog
53 Shaded sandstone cliff
54 Exposed sandstone cliff
55 Relic stand of white pines
56 Prickly pear cactus (*Opuntia compressa*)
57 Ferns on sandstone cliff
58 Beach at Trout Lake

following page 406
59 Beach at Bailey's Harbor
60 Beachgrass (*Ammophila breviligulata*) on dunes
61 Creeping juniper on dune
62 Evergreens on dunes

following page 432
63 Grazed and ungrazed prairie
64 Grazed and ungrazed forest
65 Model of three-dimensional ordination of red oak
66 Dry-mesic prairie

LIST OF FIGURES

1 Range maps, legume species, 17
2 Range maps, orchid species, 17
3 Range maps, groups of species, 18
4 Range maps, boundaries for groups of species, 19
5 Tension zone (indicated on county map), 20
6 Climatic isorithms at tension zone, 36
7 Wyalusing maps, 89
8 Red oak on southern upland continuum, 96
9 Other tree species of southern uplands, 97
10 Tree curves, combined southern ordination, 99
11 Tree curves, bimodal distributions, 100
12 Curves for major tree species on combined ordination, northern forest, 180
13 Curves for lesser tree species on combined ordination, northern forest, 181
14 Bimodal curves for two lesser tree species on combined ordination, northern forest, 181
15 Curves for groundlayer species on combined ordination, northern forest, 182
16 Curves for major trees, boreal forest, 248
17 Curves for prairie grasses, 267
18 Curves for prairie forbs, 267
19 Curves for savanna trees, 328
20 Curves for submerged aquatic species, 398
21 Typical pollen profiles, 446
22 Map of vegetation, 12,000 years ago, 449
23 Map of vegetation, 3,500 years ago, 451
24 Map of vegetation, 500 years ago, 453
25 Map of postglacial vegetational changes, 454
26 Map of Plum Lake township (1858), 470
27 Map of Plum Lake township (1929), 470
28 Map of Plum Lake township (1951), 470

29 Red oak in two-dimensional ordination, 484
30 Major trees, southern upland ordination, 484
31 Major trees, northern upland ordination, 486
32 Major trees, boreal forest ordination, 486
33 Community ordination, (A-B) axes, 490
34 Community ordination, (B-C) axes, 491
35 Community relationships, 492
36 July temperatures, 493
37 Annual temperature, 493
38 Annual snowfall, 493
39 Light intensity, 493
40 Soil pH, 495
41 Soil water-retaining capacity, 495
42 Amount of evergreen species, 496
43 Amount of shrub species, 496
44 Fern family distribution, 497
45 Composite family distribution, 497
46 Grass family distribution, 497
47 Ericad family distribution, 497
48 Community ordination; balsam fir, sugar maple, white birch and shagbark hickory, 498
49 Community ordination; white ash, white pine, white oak, and basswood, 499
50 Ordination for groundlayer species, 500
51 Ordination for groundlayer species, 501
52 Frequency distribution diagram, 507
53 Relation between presence and ubiquity, 508

LIST OF TEXT TABLES

1 Plants endemic in the Driftless Area, 14
2 Climatic factors in two floristic provinces, 37
3 Major plant communities at time of settlement, 61
4 Trees at surveyors' witness corners, 91
5 Variation in seed yield of sugar maple, 105
6 Reproduction rates of herbs in Green County, 115
7 Stand variability, Green County mesic forests, 120
8 Region variability, 121
9 Comparisons with midwestern forests, 127
10 Variation in acorn crop, 138
11 Adaptation numbers for northern forests, 179
12 Presence per cent of species pairs in mesic forests, 192
13 Geographical relations of conifer swamps, 240
14 Geographical relations of boreal forest, 256
15 Indicator species for five segments of prairie gradient, 266
16 Aggregation of mesic prairie plants, 281
17 Geographical relations of the prairies, 290
18 Relation of oak opening species to light intensity, 333
19 Differential species of sedge meadows, 368
20 Emergent aquatics by water hardness classes, 390
21 Joint occurrence groups of submerged aquatics, 397
22 Weed species on soils of same texture, 420
23 Weeds that react to soil texture, 421
24 Summary of weed communities, 421
25 Composition of grazed dry prairies, 429
26 Composition of Vilas County forest versus surface pollen composition, 441
27 Correction factors for pollen amounts, 442
28 Key for Figure 33 and other ordination diagrams, 490
29 Key to herb ordination diagrams in Figures 50 and 51, 502
30 Ubiquitous groundlayer species of terrestrial communities, 506

Acknowledgments

Perhaps no book is ever wholly original, but this one is extraordinarily dependent upon the work of others. Proper acknowledgment of my obligation to others is made difficult because it is so great. The book would have been impossible without the willing efforts and stimulating ideas of many graduate students who studied individual communities. All of them are named in the chapters which incorporate their findings. To them, my sincere thanks. Appreciation is also expressed to the many graduate students who studied other phases of ecology and contributed their views freely in the many discussions concerning the nature of vegetation, the methods for its study, and the interrelations of plant communities.

The research program of the Plant Ecology Laboratory of the University of Wisconsin during the investigation periods reported herein has received support from a number of sources. The major portion of the work was financed by grants from the University Research Committee, using funds provided by the Wisconsin Alumni Research Foundation. Without this aid, the task would have been impossible. My sincere thanks are extended to the Committee and to the Foundation for their continued interest and support. Additional financial aid for various phases of the program was received in grants from the Office of Naval Research, the National Science Foundation, the Wisconsin Conservation Department, and the Lederle Corporation. I began the initial study in 1942 during the tenure of a

John Simon Guggenheim Memorial Foundation Fellowship. Unfortunately, this investigation was interrupted by World War II. Upon completion of the investigations in 1956, opportunity was afforded for preparation of the manuscript by a renewal of the Guggenheim Fellowship, for which I am deeply grateful. Additional support was generously provided by the University Research Committee and a leave of absence was graciously arranged by Dean M. H. Ingraham, of the College of Letters and Science.

Among my colleagues of the University of Wisconsin Botany Department, Professor Grant Cottam and Dr. Henry C. Greene have played vital roles throughout the decade of study and the final preparation of the manuscript. Dr. Greene's thorough familiarity with the flora of Wisconsin and his almost uncanny ability to identify plants in their vegetative condition were of inestimable value in all phases of field sampling. Furthermore, his constant interest and willing assistance on numerous field trips permitted a degree of completeness in the study of many plant communities that would otherwise have been unattainable. Professor Cottam contributed freely of his ideas, his time, and his labor from start to finish. Especially deserving to be singled out were his efforts at improving methods of field study. Without the timesaving sampling techniques he devised, the present study would have been wholly impossible. Nearly as valuable has been his constant helpful advice during preparation of the manuscript. To Grant Cottam and Henry Greene, my heartfelt thanks. Other colleagues have also been very helpful in various phases of the field work and in manuscript revision. Professors J. W. Thomson and Hugh Iltis have given aid in the identification of difficult species, and Professors Paul B. Sears, A. D. Hasler, D. A. Baerreis, F. D. Hole, and Jonathan Sauer have offered valued criticisms of the chapters in their specialties.

Grateful appreciation is expressed to Miss Shirley Unferth and Miss Karen Valesh for secretarial assistance in the preparation of the manuscript, and to Mr. Langdon Divers for his aid in much of the photographic work. Finally, I am indebted to my wife, Jane Ann, for much help with all phases of the book, especially with the preparation of the indexes.

J.T.C.

Madison, Wisconsin
January, 1958

Part 1

Background

Introduction

The vegetation of a region consists of the total of the plants growing on its soils and in its waters. These plants are present as populations of individual species, which occur in mixtures of various sorts called plant communities. Each community is characterized by a particular structure and appearance, imparted by the numerical proportions of the particular species which compose it. Each of the component species has certain limits to the environmental variables within which it will thrive. Those species which have similar limits tend to grow together, but since the number of environmental factors which may influence the growth of plants is so very large, no two species have exactly the same limits. As a result, the communities which they form are not precise entities of fixed and unvarying composition, but rather are loose aggregations of species, whose makeup changes from place to place and from time to time in a more or less continuous fashion. The communities, therefore, and the entire vegetation which they compose, cannot be described in the exact language of physical science, but must be treated in a statistical manner as a continuous variable.

The vegetation needs to be described regardless of these difficulties in the procedure, however, since it is the major recipient of climatic forces and the prime agent in the conversion and modification of these forces, in the regulation of stream flow, stabilization of water tables, and in the production and maintenance of soils.

These interrelations between vegetation and environment have long been recognized in Wisconsin and were ably expounded by T. C. Chamberlin as long ago as 1877:

> The most reliable natural indications of the agricultural capabilities of a district are to be found in its native vegetation. The natural flora may be regarded as the result of nature's experiments in crop raising through the thousands of years that have elapsed since the region became covered with vegetation. If we set aside the inherent nature of the several plants, the native vegetation may be regarded as the natural correlation of the combined agricultural influences of soil, climate, topography, drainage and underlying formations and their effect upon it. To determine the exact character of each of these agencies independently is a work of no little difficulty; and then to compare and combine their respective influences upon vegetation presents very great additional difficulty. But the experiments of nature furnish us in the native flora a practical correlation of them. The native vegetation therefore merits a careful consideration, none the less so because it is rapidly disappearing and a record of it will be valuable historically.

Knapp, Lapham, Chamberlin, and other early Wisconsin natural scientists made an excellent beginning to our understanding of the nature, distribution, and complexities of the local vegetation. Their work was added to by Bruncken, Stout, and others in the early 1900's, and by a rapidly growing number of investigators after 1930. A definite attempt was begun at the Plant Ecology Laboratory of the University of Wisconsin in 1946 to survey the entire vegetation of the state and to learn the geographical limits, species compositions, and as much as possible of the environmental relations of the communities composing that vegetation. The present book is a summary of a decade of research directed toward that goal. It endeavors to integrate and interpret the findings and to incorporate them into the results reported by earlier investigators. It should not be regarded as a complete and final statement of the vegetation of Wisconsin, but merely as a summary of our current knowledge.

In the manifold field of conservation, the practical land manager is coming to realize that most of his activities are directed at the control and manipulation of plant communities, whether his apparent interests lie in forests, game, fish, or soils. This book is intended to help the conservationist toward his goals by providing him with a better statement of the nature of unmanaged communities than has been readily available. It is also directed to all others interested in the land and its products, be they aesthetic or practical, horticultural or recreational, or otherwise. It has been written with a minimum of technical terms, the few employed being defined at the place of first use and also in a glossary at the end of the book. One exception to this de-emphasis of technical words concerns the use of scientific names of plants. Most readers are aware that each kind of plant has only one correct name which consists of an internationally understood Latin binomial. Common or vernacular names of plants are not standardized. Frequently the same common name is applied to more than one kind of plant. Furthermore, the majority of the plants of Wisconsin have never received common names because of their rarity or inconspicuousness. As a result, it has not been possible to use common names throughout the present work. Instead, the following procedure has been adopted. Well-known trees are referred to by standardized common or English names. Scientific names are provided for these in Table IV-1 in the Appendix. In addition, the important species of each forest type are identified by both common and Latin names at their first mention in each of the chapters dealing with forests, and also whenever there is an extended discussion of a particular species in the text. Many of the truly familiar herbs and shrubs are referred to by their common names in the text, followed by their scientific names in parentheses. In the tables, however, only the scientific names are used, since these tabular lists usually contain a number of species which lack vernacular epithets. The nomenclature throughout the work follows that of Gleason in Britton and Brown (1950) see page 78, Chapter 4.

The general plan of the work consists of several introductory chapters on the flora, the environment, the nature and distribution of the vegetation, and the methods employed in the study. These are followed by a number of chapters which take up single communities or groups of communities. These employ a relatively standardized form, within the limitations imposed by the structure of the particular community. In these chapters the openings include maps showing the locations of the communities studied. For the convenience of the reader many of these maps are in the Appendix as well. In each there is an account of location, size, composition, environmental relations, geographical relations, and current use. The supporting tabular data are gathered together at the end of the book along with a brief community summary.* The recent geological history of the vegetation and the influence of man are discussed in two further chapters.

The last chapter deals with the interrelations of the various

* Text plates, figures, and tables are numbered consecutively throughout the book in Arabic numerals. Figures and tables appearing in the appendix are grouped by chapters and numbered with both a Roman numeral for chapter number and an Arabic numeral for chapter sequence.

communities and is especially concerned with the behavior of individual species in the full range of communities. It is distinctly more technical than the remainder of the book and may be omitted by those who are more interested in the applications of vegetational knowledge than in the basic nature of a plant community. The numerous references to studies on related communities in Europe and other foreign regions are included in the earlier chapters as a guide to students who wish to delve more deeply into the world literature on a particular vegetation type.

The vegetational photographs were taken by the author and have never before been reproduced.

At many places in the text, plant communities or other features are located within Wisconsin by reference to one or more counties. These governmental units serve to divide the state into an irregular grid. They may be located from the map on the endpapers which shows the name of each county.

CHAPTER 1

Flora

SOURCES AND PRESENT AFFINITIES OF THE FLORA

Any account of the vegetation of a region must take the flora of the region into consideration. Vegetation and flora differ from each other in a quantitative way. The flora is the total list of plant species present, irrespective of the numerical abundance of each species; one individual of an ultra-rare species can contribute as much to the floral list as a hundred million individuals of a ubiquitous species. The vegetation, on the other hand, has to do with combinations of species present in a given region and with the relative abundance of each species. For this purpose, the common or dominant species are far more important than the rare species; the presence or absence of the latter is of little significance.

Obviously, since the vegetation is made up of combinations of species, any attempt at description on a floristic basis calls for a detailed knowledge of the flora on the part of the investigator because he must be able to identify the species he sees in the field. Beyond this, however, a knowledge of the flora also may be of great value in giving an understanding of the origin and history of the vegetation. This knowledge is important because the geographical relations of the individual species are becoming well understood, in North America at least, whereas the geographical distribution of no single vegetation type in the same region has ever been adequately described.

Wisconsin is fortunately one of the few states whose flora is well known from this important geographic viewpoint. The information available, naturally, is not complete, but it is sufficiently complete to provide firm groundwork for vegetation studies. This fortunate situation is largely due to the efforts of N. C. Fassett of the University of Wisconsin.

The flora of Wisconsin can be divided into a number of elements, each of which shares a common type of current geographical range. Thus the Boreal element is made up of those species whose existing ranges extend from Alaska or far-northwest Canada through the northern Great Lakes region to the St. Lawrence, with a narrow extension down the Allegheny Mountains at high elevations. Many of the species are actually circumboreal, as they occur in a similar climatic belt across Eurasia (Hulten, 1937). Included in the Boreal element are trees such as tamarack (*Larix laricina*), white spruce (*Picea glauca*), and balsam fir (*Abies balsamea*), shrubs such as Labrador tea (*Ledum groenlandicum*), bearberry (*Arctostaphylos uva-ursi*), mountain cranberry (*Vaccinium vitis-idaea*), bog birch (*Betula glandulosa*), and sweet gale (*Myrica gale*), and herbs such as bunchberry (*Cornus canadensis*), twinflower (*Linnaea borealis*), shinleaf (*Pyrola secunda*), *Moneses uniflora*, bog bean (*Menyanthes trifoliata*), and many others.

Perhaps the most important floristic element in Wisconsin is the so-called Alleghenian. This group of species actually centers in the Cumberland and Great Smoky Mountains of the southern Appalachians and ranges through the area above the fall line from the Mississippi Embayment around through Georgia, up through the eastern states to New England, thence westward to the vicinity of Lake Superior, and then south near the Mississippi River. This region contains the heartland of the deciduous forest and its species are largely members of that forest. Included are widespread trees such as white pine (*Pinus strobus*), hemlock (*Tsuga canadensis*), basswood (*Tilia americana*), sugar maple (*Acer saccharum*), white oak (*Quercus alba*), yellow birch

(*Betula lutea*), white ash (*Fraxinus americana*), and ironwood (*Ostrya virginiana*), shrubs such as gray dogwood (*Cornus racemosa*), and hazelnut (*Corylus americana*), and familiar woodland herbs such as bloodroot (*Sanguinaria canadensis*), spring-beauty (*Claytonia virginica*), Dutchman's breeches (*Dicentra cucullaria*), trillium (*Trillium grandiflorum*), may apple (*Podophyllum peltatum*), trout lily (*Erythronium americanum*), and blue cohosh (*Caulophyllum thalictroides*).

The Alleghenian floristic element is a very ancient one, extending back to Tertiary times when it was part of a very widespread Arcto-tertiary flora which circled the globe in regions currently in mid- and high-latitudes (Cain, 1943). Portions of this world-wide flora persisted in the Appalachian region of North America and also in the mountains of southeast China, but most of it was wiped out in the remaining area. The former continuity is emphasized by the existence today of many of the same species both in China and in southeastern United States and by many other pairs of very closely related species, with one member of the pair in China and one in the United States (Cain, 1944). Included in the genera showing this bihemispheric distribution are the following from Wisconsin: *Thuja, Menispermum, Gymnocladus, Nyssa, Jeffersonia,* and *Caulophyllum,* while the following species have similar ranges: skunk cabbage (*Symplocarpus foetidus*), interrupted fern (*Osmunda claytoniana*), black walnut (*Juglans nigra*), and ramshead ladyslipper (*Cypripedium arietinum*).

The Ozarkian element is closely related to the Alleghenian and is probably of Tertiary origin also. It differs in that its center is in the Ozark Mountains of Arkansas and Missouri, with a range extending south and east to the fall line, west to the arid plains, and north to Minnesota and Wisconsin. Many of the genera are the same as those in the Alleghenian element but the species have become differentiated through long isolation. The Ozark flora on the whole is better adapted to drought conditions than its more eastern relative, and is thus better suited to conditions on the prairie-forest border of the mid-continent. Black oak (*Quercus velutina*), bur oak (*Q. macrocarpa*), chinquapin oak (*Q. muhlenbergii* Engelm), black maple (*Acer saccharum* var. *nigrum*), and the hickories are typical trees of this group in Wisconsin, while bladdernut (*Staphylea trifoliata*), redroot (*Ceanothus americanus*), blue phlox (*Phlox divaricata* var. *laphamii*), rue anemone (*Anemonella thalictroides*), and wild-indigo (*Baptisia leucophaea*) are characteristic groundlayer species.

The Prairie element is made up of species whose range includes all or part of the existing prairies. For the most part, Wisconsin plants of this group have ranges from the Prairie Provinces of Canada south to Wyoming, Nebraska or Kansas and from Indiana west to Colorado. A number of these plants also occur throughout a much larger region to the south and west. Characteristic members are needlegrass (*Stipa spartea*), side-oats grama grass (*Bouteloua curtipendula*), pomme de prairie (*Psoralea esculenta*), Missouri goldenrod (*Solidago missouriensis*), purple prairie clover (*Petalostemum purpureum*), and lead plant (*Amorpha canescens*). A number of species of the prairie element, like prairie sagewort (*Artemisia frigida*), pasque flower (*Anemone patens*), and June grass (*Koeleria cristata*), occur on the steppes of central Eurasia as well as on the North American grasslands.

The remaining floristic element of importance is the Coastal Plain group (Peattie, 1922; McLaughlin, 1932). For the most part, its members are of rare and local occurrence in Wisconsin and they contribute but little to the character of the vegetation. A few, however, reach significant dominance in special locations. The element has a mass range along the Atlantic Coastal Plain, from the seashore to the fall line, and from Massachusetts to Texas, with a prominent northward extension up the Mississippi Embayment to southern Illinois. It is from this last region that the important members of the group have reached Wisconsin. Included in the group are river birch (*Betula nigra*), honey locust (*Gleditsia triacanthos*), and partridge-pea (*Cassia fasciculata*).

The rest of the native flora of Wisconsin can be considered epibiotic, in the sense that it contributes almost nothing to the major biotic communities and exists as isolated or relic populations in unusual habitats like rock cliffs and lake shores. These rare plants do have geographical affinities with other regions and can be considered as minor floristic elements. Thus there is a trace of an Arctic-Alpine element, with such species as *Rhododendron lapponicum,* existing only as a few individuals in one location, *Euphrasia artica,* and *Pinguicula vulgaris,* both of rare occurrence on the Lake Superior shore. A few species may be ascribed to the Western Mountain element, including *Rubus parviflorus, Goodyera decipiens, Asplenium viride, Adenocaulon bicolor, Osmorhiza divaricata,* and possibly, *Dodecatheon amethystinum.* The existence of a local Preglacial element which survived in the Driftless Area is discussed in more detail below.

In addition to these groups of native plants of different geographical affinities, there is a large segment of our existing flora, exotic in origin, which has entered Wisconsin as a result of man's activities. The most important portion of this group comprises the weeds. As will be seen later, the majority of the weeds are of Eurasian origin, although some are of tropical derivation, and another important group comes from the plains and deserts of western United States. As in the case of native species, the place of origin of the weeds is of great importance in determining the type of habitat they will occupy in Wisconsin.

POSTGLACIAL MIGRATIONS

The bulk of the land surface of Wisconsin was covered by one or more of the continental ice sheets. In their passage over the state, the glaciers destroyed all existing vegetational cover. Upon their retreat, they left an assorted blanket of glacial till, outwash sand, or other finely comminuted sediments. This blanket was essentially mesic in its water-holding properties and must have been a favorable surface for immediate occupation by many plants. As more and more of the state was uncovered by the melting ice, greater and greater opportunities for inward migration of various floristic elements arose. The exact sequence of this invasion is in doubt; such information as is available will be discussed in detail in a later chapter. Here it is essential to point out only the various migration routes that may have been used by the invading flora.

The Prairie element, the Ozarkian element, and the Mississippi-Coastal Plain element could have advanced directly from the south and west, either from the Driftless Area or from regions beyond it. The river valleys and their adjacent hills provided a variety of continuous terrains that could have been occupied in a direct, unbroken sequence, without calling on any special mechanisms of seed dispersal. The north-south orientation of Lake Michigan, however, apparently provided a barrier to similar migration from the east. It has been postulated that some of the Alleghenian element entered Wisconsin by the northern route around the upper end of Lake Michigan while other species, like the beech (*Fagus grandifolia*), moved both around the north and the south and converged on the state in a sort of pincers movement. This supposition overlooks two facts of great significance; first, that fossil beech pollen has been found in peat deposits of southern and western Wisconsin and northern Iowa while fossil hemlock pollen occurs in Driftless Area bogs, and second, that Lake Michigan had dwindled to scarcely more than a river with a few widespreads or lakes along its length at the time of Glacial Lake Chippewa in

3000 B.C. There would, therefore, have been ample opportunity for a direct westward march of the Alleghenian element and the fossil record indicates that this may actually have occurred.

In view of the presence today of the full range of floristic groups in the Driftless Area, the most reasonable hypothesis is that all of our major community dominants survived the glacial advances in or near the non-glaciated region and subsequently spread from that center. Any other view is more tenuous and should be accompanied by definite proof of an alternative migration route. In some of the more specialized communities, however, it appears probable that their floras are the result of an inward, postglacial migration. This seems particularly clear in the case of the beach flora (Chapter 20), most of whose members are found on the ocean shores (*Cakile edentula, Ammophila breviligulata, Lathyrus maritimus,* etc.). These plants could find congenial habitat conditions only on the shores of large bodies of water and beyond doubt would not be able to exist in the Driftless Area. It has been supposed (McLaughlin, 1932) that they migrated along a continuous pathway up the Mohawk outlet of certain stages of the Glacial Great Lakes or up the St. Lawrence at a time when the sea extended much farther inland than it now does. It is true that their habitats favor long distance dissemination by ducks or shore birds. The varietal endemism in *Cakile* and *Lathyrus* indicated by distinct morphological sub-speciation along the Great Lakes, however, points toward the former conclusion as the correct one, since there appears to be no reason why birds would not continue to carry the seeds and thus prevent maintenance of isolated populations which would be needed to build up the differentiation.

PREGLACIAL RELICS

In the southwestern part of Wisconsin lies an extensive area that has never been covered by glacial ice, although it is surrounded by lands that have been glaciated. Upon the usual glacial map it appears as an island, but in actual fact it never was surrounded completely by ice at any one time, because the glaciers to the east and to the west of it were present in different epochs. This area, the "Driftless Area," thus was open to direct colonization by plants at all times, but the fact that it was available is in itself no proof that plants actually did exist there during glacial times. The entire question of periglacial climates is involved. Without doubt the land surfaces immediately adjacent to the ice front were influenced by the temperatures of the ice, but considerable uncertainty exists as to the distance over which such an effect might persist (Braun, 1951). Some portions of the Driftless Area were at least 100 miles from the ice front, regardless of which sheet was present, and this distance is thought by many geologists to be far removed from direct effects of the ice. If recent theories on the climate necessary for glacier formation are correct, then the prevailing temperatures of the region may have been similar to those existing now or were perhaps slightly warmer. Wolfe (1951) reached similar conclusions concerning the climate at the glacial border in Ohio. Under such conditions of a warm, moist, growing season there is every reason to believe that a vegetational cover existed over most of the Driftless Area and furthermore, that the composition of the vegetation was very similar to that found today.

Unfortunately, no fossil record of any kind has been discovered which would prove or disprove this conclusion. N. C. Fassett of the University of Wisconsin was deeply interested in the problem and from 1925 to 1950 devoted much effort to the study of the flora of the Driftless Area (Fassett, 1931, 1943). He and his students succeeded in finding a number of species and varieties that were either endemic to the region, or were found there and in other places beyond the edge of the continental glaciation, but not in places which had been covered by ice. Among them were the species listed in Table 1. The rather remarkable distributions of *Aconitum noveboracense* and *Dodecatheon amethystinum* are best explained by assuming that the plants have persisted within the Driftless Area from some past time when their range was continuous through the area now covered with drift. Both of these species and a number of others on the list have rather specialized habitat requirements, growing best on shaded, dripping cliffs. Some investigators have considered that the plants are confined to the Driftless Area because of the prevalence there of these cliffs. Such geological formations are naturally more common in areas not subject to glaciation but they are also present in glaciated regions. In addition, other species on the list such as *Muhlenbergia cuspidata* and *Psoralea esculenta* are prairie plants of open places and are in no way dependent upon special rock formations so the explanation would not hold for them. It would appear, therefore, that these species rightly may be called preglacial relics.

One of the peculiarities of the plants in Table I is that they have remained confined to the Driftless Area. Of course, this characteristic was instrumental in their original discovery but the point remains that they have shown an extraordinary tendency to a sedentary ex-

Table 1
Plants endemic in the Driftless Area or whose range in Wisconsin is restricted to that region

Aconitum noveboracense var. quasiciliatum
Adoxa moschatellina
Anemone caroliniana
Asplenium platyneuron
Azolla caroliniana
Callirhoe triangulata
Cassia marilandica
Cheilanthes feei
Chrysopsis villosa
Commelina erecta var. greenei
Dasistoma macrophylla
Dodecatheon amethystinum
Elatine triandra
Eleocharis wolfii
Elymus riparius
Gerardia skinneriana
Gnaphalium saxicola
Hibiscus militaris
Hypericum mutilum
Linaria canadensis
Lycopodium selago var. patens
Muhlenbergia cuspidata
Myosotis laxa
Onosmodium occidentale
Pellaea atropurpurea
Penstemon gracilis var. wisconsinensis
Psoralea esculenta
Rhamnus lanceolatus
Rhododendron lapponicum
Rhus trilobata
Solidago castrensis
Solidago sciaphila
Sullivantia reniformis
Talinum rugospermum

istence. Many other species may have persisted similarly within the Driftless Area during glaciation but have long since spread out and are no longer recognizable as to origin. A few species have rather small total ranges which center on the Area and are presumed to have spread in postglacial times. Included here is Hill's oak (*Quercus ellipsoidalis*).

On the basis of the relative level in peat bogs at which certain fossil pollens appeared, Sears (1942a) postulated that several trees moved eastward from a *refugium* in the Driftless Area. He included hickory and basswood in this group. E. Lucy Braun (1950) considered that the maple-basswood forest community persisted in the Driftless Area throughout glacial times, while O. Anderson (1954) concluded as a result of his studies of dry prairies that such communities were present as preglacial relics throughout the Area. Hansen (1939, see Chapter 22) found fossil pollen of mixed hardwoods all the way to the bottom of bogs he examined within the Area, but these bogs are all of postglacial origin and do not definitely indicate the nature of the vegetation during glaciation.

A variety of geological, climatological, and ecological evidence, therefore, strongly supports the hypothesis that the Driftless Area was at least partially covered with vegetation at all times and that it formed the source for the bulk of the plant cover which later spread out over the remaining parts of the state as these were deglaciated.

TENSION ZONE AND FLORISTIC PROVINCES

The various floristic elements are not distributed uniformly over the entire state. Rather, there are two distinct floristic provinces. One of these, in the southwest half of the state, contains the Prairie element, and the other, in the northeast half, contains the Boreal element. The Alleghenian element is generally distributed throughout both provinces. The two provinces have been called the prairie-forest province and the northern hardwoods province, respectively (Curtis and McIntosh, 1951).

The two provinces are separated by a narrow band or zone which contains some members of each. The location of this zone has been recognized for many years. Knapp, in 1871, published a map of Wisconsin on which he showed the northern limit for the successful cultivation of dent corn and Concord grape. This line passed from St. Croix Falls in the northwest to Chippewa Falls, Wisconsin Rapids, Berlin, and Union Grove in the southeast. His paper attempted to explain the position of the line by the relative position of polar and subtropical air masses, with consequent differences in climate on the two sides of the line. L. S. Cheney, in 1894, again defined the line or zone and pointed out the major differences in vegetation in the two provinces.

A more precise method for locating the boundaries of floristic provinces was proposed by Clements in 1905 following the earlier studies of Livingston (1903). He suggested, "The limiting line or ecotone of a . . . province is a composite obtained from the limits of principal species and checked by the limits of species typical of the contiguous vegetations." To put this method into practice, it is necessary to have accurate range maps for the principal species. In Wisconsin, we are fortunate to possess such maps for the trees. They were prepared by L. S. Cheney in the course of his travels through the state in connection with a wheat rust control program, which took him to nearly all townships. He intended to prepare a book on the trees of Wisconsin and so he took complete notes on the species of trees that he saw wherever he went. From these notes, he made maps which show not only the range but also the relative abundance of the species.

The situation is not so satisfactory for the non-tree species. The *Preliminary Reports on the Flora of Wisconsin* prepared by Fassett and his students, which were begun in 1929 and continued through

1953, contain distribution maps for all members of families so far treated, but they are based solely on herbarium specimens. Taxonomists, like other citizens, are more impressed with the unusual than the common. As a result, herbaria tend to have an excess of specimens from the borders of a range where a plant is rare and unusual but little representation from areas where the same plant is to be seen at every turn. A more usable map could be prepared if some indication of actual abundance were available, but such information is possible only on the basis of sight records and these are understandably not acceptable for taxonomic purposes.

Within these limitations, however, it is possible to prepare composite maps which show the approximate range limits of a large number of species. When these maps are drawn for objectively chosen groups of species, such as all trees, all orchids, or all members of other families or genera, a great concentration of range limits fall in a narrow band in nearly the exact position described by Knapp 80 years ago. This band includes the southern limits of many northern species and the northern limits of many southern species and conforms to the criterion for provincial boundaries set by Clements.

A selection of the many maps that have been prepared is given in Figures 1 and 2, while summation maps are shown in Figures 3, 4, and 5. This tension zone (Griggs, 1914) also marks the range limits of many animals, including such birds as the bobwhite and the mourning dove, and a number of fishes, including the central stone roller and other small, stream-inhabitating forms (Greene, 1935). The zone width is variable. When it approximately coincides with a sharp change in soil type, as in Jackson County or Burnett County, it may be as narrow as 10 to 15 miles, while over areas of uniform topography and soil, it may be 20 to 30 miles wide. The zone is shown by the shaded band on the map at the opening of this chapter.

The same tension zone can be traced through Minnesota and Michigan (Livingston, 1903). It is very clear in northwest Minnesota where the presence of special soils in the bed of Glacial Lake Agassiz may be influential, but it becomes increasingly diffuse to the eastward. As it turns south in eastern Ohio, the zone is very wide and indistinct (Griggs, 1914).

The significance of floristic provinces in the study of regional vegetation lies in their influence on the possible floristic composition of the individual communities (Cain, 1947). Obviously, a forest in Grant County, in the extreme southwest corner of Wisconsin, cannot be expected to contain balsam fir, white spruce, or balsam poplar

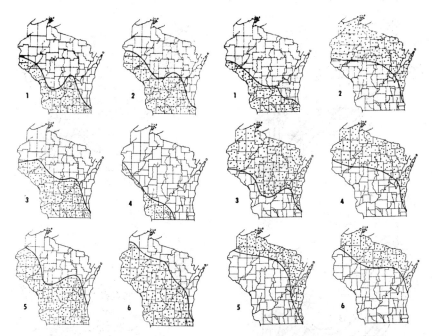

Figure 1.—Distribution maps for legume species. 1, *Cassia fasciculata;* 2, *Baptisia leucantha;* 3, *Baptisia leucophaea;* 4, *Psoralea esculenta;* 5, *Amorpha canescens;* 6, *Petalostemum purpureum.* Species are present throughout the shaded areas.

Figure 2.—Distribution maps for orchid species. 1, *Cypripedium candidum;* 2, *C. arietinum;* 3, *C. acaule;* 4, *Habenaria dilatata;* 5, *H. hyperborea;* 6, *H. obtusata.*

(*Populus balsamifera*), since these boreal species do not occur anywhere within the province. On the other hand, Grant County contains practically all members of both the Prairie and Alleghenian elements; hence any stand of vegetation has had equal opportunity of receiving propagules of all of these species. The fact that they have sorted themselves into different combinations must be due to varied historical and environmental factors. The communities they form and the dynamics of their interrelations thus become the subject matter of ecological investigation.

ECOTYPIC DIFFERENTIATION

The above discussion of the flora, floristic elements, and floristic provinces is all based on the assumption that the species concerned are discrete entities with definite characteristics. There is every reason to

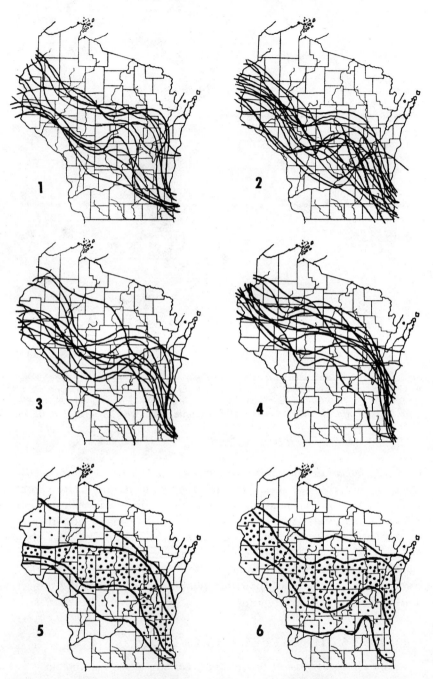

Figure 3.—Range boundaries (either northern or southern limits) for groups of species. 1, *Geraniales*, 12 sp.; 2, *Gramineae*, 16 sp.; 3, *Leguminosae*, 11 sp.; 4, *Ericaceae*, 12 sp.; 5, *Orchidaceae*, 18 sp. (light shading, 1 to 5 species limits per county, heavy shading, 6 to 10 species per county); 6, trees, 30 sp. (light shading, 5 to 10 species limits per county, heavy shading, 11 to 15 species per county).

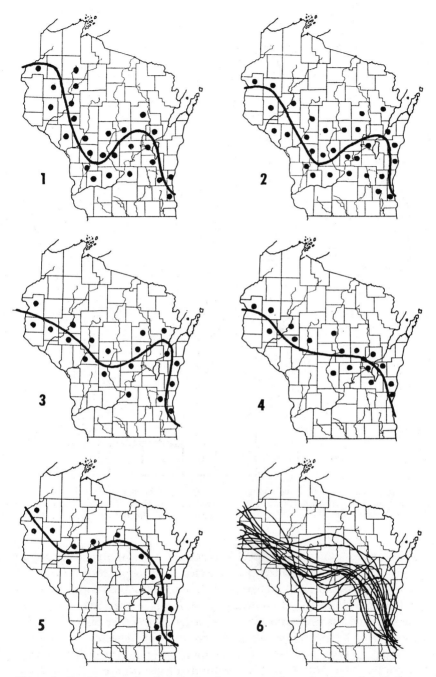

Figure 4.—Combined range boundaries for groups of species. The dots indicate counties with greatest numbers of species limits. 1, *Cyperaceae,* 15 sp.; 2, *Scrophulariaceae,* 13 sp.; 3, *Liliales,* 13 sp.; 4, weeds, 13 sp.; 5, *Caprifolaceae,* 8 sp.; 6, average range limits for 182 species in 12 families or orders.

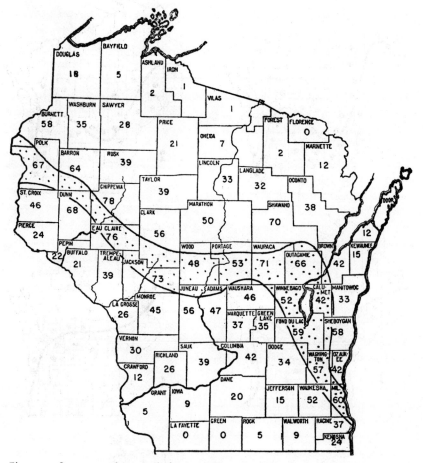

Figure 5.—Summary of range limits for 182 species. The figures in each county indicate the number of species attaining a range boundary there. The shaded band is the tension zone. Its exact location was determined from the densest concentration of individual range lines.

believe that most of the species, certainly most of the wide-ranging species, are actually made up of a series of subpopulations, each with its own combination of characters. These subpopulations may have minor morphological differences, in which case they may have been distinguished as taxonomic forms or varieties. In other cases, the differences may be physiological only and may be reflected in different seasonal growth patterns or in adaptations for growth in different environmental situations. These subsidiary groups are called ecotypes (Turesson, 1925; Gregor, 1946). They have been found in nearly every instance where methods of study suitable for their discovery have been employed. It seems probable that they occur in every species of wide distribution.

The significance of ecotypes in the description of vegetation is obvious. If two communities of widely different environmental requirements are described upon a floristic basis, then the presence of several species in common between the two communities will indicate a degree of similarity which may be highly misleading if the particular plants concerned actually belong to two different ecotypic populations (McMillan, 1956). Recognition of this situation and the solution of the problems it raises are among the greatest questions faced by vegetation ecologists today.

Ecotypes may originate in a variety of ways (Stebbins, 1950). Simple migration outward from an original center will often suffice to slow down gene exchange between peripheral individuals and central individuals. If the direction of migration is along a gradient of changing moisture conditions, for example, Darwinian natural selection will tend to increase the success of those individuals which happen to have preadaptations to the new conditions. Close breeding among these peripheral plants is more likely than outbreeding to the parent stock. In time, therefore, a variety of different strains may become more or less purified, with each strain adapted to the particular conditions of its portion of the environmental gradient (Clausen, Keck, and Heisey, 1940). Eventually, ecotypes may be evolved which are so differently adjusted to their environment that they are incapable of surviving in their original habitats. When this situation occurs, the indicator value of the species in describing a particular community may be nil. It is probable that such severe deviations are very rare in nature, except when accompanied by observable changes in morphology.

The work of Edgar Anderson (1948) and others has shown that the initial variability which is essential to ecotype development has often come about through hybridization between different species. This is especially likely to be permanently effective when the hybrid progeny have a disturbed environment in which to grow. An example might be the centripetal migration of two related taxa across recently deglaciated lands. Upon coming in contact, hybrids would be produced, and these hybrids would have a great variety of habitats in which to become established, perhaps on substrate types not favored by either parent. Continued backcrossing from such hybrid populations to either or both parents might introduce the adaptive genes characteristic of one into the other. Further migrations then might produce the ecotype development outlined earlier.

One of the best examples of this process in Wisconsin was worked out by Desmarais (1952). He studied the sugar maple (*Acer saccharum saccharum*) and the black maple (*Acer saccharum nigrum*) over their entire ranges. He found that pure black maple was confined to the Ozark region while pure sugar maple was present from the northern Great Lakes eastward to the St. Lawrence and southward along the Appalachians. In the region between Missouri and central Wisconsin, the populations were mixed with some individuals showing characters of both varieties. In an extensive statistical survey, he showed that the proportion of black maple characters gradually decreased along the gradient. In Wisconsin, the black maple characters were in a minority and were confined to the prairie-forest province in the southwest. He undertook no work on physiological characters, but it is presumed on the basis of their ranges that the black maple is more drought resistant and less cold tolerant than the true sugar maple. If this is really the case, then statements as to the composition of forests containing maple which fail to differentiate between the two varieties do not give a sufficiently accurate picture of the situation.

The great difficulty lies in the application of this knowledge in the field. An ecologist using ordinary methods must classify each plant he sees in his samples. If individuals are actually mixtures of two entities, then a simple yes or no classification is improper and misleading. In cases where definite morphological characters are involved, then perhaps an index number could be assigned to each individual and the average degree of intermixture determined for each stand. In cases that are probably more common where no morphological differences are known, the ecologist has no recourse except to lump everything under one name.

The entire question of ecotypes and their role in community study needs much further intensive investigation (Turrill, 1946). For the present, we can merely recognize that the problem exists and try to point out the cases where the ecological data seem to imply that more than one population is involved. The phenomenon is exceedingly important in Wisconsin forests, since the number of dominant trees is relatively small and since hybridization and introgression are known or suspected in the majority of them, including: black maple-sugar maple, silver maple-red maple, white ash-red ash, black ash-blue ash, black spruce-white spruce, shagbark hickory-yellowbud hickory, American elm-slippery elm-rock elm, all of the oaks, beech, basswood, hackberry, ironwood, and all of the birches, aspens, willows, and cherries. In fact, a few conifers and the leguminous trees seem to be the only forest dominants that are not suspect in certain places. This should mean that the information which follows in this book is subject to refinement and reinterpretation as further information becomes available. It should also strengthen the opinion held by many interested citizens that detailed botanical studies of our major forest trees are in need of much greater support, both on a state and national basis.

HISTORY OF KNOWLEDGE ABOUT WISCONSIN FLORA

The earliest comprehensive list of Wisconsin's flora (213 species) was published by Increase Lapham in 1836, although descriptions of a number of species collected in Wisconsin had been published earlier by Nuttall, Torrey, and de Schweinitz. Lapham rapidly increased knowledge about our flora and was able to publish a list of 933 species in 1852. Further additions were made by Lapham and T. J. Hale, a student at the University of Wisconsin (1860), and by T. A. Bruhin, a Swiss priest living near Milwaukee (1876). In 1883, Swezey gathered together all available information and printed a list of 1471 species in the *Geological Survey of Wisconsin,* Volume 1. No further listings were attempted until Russel, as chairman of the Botanical Committee of the Wisconsin Natural History Society, issued a mimeographed list of the flora in the years 1914 to 1916. This list contained 1605 taxa and was largely based on herbarium specimens.

No complete listing of the flora has been made since 1916. N. C. Fassett of the University of Wisconsin and his students began publication of the "Preliminary reports on the flora of Wisconsin" in the *Transactions of the Wisconsin Academy of Sciences, Arts, and Letters* in 1929. This portion of the series was continued until 1953. It consists of treatments of individual families or orders and contains distribution maps of each species and frequently keys to the species. Thirty nine families or small orders have been covered. In addition, larger treatments of the orchids (Fuller, 1933), legumes (Fassett, 1939), ferns (Tryon *et al.*, 1940), and grasses (Fassett, 1951) have been published separately. The remaining reports still to be published include the *Rosaceae, Labiatae,* the genus *Carex* of the *Cyperaceae, Compositae,* and *Cruciferae,* all with difficult sections or genera and many species. In spite of incompleteness, the "Preliminary report" series has proved highly useful in the recent studies of the vegetation of the state. It is hoped that the reports will be completed and a general work on the flora of Wisconsin issued in the near future, since such information is basic to all work in field biology and conservation.

In 1955, Margaret Bergseng, herbarium assistant at the University of Wisconsin, made a list of all species of seed plants from Wisconsin as represented in the University herbarium at Madison. It is not a complete listing since some rare species are represented only at the Milwaukee Public Museum or at other institutions. However, its total of 1829 species is probably within 50 or 100 of the actual total number of native and naturalized seed plants in the state. There are 123 families and 535 genera listed, for an average of 14.8 species per family and 3.4 species per genus. Far and away the largest genus is *Carex* with 141 species. Other large genera are *Potamogeton* (29 sp.), *Aster* (26 sp.), and *Polygonum* (24 sp.). The ten largest families, with their number of species, are: *Compositae*—326, *Gramineae*—227, *Cyperaceae*—184, *Leguminosae*—74, *Rosaceae*—65, *Cruciferae*—62, *Labiatae*—52, *Scrophulariaceae*—50, *Orchidaceae*—43, and *Ranunculaceae*—42.

CHAPTER 2

Environment

PHYSICAL GEOGRAPHY
Bedrock geology

Apart from the Driftless Area, the surface of Wisconsin is mostly covered by glacial drift. The underlying bedrock, therefore, has relatively little direct effect on the vegetation except at the infrequent outcroppings in escarpments and ravines. Indirectly, however, the effect may be very great, since the lithological composition of the drift is largely determined by the nature of the immediate bedrock and is only slightly influenced by rocks that are more remote. A knowledge of the subsurface geology thus is important in understanding the environment of the vegetation.

Wisconsin is largely a plain, with a maximum elevation differential of only about 1400 feet. It is underlain throughout by ancient pre-Cambrian rocks including quartzites, granites, trap rocks, and other igneous rocks and their metamorphic equivalents of Archaean to Keweenawan ages. These pre-Cambrian rocks form the surface layer in the north central and northwestern third of the state at a general elevation of 1000 feet (300 meters) above sea level or higher. Southward, they are buried beneath one or more layers of Cambrian and post-Cambrian rocks, at increasingly deeper levels. Thus the pre-Cambrian rock surface is at sea level near Madison, but is more than 2000 feet (600 meters) below sea level at Racine and Kenosha in the southeastern corner of the state. An extremely useful discussion of the relations of geology to surface physiography is given by Martin in his book on the physical geography of Wisconsin (1932). This work should be consulted by all who are interested in further details.

The covering bedrock layers are all of Paleozoic age. They include Cambrian sandstones, Ordovician dolomites, sandstones and shales, Silurian dolomites and Devonian dolomites and shales, in order from bottom to top. These sedimentary rocks form a series of layers, all of which are present in a vertical section in the vicinity of Milwaukee. The Devonian rocks, on top, have only a slight areal extent. A few miles west of Milwaukee, they disappear, and the surface rock becomes Silurian dolomite. This extends westward on a gently rising slope to an abrupt clifflike edge, the Niagara escarpment. West of this escarpment, surface rocks are the Platteville-Galena dolomites of the Ordovician, which rise to another escarpment. The Prairie du Chien dolomites, also of the Ordovician, are exposed next. Finally, the Upper Cambrian sandstones become the surface rock. The Ordovician shales and sandstones are relatively weak; they are found mostly at the bases of the escarpments or in deeply recessed valleys. The major portion of the area of southern and eastern Wisconsin is thus covered with limestones, while sandstones without a limestone cover are mostly restricted to central Wisconsin, in the region between the pre-Cambrian igneous rocks and the Prairie du Chien (frequently called Lower Magnesian) escarpment. The acidic character of these Cambrian sandstones and the pre-Cambrian granitic rocks has a considerable influence on the chemical nature of the ground water. This is expressed in the predominately soft water of the lakes of central and northern Wisconsin and in the podzolic soils, in contrast to the hard water lakes and more alkaline soil materials of the calcareous regions of Ordovician, Silurian, and Devonian rocks.

Wisconsin has had a very ancient and stable geological history. There were apparently no geologic revolutions or violent changes from the Devonian until the Pleistocene, a period of more than 200 million years. The Driftless Area (the shaded portion of the map at the opening of this chapter), not being affected by the Pleistocene glaciations, has been continuously available for plant occupation throughout the entire period of evolution of the higher land plants. Whether or not it was actually so occupied is unknown, since no fossil remains of ancient plants or animals have been found there. The entire area was once covered by the Niagara dolomites as is shown by the isolated outlier at the Blue Mounds. Erosional processes have worn away 1000 feet or more of the overlying sediments and have removed any evidence that might formerly have existed.

Outcrops of the various kinds of bedrock occur largely on the escarpment faces, where they sometimes make imposing cliffs with a sheer drop of 100 or 200 feet, in the gorges and valleys cut by the drainage system, and especially along the shore lines of the larger extinct glacial lakes where the action of the water has produced many wave-cut cliffs. As will be seen in a later chapter, these cliffs and rock exposures provide microhabitats for a number of plant species which do not occur elsewhere.

Glacial geology

During the Pleistocene epoch, Wisconsin was invaded several times by major continental ice sheets. These huge masses of ice built up in Canada and spread southward over the northern region of the United States. They exerted a very great effect on the topography of the regions they covered, not only by their action in depositing a deep layer of glacial drift, but by their scouring and planing effects on the bedrock. The resulting glacial landscape is usually a rolling plain, with numerous abrupt local topographic variations such as those produced by end moraines and interlobate moraines. The more nearly level ground moraine may be broken by pits or depressions. When these extend below the local water table, lakes occupy the pits, as they do in Vilas County. Both glacial ice and glacial drift at various times dammed former drainage ways, creating Glacial Lake Wisconsin in central Wisconsin, Glacial Lake Oshkosh in Winnebago County, and similar lakes in other areas.

There is some doubt as to the age of the oldest exposed glacial drift in Wisconsin. Some old glacial material covers a small area in Pierce and Dunn counties, just to the north of the Driftless Area. It has been assigned by various geologists to Nebraskan, Kansan, and Illinoian ages, but may be younger. The glacial drift in other portions of Pepin, Pierce, and Dunn counties is either Illinoian or early Wisconsin in age.

This drift is relatively thin and its exact geological status is difficult to determine.

The major drift sheet of definitely known origin is called the Wisconsin. It came into being after the Sangamon interglacial period, an ice-free interval of 50,000 to 100,000 years duration, which followed the Illinoian stage. The Wisconsin glaciation was made up of a series of advances and retreats of a single ice sheet. Six of these substages may have been present in Wisconsin (Leighton, 1957; Wright, 1957). The first of these is the Farmdale substage, which covered most of Green County and a portion of Rock County. This drift has been assigned to the Illinoian in the past, but the lime in its till is leached only to a depth of a few feet, which is irreconcilable with this conclusion. Alden (1918) believed it to be identical with the drift in adjacent counties of Illinois, which definitely has been shown to be of Farmdale age (Shaffer, 1956). No exact dating of this sheet is available at present, but it may have existed about 25,000 years ago (Flint and Rubin, 1955).

One of the major effects of the Farmdale substage in the Green County region was the creation of temporary lakes in the Pecatonica and Sugar river valleys. The ice sheet moved into the region from the southeast and created dams near Browntown and Verona. The waters backed up by these obstructions sought an outlet through the Apple River of Illinois, but during their high-water stages a considerable lake was formed in Lafayette, Iowa, Green and Dane counties. This lake filled the preglacial dendritic drainage valleys of the area and served to steepen the sides by wave-cutting action. These cliffs form the major habitat for the pine and hemlock forest relics to be found there now (see Chapter 11).

The second substage is the Iowan, which extended into Iowa and may have been present in the west central portion of Wisconsin about 22,000 years ago. This was followed by the Peorian interstadial period. The next substage was the Tazewell, which began about 19,000 years ago, according to recent radiocarbon dates. The only known exposure of the Tazewell in Wisconsin is in southern Walworth County. A large area in Marathon, Portage, Wood, and Clark counties is possibly of this age, although recent studies by Hole (1943) suggest a Cary age for the till. After a brief recession beginning about 18,000 years ago, the Cary substage (possibly the equivalent of the Pomeranian substage in Europe—Flint and Deevey, 1951) advanced between 14,000 and 15,000 years ago. The Cary stage left the most extensive fields of glacial drift, in terms of area exposed, of any of the glaciers. It covered all of the northern two or three tiers of counties and the eastern two or three tiers. There were two main lobes of the Cary ice in eastern Wisconsin, the Green Bay and Lake Michigan lobes. Their line of contact is marked by the spectacular Kettle Moraine which extends from Walworth to Manitowoc counties.

The Cary ice dammed the Wisconsin River in the vicinity of the Baraboo Bluffs, with the resultant creation of Glacial Lake Wisconsin which covered most of Adams and Juneau counties and parts of Monroe, Sauk, and Wood counties. Wave action on this lake was apparently responsible for the castellated buttes and mounds so characteristic of the region today. Where the sandstone is soft, the mounds have wasted rapidly since the glacial period, but have maintained steep sides. Sedimentation of outwash sands and gravels from the melting ice served to level the bottom of this lake and to produce the flat, poorly drained marshlands of today.

Following the retreat of the Cary ice, the next substage was the Mankato, which may have been only a minor readvance of the Cary (Wright, 1957). It is tentatively dated at 12,000 to 13,300 years ago. It influenced only the northwestern counties of Wisconsin, but the boundaries of its activities there are uncertain. The Cary and Mankato sheets had receded 11,400 years ago to a point beyond the Straits of Mackinac outlet of Lake Michigan, which allowed that lake to drain out to the north. This interstadial period is called the Two Creeks interval. It is world famous because of the extensive forest beds dating to that time which have been found at Two Creeks, Wisconsin. It appears to have been nearly contemporaneous with the Alleröd period of Europe (deVries et al., 1958). The botanical significance of the beds will be discussed later (Chapter 22). A further readvance of the Wisconsin glacier about 11,000 years ago is called the Valders substage. This extended inland westward as far as Shawano County and southward as far as Milwaukee. At its maximum in Shawano and Outagamie counties, it dammed the Fox-Wolf River system, creating Glacial Lake Oshkosh to the north and west of present Lake Winnebago. Lake Michigan was also affected by the Valders ice; the waters were raised to form one of the stages of Glacial Lake Chicago around the south end of the present lake. The shores of this lake were some distance inland from those of the present lake in Racine and Kenosha counties. The maximum extent of Glacial Lake Chicago occurred about 10,200 years ago. Both the Mankato and the Valders sheets left tills which were red in color from iron compounds, and rich in lacustrine clay minerals, with consequent poor internal drainage.

As the Valders substage receded, the glacial lakes at the ends of the basins of present Lakes Michigan and Superior increased in area, forming the Glacial Lake Algonquin stage of the Great Lakes about 8,500 years ago. This stage was at a much higher level than the current lakes and therefore submerged much of the area near the present shores.

A minor readvance of the Wisconsin glacier, called the Cochrane substage, took place in Canada about 5,800 years ago. This sheet never reached Wisconsin and exerted no direct physical effects on the state. Shortly thereafter, however, the Great Lakes reached a very low level, presumably due to the depression of the outlet by the weight of the ice. This stage in the Lake Michigan basin was called Glacial Lake Chippewa (Quimby, 1954). It occurred about 5,000 years ago and during it, the waters receded to levels 350 feet below the present lake, thus making a very narrow lake in the center of the present bed. The exact location of the shore lines are uncertain but it appears that direct plant migration in either direction across Lake Chippewa would easily have been possible.

A rise in the level of the land following complete recession of the Wisconsin ice caused another rise in water level, and created the Glacial Lake Nippissing stage of the Great Lakes. In Wisconsin, the Lake Nippissing waters covered much land that had previously been exposed, including numerous peat beds. The Glacial Lake Nippissing stage presumably lasted from 4,200 to about 3,000 years ago, after which the waters receded to approximately their present levels.

These changes in the lake levels and shore lines of Lakes Michigan and Superior had profound influences on the vegetation near the shore. Much of the land there has been available for colonization only for about 3,000 years; the current vegetation, therefore, can scarcely be used to indicate migration routes or other phenomena concerning plants or communities which are themselves much older.

Influence of glaciation on local environment

Glacial geologists do not agree as to the causes of glaciation, but some of their theories seem more useful from a vegetational standpoint than others. This is obviously not a valid reason for accepting the theories, but it may justify the following presentation. Several of the recently proposed theories (Stokes, 1955; Ewing and Donn, 1956) concerning the origin of the continental ice sheets of the Pleistocene and their cyclic fluctuation have involved climatic changes related to the temperature of the oceans. It is supposed that conditions favorable for glaciation in the Pleistocene were correlated with a shift of the poles from a position in the open oceans of the north Pacific and the south

Atlantic to their present positions in the nearly landlocked Arctic Ocean and on the Antarctic land mass. Changes in the earth's crust associated with the continuing rise of the Rocky Mountains are presumed to have provided the unbalance in the earth's crust which led to this shift of the poles. When the poles reached their present position, then the stage was set for a differential fluctuation in temperature of the Arctic Ocean in relation to that of the Atlantic Ocean.

According to the theories, the initial formation of a continental ice sheet was occasioned by an abundant supply of moisture from warm tropical oceans; this moisture became locked in the form of snow and ice in the polar regions, thereby lowering the ocean levels and disturbing circulation patterns there. As the ice sheets pushed toward warmer regions nearer the equator, their leading edges provided increasing volumes of cold meltwater during the summer seasons. This caused a gradual lowering of the ocean temperature; hence there was a decrease in the atmospheric moisture supply. Eventually this resulted in stagnation of the glaciers and their final disappearance.

If these general theories of glaciation are borne out by further evidence, they can be used to interpret the probable climatic conditions in Wisconsin during and between glaciations, particularly in the Driftless Area. It is possible that conditions in Wisconsin during the buildup of a glacier were warmer and wetter than they are at present, due to the high rate of flow of tropical air. At the time of a glacial maximum, both temperatures and precipitation may have been slightly lower, but the drop could not have been excessive, or the glaciers would not have halted where they did.

During periods of glacial disintegration, atmospheric moisture supplies would be at their lowest. It is supposed that conditions of aridity in mid-continent were severe enough to cause the destruction of the vegetational cover on the high plains, with consequent periods of loess production and transport onto the newly exposed glacial lands and adjacent non-glaciated regions. The extensive loess deposits of the Sangamon and later interglacial periods are in accord with this interpretation (Beavers, 1957). No evidence exists to support the conclusion that the entire vegetal covering of the Driftless Area was similarly destroyed by desiccation. In fact, the protection offered by the steep valleys would make such an event highly improbable. The level, upper peneplain surfaces, however, might well have been nearly denuded. Deposits of gravelly colluvium on valley sides at 150 to 190 feet above the Wisconsin River represent movements during an early period of erosion (Hole, *et al.*, 1952). Such a period would account for the nearly total absence of deep beds of residual soil on such areas today, beds which would be expected to exist in view of the extremely ancient age of these surfaces.

The evidence, therefore, seems to support the contention that the Driftless Area provided suitable conditions for growth of plants throughout the Pleistocene, conditions which did not differ significantly from those existing in the region today.

Another interesting aspect of the relations of vegetation to glaciation is concerned with the hypothesis that the surfaces of the ice sheets were covered by forest. This idea was first advanced for the Wisconsin region by H. H. T. Jackson (1914). He based his arguments largely upon the observations of Russell (1893) who described the dense spruce-alder forests growing on the surface of the great Malaspina Glacier in Alaska. In describing the tamarack-black spruce bogs of northern Wisconsin, Jackson said:

In view of these observations of Russell . . . it seems not improbable that during the glaciation of the Wisconsin ice sheet a dense plant growth might have existed on top of the border of the ice lobe itself as well as a short distance below the edge of the ice. The cold damp substratum of such an habitat would lead one to suspect that the plant growth occupying it would be . . . not essentially dissimilar to the flora of the Cassandra [*Chamaedaphne*]-Tamarack-Spruce Associations of Ridgway Bog . . . there is no conclusive evidence that the major part of it [the Tamarack-Spruce community] ever occupied the area south of the Wisconsin drift.

This is a very attractive idea, and as will be seen later, it serves to explain much information from fossil pollen analyses that is otherwise difficult to understand. The existence today of large areas of boreal forest in Canada under conditions of permafrost and with much colder and shorter growing seasons than those prevailing in Wisconsin certainly lends support to Jackson's hypothesis.

Physiography

As indicated earlier, Wisconsin is essentially a rolling plain with no great or abrupt differences in elevation. Physiographic features which are of importance to vegetation are thus largely ones of microrelief. In most cases, elevation is of less significance than degree of slope or direction of slope. A deep stream-cut valley or gorge in the Driftless Area may show a total elevational difference of 300 feet, rarely more, but the environmental differences between top and bottom on one side of the valley are rarely as great as the differences between opposite sides. This is most marked when the valley trends east and west so that one slope faces south and the other north, but it is evident in all valleys, as both south and west exposures tend to be much more xeric than north and east exposures (see Figure 7).

This influence of exposure is most readily observed in the Driftless Area. Other regions with similarly abrupt micro-relief are to be found in the Kettle Moraine of eastern Wisconsin, along the Niagara escarpment, in the Baraboo Bluffs, the Barron Hills and the Penokee Range, and on the Washburn Hills of Bayfield County. Such isolated peaks as the Blue Mounds, Sinsinawa and Platte mounds, Rib Mountain, and Powers Bluff are additional examples.

Most of the glaciated area is lacking in rises or hills of sufficient magnitude to influence the vegetation. The opposite micro-relief type, the kettle-hole depressions, are well developed in the glaciated areas, where the lake regions of Waukesha, Dane, Vilas, and Washburn counties are notable examples. Many other parts of the state contain similar depressions which were more shallow and have long since been converted to marshes and swamps. The major wet soil areas of Wisconsin are confined largely to the glaciated region, and the vegetation adapted to such habitats is naturally best developed in that terrain.

CLIMATE

Climatogeny

The climate of any particular geographical region is largely controlled by the air masses which pass over the region. These masses are huge bodies of air which have acquired more or less definite physical characteristics during a period of stagnation over a particular region of the globe and which retain those characters after they have begun to move. In Wisconsin, there are three important types of air mass, Arctic, Subtropical, and Continental. The Arctic mass becomes very cold and dry as it builds up in polar regions. The Subtropical mass, on the other hand, becomes warm and very moist during its stay over the tropical Atlantic Ocean. The Continental air becomes very dry during its inception over the Great Basin.

In the normal course of events, a Subtropical air mass moves northwestward across the Caribbean and then northward up the Mississippi valley. In mid-latitudes, it turns and moves eastward, usually paralleling the Ohio valley. The Arctic masses flow southward from the polar regions down the lowlands of the McKenzie trough just east of the Canadian Rockies. Upon reaching mid-latitudes, they, too, move eastward, parallel to the Ohio valley. Where these masses come in contact, the cold air of the Arctic mass flows under the warm subtropical air, causing it to rise and to cool off to at least the dew point, when the abundant moisture of the subtropical mass is squeezed out in the form of rain. The great bulk of the rainfall in the northern Middlewest has this origin. The flow of Continental air from the western mountains sometimes appears to drive a wedge between the other two masses, postponing their contact to a later, more easterly, location. During periods or seasons when the westerlies are strong, drought conditions tend to prevail on the plains and in the Upper Mississippi and Ohio valleys while heavier-than-normal rains fall in northeastern United States (Borchert, 1950). When the westerlies are weak or lacking, the Arctic and Subtropical masses meet over the plains or the Upper Mississippi. The rainy periods there are matched by drought conditions in the Northeast.

For the most part, Wisconsin's weather is governed by the relative duration of each of these three types of air and also by the frequency of their shifts. On relatively rare occasions, cool moist air from the North Atlantic penetrates this far inland, and at still less frequent intervals, similar air arrives from the North Pacific. Under more usual conditions, the dominant air mass in the winter months is the Continental. This begins to weaken in the spring and by summer the mean air flow over the state originates in the Subtropical air mass. During drought years or in more extended drought periods, the mean air flow from the western mountains continues throughout the year. The area of United States and Canada most under the influence of these westerly winds lies in the region of plains and prairies, with relatively steep gradients around the borders of these grasslands. Thus in Wisconsin, the average flow is from the west for seven months of the year in the extreme southwestern part of the state, but this falls to less than five months in the northeast. The location of these average patterns of air mass movement were used by Borchert in the presentation of his map of the climatic regions of Eastern America. According to Borchert's classification, northeastern Wisconsin is in Region I, which is characterized by deep winter snows and reliable summer rains. The extreme southwestern corner is in Region IV, which has relatively dry winters and frequent severe summer droughts. The intervening region, including most of the southwestern half of Wisconsin is thought to be transitional. It is of interest that the boundary of Region I coincides almost exactly with the floristic tension zone discussed earlier. The transitional region includes the prairie-forest floristic province.

Major features of climate

Many measurements have been made of various features of the climate of Wisconsin. Probably not all the measured phenomena are of equal significance to the vegetation, but there is little information concerning the relative importance of each (Table IV-3). The fact that winters are controlled by different air masses than summers tends to reduce the biological validity of many averages made on an annual basis. In the case of temperature, for example, the isotherms in winter show parallel trends from southwest to northeast, while the summer isotherms run at right angles to these, with the trend from southeast to northwest. As a result, the average temperatures, on an annual basis, show an intermediate pattern trending from east to west. No such pattern of actual temperatures is ever present at any given time.

Perhaps the most significant climatic factors are those which are correlated with the biological tension zone discussed earlier and which are involved in Borcherts' climatic types. A number of meteorological variables show a pattern similar to that given by the distribution of individual species and groups of species. Since many of these factors are correlated with each other there is no way to choose between them as to their relative importance in influencing vegetation. No single factor so far mapped shows a markedly steeper gradient in the region of the tension zone than elsewhere, but this does not rule out the possibility that certain combinations of these factors, unrecognized at present, actually do unite to form a barrier in the critical zone.

On the basis of correlations between isorythms and plant range boundaries, it appears that the tension zone is marked by shifts, from south to north, to levels of July evaporation below 5 inches, to average summer temperatures below 67°F., to a total winter snowfall above 48 inches, to a decrease in days with average temperature above 68°F. to less than 60, and to an increase in rainy days above 95 (Figure 6). Many other factors also show a good correlation with the zone. The phytoecological climate in the prairie-forest province in comparison with that of the northern conifer-hardwood province can be characterized by the following summary of certain of these factors; data were obtained in part from the maps of Visher (1954). The prairie-forest province has a more variable precipitation pattern, with an average of over 15 per cent variability from the normal. The number of thunderstorms per year (more than 20) is greater than north of the zone, as is the number of hailstorms, and excessively heavy rains, including storms with more than 1 inch of precipitation per hour, and more than 5 inches in 16 hours. More than 5 per cent of the annual rainfall occurs on days with 2.5 inches or more of rain. The daily range of temperature exceeds 20°F. in the prairie-forest province. There are more than 10 days per year with a maximum over 90°F. south of the tension zone, while the highest July temperatures recorded are over 105°F. Spring comes early in the prairie-forest region, with a mean of 50°F. reached by May 1. Although the annual precipitation is greatest south of the tension zone, the higher temperatures and lower relative humidities (July mean at 8 p.m. less than 65 per cent) lead to greater evaporational stress. Growing season evaporation from open-pan evaporimeters is over 30 inches. Among other factors contributing to this

Figure 6.—Isorithms for certain climatic factors which are correlated with the tension zone. 1, average summer temperature of 67°F.; 2, average annual snowfall of 48"; 3, average evaporation during July of 5"; 4, sixty days per year with average temperature above 68°F.; 5, ninety-five days per year with 0.01" of rain or more.

atmospheric dryness in the south are a greater number of hours of sunshine per year (more than 2500), a greater normal daily total solar radiation (more than 150 langleys in January and more than 450 langleys in August), and a smaller number of days (fewer than 95) with light rains of 0.01 to 0.25 inch.

Winter conditions are quite different in the prairie-forest province, with less than 40 days on which snow falls, and a mean total winter snowfall of less than 48 inches (Waite, 1958). These appear to be quite critical values, because they frequently result in the lack of a continuous snow cover south of the tension zone, while there is almost invariably a solid cover north of the zone during the winter. Snow blankets in the south are gone before April 1 on the average, whereas they persist much longer in the north. The alternate freezing and thawing brought about in the south by this fluctuation in snow cover creates very severe conditions for overwintering plants, particularly those in the seedling stage. In the north, seedlings and other small plants are continuously protected by a deep layer of insulating snow which falls regularly before severely cold weather sets in, and keeps the ground from freezing, especially in the forests.

It should not be assummed from the above that all factors of the climate run parallel to the tension zone. A considerable number of phenomena show isorythms that run at right angles to the zone, especially those concerned with winter temperatures and with winter and spring precipitation. Certain plant species are known that have ranges trending in this southwest-northeast pattern, but they do not include

Table 2
Important climatic factors in the two floristic provinces

Factor	Prairie-forest province	Northern conifer-hardwood province
January mean temperature	15.1° F.	11.4° F.
July mean temperature	71.1°	68.0°
Annual mean temperature	44.6°	41.5°
Length of growing season	148 days	126 days
Annual precipitation	31.5 inches	30.1 inches
Annual total snowfall	41 inches	51 inches
No. of days with 0.01″ ppt.	90	105
Average evaporation in July	5–6 inches	3–5 inches
No. of days per year with mean above 68° F.	60–90	0–60

any of the community dominants and are infrequent in comparison with those which are related to the tension zone. The existence of these divergent isorythms means that the two major floristic provinces do not have entirely homogeneous climates. A selection of those meteorological phenomena which do appear to show provincial homogeneity are summarized in Table 2. They are based on the mean of records from weather stations located within the main body of each province.

Phenology

One of the interesting correlations between climate and plant behavior is the object for study of that branch of biology known as phenology. This science is concerned with the seasonal march of observable biological events; the date of arrival of the first bluebird, the date of flowering of the first trillium, or the date of ripening of wild strawberries. Such events are related to the weather, especially to the number of degree days or other measures of accumulated heat, and to seasonal changes in length of the days. Many phenological events have crept into local folklore, especially those related to farming practices. Thus in Wisconsin it is said that the corn should be planted when the young oak leaves are as big as squirrels' ears and that the resulting corn crop should be knee-high by the Fourth of July. No mention is made of the variety of corn or the species of squirrel. More scientific approaches to phenology have been employed by biologists in recent years. The contribution of Leopold and Jones (1947) on the phenology of Sauk and Dane counties in Wisconsin is a good example of this modern work. They used a great number of botanical and zoological events to create a measure of the effective weather of a given year. For example, they found that the seasonal progression of the year 1945 was early from March 1 to May 15 and late from May 15 to August 20, in comparison with the average.

Phenological records are of great value in many applied phases of field biology from agriculture to conservation. In vegetation ecology, they enable us to predict when the major floral display of any particular plant community will occur in any region. The predictions are made on the basis of Hopkins' Law, which states that phenological events vary at the rate of 1 day for each 15 minutes of latitude, 1.25 days for each degree of longitude, and 1 day for each 100 feet of altitude, being later northward, eastward, and upward (Hopkins, 1918). This law appears to hold almost exactly along a north-south axis in Wisconsin. Thus a comparison of events at Brodhead, a town on the Illinois border, with the same events in Boulder Junction, a town directly to the north on the Michigan border, shows that there should be a 22-day-interval between the two stations, according to Hopkins' Law. The full bloom of duchess apples in 1911 showed a 22-day lag according to Whitson and Baker (1928). On the other hand, a comparison of Hudson, a town on the extreme west, with Sturgeon Bay, a town at the same latitude but on the easternmost edge of the state, showed an expected interval of 5 days, but the actual value was 10 days. This was due to the influence of Lake Michigan in retarding the spring warmup.

In the spring of 1957 an additional test of Hopkins' Law in the central region was made through the coöperation of Philip Jones, who was conducting extensive field investigations in Vilas County at the time. The weather was exceedingly irregular with long, hot, dry periods alternating with record-breaking cold and wet spells. In spite of this, the average deviation of date of first bloom of ten early species of woodland plants was 21 days between Vilas County and Green County, as compared to the 22 days expected by the law. Additional observations on seven species of maple forest herbs near Pelican Lake in southeastern Oneida County showed an average 19 day difference from Green County, which was exactly in accord with expectation.

The altitudinal retardation of phenological events is one of the few ways in which the elevational differences in Wisconsin are of significance to vegetation. The maximum difference of just over 1400 feet makes a phenological difference of 14 days. Because the higher lands are all in the north, the combination of altitudinal and latitudinal effects becomes quite marked. On the average, events are one day later for every 12 to 15 miles in the southernmost three-fifths of the state and one day later for every 8 to 9 miles in the northern two-fifths, for an over-all change of one day later for every 10 to 11 miles northward. These values apply in the center part of the state. Along Lake Michigan, they are increased to one day for 9 miles, whereas along the Mississippi and St. Croix rivers on the western border they are reduced to one day per 13 miles.

The figures above are probably valid only for the spring season up to June 1. Adequate information is lacking concerning the behavior of summer and autumn events, but scattered observations indicate that many members of the aster family bloom earlier in the north than they do in the south, including species of *Liatris, Solidago, Aster,* and *Helianthus*. The same is known to be true of *Spiranthes cernua* and *Gentiana crinita* and may apply to the majority of autumnal species. One possible cause of such a phenomenon would be the more rapidly decreasing day lengths present in the north after the summer solstice.

Hopkins' Law is an empirical statement originally propounded upon the basis of extended observations; the above confirmation in Wisconsin is similarly observational in nature. The delay in phenological events in the spring with increasing latitude appears to be a reasonable response of plants to a delayed warming of the northward regions, but the speedup of events with increasing longitude has no similarly obvious basis. Recent studies of McMillan (1956, 1957) in Nebraska indicate that phenological changes with distance may be enormously more complicated than the simple climatic control usually postulated in explanations of Hopkins' Law. He found that plants of various prairie grasses did indeed bloom earlier in the west than in the east, but that these differences were maintained or magnified when all the plants were grown together in one transplant garden. In other words, the early blooming habit of the western strains had become genetically fixed as an ecotypic response and was no longer under the direct control of the local weather pattern. Other studies with prairie grasses collected along a north-south gradient (Olmsted, 1944), indicate an inherited response to length of day rather than to temperature alone. These investigations of observed phenological patterns do not detract from the validity of Hopkins' results, but merely alter and expand our explanations of the phenomena.

Microclimates

The previous discussion of the climate of Wisconsin was based entirely upon weather bureau records which are made for the most part in cities and towns. The actual climate in nearby stands of native vegetation may be markedly different from that in the urban areas. The climates of these restricted, local areas are called microclimates. They have been studied very little but enough is known to hint at their very great importance in interactions with vegetation. One of the most thorough investigations was that of Wolfe et al. (1949) in the nonglaciated regions of Ohio. There the microclimate in sheltered ravines and valleys was found to be greatly different from the macroclimate as measured at a nearby weather bureau. There were such temperature differentials as a 26°C. difference in minimum winter values. Evaporation conditions also varied greatly, with a tenfold difference between valley bottom and adjacent cliff top.

In Wisconsin, Van Arsdel (1954) found a similar situation in the deep valleys of the Driftless Area. He was studying the conditions necessary for the transmission of blister rust disease from gooseberry (*Ribes*) bushes to white pines. Among the factors needed, a period of saturated atmosphere of more than 24 hours duration was most critical. These periods were of frequent occurrence in the area of Wisconsin north of the tension zone but were rare or lacking south of the zone except in the deep, rock-walled valleys with a northern exposure. These sites frequently supported relic, isolated colonies of white pine. They were the only places in the prairie-forest province where blister rust was prevalent. McIntosh (1950a) also studied the climatic conditions in the pine relics on cliffs in the Driftless Area. He found the summer soil temperatures at a depth of 5 cm to average 1.7°C. lower than on the adjacent uplands.

A closed plant community naturally exerts a considerable influence on certain factors of the microclimate, such that conditions inside of the community are very different from those outside. Cottam (1948), for example, found in an oak forest in Dane County that the evaporation from standard Livingston evaporimeters during the summer months was 0.33 ml. per hour, as compared to 0.83 ml. in the open or outside the forest. The soil temperatures at a depth of 7.5 cm averaged 15.3°C. in the forest during the same period, as compared to 16.8°C. in the open.

Comparative studies were made in the University of Wisconsin Arboretum in Madison between the microclimates of a prairie and two adjacent forests of pine and oak. The forests were similar in their temperature and moisture conditions, and were cooler and less dry than the grassland. A major difference between forest and open was seen in the rate of advance of the spring season, where snow-melting and ground-thawing regularly took place three to four weeks later in the forest than in the prairie.

SOILS

Parent materials

The soils of Wisconsin have been derived from a variety of parent materials. In a few places, particularly on steep slopes in the Driftless Area, on escarpments, and rock hills in the glaciated area, the soil parent materials are produced directly from the weathering of the bedrock (Plate 1). In the great bulk of the state, however, coverings of one sort or another have been deposited over the underlying rocks and the soils are derived from the covering materials. The most common covering is a blanket of glacial till, but wind- and water-borne sediments are also very important in some areas. The glacial till is greatly influenced in its lithological composition by the nature of the local bedrock, but the comminuted condition of the till presents very different opportunities for soil development than does the solid rock beneath it. The water-holding capacity and nutrient supplying power of the till present a favorable medium for plant growth from the date of initial exposure. The weathering of till and the transformation of the upper part into soil, therefore, has taken place under the continuous influence of a vegetational cover. Bare rock, on the other hand, has gone through a slow process of direct physical weathering and a gradual succession of various simple plants like lichens and mosses before higher vegetation and true soils could develop. These processes slowly produced soils on sandstones, shales, and dolomites; however, some of the quartzites in Wisconsin are so hard and so resistant to decomposition that soil production has been literally impossible. The striations and chattermarks left 12,000 years ago by the passage of the Cary ice over the quartzite of Observatory Hill in Marquette County show no indication of surficial weathering whatever. The present vegetation of other quartzite prominences, such as Rib Mountain, the Baraboo Bluffs, and the Barron Hills, is rooted largely in weathered loess which was deposited by wind in cracks of the unyielding bedrock.

The glacial lakes frequently accumulated a considerable layer of lacustrine clays on their beds, which became sites for soil development upon retreat of the water. On their immediate shores as well as the shores of the present Great Lakes, sand deposits accumulated as a result of wave action. Movement of this sand in the form of dunes provided another type of soil parent material. Another wind deposit was the blanket of fine silt called loess which covers much of the southwestern-half of the state. The loess, sand, and lacustrine clay, like the till, can support a vegetation directly without prior physical weathering and the soils on these materials are all developed under the influence of vegetation.

Soil-building processes

Given one of these comminuted parent materials, the kind of soil which developed was largely a function of the type of vegetation which became established and the type of climate which prevailed in the region (Wilde, 1958). Since vegetation is related to climate, the effects of these two factors cannot be readily separated but must be considered as a unit. Upon recession of the Cary or Valders ice, the exposed till was invaded by a forest composed predominantly of coniferous trees, especially spruce and fir. For purposes of this discussion, it does not matter whether this was accomplished by an invasion of bare till following a receding ice front or whether, as is more likely, it came about through the stagnation and gradual wasting away of an ice sheet which supported such a forest on its surface and lowered the forest to solid ground as it disappeared. In either case, the climatic conditions corresponded to Borchert's type I, which means relatively low temperatures, uniform moisture supply with a large fraction in the form of snow, and, especially, low intensities of evaporation. Under such conditions, an excess of moisture was present which percolated downward through the raw till. The surface of the till was covered by a thick layer of undecomposed and partially decomposed conifer needles. This needle mat was low in basic nutrient salts because of the nature of the metabolic processes of conifers, but was rich in organic acids and acidic inorganic compounds such as silica. The percolating waters, therefore, were in the nature of weak acids. As they moved through the upper layers of the till they gradually dissolved out the basic elements, especially calcium, magnesium, and potassium salts. In addition, iron and aluminum gradually were removed by an interaction with organic chelating agents derived from the conifer needles (Swindale and Jackson, 1956). All of these materials displaced from the surface layers became deposited at lower levels, often in the form of cementing layers between existing fragments of till. In this way, a soil finally was formed which was composed of a series of horizontal layers, called soil horizons, each of which had its own particular chemical and physical characters. This process is called podzolization, and the mature soil is Podzol. It has a mat of partially decomposed plant remains on the surface (the A_o layer, according to the current designation by soil scientists) which is usually quite sharply separable from the next lower thin layer containing an incorporated mixture of decomposed organic matter and mineral particles (the A_1 layer). Next below is the A_2 layer, from 1 to 8 inches thick, of sub-pure silica which may be ashy gray or purple gray in color in extreme cases. Below this and extending for a considerable depth is a layer which has been enriched by the materials leached out over the years from the upper layers (the B layer). In Wisconsin, this layer is usually a bright coffee-brown color. Still deeper in the profile, the unaltered parent material may be found.

Soils with these characteristics are found today in Wisconsin in the northern and northeastern counties which support a coniferous forest. They are particularly well developed under spruce and fir and in the hemlock forests. Formerly they must have been distributed widely throughout the glaciated region except on clay or clay-loam parent materials. However, climatic changes associated with postglacial times eventually led to the present prairie-forest climatic pattern in the southwestern-half of the state. As the conifer forests were invaded by hardwoods under the influence of a drier, warmer climate, certain phases of the soil-building process were changed or even reversed. Many of the new deciduous trees, especially birches, maples, and basswoods, had leaves which were rich in calcium and other bases at the time of shedding. These bases were originally obtained from deep layers of the soil or parent material by the tree roots and were literally pumped up into the foliage of the trees and then deposited in the form of leaf litter on the surface of the soil several feet above the sites of origin. Downward percolation was much slower than in the conifer forest because of changed conditions of temperature, moisture, and evaporation. Under these circumstances, the uppermost layers of the soil became enriched rather than depleted. The layer of incorporated organic matter also became deeper, while the top layer of undecomposed organic materials became thinner. In fact, in many situations, the rate of decomposition was so speeded that a year's crop of leaves was gone before a new crop was produced.

Forest soils in Wisconsin produced by this process of depodzolization are known as Brown Forest soils. They largely support stands of southern mesic hardwoods. Over large areas, however, the processes of calcification (soil enrichment) and podzolization (topsoil impoverishment by leaching and translocation) were simultaneous under xeric deciduous or mixed coniferous-deciduous forest, and a "hybrid" group of soils resulted, called the Gray-Brown Podzolic soils. These soils have surface layers enriched in organic matter, but are acid in the subsurface soil, and have a subsoil horizon somewhat enriched with iron and aluminum compounds.

When the mesic forests were replaced by prairies (Chapter 14), still greater changes were induced in the Brown Forest soils. The pumping activity of the prairie grasses exceeds that of the deciduous trees, so that an intensification of the accumulation of bases (called calcification) took place. In addition, the massive development of fibrous roots of the prairie grasses and their decomposition in place led to a greater and deeper incorporation of organic matter into the soil. In contrast to the Podzols and Gray-Brown Podzolic soils the Prairie soils (Brunizem in Hole and Lee, 1955) are almost free of discernible layers. Rather there is a gradual reduction in content of bases and organic matter from the surface downward that results in a soil which is rich and black on top, but which fades out to the parent materials at depths of 2 or 3 feet or more.

It should not be assumed that the plants alone are responsible for these soil-building processes. The total biota of plants and animals is involved and the microscopic forms of both kingdoms are of great importance. Obvious soil movers such as earthworms, ants, and burrowing mammals clearly exert a considerable influence on the rate of soil-stirring. It is significant that all animals with these habits are rare or absent in true Podzol regions while all are at their maximum, in Wisconsin at least, in the prairie areas. The relative degree of distinctness of the soil horizons in the two areas is in large part dependent upon this distribution.

Loess cap

One of the most significant parent materials of soils in Wisconsin is an extensive loess cap over the southwestern-half of the state. The loess is an aeolian sediment made up of silt-sized particles; the loess lies as a blanket over all other soil parent materials in depths varying from 16 feet or more near the Mississippi, to a thin layer of several inches in the central counties. Its source and time of origin are in doubt. A commonly expressed opinion of some glacial geologists is that the Wisconsin loess was derived from the glacial flour exposed by the receding meltwaters in the outwash valley trains of the Mississippi River. The loess in Grant County is leached to a depth of 7 feet and has been estimated to be 20,000 years old (Robinson, in Hole, *et al.*, 1952). If correct, this date would relate the origin to the Iowan substage of the Wisconsin glaciation. However, extensive deposits of loess followed the Cary substage, as shown by the aeolian cap over till of Cary age in Dane and Columbia counties. Such deposits would necessarily be less than 14,000 years old, and their addition to the surface of Iowan loess might complicate calculations based on depth of leaching. The absence of weathering on the surface of the Cary till beneath the loess cap tends to support the glacial outwash theory of origin as far as the earliest layers of loess are concerned. It is probable, however, that additional layers were produced at later times and especially during the period of advance of the prairies around 2000 B.C. from sources on the drought-denuded high plains far to the west of the Mississippi. Robinson found that the soils developed on the loess of late Cary terraces in Richland County were only 8000 years old. The great dust storms of the 1930's and the less frequent storms in more recent years testify that the loess blanket is still being increased.

One of the interesting characteristics of loess is its uneven thickness. It apparently behaves very much like snow in drifts on open fields, with deep deposits on the lee side of obstacles to the prevailing wind and thin layers or none at all on the exposed windward slopes (Plate 2). Because snow does not behave this way in extensive forests, it appears that the loess must have been deposited in the open during a period of prairie occupation, or, at least, have been redistributed during such a period.

The loess in Wisconsin is uniform in chemical characteristics. Even the leached and weathered loess is rich in bases, particularly calcium. The gradual deposition of loess on the surface of a spruce-fir or hemlock forest would do much to counteract the normal processes of podzolization. It is possible that the reversal of podzolization which must have occurred in southern Wisconsin in postglacial times was caused as much by the addition of loess as by the activities of a changed vegetation.

Perhaps the most significant effect of the loess on vegetation is its masking of various soil parent materials beneath it. It would be expected that the Driftless Area, with its ancient soils and exposed bedrocks, would have markedly different assemblages of plants than the adjacent glaciated areas with young tills. As will be seen later, however, no such vegetational differences actually exist. The fact that the vegetation has developed on a constant substrate of recent origin in both regions is largely responsible for this lack of differentiation.

Major soils of Wisconsin

Although soil scientists realize that soils in nature are subject to continuous geographic variation correlated with topographic position, parent material, internal drainage, and other factors, nevertheless the apparent existence of relatively discrete areas of fairly homogeneous composition has led to the use of an arbitrary classification system for the description of soils. In this system a binomial nomenclature, related to that used in biological taxonomy, is employed. Thus, a particular kind of soil is indicated by the name Miami silt loam. The word "Miami" is comparable to a generic name in biology. It is called a series instead of a genus and is based on such factors as the arrangement of layers in the profile, their origin, and chemical composition. The words "silt loam" are comparable to a species name but are called types. The types are based on texture or relative content of clay, silt, and sand. There may be several types of soil in the Miami series, depending on their textures. The biological category of family is as yet little used in soil classification, but the ecological concept of plant community is represented by a soil association, which is a group of two or more kinds of soils occurring together in a particular landscape. Hierarchic units comparable to order or class in biology are perhaps represented in soil science by the Great Soil Groups or Zonal Soil Groups, which are differentiated by their method of origin. Thus, the Great Soil Group called Podzol includes a number of soil series which are related in the sense that they have all been formed by the podzolization process, although their parent materials, age, slope, and other characteristics may be very different.

The various soil associations may be arranged in a variety of sequences to show different relationships (Hole and Lee, 1955). Perhaps the most significant sequence from the standpoint of relations to vegetation is based on natural drainage condition or degree of aeration. Such a sequence is termed a catena (Milne, 1935; Bird, 1957). It is an association of soil series arranged in order of increasing wetness, from excessively well-drained to very poorly drained. In an ideal situation, such a drainage catena may be found on a single hill slope, with the best-drained soil bodies at the crest and the worst-drained at the foot of the slope. Under such conditions, the assemblage of soil bodies most nearly approaches a complete continuum, whose components are distinguished only by means of arbitrary soil type boundaries established by man. On a given hillside, the catena may be present only in part or it may be interrupted by local areas of higher elevation, as in regions with steeply rolling topographies.

From the standpoint of vegetation, certain significant factor gradients are associated with these "drainage" catenas. The best drained terminus of the catena naturally tends to be the driest, since direct rainfall is the only source of moisture and this tends to pass through the soil quickly. The rapid and frequent percolation of water leaches out the mobile salts, removing the bases and the finer colloidal clay particles. In soil bodies farther down the slope, moisture conditions improve, both because of slower movement of water through the profile and an increased moisture supply due to surface runoff from higher regions. These trends become accentuated toward the base of the slope. Here the water supply may become excessive, due to runoff from above. If surface drainage is poor, the excess water may accumulate for all or part of the year, either on top of the soil or at a shallow depth below the surface. Mobile salts and colloidal particles from up-slope accumulate in these lower soil bodies. The colloids, particularly, contribute to the poor internal drainage. The displacement of air in the interstices between soil particles by the standing water results in an anaerobic or reducing condition some place in the profile, thus greatly affecting the ability of most plants to secure sufficient nutrients. The high base content of depressions frequently results in a neutral or alkaline soil condition, particularly in dolomite country, whereas the drier soils of the catena are usually acid. This combination of low oxygen and low acidity in depressions tends to produce a black or dark color in the organic matter. Under extreme conditions, peat may form on the lowest areas.

The complete array of soils of a catena can be found on uniform parent materials throughout the slope, although this is not always the case. Under such conditions, there may be a corresponding continuum of vegetation from top to bottom. The species making up the vegetation may be responding to different factors of the catena so that over-all correlation can be related only to the slope. Wilde and Leaf (1955) have shown the changes in ground-cover vegetation associated with a catena varying from melanized sand to sandy gley Podzols in central Wisconsin. Blueberry (*Vaccinium* sp.) grew well over the entire gradient, while redroot (*Ceanothus ovatus*) was confined to the well-drained portion, and bunchberry (*Cornus canadensis*) was confined to poorly drained areas. Similar studies on other catenas are needed for other regions of Wisconsin.

From this discussion of catenas, it should not be assumed that a particular soil characteristic induced by topographic position on a slope is immutable. The process of podzolization, for example, may increase the vegetational carrying capacity of porous sands originally present at the top of the catena. According to Wilde *et al.* (1949):

First of all, the soil accumulates a substantial layer of highly nutritive and water-holding litter. Second, some mineral colloids and humus suspensions are removed from the soil surface and redeposited at a depth of about from 1 to 2 feet, materially increasing the water-holding capacity of the soil. In this manner, a coarse sand is converted into a soil which is equivalent to a sandy loam.

This process of amelioration is accentuated by prairie vegetation and by certain hardwoods. A particular catena, therefore, is the result of interaction of vegetation, time, and slope; and its characteristics may be expected to change in the future as the vegetation changes.

A detailed description of even the major soil series in Wisconsin is not possible here. Relatively complete accounts may be found in the papers of Hole and Lee (1955), Wilde, Wilson, and White (1949), and Kellogg (1930). Some details will be given later in the accounts of the individual vegetation types of the state. The Great Soil Groups in Wisconsin cover a considerable range, including the climatic zonal groups of Podzol, Gray Wooded, Gray-Brown Podzolic, Brown Forest, and Prairie, as well as several intrazonal groups (Table IV-4).

Several specialized soil types which are associated with unusual plant communities in other parts of the world are missing in Wisconsin. For example, no truly saline soils are present. The closest approach to salinity is shown by certain soils very high in lime, especially those developed on marl beds around alkaline lakes in the southern part of the state. Another simulated saline habitat is the very specialized environment created by the cinders used as ballast on railroads. A number of the weeds found only on such sites are widespread in the western states on naturally saline soils. Serpentine soils with their high content of toxic heavy metals like chromium and nickel and their high degree of plant endemism are likewise lacking in Wisconsin.

CHAPTER 3

Plant communities and their distribution

NATURE OF VEGETATION

Within the two floristic provinces of Wisconsin there is a more or less homogeneous set of species which is found distributed more or less uniformly throughout the area of each province. Any particular small location within a province, however, is likely to have a very different assortment of species than another location within the same province. These local assemblages of species are called plant communities. They may differ from one another in the kinds of species they contain, in the relative amounts of the same species, or in both ways. Collectively, the plant communities of a region are called vegetation. The vegetation of an area differs from its flora in that the former is quantitative and the latter is qualitative.

The flora of a given floristic province consists of a great number of species of diverse origins and histories, as discussed in Chapter 1. The species further differ in their tolerance of the various environmental factors discussed in Chapter 2, although all must be able to tolerate certain features of the macroclimate characteristic of the province, such as day length, general seasonal pattern of temperature changes, and the like. Because of these inherent differences in the flora, the various members react differently to local microclimates and local soil conditions; thus they produce aggregations of species which differ in composition from place to place.

The nature of these compositional variations in community makeup can be understood by considering an area of bare soil of some magnitude suddenly made available for plant colonization somewhere in the middle of a floristic province on a site that is not extreme with regard to the regional environment. The probabilities are high that this new area will be invaded at first by members of the adjacent plant communities, with those species which produce the greatest number of easily dispersed seeds having greater chances of success than those which produce few seeds which are disseminated with difficulty. Of the plants which reach the new soil, only those will succeed that are adapted to the microenvironment of the site at the time. Since this is a bare area, the main feature of the environment is its great variability, with rapid diurnal changes in temperature and often with great changes in soil moisture over a period of several days. Plants adapted to such a fluctuating environment are called pioneer species. The bare area may be covered almost completely within a year by the immigrating pioneers, but their mere presence brings about great changes in the local environment. Soil erosion and surface movement are greatly reduced because of the protection against raindrop impact offered by the plants. A portion of the radiant energy formerly lost as heat from the area is now stored in the form of organic matter as a result of photosynthesis and eventually is added to the soil, improving its physical and chemical properties. The new conditions may become less favorable for the original immigrants and more favorable for other species which were excluded by the initial environment, particularly by changes in light intensity and soil stability.

A continued reaction of the plants on their environment produces further change which is favorable for some species and unfavorable for others. The exact species composition at any given time is a resultant of the preadaptations of the nearby flora to the microenvironment at that time and of the chance factors which allow some of the adapted species to enter and which prevent other equally well-adapted species from reaching the spot. The chance factors are of great importance; they include such unpredictable things as wind direction and velocity at the time of seed ripening, drought periods, periodic insect infestations of seed crop or seedling stand, distance of seed plants from the colonizable area, temporarily high or low population levels of rodent enemies, and a host of other infrequent events. Because of these chance factors, the exact species to enter an area and the exact number of individuals of each species which succeeds are unique phenomena never to be repeated at any two places or any two times.

The above is a brief statement of the individualistic hypothesis of plant communities as expounded in detail by Gleason (1926, 1939); all of the information to be given later about the detailed composition of the plant communities of Wisconsin is in full support of this hypothesis. It must not be assumed, however, that the vegetation of Wisconsin is a chaotic mixture of communities, each composed of a random assortment of species, each independently adapted to a particular set of external environmental factors. Rather, there is a certain pattern to the vegetation, with more or less similar groups of species recurring from place to place. The main reason for this is the great potentiality for dominance possessed by a relatively small group of species. These are plants that are well adapted to the over-all climate and soil groups of the province and which have the ability to exert a controlling influence on the communities where they occur because of their size or their high population densities. Over most of Wisconsin, these dominants are trees, but only 12 or 15 of the tree species in the flora commonly achieve leading positions. The interactions of this tiny group of plants with the general climate and the regional soil groups produce a series of microenvironments which differ according to the biological characteristics of the dominants. Most of the remaining species of the flora must grow in these modified conditions and they tend to be sorted out in groups aligned with the particular dominant concerned. The groups are not discrete and separate from one another but gradually shift in composition because the dominants themselves rarely grow in pure stands but occur rather in mixtures of varying proportions. The mixtures of dominants vary according to a pattern commonly associated with a soil-moisture gradient. Thus, in the prairie-forest floristic province of southern Wisconsin, non-aquatic, non-burned sites are likely to be occupied by hardwood forest. This forest may be dominated by willow, cottonwood, American elm, or silver maple in places of abundant moisture; by maple and basswood on well-drained mesic sites; and by a series of oak species on progressively drier sites. Intermediate places have an intermediate mixture of dominants. The lesser plants making up the remainder of the communities are similarly found in shifting assortment, with the set adapted to the wettest conditions gradually losing some species and gaining others as the habitat becomes drier, so that the set in the most xeric oak forests has very few species in common with the wet group. The entire series of communities whose floristic composition gradually changes along an environmental gradient has been termed a "vegetational continuum" (Curtis and McIntosh, 1951), to emphasize the fact that no discrete divisions, entities, or other natural discontinuities are present.

Not only do the existing stands of a particular vegetation type, such as the hardwood forest, form a continuum, but the types themselves frequently blend gradually into one another. In southern Wisconsin, the forests in many places gradually become thinner until the spaces between the trees are greater than the spaces occupied by the trees, thus forming a savanna. The savannas themselves may show a gradually decreasing density of trees until the few remaining ones are so far apart that the area becomes a grassland rather than a savanna.

It must not be assumed that this gradual blending of one community into another or one vegetation type into another is always expressed in the field. On the contrary, there are many examples of abrupt shifts from one assemblage to another, sometimes along a line so sharp that it may be crossed at a single step. The relic tamarack swamps to be found in glacial kettle holes in southern Wisconsin offer a good case in point. It is not uncommon to find an oak woods on the surrounding hillsides with a distinct boundary at the level of the water table in the kettle hole. Across the boundary on the surface of the swamp there may be a completely different assemblage of plants, with no species whatsoever in common between the swamp community and the hillside deciduous forest.

In other places, a prairie community may give way to a closed forest within the space of 15 or 20 feet. Usually these abrupt boundaries between dissimilar communities are associated with equally abrupt changes in the habitat, mostly due to topographic features. They must not be taken as evidence of the discrete nature of plant communities, because an extensive study of the entire region will quickly show that the prairies in different sites differ from each other in a continuum pattern, as do the oak forests and the conifer swamps. Thus, communities largely or totally different in composition may exist side by side, yet they cannot be categorized, because more remote examples of each of the locally distinct types will differ slightly from the initial stands and further examples can be found which will show much less differentiation, and finally culminate in stands which are intermediate between the original pair.

This situation means that it is not possible to erect a classification scheme which will place the plant communities of any large portion of the earth's surface into a series of discrete pigeonholes, each with recognizable and describable characteristics and boundary limitations. The plant communities, although composed of plant species, are not capable of being taxonomically classified, as are the species themselves. The communities are the result of chance historical happenings, accidents of migration, and partial catastrophic destructions, with no two stands ever of the same composition. They have no method for self-duplication, such as is provided by the chromosome mechanism in their component species, and they are not genetically related to one another in either time or space. Their lack of an inherent discreteness, however, does not prohibit their orderly arrangement into groups for purposes of study and discussion. The groups may be of such size as is convenient for the purpose at hand. When considering small or local areas, the resulting groups may actually have a certain degree of discontinuity, since the bridging forms may be absent in the area. This idea is familiar in taxonomy, where very distinct and apparently unrelated forms in a particular region are actually related by a series of intermediate types in other regions, as in the case of the creeping snowberry, formerly classed as *Chiogenes hispidula*, and the wintergreen (*Gaultheria procumbens*). Recent research has shown that these two genera, although apparently distinct on the basis of the Wisconsin representatives, are actually connected by a series of intermediate forms in other lands and must be combined into a single genus, *Gaultheria*.

This concept of the nature of plant communities is fully supported by the data to be provided in later chapters and is equally applicable in the prairie-forest floristic province and in the northern hardwoods province of Wisconsin. However, the concept did not originate with the current work and is definitely not limited to the vegetation of Wisconsin. It appears in the ideas of many of the plant geographers of the early nineteenth century and received definite statement by many field scientists (Chamberlin, 1877) in the late years of the same period. The first detailed treatments of the individualistic concept were provided by Ramensky (1926, 1930, 1953) in Russia, and Gleason (1926, 1939) in the United States. Support based on quantitative analyses, in addition to that reported for Wisconsin, was given by the findings of Kulczynski (1927) in Poland, Hansen (1930) in Iceland, Tuomikoski (1942) in Finland, Motyka (1947) and Matuszkiewicz (1948) in Poland, Sorensen (1948) in Denmark, Whittaker (1951) in Tennessee, Goodall (1953) in Australia, Parmalee (1953) in Michigan, deVries (1954) in the Netherlands, Horikawa and Okutomi (1955) in Japan, and many other workers in more recent years. The concept has been reviewed by Goodall (1954a), Greig-Smith (1957), Whittaker (1954, 1957), and Webb (1954).

Similar concepts of the essential continuity of many natural phenomena have become well established in soil science, where the idea of a catena (Milne, 1935; Bushnell, 1942) is directly comparable to a vegetational continuum, and in taxonomy, where the idea of a cline (Huxley, 1938) and an ecotype as a particular range on an ecocline (Gregor, 1946) are clear parallels.

Thus there is widespread support for the idea that vegetation must be studied as a continuous variable. However, the tremendous range in kinds of plant combinations to be found in an area as large as Wisconsin makes such a study very cumbersome if attacked in its entirety. Accordingly, it is convenient to initially break the vegetation into major units, recognizable by simple criteria, and study the variations in each unit, after which the units may be recombined according to their degree of similarity, as shown by the detailed studies. This is the plan to be followed in this book.

The first step in grouping the communities of any vegetational region, such as one of the floristic provinces of Wisconsin, is to erect the groups on the physiognomy or gross appearance of the community. Following this custom, the major portion of the vegetation of Wisconsin may be divided into three types, the forests, the savannas, and the grasslands. These three may grade into one another, as indicated above, but arbitrary and specified criteria may be used to assign any particular stand to one of the three types. A further separation is possible on the basis of the floristic provinces, thus giving northern and southern subtypes of the forests, savannas, and grasslands.

Finer subdivision of the vegetation is usually based on floristic composition rather than physiognomy. In North America, Australia, and elsewhere, major emphasis in this procedure is placed on the dominants, whereas in Europe it is common to take the lesser plants into account, sometimes to the exclusion of the dominants. In this book a partial compromise is attempted; the major subdivisions of the forest communities are based on the dominants although the groundlayer is also considered; and in the non-forest communities the most prevalent species of the entire assemblage are used, regardless of whether or not they are dominants.

In general, the system employed for a particular community was that which seemed best adapted to both the community and to the type of information available. There was no attempt to follow blindly a standardized or fixed system based on dominants, characteristic species, constant species, or other single criteria, since there was no prior information which could be used to justify such a procedure. In a few minor communities, the finer subdivision was done on physiographic grounds (as the lake beaches, lake dunes, and cliffs), or on growth form (as the emergent aquatics versus submerged aquatics).

Within the major or minor units of vegetation delimited by these varied methods, an attempt was made to collect comparable information whenever possible. The only standard information which was collected in all of the communities was presence—the number of stands in a community in which each species occurred, expressed as a percentage of the total number of stands studied. The comparisons of communities given in the last chapter are based on these presence values.

When the communities of a particular vegetation type, such as the southern forests or the prairies, are shown to form a vegetational continuum, then a subdivision can be made by dividing the entire range into arbitrary segments. Purely for purposes of convenience, this has been done by erecting five segments, termed wet, wet-mesic, mesic, dry-mesic, and dry, respectively. This terminology emphasizes the correlation between community composition and a major environmental gradient, but the causal mechanisms are far more complicated than simple changes in soil moisture. The terms, like the segments, are purely arbitrary and are used solely for convenience and uniformity. Exactly similar divisions have long been used in Europe (Vorobiev, 1953).

Following the above procedures, it has been possible to divide the bulk of the native vegetation of Wisconsin into 21 major communities. These include the five segments (wet to dry) of the southern forest, the five of the northern forest, and the five of the southern grasslands or prairies. In addition, there are four kinds of savannas, a northern grassland, and a boreal forest in the far north. Lesser communities of minor areal importance comprise another 13 types, for a total of 34. These are separated by certain diagnostic features in the key which follows. Their salient characteristics are described in the community accounts in the following chapters.

KEY TO WISCONSIN PLANT COMMUNITIES

A. Mature trees present (more than one per acre)
 B. Scattered trees only (less than 50 per cent canopy) . . . . . Savannas
  C. Hardwood trees dominant
   D. Major tree species are black oaks (*Quercus velutina* or *Q. ellipsoidalis*) . . . . **Oak Barrens**
   DD. Major tree species are bur (*Q. macrocarpa*) or white oak (*Q. alba*) . . . . **Oak Opening**
  CC. Coniferous trees dominant
   E. Major tree species is jack pine (*Pinus banksiana*) . . **Pine Barrens**
   EE. Major tree species is red cedar (*Juniperus virginiana*) . . . . **Cedar Glade**
 BB. Trees form closed stands (more than 50 per cent canopy) . **Forests**
  F. Trees hardwoods only; forests south of tension zone
   G. Sugar maple (*Acer saccharum*) or beech (*Fagus grandifolia*) is major dominant; canopy usually more than 90 per cent . . . . **Mesic Southern Hardwoods**
   GG. Other species dominant; canopy usually between 50 per cent and 90 per cent
    H. Communities of dry, upland sites; oaks usually dominant
     I. Bur (*Q. macrocarpa*), black (*Q. velutina*), or white oak (*Q. alba*) dominant . . . . **Dry Southern Hardwoods**
     II. Red oak (*Q. borealis*) or basswood (*Tilia americana*) dominant . . . . **Dry-Mesic Southern Hardwoods**
    HH. Communities of wet, lowland sites; oaks rarely dominant
     J. Willow (*Salix nigra*) or cottonwood (*Populus deltoides*) dominant . . . . **Wet Southern Hardwoods**
     JJ. Silver Maple (*Acer saccharinum*), American elm (*Ulmus americana*), or ash (*Fraxinus* sp.) dominant . . . . **Wet-Mesic Southern Hardwoods**
  FF. Coniferous trees usually present; forests north of tension zone
   K. White spruce (*Picea glauca*) and balsam fir (*Abies balsamea*) dominant; forests of far northern counties . . . . . **Boreal Forest**

KK. Other conifers dominant, with or without hardwoods

 L. Communities of mesic sites; sugar maple (*Acer saccharum*), beech (*Fagus grandifolia*), hemlock (*Tsuga canadensis*), or yellow birch (*Betula lutea*) dominant; canopy over 90 per cent . . . . **Mesic Northern Hardwoods**

 LL. Communities of wet or dry sites; other species dominant; canopy 50 per cent to 90 per cent

  M. Communities of dry upland sites; pines usually one of major dominants

   N. Jack pine (*Pinus banksiana*), red pine (*P. resinosa*), or Hill's oak (*Q. ellipsoidalis*) dominant . . **Dry Northern Hardwoods**

   NN. White pine (*Pinus strobus*), red maple (*Acer rubrum*), or red oak (*Q. borealis*) dominant . . . . **Dry-Mesic Northern Hardwoods**

  MM. Communities of wet lowland sites; soil usually peaty

   O. Tamarack (*Larix laricina*), black spruce (*Picea mariana*) dominant . . . . **Wet Northern Forest**

   OO. White cedar (*Thuja occidentalis*), balsam (*Abies balsamea*), black ash (*Fraxinus nigra*) dominant . . . . **Wet-Mesic Northern Forest**

AA. Mature trees absent; woody plants, if any, are bushes or shrubs

 P. Terrestrial communities (water table below soil surface during most of growing season)

  Q. Community dominated by shrubs

   R. Low, evergreen, ericaceous shrubs; sphagnum understory . . . . **Open Bog**

   RR. Tall deciduous shrubs; no sphagnum

    S. Alder (*Alnus rugosa*) dominant; north of tension zone . . . . **Alder Thicket**

    SS. Willows (*Salix* sp.) and dogwoods (*Cornus* sp.) dominant; south of tension zone . . . . **Shrub-Carr**

  QQ. Community dominated by herbaceous plants

   T. Essentially closed communities, usually with 90 per cent to 100 per cent coverage

    U. Over one-half of dominance contributed by sedge-like plants

V. Communities north of tension zone; *Scirpus* prominent . . . . **Northern Sedge Meadow**

VV. Communities south of tension zone; major dominance by *Carex* . . . . **Southern Sedge Meadow**

UU. Over one-half of dominance contributed by grasses (*Gramineae*)

W. Communities south of tension zone; ferns never codominant

X. Communities of wet sites, with deep soil and a gley layer

Y. Spring-fed supply of internally flowing, calcareous waters . . . . **Fen**

YY. Water supply from rain and surface drainage

Z. Sites commonly inundated in spring season; *Calamagrostis* and *Spartina* dominant . . . . . **Wet Prairie**

ZZ. Sites rarely inundated . . . . **Wet-Mesic Prairie**

XX. Communities of mesic or dry sites; no gley layer

a. Communities of mesic sites; deep soils . . **Mesic Prairie**

aa. Communities of dry sites; soils thin or sandy

b. Thin soils, over bedrock or gravel; *Bouteloua* and *Andropogon scoparius* dominant . . . . **Dry Prairie**

bb. Deeper soils, often sandy . . . **Dry-Mesic Prairie**

WW. Communities north of tension zone; bracken fern (*Pteridium*) is codominant . . . . **Bracken-Grassland**

TT. Relatively open communities of rocks or sandy soils; often less than 50 per cent coverage in ground layer

c. Communities of cliffs and rock outcrops

d. Communities exposed to full sun . . **Open Cliff Community**

dd. Communities shaded by trees or other cliffs . . . . **Shaded Cliff Community**

cc. Communities of essentially level topographies

e. Inland communities of sand plains . . . . **Sand Barrens**

ee. Lake shore communities—Great Lakes and larger inland lakes

f. Communities of lake strands; *Cakile* often dominant . . . . **Beach Community** *ff.* Communities of lake dunes; *Ammophila* and *Lathyrus* often dominant . . . . **Lake Dunes**

PP. Aquatic Communities (water table above soil surface during most of growing season)

    g. Dominated by emergent aquatics, such as *Scirpus, Typha* . . . . **Emergent Aquatic Community**

    gg. Dominated by submergent aquatics, such as *Potamogeton, Anacharis* . . . . **Submerged Aquatic Community**

VEGETATION MAPS

A number of maps of the vegetation of all or part of Wisconsin have been prepared by several investigators in the past. The first, presented by J. W. Hoyt in 1860, is remarkably accurate considering the nature of the information available to him. Areas of dense forest, according to Hoyt, are shown by the shading on the map at the opening of this chapter. The second map was drawn by Judge Knapp in 1871 to accompany his article on the vegetation of Wisconsin. It is very generalized, designating several "belts" of vegetation. The first detailed map was prepared by Chamberlin as Plate II-A in the *Atlas for the Geology of Wisconsin* (1877). This colored map was based on land surveyors' data for prairies, marshes, and swamps, and upon observations of members of the Geological Survey for the various upland forest types. Its accuracy is extremely variable, apparently depending upon the botanical knowledge and interests of geologists who examined different regions of the state. Thus, the St. Croix area and the region to the north of it is presented in great detail, based on the colored map of Wooster in Volume 3 of the *Geology of Wisconsin,* and on the detailed field notes of Moses Strong in Volume 4 of the same work. The eastern counties are well depicted, apparently based upon the combined field work of Chamberlin and Lapham. The north central region is the poorest, with distinctions between pine and hardwood forests seemingly having been made by exercise of the imagination. Another serious defect of the map is the failure to differentiate oak savanna from oak forest in the south, although pine savannas are separated from pine forests in the north.

This map by Chamberlin has been copied by many subsequent workers, usually with little or no modifications. Relatively generalized maps of the vegetation of Wisconsin were prepared by several writers as parts of larger maps of the vegetation of all or part of the United States. These maps included those of Sargent in his *Tenth Census Report* (1884), Shantz and Zon in the *Atlas of American Agriculture* (1924), and Braun in her book, *Deciduous Forests of Eastern North America* (1950).

The records of the land survey were used to provide data for detailed vegetation maps of counties or smaller areas as published by a number of workers. These include a colored map of the vegetation of the Brule River basin in Douglas County by Fassett (1944), a colored map of the vegetation of Dane County by Ellarson (1949), and a black and white map of the vegetation of Racine County by Goder (1956). In addition, manuscript maps of the same type are available for Barron County by M. A. Fosberg, Dodge County by Herbert Neuenschwander, Green County by H. C. Greene and J. T. Curtis, Jefferson County (Zicker, 1955), Kenosha and Milwaukee counties, by H. A. Klahorst, Vernon County by J. T. Curtis and N. C. Fassett, Vilas County by J. T. Curtis and N. C. Fassett, and for the entire state (Finley, 1951).

All of the maps based on surveyors' records suffer from certain defects inherent in the data. The surveyors were primarily interested in establishing a grid system of land measurement (see Chapter 4) and their use of botanical information was distinctly secondary. Thus, in forested regions, they used trees as witness markers for corners and the recorded information on the trees may be used to gain knowledge about the composition of the forest. On the other hand, in treeless areas such as prairies, meadows, or barrens, they built a mound of earth at the corner point and simply stated that the corner was in an open area. There is, therefore, no way by which the varying composition of these non-forested areas may be determined from the survey records. Because there are more non-forest communities than forest communities in Wisconsin, any maps based on survey data are necessarily very incomplete. In spite of this inadequacy, however, the surveyors' records remain the best source of information available on the nature of Wisconsin's vegetation before the coming of major European settlement.

The vegetation maps in this book are based upon the existing maps of survey origin as mentioned above, checked by reference to the actual records in cases of doubt, or by field inspection of questionable areas. It is hoped that the resulting maps will give a fair impression of the regional distribution of the major vegetation types but it is recognized that only a low degree of precision exists. This is due to two important difficulties.

The first difficulty concerns the continuum nature of the vegetation with its gradual blending from one type to another. This is particularly important when making decisions on the location of the boundary between forest and savanna or between savanna and grassland. The difficulty has been partially surmounted in some of the individual maps (Figures VII-2 and XI-2) by showing the related community in stippling. On Figure VII-2, for example, the oak forests are shown in black, mostly surrounded by oak savanna in stippling. The combined representation may be interpreted as a map of the area within which closed forests of oak may be found, with the chances of such a forest being much higher in the black area than in the stippled area.

The second difficulty is concerned with the subjective reduction of types as made necessary by a small-scale map. In general, areas of a particular community which are smaller than a township (36 square miles) are omitted from the map unless they are contiguous with other areas of the same type. This is most serious in sedge meadows, open bogs, and conifer swamps, which frequently occupy glacial kettle holes much smaller than this minimum.

The end-paper map in this book shows the general distribution of the major plant communities of Wisconsin as they existed about 1840, at the time of the land survey. The communities dominated by trees are shown in several shades of green, while the communities dominated by grasses are shown by stippling. The combined symbols of color and stippling thus represent the savannas.

The acreages of the major types were obtained by cutting out the designated areas of each type and calculating their area by comparing

Table 3
Areas of major plant communities at the time of settlement (1830–1850)

Community	Acres	% of total land surface	Community	Acres	% of total land surface
Southern Mesic Forest	3,432,500	9.81	Northern Lowland Forest	2,241,500	6.40
Southern Xeric Forest	1,386,500	3.96	Boreal Forest	672,500	1.92
			Prairie	2,101,500	6.00
Southern Lowland Forest	420,000	1.20	Oak Savanna	7,257,000	20.74
			Pine Barrens	2,340,000	6.68
Northern Mesic Forest	11,741,000	33.55	Sedge Meadow	1,135,000	3.24
			All forest	22,163,500	63.32
Northern Xeric Forest	2,269,500	6.48	All savanna	9,597,000	27.42
			All grassland	3,236,500	9.24

their weight to the total weight of the state in outline. Because of the above mentioned limitations on the precision of the map, the resulting acreage figures are only approximations. For those communities not shown on the map, such as the various subtypes of prairie, an estimate was made from a consideration of the existing relics of each type. These estimates obviously have a very low precision. The data on acreages are summarized in Table 3.

CHAPTER 4

Vegetation study methods

SOURCE OF DATA
Surveyors' records
Our knowledge of the vegetation of Wisconsin is derived from a number of different sources which includes generalized surveys and detailed local studies. One of the most complete sources is provided by the records of the federal land surveyors as mentioned in the last chapter (Bourdo, 1956). The surveyors' task was to lay out a grid system that would enable the incoming settlers to find and describe the parcels of land they wished to occupy. This grid consisted of rectangular coordinates spaced 6 miles apart in both north-south and east-west directions. The areas within the grid lines were called townships. Each of them in turn was divided into 36 square sections, each 1 mile on a side. In the actual operation of surveying, each line was measured on the ground. At every half-mile interval on the lines in each direction,

a corner point was established. These points were described by giving their bearings, in direction and distance, to two or four blazed trees nearest the corner. The species of each tree was listed as was its diameter. In this way, settlers could find the exact spot with relative ease and locate their claim.

Botanically, these surveyors' records constitute an unbiased sample of the vegetation as it existed in presettlement times. By proper calculations, it is possible to determine the number of trees per acre, their average size, and their frequency of occurrence from the surveyors' data. In addition to these figures on the trees, the surveyors also listed other species they saw along their traverse and gave a brief summary of the understory vegetation. When trees were lacking, as on prairies or on marshes, this fact was clearly indicated. Swamps were distinguished from uplands, and soil quality was estimated, since the surveyors also were charged with assessing the agricultural possibilities of the lands they saw. For each township they drew a map showing prairies, swamplands, lakes, streams, and other features.

The original record books and maps of the survey in Wisconsin are on file in the office of the Commissioners of Public Lands, in the Capitol at Madison. From these records it is possible to obtain a quantitative picture of the vegetation at the date of the survey. From the surveyors' own statements as to the nature of the vegetation and from their maps, areas can be delimited which appear to be relatively homogeneous in composition. Data from all corner points in these areas can then be tabulated and the composition of the forest in terms of quantitative measurements of density, dominance, and frequency for each species of tree can be calculated. Such a procedure was followed in the investigations of Stearns (1949), Cottam (1949), and Ward (1956).

A less satisfactory, but still very valuable use of the surveyors' data is in the construction of vegetation maps which have no quantitative basis but rather are subjective interpretations of the vegetation as gained from a simple reading of the surveyors' notes. This approach was used for many of the maps described in Chapter 3, such as those by Fassett (1944) in his study of the Brule River Basin, Ellarson (1949) in Dane County, and others by Fassett's students who prepared manuscript maps of various counties.

The geographer Finley (1951) prepared a manuscript map of the vegetation of the entire state, based largely on a combination of floristics and apparent physiognomy, rather than on the phytosociological information contained in the records. His map lists 32 combinations of species, which are shown in great detail by different colors and shades of color. It is perhaps the most useful map ever made to convey at a glance the tremendous complexity of the vegetation on a large area.

In addition to the information on distribution of the vegetation that can be obtained from the surveyors' records, data are also available concerning the quantitative changes that have taken place in the hundred or so years that have elapsed since the surveys were made. Stearns, Cottam, and Ward used the data for these quantitative changes, but there are many other situations where such use would be extremely valuable.

Other non-ecological sources of data

The surveyors' botanical information was incidental to their main activity and obviously was not the result of an ecological investigation. A number of other sources of information are available in which the vegetational data also are incidental to the main purpose of the work. Included here are reports and accounts of explorers and early travelers in the state. For the most part, these are fragmentary, but they frequently give a good picture of a particular region. Their main value lies in the long period involved, which spans more than three hundred years and goes back to 1636.

In the 1800's, a number of more detailed descriptions of the vegetation began to appear. Perhaps the best of these are included in the accounts of the geology of various sections of the state, as given in the four volume *Geology of Wisconsin* (1881–83). At about this time, a volume on the local history was produced for almost every county; these frequently included detailed descriptions of the vegetational cover of the county (Greene and Curtis, 1955).

Another source of information prepared on a county basis was the material included in the books accompanying the soils maps that were produced by the Wisconsin Geological and Natural History Survey. These were largely restricted to the southern half of the state, and the coverage was incomplete even there. Nevertheless, they are valuable because they are based on detailed inspection by well-trained field men.

A more ambitious source of similar information is to be found in the *Wisconsin Land Economic Inventory* of the Wisconsin Department of Agriculture. This agency mapped the entire state on the basis of data derived from a series of transects one-quarter mile apart. The maps show the location of wood lots and forests and give the only available information on the distribution of non-forest lands, such as the various kinds of marshes, bogs, and thickets. This inventory was conducted during the 1920's and 1930's.

A more recent survey of the northern portion of the state, called the Wisconsin Forest Inventory, was carried out in the 1950's under the auspices of the Wisconsin Conservation Department. This survey was based on the use of infrared aerial photographs and frequent ground controls. Of greatest vegetational interest is the series of maps that resulted, one per county, which showed the ground cover according to broadly standardized forest types. A comparison of these maps with similar ones prepared from the early land survey adds greatly to an understanding of the changes that have occurred in the last century (see Figures 26, 27, and 28).

Non-University of Wisconsin ecological studies

Studies of plant communities by investigators directly interested in ecological aspects are restricted to the current century. For convenience, the studies will be divided into those with no connection with the botanical research program of the University of Wisconsin and those associated with the University. In the first category, the oldest investigations are those of Bruncken who wrote a series of papers called "Studies in Plant Distribution" from 1902 to 1910. These were concerned mostly with the region around Milwaukee; they included quite detailed analyses of the relations between vegetation types and soil or climatic factors and also included discussions of the successional history of the communities.

Pammel (1907) gave similar information for the vegetation of western Wisconsin, centering on La Crosse County. He included many lists of species typical of the various communities. The prairies of the Lake Michigan shore in Kenosha County were included in a study by Gates (1912), again with floristic lists as the main emphasis. In 1914, H. H. T. Jackson described the floristic aspects of the various stages in bog succession in Oneida County in connection with his study of land vertebrates.

Apparently the first attempt to use quantitative measures of vegetation by means of quadrats was made in 1906 by Ruth Marshall (1910), when she described the plant communities of Twin Island in Lake Spooner, Washburn County. Miss Marshall, then a student at the University of Nebraska, came into contact there with the new methods that had recently been published by F. E. Clements in his book on quantitative measures (1905). Bruncken also used the quadrat method in his study of Lake Michigan beaches in 1910. Another quantitative study of vegetation appeared in 1936 when D. F. Costello, a student at the University of Chicago, published on the tussock meadows of southeastern Wisconsin. Costello gave quantitative data with quadrat frequencies for the plants in a number of stands of sedge meadows dominated by *Carex stricta*.

The next use of the quadrat method was by W. A. Eggler in 1938. Eggler was a student in the famed Plant Ecology Seminar of Professor W. E. Cooper of the University of Minnesota. He studied the maple-basswood forest of Washburn County, Wisconsin, and published detailed phytosociological information according to modern methods. He also reported on the vegetational history of the region as revealed by age determinations of living trees and stumps. This was the first complete study in forest ecology to be made in Wisconsin.

In the same year, 1938, Kittredge reported on the aspen community in Wisconsin, with detailed statistical treatments of the understory plants of this forest type. Since 1938 there have been a number of other non-University of Wisconsin studies, both quantitative and otherwise, including those of Stout (1944) on the oak openings, Scully (1942) on the oak-maple forest at Lake Geneva, and Russell (1953) on the plant communities of Apple River Canyon in St. Croix County.

University of Wisconsin ecological studies

The first ecological study by a botanist from the University of Wisconsin was the amazingly detailed, quantitative, phytosociological investigation of a sedge meadow near Madison by A. B. Stout. This was begun in 1907 and appeared in print in the *Transactions of the Wisconsin Academy of Science* in 1914. It consisted of a complete analysis of frequency, density, and weight dominance by means of the quadrat method. Among the results was the separation of the meadow into three subcommunities which were termed *Caricetum, Lycopus-Caricetum,* and *Calamagrostis-Caricetum*. Stout also conducted comprehensive studies of the autecology of the major dominants of the assemblage. No equally thorough ecological research on a single plant community in Wisconsin was reported between the time of this pioneer investigation and the work of Cottam in 1948. In fact, it would appear that little emphasis was placed upon an ecological approach to the plants of Wisconsin by university botanists prior to 1930, although a few general descriptions with floristic lists of plant communities were produced during the Stout era in 1910 in three theses by Davis, H'Doubler, and Heddle. Nothing further appeared until students of

N. C. Fassett of the Botany Department began publishing on various communities of the state, beginning with the bog studies of Hansen and of Rhodes in 1933. Most of these studies had a floristic-plant geographical basis, which was in keeping with Fassett's interests in taxonomy and plant geography, but they were centered on a particular community as the unit of study. Among the many such studies, reference can be made to those which saw the light of publication; these included Thomson's studies on relic prairies (1940) and succession on abandoned fields (1943), Gould's paper on indicators of original prairie (1941), and Fassett's (1944) work on the vegetation of the Brule Basin.

A number of investigations were conducted from the University of Wisconsin Biological Station at Trout Lake in Vilas County, then directed by E. A. Birge and Chauncey Juday and operated at the time largely as a limnological research station. Among the plant ecological papers resulting from these investigations were those of Wilson (1935) and Potzger (1943) on the aquatic plant communities of northern lakes and those of Potzger (1946) on the hemlock-hardwood forests of Vilas and Price counties.

During this same period a number of studies of fossil pollen as preserved in peat bogs were made by Fassett's students. These are referred to in more detail in Chapter 22.

Plant Ecology Laboratory studies

All of the foregoing sources of information have been drawn upon to provide a portion of the basic data on the plant communities discussed in succeeding chapters. By far the greater part of the data, however, has been taken from the files of the Plant Ecology Laboratory (hereafter abbreviated as P.E.L.) of the Botany Department of the University of Wisconsin. Specific early works of the present author on various phases of plant reproduction (Curtis, 1932, 1936, 1939, 1941, 1943) served to emphasize the importance of a knowledge of the ecological life histories of native plants in any explanation of their behavior in the field or of the behavior of their collective community groups. The great research inefficiency of the accustomed method of studying one phase of the behavior of one species at a time led to the conviction that the necessary information could be obtained only by studying related groups of species simultaneously and by studying all aspects rather than a single one of those species which appeared to be significant. Such an approach demanded an intimate knowledge of the phytosociological relations of the flora, the species which commonly grew together, the conditions under which each grew at optimum levels, and the geographic locations where each combination could be found. Information on these points was almost completely lacking in 1946, so the decision was made to concentrate on collecting this information before intensive work on behavior studies was undertaken. During the next ten years, all major communities in Wisconsin, and most of the minor ones, were subjected to study. The present book is largely an attempt to summarize this information and to integrate it with the results of other workers.

A total of 1420 separate stands were investigated, including 1045 in native terrestrial communities, 187 in native aquatic communities, and 188 in weed communities of disturbed places. Distribution maps showing the locations of the several groups of stands are given in the summaries in the Appendix, and all of the stands in the entire P.E.L. study are shown in the figure reproduced at the opening of this chapter. On the average, there were 2.59 stands per 10 square miles, but they were not uniformly distributed, with greater than normal concentrations in the vicinity of the University, in and near Dane County, and in the vicinity of the Trout Lake Biological Station in Vilas County, and less than normal concentrations in the general region of the tension zone. In spite of these deviations from uniformity, it is believed that the coverage is adequate to justify the treatment of the vegetation of the entire state as given in the following chapters.

METHODS OF P.E.L. STUDIES

Field methods

Early in the course of the P.E.L. studies, it became apparent that the quadrat method then in common use would be inadequate for an investigation of the scope contemplated. The tremendous amount of time needed in the field to sample properly a single forest stand by the use of 10 meter by 10 meter quadrats meant that only a few stands of each type could be studied without the expenditure of inordinate amounts of money. Accordingly, much attention was devoted to the development of a new method which would give similar results in a much shorter time. As a result, a whole new approach to the sampling problem was developed (Cottam and Curtis, 1949). In brief, this consisted of the use of point samples, rather than samples of a fixed area, and involved the determination of the space occupied per plant, rather than the number of plants per unit area. Certain details of these new methods were improved during the course of the community studies (Cottam and Curtis, 1956), but the same basic technique was used throughout, so that all of the results are on a comparable basis. In addition, the methods give results very similar to those which may be obtained by proper calculations of the figures of the original land survey; thus then-and-now comparisons can be made with little difficulty.

As an aid in understanding the results to be presented later, and to assess their limitations, it seems advisable to present a complete description of the procedures followed in both the field and the laboratory. Forests are the most common type of community in Wisconsin today and more detailed work has been conducted in them than in any other community, so the procedures used in the study of a forest will be given first.

It has been said that the most important decision made by an ecologist is that made when he stops his car (Ashby, 1948). In other words, the choice of a place to study is more likely to affect the results than anything the ecologist does subsequently. There is no feasible way whereby this subjective judgment can be completely avoided. It is possible, however, to greatly minimize this influence by the use of a properly chosen set of criteria for the recognition of a piece of vegetation suitable for study. Given these criteria, then any stand which does not pass the test is omitted, while all of those which do (or as many of them as possible) are included. In the P.E.L. forest studies, the following criteria were used throughout.

Each stand was 15 acres or more in size, to reduce the edge effect from wind movement which is excessive in very small stands. Each stand was upon a uniform topographic site. In the case of upland forest studies, this meant that rain water never accumulated on the surface, while in lowland forests, it meant areas which were occasionally inundated by runoff water from adjacent upland or by rivers. Each stand was free from serious disturbance during the life of the current generation of trees. This was interpreted to mean that no visible signs of past grazing by domestic animals could be present, nor could the area have been selectively logged in the past to an extent greater than a 10 per cent decrease in canopy coverage. Because almost all the stands had been subjected to occasional ground fires, total absence of past fire could not be used as a criterion for choice.

The most important criterion used was that of homogeneity. This is a difficult concept to understand and to employ in the field. It has to do with the uniformity of distribution of the species throughout the area sampled. In southern Wisconsin, most forest stands exist as farmer's wood lots, usually bounded on all sides by cultivated fields. Frequently these wood lots have straight sides and give the appearance of a unit. Common ownership, however, is not a sufficient indication of ecological uniformity. Many wood lots contain portions of two or more forest types, due to differences in topography or past history. Nothing of value is to be gained by sampling them together in one study and presenting the average results. In an extreme example, the averages might show a forest composed of equal quantities of white oak and sugar maple although in actuality one half of the woods was pure oak and the other pure maple. To guard against such occurrences, all of the P.E.L. stands were tested for homogeneity. In this way, the results applied to a particular, uniform portion of a forest, regardless of the fact that a farmer's wood lot might contain, in addition, several other types.

The test applied was the Chi-square homogeneity statistic, used on all of the tree species with a relative density of 25 per cent or more (Curtis and McIntosh, 1951). The data taken in the field were divided into four equal portions in the geographic order in which they were recorded. Each portion was required to be approximately equal to each other portion, within the limits of expectancy set by Chi-square, for these major trees. If a stand proved homogeneous by this test, it was included in the study, because it was assumed that the major trees were most important in determining microclimate and other conditions in the forest. It would be impossible, of course, to demand complete homogeneity from all members of the forest since this condition never occurs in nature nor is it ever very closely approached. After a little practice with this homogeneity test, an investigator soon learns to recognize non-homogeneous areas in the field and to avoid their initial sampling.

Given a forest stand which fits the above criteria, a relatively standardized procedure was followed in the collection of the necessary field data. A reconnaissance trip was made through the stand to check condition and apparent homogeneity. The portion selected for study was examined for species complement, and every species seen was recorded on a printed presence list which included the names of all of the common species of the type to be studied and which also included spaces for rare species. The size of the area to be sampled was then estimated and the necessary distance between samples determined. The first sampling point was determined by a random toss of a quadrat stick which is usually used for a marking point. Subsequent points were determined by pacing a compass line for the requisite distance. This line was usually predetermined in the form of a square U, so that the investigator would return somewhere near his starting place. The compass lines were located at least 100 feet inward from the edge of the wood lot, if such an edge were present.

At each sampling point selected along the line, a quadrat was laid out to sample the presence or absence of herbs and shrubs, herein called the groundlayer. The square quadrat was ¼ milacre in size, or 1.008 meters on edge, and was centered directly over the sampling point. The plants present were checked off on the printed sheets, one per quadrat, which contained the names of all common plants likely to be encountered. Following this, the trees were examined. Most of the information about the trees was required to be on an area basis such as, so many trees per acre, so much basal area per acre, or the like. The usual methods for securing this information call for samples of fixed area, actually laid out upon the ground, within which the trees are counted or measured. Similar results may be obtained by measuring the distance between the trees and determining the average amount of space per tree, rather than the number of trees per unit of space.

In early studies, the random pairs method (Cottam and Curtis, 1949, 1955) was used, while in later work, the quarter method (Cottam and Curtis, 1956) was employed. Both methods are similar because they determine the average distance between the trees in the forest. The random pairs method measures the distance between the tree nearest to the sampling point and a second tree which is the closest to the first but which lies outside of an exclusion angle of 180°. The line between the sampling point and the nearest tree is considered the bisect of a straight angle through the point, extending along an imaginary line 90° to the right and to the left of the point. In simple terms, all trees in front of the investigator as he stands on the point with his arms outstretched and faces the closest tree to him are excluded. The distance from the closest tree to the one nearest it but behind him is measured. The average of all of these distances for all sampling points gives the average distance between all trees in the forest. This distance, when squared, is equal to the average space occupied per tree. From this it is an easy step to calculate the number of trees per acre, per hectare, or other unit area. In addition to the distance between this pair of trees, the species and basal area of each tree of the pair at a level 4.5 feet above the ground is recorded on a special blank.

In the quarter method, the procedure is similar but differs in the choice of trees. At each sampling point, the direction of the compass line is used as a bisect of the space around the point and this space is further divided by another imaginary line erected at right angles to the first, with the point as center. Within each of the quarters thus demarcated, the closest tree to the point is chosen and its distance from the point determined. These measurements are made with an optical range finder by the investigator standing on the point or by means of a conventional tape. All of the distances thus determined at all sampling points are summed and the total divided by four times the number of points used. The result equals the average distance between trees and is treated as in the random pairs above to obtain the average space occupied per tree. Species and basal areas are determined for each tree as before. A given number of points gives twice as many trees in the sample as in the random pairs method; the advantages of the quarter method largely stem from this fact. These methods are discussed in detail by Greig-Smith (1957).

In the forest studies, certain environmental measures were made in each stand. These included an ocular estimate of the percentage canopy coverage at each sampling point (in many cases a photoelectric measurement of light intensity was made as well) and the soil permeability as measured with the Wilde soil permeameter. Soil pits were dug at three places along the line of transit; in each the soil profile was described and soil samples of the A_1 soil horizon were collected. These collections were pooled to give a single sample for each stand and were later analyzed for water retaining capacity (herein abbreviated as w.r.c.), soil reaction (pH), and available nutrients in the laboratory. In most cases, the latter analyses were simple colorimetric tests based on standard soil extracts; they included calcium, magnesium, potassium, phosphorus, nitrate nitrogen, and ammonia nitrogen. The results are expressed as parts per million (p.p.m.) of the dry weight of the A_1 horizon and thus represent concentrations rather than total amounts, since the latter figure is greatly influenced by the depth of this layer. The water retaining capacities were determined on samples which were ground to pass a 2 mm. screen and placed in a standard cup in a layer 1 cm deep. The pH determinations were made with a glass electrode on water suspensions of the soil. In the case of prairie soils, sedge meadow soils, and other types with poorly defined horizons, the sample consisted of a standard volume taken to a depth of 4 inches.

In the major non-forest communities, the only phytosociological measurement made was quadrat frequency, although similar environmental studies were made in many cases. For the lesser communities, the only record taken was species presence. Special techniques were necessary in the studies of soil fungi and corticolous bryophytes, in keeping with the size and difficulties in identification of these organisms, but the results in all cases were made in terms of presence or quadrat frequency, so they are comparable to the results on herbaceous plants in the other studies.

Laboratory methods

Upon return to the laboratory, the data of the field studies were tabulated and reduced to average values. In the case of forests, the averages included: number of trees per acre, total basal area per acre, average basal area per tree; frequency, density, and basal area per species; relative frequency, relative density, and relative dominance per species. These last three values were summed to give importance values (I.V.) for each species (Curtis and McIntosh, 1951; Cain, Castro, and Pires, 1956). All of these figures were reported upon special tabulation sheets. The groundlayer quadrat records were similarly transformed to frequency percentages and tabulated on special forms. All of the tabulations plus the original field sheets were then filed, one stand per folder, according to the number of the stand.

After all stands had been examined, it was frequently desired to arrange them according to some gradient by an ordination method suitable to the data. These methods are discussed later under the several community treatments. For the arrangement, all of the important data for a given stand were copied to a Post-Index Visible File system. In this system, all variable figures were recorded along the extreme edge of the sheet, and the sheets were arranged in a tray in a staggered or offset fashion, so that only the edges with the recorded values were visible. The sheets were placed in the tray in the numerical order called for by the gradient under study. In this way, it was easily possible to make various syntheses of the data and study the behavior of each species in relation to any particular gradient. The use of the visible index meant that the data were copied only once, rather than frequently as called for by other systems, and the clerical errors involved were thus reduced to a minimum.

In certain studies, the data from individual quadrats or from individual stands were transferred to punched cards of the IBM system (Cottam and Curtis, 1948). By the use of appropriate sorting equipment and electronic computers, the occurrence of any one species with each other species could be checked against random expectation, and the deviants checked for significance by Chi-square tests.

Structure of forest stands

In each of the major kinds of forests, an attempt is made to portray the numerical structure of the forest by presenting a structural analysis of one or two typical stands of each type. These analyses were obtained by a sampling procedure, similar to that used in all stands, which surveyed a sufficient number of individual trees so that the data could be presented by size classes. Special quadrat studies were employed to obtain figures on density of seedlings and saplings. The data in each case are presented in the standard form usually used for such information (Eggler, 1938). A seedling is defined as a young tree less than 1 inch in diameter at breast height (d.b.h.), with a subdivision into those individuals less than 1 foot tall and those individuals more than 1 foot tall but less than 1 inch d.b.h. Saplings are defined as young trees between 1 inch d.b.h. and 4 inches d.b.h., and mature trees are those more than 4 inches d.b.h. The latter are subdivided into those between 4 inches and 10 inches d.b.h., those from 10 to 20 inches d.b.h., 20 to 30 inches, etc. In all of the structure tables, the figures in the body of the table are densities of indicated size class and species in numbers per acre; they may be multiplied by 2.47 to get numbers per hectare.

Statistical reliability of the results

At the outset if may be said that no study of a single stand in the entire P.E.L. series gave statistically reliable results for all species encountered. However, this same statement can be made of all ecological studies ever reported by all investigators in all countries. This conclusion is inherent in the nature of plant communities. A normal plant community contains a few species which are present in great numbers or great amounts, more species which are intermediate in density, and a great majority of species which are rare or sporadic in occurrence. The sampling process can easily be designed to accurately measure the common species and, with greater expense, to measure satisfactorily the intermediate species, but the rare species require such an intensity of sampling that their precise measurement is beyond practicable attainment. The situation is not as unsatisfactory as it may sound, however, since greatest emphasis in interpretation of community composition and behavior is always placed on the common species, and these are exactly the ones most accurately measured by the usual sampling methods.

Within these density-dependent limitations, it is instructive to examine other possible sources of error in the data and to assess their magnitude where feasible. A number of tests, in which the same community was measured by different investigators at different times, indicated that the importance values of the individual tree species were duplicated within 12 to 15 per cent on successive trials. The compositional index of forest stands (see Chapter 5) was approximated within 5 per cent. Frequency values for groundlayer species were accurate to within 10 per cent on the average. The variability in the case of forest trees is known to be greatly influenced by the degree of aggregation of the species, with those kinds which tend to grow in clusters being sampled with a much higher error than those which are dispersed at random throughout the stand. Frequency values of the groundlayer as obtained from quadrat studies appear to be independent of degree of aggregation. This is understandable if frequency is thought of as the chance of finding a given species within a distance of one-half of the square root of the quadrat area from any random point chosen in the stand. The main factor affecting the accuracy of quadrat studies is that of density-dependence mentioned earlier. Rare species are encountered an insufficient number of times to give statistical reliability, and the sum of the resultant errors tends to reduce the over-all reliability of the entire complement.

One aspect of the sampling question that is of importance in assessing the reliability of results is the reinforcing factor that comes from replication. If a single stand were all that were available, then the results for all rare species would remain doubtful. However, when a number of similar stands are studied and a particular species shows low values in all of them, then reliance can be placed on conclusions drawn from the combined data, even though each individual value is in some doubt. In the P.E.L. studies, at least one hundred stands each were examined for all major communities and for several minor communities. The reinforcing factor thus strengthens the results for all but the very rarest species.

Taxonomic problems

If the statistical errors resulting from the sampling methods seem to be within reasonable limits, the same cannot always be said for another major source of error in phytosociological investigations. This may be termed the taxonomic error. It is brought about by the difficulties in correctly identifying every specimen encountered in the sample. The error has two main components, a human error and a miscegenation error in the plants. The first is due to lack of experience on the part of the investigator. Only very long-term familiarity with all of the species likely to be encountered can provide the field man with the necessary background to be able to identify every plant, regardless of the stage of its growth. It is doubtful whether the normal human life expectancy is sufficient for this purpose, if a total flora the size of that found in Wisconsin is involved. In more complex regions towards the tropics, one lifetime is certainly not enough.

The plant error involves the entire question of species delimitation, of hybrid introgression, and of ecophene and ecotype development. As pointed out in Chapter 1, hybrid swarms between two or more species are frequently encountered in the field, with no fully satisfactory way of handling the intermediates available at present. The problem of species demarcation includes not only the uncertainties involved in the questionable taxonomic status of many entities, but also the literal impossibility of identifying members of many groups when they are not in flower or fruit. This last applies with full force to the sedges and other members of the *Cyperaceae* and *Juncaceae* and is present in many other groups as well.

It has been suggested by some well-meaning but inexperienced taxonomists that ecologists should collect a voucher specimen for every species they encounter each time they encounter it, so that any possible errors in identification may later be checked by competent authorities. This suggestion has not one but two fatal weaknesses. In the first place, the labor necessary to collect, press, label, and ship the specimens would more than double the cost of the field work. In the second place, the number of voucher specimens required for a study such as the P.E.L. series would stagger even the most ardent herbarium curator. On the basis of one specimen for each species in each stand, the present investigation would have required 54,015 collections. If the voucher idea is extended to the individual quadrats in those communities where this method was employed, then the number of encounters with separate species amounted to 93,320. This number compares to the 75,000 Wisconsin sheets currently present in the University of Wisconsin herbarium and would mean an expenditure of nearly $50,000 merely to house the vouchers.

There is no intent to minimize the taxonomic error, as it is certainly the greatest to be faced in any extensive ecological study. A much more adequate series of keys for the identification of plants in a vegetative condition are needed as well as a great increase in the number of ecological herbaria which contain known species at all stages in their life cycles.

In the P.E.L. studies, especial difficulty was encountered in the following groups, which are listed to give the reader some basis for assessing validity of the results for any group which is of particular interest to him. The greatest source of trouble was the genus *Carex*. The genus is currently being studied in Wisconsin by James Zimmerman who identified all flowering or fruiting specimens which were brought to him. As a result, the over-all occurrence of *Carex* species in each community is probably fairly well understood. On a quantitative basis, however, the situation is very different. Most species of *Carex* flower and fruit in late June and early July. Individuals encountered at other seasons in quadrat samples are practically undeterminable. All quantitative values reported for *Carex*, therefore, are minimum values. In a less extreme way, the same comments apply to species in all genera of the *Cyperaceae* and *Juncaceae*. In some notoriously difficult groups like *Amelanchier, Crataegus, Rosa*, and certain sections of *Rubus*, no attempt was made to record species determinations, since it is doubtful if species in the usual sense exist. Other groups in which a certain element of doubt is present in some of the P.E.L. studies include *Aronia, Bromus ciliatus–B. kalmii, Cerastium arvense–C. vulgatum, Chenopodium, Digitaria ischaemum–D. sanguinalis, Eriophorum, Gerardia, Helianthemum, Hypericum, Lonicera canadensis–L. oblongifolia, Lysimachia quadrifolia–L. quadriflora* (not because of similarities of the plants but because of the confusion of the names), certain sections of *Panicum, Polygonum, Ribes, Scrophularia, Scutellaria leonardi–S. parvula, Sisyrinchium, Solidago juncea–S. missouriensis, Sparganium, Stachys*, and *Viola pallens–V. incognita*. In all of these cases, positive identifications for each community of occurrence were made on flowering specimens wherever possible, but quadrat frequencies based on non-flowering material may be in error. The *Lonicera* and *Scutellaria* troubles were due to a faulty understanding of these taxa in the early years of the investigation.

The problem of correct nomenclature was a difficult one in the P.E.L. investigations. During the early years of the study, the accepted standard was the 7th edition of Gray's *Manual of Botany*, although many of the names included in that work had been changed as a result of monographic revisions in the years since the first publication. In the later phases of the P.E.L. study, the 8th edition of Gray's *Manual*, as prepared by Fernald, was published and this was quickly followed by Gleason's revision of Britton and Brown's *Illustrated Flora*. These two manuals employed the same nomenclatorial system and they were in agreement for most species. In some of the difficult groups, however, there was a considerable divergence in treatment. Careful study of these differences in relation to the actual Wisconsin populations showed that the more conservative treatment by Gleason was more useful than the atomistic splitting by Fernald, probably because Gleason had had a long personal experience with the flora of the Midwest as well as the East and was better able to judge the significance of minor variations. As a result, it was decided to follow Gleason's nomenclature in the P.E.L. studies. The species names in that book are given without authorities, since these may be ascertained easily from Gleason's *Flora*. In the very few cases where other names are used, these are accompanied by authorities. The main exception is the case of the yellow ladyslippers (*Cypripedium parviflorum* Salisb. and *Cypripedium pubescens* Willd), where Fuller's treatment (1932) is clearly more in accord with the local facts than that followed by either Gleason or Fernald.

Summation of data on community basis

As indicated in Chapter 3, the vegetation has been divided into communities on the basis of objective segmentation of compositional gradients, or by consideration of physiognomy, or of topography. The total flora in each of these communities is rather large, whereas the type of information available about the flora differs greatly from community to community, with detailed quantitative figures on density, frequency, and dominance for the trees in the forest and savanna stands, and on frequency for the groundlayer species in the forests and many of the grassland stands. In most of the lesser communities, the only measure available is presence, which is the number of stands of occurrence of a species, expressed as a percentage of the total number of stands examined. Since presence values are also at hand for all communities in which more detailed measures were made, it seems best to standardize on presence as the basic guide in the description and comparison of communities. This is the usual practice in most European phytosociological studies and is frequently employed in America as well.

The following procedure was used for all communities. First, the average number of species per stand (called the species density) was determined. This was most easily done by summing the presence percentages of all species and dividing the sum by 100 (Raunkiaer, 1934). Then the species were arranged in decreasing order according to their presence values, beginning with the species of highest presence. A group of the most widely distributed species was marked off, to be used in later comparisons between communities. The limits of this group were determined objectively by use of the value for species density. For example, if there were an average of 35 species in each stand of a given community, then the top 35 names on the presence list were included in the group. This method has been employed by Raabe (1952) who termed the group the "floristic-characteristic species combination." In this book, the more descriptive name of "prevalent species" will be given the group. It contains the species most likely to be encountered in any stand of the community and these species are also the ones present in highest densities within the stands.

Some idea of the homogeneity of the community (the amount of stand to stand variability) can be obtained by comparing the sum of presence of the prevalent species with the total sum of presence of all species. Similar methods were suggested earlier by Raunkiaer (1934), Curtis and Greene (1949), and Raabe (1952). If the same species, and only the same species, were found in each stand of a community, then the prevalent and total floras would be the same and the ratio of their sums of presence would be 1.0. On the other hand, if only a few of the species were found in all or most of the stands and the remainder were confined to one or a few stands, then the ratio would be some place between 0 and 1.0. In the actual P.E.L. studies, the homogeneity index varied from 34.1 per cent for the beach community to 70.3 per cent for the dry prairie, with the ratio expressed as a percentage (Table IV-2).*
One advantage of this index is that it is more or less independent of the number of stands (Raunkiaer, 1934), since the species density quickly reaches a constant value after the first few stands are studied.

Some investigators (Dahl, 1957) have used the index of diversity (Williams, 1944) as the basis for a measure of homogeneity. This index gives a measure of the floristic richness of a community. It is calculated by a formula which divides the difference between the total number of species in all of the stands and the average number of species per stand by the Naperian logarithm of the number of stands. In the P.E.L. study, the index of diversity ranged from 8.87 for the submerged aquatic community (Chapter 20) to 70.40 for the boreal forest (Chapter 13). The indices for all communities are given in Table IV-2. Dahl suggested that the ratio of the species density to the index of diversity might be a good measure of homogeneity, with high values indicating homogeneity. Calculations of these ratios for the P.E.L. communities showed them to be closely correlated with the index of homogeneity values described above, with a linear relation between the index of homogeneity and the logarithm of Dahl's ratio. The exact significance of this relation is difficult to assess.

* Figures and tables appearing in the Appendix are grouped by chapters and numbered with a Roman numeral for chapter number and an Arabic numeral for sequence.

In addition to the lists of prevalent species given for each community in the following chapters, another group of species is indicated for each, namely, the "modal species." These are the plants which have their maximum presence value in the given community. When such species are also prevalents, this fact is indicated by an asterisk after the name on the prevalent list. Many of the modal species, however, are infrequent or rare. They are listed in a separate paragraph. Since all species encountered in the P.E.L. study necessarily have a maximum value in one community or another, the combined lists of modal species include the total understory flora found in the study, or 1105 species. The determination of modality was based solely on the arithmetic values for presence. If this number for a given species was but slightly higher in one community than in another, then the species was called modal for the community where it had the highest value. No attempt was made to determine the statistical significance of this superiority. The modal species, therefore, are not fully equivalent to the "indicator" species of Goodall (1953) which were based on a more sophisticated measure of the degree of difference between the presence values in the several communities. In the Species List at the back of this book, the community of modality is indicated for each species by an abbreviation of the community name, according to the key given in Table 28.

In the communities for which both presence and frequency data are available, it is possible to rank the species by an index based on the product of these two figures. For example, a species may be present in 70 per cent of the stands of a given community and may have an average frequency of 25 per cent in all of the stands. The product of 70 × 25, or 1750, is an index of the commonness of the plant and is directly related to the chances of finding that particular species at any given point in any stand of the community. This frequency-presence index has a maximum value of 10,000 where a species is found in all quadrats in all stands and a minimum value approaching zero for a rare plant found in only one quadrat in only one stand, with the exact value depending upon the number of stands in the study. The index was first employed in the Wisconsin investigation by Anderson (1954) for the species of the dry prairies. It was developed independently by him but appears to be a modification of a somewhat similar approach used by Etter (1949) in Switzerland. In the following chapters, the frequency-presence index will be used for many of the major communities of forest and prairie, but not for the lesser communities, since frequency values are not available for them.

Summaries are presented for each community in the Appendix.

Whenever possible, these summaries include the following information, gathered together in more or less standardized form: average tree composition of forests, showing average importance value and constancy per cent for each species; structure of typical forest stands, showing density per acre for each species by size classes; understory composition, with percentage presence and percentage frequency for the prevalent species and percentage presence for the modal species; map of stations in P.E.L. study; map of original area of the type in presettlement times; and a community summary table giving (when available) the area, major dominant species, leading prevalent species, species density, homogeneity, related communities, climatic averages, typical weather stations located in the type, soil series and average soil analyses, location of typical examples under public control, and major publications dealing with the community in Wisconsin. Comparative climatic data for the major communities are tabulated in Table IV-3, while a summary of average soil analyses is given in Table IV-4.

Comparison of communities with each other

A plant community may be very homogeneous, as shown by a high index of homogeneity; this means that any given stand is very similar to any other stand. Homogeneity, however, does not tell anything about the distinctness of the community, about how much it differs from other kinds of communities. One approach to the measurement of distinctness is offered by the modal prevalent species. If a large number of the most common species have optimum presence values in a particular community, then that community is recognizably more distinct than another assemblage in which the most common species are actually modal elsewhere. This measure is best expressed by an index number showing the number of modal prevalents as a percentage of total prevalents. A low value indicates that the community is intermediate or hybrid in nature, with its most important species really growing best in some other community. On the other hand, a high value indicates a more distinct assemblage, dominated by species which achieve their optimum there. Calculation of this index of distinctness for the Wisconsin terrestrial communities revealed a tremendous range in values, from a low of 2.1 per cent for the oak opening to highs of 56.3 per cent for the bracken-grassland and 57.5 per cent for the dry prairie. Still higher indices were attained by the aquatic communities, with values of 85.7 per cent for the submerged and 90.8 per cent for the emergent groups (Table IV-2).

Many methods exist whereby the actual degree of similarity of one community to another may be assessed. Most of these are based on the "coefficient of community" of Jaccard (1902, 1928) or on its modification for quantitative data by Gleason (1920). The method chosen here is the one which has long been used by a group of Polish ecologists. It apparently was first proposed by the statistician Czekanowski (1913) and developed with modifications by Kulczynski (1927), Henzel (1938), Matuszkiewicz (1947), and Motyka (1947). It was later used by Sorensen (1948), Clausen (1957), Dahl (1957), and others in Scandinavia; and by Culberson (1955a), Bray and Curtis (1957), and others in this country. Basically, it is a correlation coefficient for use when a number of different, quantitatively expressed measures or attributes are available for the two entities being compared. The formula used in its calculation is $2w/a + b$, where a is the sum of all measures for one entity, b is the sum of all measures for the other entity, and w is the sum of the lower values for each measure. In effect, $2w$ represents the degree of coincidence between the two sets of measures, the amount which they have in common. The resulting index of similarity, therefore, may vary from zero for two entities which have no measurements in common to 1.00 (or 100 per cent) for two quantitatively identical entities. In phytosociological work, the entities are plant assemblages (quadrats, stands, or communities) and the measures or attributes of each assemblage are the component species or a selection of those species. The quantitative expression may be density or dominance for individual quadrats, frequency for stands, presence for communities, or any other statistic available. These values may be used directly or they may be weighted by various means (Demianowiczowa, 1952; Motyka and Zawadski, 1953).

Indices of similarity were calculated for all combinations of the native terrestrial communities in the P.E.L. study. For this purpose, a master list of all groundlayer species occurring in the prevalent groups of each community was compiled. There were 387 species on this master list, and their presence values in each community were used in a direct $2w/a + b$ calculation without weighting, resulting in 496 indices. The two most closely related communities (dry prairie and dry-mesic prairie) had an index of 78.9 per cent, whereas the two least closely related (open bog and dry prairie) had a value of only 0.4 per cent. Further discussion of these similarity figures is given in Chapter 24.

Part 2

Southern forests

CHAPTER 5

Southern forests—general

NATURE AND GENERAL DISTRIBUTION

The forests of the prairie-forest floristic province are located in the southwestern-half of the state, south and west of the tension zone, in the shaded region shown on the map at the opening of this chapter. They occur on a full range of moisture sites, from very wet places along streams and lakes, through mesic sites with deep soils, to very dry places on the thin soils of exposed hills and bluffs. It is estimated that 5,200,000 acres were covered by this forest in presettlement days, and an additional 7,250,000 acres were in the closely associated oak savanna type. As mentioned in Chapter 3, estimates are necessarily of low precision because of the difficulty of separation of savanna from forest.

Early descriptions of the southern forests by Hoyt, Lapham, Knapp, and others indicated a widespread occurrence of oak forests on hills and exposed sites, a dense forest of mixed hardwoods like sugar maple (*Acer saccharum*), and basswood (*Tilia americana*), on north slopes and other protected places, and a floodplain forest of elm (*Ulmus americana*), ash (*Fraxinus* sp.), and silver maple (*Acer saccharinum*) along the major streams.

In some areas with non-extreme topography, a fairly uniform combination of species extended over many square miles, while in regions of greater relief, the mixture of trees varied greatly within short distances. The latter condition is seen in most pronounced form in the Driftless Area. In Figure 7, for example, the forest groupings on the 250-acre Scientific Area at Wyalusing State Park in Grant County are shown, together with a sketch map of the topography and geology. The same area is shown in the photograph in Plate 10 taken from a hill to the west. The mesic maple-basswood forest is largely confined to the steepest portion of the north face of the hill, while the much more xeric stands of chinquapin oak (*Quercus muhlenbergii* Engelm.), black oak-shagbark hickory (*Q. velutina–Carya ovata*), and red cedar (*Juniperus virginiana*) are on the southwest slope. There is a maximum elevational difference of 380 feet within a lateral distance of only 500 feet. Such extreme topography naturally accentuates the microclimatic differences ordinarily associated with the north and south sides of more moderate hills in the vicinity.

Chamberlin summarized the knowledge available in 1877 with an account of the forests which he divided into four types. This now classical work is remarkable for the insight shown by Chamberlin as to the nature of plant communities and their relationships. The four forest types were called the Oak Group, the Oak and Maple Group, the Maple Group, and the Maple and Beech Group. The first was dominated by bur oak (*Q. macrocarpa*), white oak (*Q. alba*), red oak (*Q. borealis*), and Hill's oak (*Q. ellipsoidalis*) and was said to be closely related to the prairies, because it included both oak savanna and denser oak forests. Other associated trees were trembling aspen (*Populus tremuloides*), shagbark hickory, Iowa crab (*Pyrus ioensis*), and black cherry (*Prunus serotina*). Hazlenut (*Corylus americana*) and gray dogwood (*Cornus racemosa*) constituted the chief underbrush, along with raspberries and blackberries (*Rubus* sp.). The Oak and Maple Group was thought to be transitional. As Chamberlin said, "It is difficult to draw sharp lines of demarcation between the several classes of heavy forests, and to circumscribe the areas occupied by each. The fact is that no abrupt line of separation exists." The Oak and Maple Group is not characterized by the exclusive presence of any prominent plant,

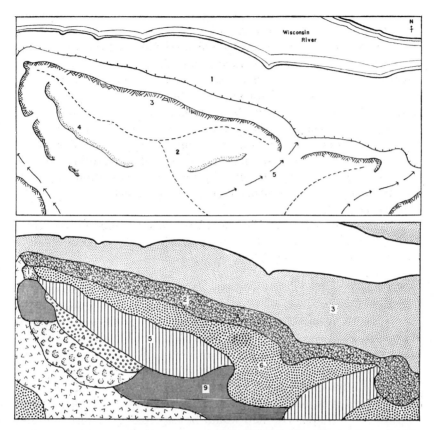

Figure 7.—Wyalusing Scientific Area in Wyalusing State Park, Grant County. See photograph in Plate 10. The upper map shows the topographical and geological features of this 250-acre area of virgin forest. 1, floodplain of Wisconsin River (640-foot elevation); 2, crest of hill at 1020-foot elevation (Platteville-Galena dolomite); 3, escarpment of Prairie du Chien dolomite (750- to 850-foot elevation); 4, cliff of St. Peter sandstone (910-foot elevation); 5, the arrows indicate drainage channels which carry water only during rains. The lower map shows the present vegetation of the area. 1, cedar glade; 2, southern mesic forest of maple and basswood; 3, lowland hardwood forest of silver maple, American elm, and green ash; 4, dry forest of chinquapin oak; 5, dry-mesic forest of white oak and red oak; 6, dry-mesic forest of red oak and basswood; 7, mixed forest of red oak, black walnut, sugar maple, and honey locust; 8, dry forest of black oak and shagbark hickory; 9, dry-mesic forest of nearly pure red oak.

but by a distinctive association of plants common to several classes. The group was said to include white, red, and bur oak, sugar maple, red maple (*Acer rubrum*), American and slippery elms (not prominent), basswood, and white ash (rarely seen). Round-leaved dogwood

(*Cornus rugosa*) tended to replace gray dogwood in the understory, whereas hazelnut was not prominent.

The Maple Group included the densest forests of the state. It was thought to be a fairly well-marked group, dominated by sugar maple, basswood, white ash (*Fraxinus americana*), both elms, and ironwood ("abundant and highly characteristic"). Black walnut (*Juglans nigra*), butternut (*J. cinerea*), yellowbud hickory (*Carya cordiformis*), and large-toothed aspen (*Populus grandidentata*) were also present. The round-leaved dogwood was the most conspicuous shrub. The Maple and Beech Group was "essentially the same as the preceding, with the addition of the beech (*Fagus grandifolia*)." The separation was made largely because of the restricted range of the beech in the counties along Lake Michigan.

In common with the other early observers, Chamberlin had little to say about the forests on the lowlands, other than to point out that black ash (*Fraxinus nigra*) was to be found in the contiguous wet grounds.

An examination of the records of the government land survey substantiates the observations of the early botanists. By tabulating the species which occurred together as witness trees at section corners, it is possible to ascertain the degree to which certain species are associated. For example, the data for all corners with at least one bur oak have been examined, and the number of times the second tree was a bur oak, a black oak, a sugar maple, or other species determined. From these figures, it is possible to determine which species occurred with bur oak more frequently than would be expected by chance and also those species pairs which were less frequent than expected. A similar method was used by Shanks (1953) in studying certain Ohio forests. A summary of information derived in this way for Green County is given in Table 4. As can be seen, there is a certain order to the arrangement, with bur oak at one end and sugar maple at the other. Bur oak is most likely to be associated with other oaks, in order from black to white to red, while sugar maple is most associated with basswood, slippery elm, and red oak. The surveyors' records usually indicate a gradual spatial transition on the land from the bur oak end to the sugar maple end, although the shift may be abrupt if some abrupt topographic boundary is present.

The general conclusions to be drawn from all of these early records may be easily summarized. The closed forests of the prairie-forest province were adjacent to oak savannas and gradually blended with them in most places. The forests on the drier sites were dominated by

Table 4
Associations of tree species at surveyors' witness corners
(Values are percentages of total stations with first species which have the indicated species as the second tree of the pair)

	Second species							
First species	Sugar Maple	Basswood	Slippery Elm	Red Oak	White Oak	Aspen	Black Oak	Bur Oak
Sugar Maple *Acer saccharum*	29.4	23.6	16.6	14.9	7.7	5.9	....	1.9
Basswood *Tilia americana*	31.3	12.2	12.2	15.6	22.2	....	....	6.7
Slippery Elm *Ulmus rubra*	25.4	15.4	30.8	....	13.4	7.7	....	7.7
Red Oak *Quercus borealis*	4.1	8.0	....	20.8	51.4	4.1	3.2	8.4
White Oak *Quercus alba*	1.7	3.4	1.7	11.2	57.7	3.4	5.2	13.8
Aspen *Populus tremuloides*	9.1	....	9.1	9.1	33.3	18.1	3.0	20.0
Black Oak *Quercus velutina*	....	....	....	2.1	40.2	3.0	5.7	49.0
Bur Oak *Quercus macrocarpa*	1.7	3.4	1.7	5.2	27.6	3.4	45.0	12.1

oaks, including black, white, and red oaks, while the mesic sites on deep soils in protected places contained sugar maple as their most important tree, with appreciable amounts of basswood and red oak. The low ground forests were of limited extent and confined to stream banks and lake shores. They, too, adjoined oak savannas on the landward side. The major species of the closed forests were swamp white oak (*Q. bicolor*), American elm, silver maple, and green ash (*Fraxinus pennsylvanica* var. *subintegerrima*). The entire provincial forest contained many other species of trees. A number of these were confined to this region of Wisconsin, including Kentucky coffeetree (*Gymnocladus dioica*), honey locust (*Gleditsia triacanthos*), shagbark hickory, black walnut, chinquapin oak, black oak, blue ash (*Fraxinus quadrangulata*), sour gum (*Nyssa sylvatica*), sycamore (*Platanus occidentalis*), redbud (*Cercis canadensis*), and boxelder (*Acer negundo*).

P.E.L. STUDY OF UPLAND FORESTS

The first detailed studies of Wisconsin plant communities made by the Plant Ecology Laboratory of the University of Wisconsin were conducted on the upland forests of the prairie-forest region. These uplands were defined on a topographic basis as sites upon which rain water never accumulated, due to runoff and percolation. The first investigations included a detailed examination of one region by Grant Cottam (1948) and comparative studies of series of stands by P. W. Whitford (1948) and by M. L. Partch (1949). Later, R. P. McIntosh (1950) combined a detailed study of one stand with an extensive survey of a large number of stands throughout the province. In subsequent years, other investigators continued the extensive studies; these included M. L. Gilbert (1953) and W. E. Randall (1952), who concentrated on the groundlayer species, M. E. Hale (1955), who examined the lichens and mosses, H. E. Tresner (1952), who investigated the soil microfungi, and R. R. Bond (1955), who studied the associated bird communities. The combined results of these and other studies are the basis of the discussion which follows. Since the methods of presentation of data to be used throughout this book were largely developed by these P.E.L. studies of the southern upland forests, they will be described in some detail here.

Objective criteria were used in the selection of stands. In addition to the upland nature of the site, each stand was required to be at least 15 acres in size, to be located southwest of the tension zone, and to be relatively undisturbed during the lifetime of the dominant trees. No criterion concerning species composition was employed. A mixture of royal palms and magnolias would have been acceptable, had it occurred. The only attention devoted to composition was that the forest be homogeneous, as explained in Chapter 4. The question, whether a given area was a forest and not a savanna, was decided on the basis of cover, with an arbitrary value of 50 per cent canopy being the deciding point.

In each stand, measurements were made which resulted in importance values for all tree species and frequency percentages for all groundlayer plants. Altogether, such information was available for 170 upland stands. The study of the relationships of such a large number of sets of values is a difficult task in any science, but is especially troublesome in ecology because of the extreme variability in biological behavior exhibited by any large series of plant species. As a first step in ordering the data so that relations might be made more apparent, certain over-all averages were determined for the trees. These included percentage presence, average importance value for all stands, average importance value for stands of occurrence only, and maximum importance value in any stand. When the tree species were ranked in

Plate 1.—Thin layer of residual soil over limestone bedrock. Dry prairie in Dane County.

Plate 2.—Blanket of loess over glacial till, showing increase in depth on northeast side of a hill. Arlington Prairie in Dane County.

LEFT: *Plate 3.*—Colony of Dutchman's breeches (*Dicentra cucullaria*), a spring ephemeral in the southern mesic forest. RIGHT: *Plate 4.*—Spring-beauty (*Claytonia virginica*) in early spring.

LEFT: *Plate 5.*—Bloodroot (*Sanguinaria canadensis*), a spring-flowering herb that retains its leaves during the summer. RIGHT: *Plate 6.*—Hepatica (*Hepatica acutiloba*), a spring-flowering herb with evergreen leaves.

Plate 7.—Southern mesic forest in midsummer, showing sun flecks and lack of groundlayer species. Wollein's Woods, Jefferson County.

Plate 8.—Southern mesic forest in early spring, showing light-colored leaf litter. The large tree in the left foreground is hackberry. Abraham's Woods, Green County.

Plate 9.—Circle of basswood trunks formed by vegetative reproduction. Southern mesic forest in Green County.

Plate 10.—View of Wyalusing Scientific Area, as seen from the west. This is the area shown on the maps in Figure 7.

the numerical order of each of these measures, it was seen that four species separated as a distinct group. Since all of these figures gave an indication of the ecological importance of the species, the top group of four clearly included the most important trees of the region. These four were white oak, sugar maple, red oak, and black oak. They were termed the leading dominants by Curtis and McIntosh (1951) who found the same four species by similar methods in the data from the first 95 stands which were studied.

The leading dominants may be used in a further reduction of the data to a greater understandability. If all stands in which each of these trees is actually the top ranking species are segregated into four groups, it is possible to calculate the average importance value of all other species, in much the same way that the surveyors' combinations of species pairs were treated. When this is done, it is seen, that in the stands dominated by black oak, the second most important tree is white oak, red oak is of much less importance, and sugar maple is negligible. In contrast, in the stands dominated by sugar maple, red oak is second, white oak is third, and black oak a poor fourth. An extension of the method to the leading pairs of species adds precision to the findings and results in a phytosociological order of species which indicates their propensities for coexistence. The species close together in the order are more likely to be found together in a given forest stand than species which are remote in the order. The actual sequence for the major species of the region as determined by this method of leading dominants is bur oak–black oak–black cherry–white oak–shagbark hickory–red oak–red maple–white ash–basswood–yellowbud hickory–slippery elm–ironwood–sugar maple.

Another approach to an understanding of the ecological behavior of the trees is based on a study of their reproduction habits. For this purpose, stands are grouped according to their leading dominants, and the density of saplings and seedlings of all species determined for each group. For example, in all stands dominated by oak species, the most numerous saplings are never of the same species as the dominant tree. In black oak forests, the saplings are likely to be white oak, red oak, or black cherry; in red oak forests, the saplings are usually ironwood, basswood, or sugar maple. On the other hand, forests dominated by sugar maple have sugar maple as the most common sapling, with lesser representations of saplings of the other species found in the canopy along with the maple, including basswood, butternut, hickory, and slippery elm. A consideration of these reproductive habits can result in an order of species based upon a replaceability sequence, where any given species is capable of replacing those species lower in the sequence, but can be replaced by those species that are higher.

By combining the information from early observers and from the surveyors' records with that derived from the P.E.L. studies, it is possible to rank all of the tree species in the upland forests of the prairie-forest floristic province according to their ecological behavior. For convenience, the resulting rank has been assigned a series of numbers from one to ten, with bur oak receiving number one, sugar maple number ten, and the other species appropriate intermediate numbers, which depend upon their degree of association with either the oak or the maple and upon their reproductive habits (Table V-1). These numbers are known as adaptation numbers (Curtis and McIntosh, 1951), because they can be interpreted as a measure of the biological adaptations of a species toward successful survival in the complex of environmental conditions found in a forest dominated by sugar maple. The higher the number, the greater is the possibility that that species will be able to exist in company with sugar maple. The value of 5.5 for red oak, for example, indicates that this species is better adapted to the low light, low evaporation, mesic soil moisture, and high nutrient levels of a maple forest than is black oak at 2.5, but less well adapted than basswood at 7.5. The use of the word, adaptation, here is purely interpretive, since the numbers are primarily based on observed joint occurrence of the species. The fact that red oak is actually found more frequently with sugar maple than is black oak and less frequently than basswood is taken as evidence of its adaptations, even though the detailed morphological and physiological characteristics responsible for the behavior are largely unknown.

ORDINATION OF UPLAND FOREST STANDS

One of the widely held principles in forest ecology states that the dominant plants of a community are the most important for the classification of that community, because they receive the brunt of the external environmental forces, such as rain, wind, and sunshine, and modify and temper these forces before they act upon the subordinate members of the community. In a forest, the dominant trees greatly influence not only the light and moisture conditions, but the processes of soil development as well. According to this concept, it should be possible to arrange a given series of forest stands into discrete groups or along a vegetational gradient or cline on the basis of the trees alone. When this is done, it is expected that the understory plants will exhibit similar patterns of distribution. The major problems in this approach have been concerned with the method of arrangement, the questions of which trees to use and the measurements which are best suited to demonstrate the actual importance of the species chosen. In the P.E.L. work, the measure adopted was the importance value (I. V.) which integrates the size of the individuals, the number of individuals, and their ubiquity of distribution. The problem of choice of species was eliminated by considering all of the trees present in the stand in proportion to their importance values.

The actual field measurements of all species in each stand were utilized to arrive at an index number which would reflect the total composition of that stand. This was done by weighting the measured importance value of each species through a procedure of multiplying the value by the adaptation number of the species and then summing the weighted values of all species in the stand. For example, a given forest might have an importance value of 120 for red oak, 80 for basswood, 50 for white oak, 30 for sugar maple, and 20 for black walnut. Multiplying 120 by 5.5, 80 by 7.5, 50 by 3.5, 30 by 10, and 20 by 6.5 and summing these weighted values gives a total of 1865. In contrast a stand with bur oak at 200 and black oak at 100 would have an index of 450 and a stand with 200 sugar maple and 100 basswood would have an index of 2750. The range in possible index numbers is from 300 for a pure stand of bur oak to 3000 for a pure stand of sugar maple. Any intermediate combination will have an intermediate number, the magnitude of which is a measure of the relative adaptations of the included species with respect to sugar maple. In this way, each species in the stand contributes to the compositional index number of that stand in joint proportion to the actual amount of its occurrence and its biological behavior and potentialities as shown by its adaptation number. This method of using weighted averages, combining a measured value of plant quantity with an estimate of plant behavior, has been used by a number of other ecologists in recent years (Whittaker, 1952; Ellenberg, 1952) and has been discussed by Whittaker (1954).

The original mass of data concerning the trees in the 170 stands of the study was thus reduced to 170 index numbers which indicated the phytosociological relationships of the stands. Further treatment of the data made possible a more detailed study both of the behavior of individual species and of environmental correlations. This was accomplished by arranging all stands in the numerical order of the compositional index and plotting the measurements of single species or single environmental factors in each stand against the ordination.

Figure 8 shows the results for red oak. Each dot represents the measured importance value of this species in each stand of occurrence. As can be seen, the outer periphery of these dots forms a curve very similar to the Gaussian normal curve, with maximum values of importance located in a narrow region in the center of the curve. Even in this region, however, some stands occur which have very low values for red oak. The curve can be looked upon as permissive, with stands

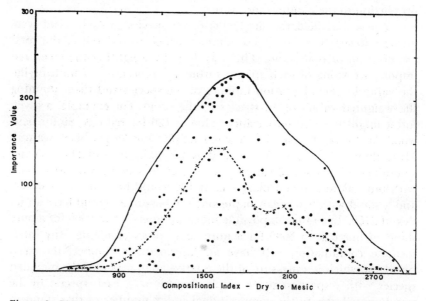

Figure 8.—Behavior of red oak (*Quercus borealis*) on the compositional gradient of the southern upland forest. Each dot indicates the importance value of this species in a separate stand, with the stands located along the abscissa according to their compositional index. The solid line includes all of the stands containing red oak. The dotted line is the average importance value of this species as determined for sequential groups of stands of 100 compositional units each.

below 900 or above 2600 on the ordination having conditions which prevent the successful growth of red oak while stands from 1400 to 1800 may allow maximum development but do not necessarily do so. Solid normal curves of this type, with points scattered throughout an area subtending a true normal curve, may be reduced to a line by means of a moving average. In Figure 8, such a line is shown which was based on a direct average of all points within each 100 unit segment of the ordination and a moving average of these segment averages.

When the procedure was carried out for all tree species in the southern upland forests, the result was a family of curves (Figure 9) which covered a wide range in amplitude and in location of peak or modal values. In fact, no two species showed identical or nearly similar curves and there were no groups of stands which had the same or even closely similar combinations of species at similar levels of importance. Rather, the 170 stands of upland forest formed a vegetational

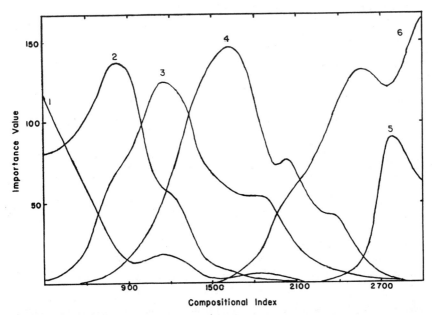

Figure 9.—Behavior of major tree species on compositional gradient of southern upland forest. The curves were obtained in the same way as those of red oak shown in Figure 8. 1, bur oak (*Quercus macrocarpa*); 2, black oak (*Quercus velutina*); 3, white oak (*Quercus alba*); 4, red oak (*Quercus borealis*); 5, beech (*Fagus grandifolia*); 6, sugar maple (*Acer saccharum*).

continuum, with no evident breaks and no discrete communities dominated by uniform combinations of species (Curtis and McIntosh, 1951). On the other hand, the distribution of species was not a random one. A distinct pattern was present, related to the proportionate representation of a continuously changing series of species combinations.

When certain environmental factors are compared with the ordination, clear-cut trends are evident. The amount of light decreases with increasing index number, whereas soil calcium, soil pH, and soil water-retaining capacities are positively correlated with the phytosociological gradient. Such comparisons merely indicate that the different combinations of trees are accompanied by different environmental conditions but do not, by themselves, indicate any clear cause and effect relationships.

COMBINED ORDINATIONS OF UPLAND AND LOWLAND FORESTS

A P.E.L. study of the lowland forests of the region was made that was similar in scope to the one described for the uplands. This study, largely the work of G. H. Ware (1955), with the aid of investigations by R. A. Dietz (1950), and several other workers, was based on 117 stands. The lowlands were chosen by the same criteria emphasized before, except that the topographic position was one on which runoff waters or stream flood waters accumulated, so that an excess of soil moisture beyond that predictable from direct rainfall was present for all or part of the year. Similar analysis of the data resulted in a series of adaptation numbers for the trees encountered in the lowlands (Table V-1). When the stands were arranged in order of the index numbers derived from the use of these adaptation numbers, it was possible to plot the behavior of individual species and factors against the ordination as before. Without giving the results in detail at this time (see Chapter 8), it is sufficient to point out that the species formed a series of overlapping curves, with willow and cottonwood on the wettest sites and American elm, silver maple, and green ash at intermediate ranges on the compositional gradient. The stands on the most mesic sites were dominated by sugar maple, with high values of the accompanying basswood, ironwood, yellowbud hickory, and slippery elm. These mesic stands could scarcely be distinguished from the comparable upland stands except by gross land form of the site. It seemed, therefore, that a more useful over-all picture of the forests of the prairie-forest province could be obtained by combining the two ordinations into a single one, with a concomittant moisture gradient from wet through mesic to dry.

This was done mechanically by simply reversing the order of the compositional index numbers on the graphs of the upland ordination, so that they began on the left with the high index number representing mesic sites and decreased toward the right to the low indices representing the driest places most different from the sugar maple stands. The two ordinations were then combined onto a single graph. The compositional index numbers began at 300 on the left (representing the wettest

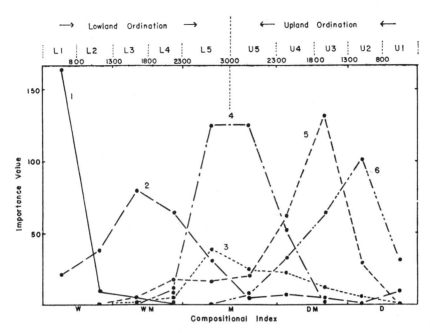

Figure 10.—Behavior of major tree species on the combined ordination of southern upland and lowland forest. 1, black willow (*Salix nigra*); 2, American elm (*Ulmus americana*); 3, slippery elm (*Ulmus rubra*); 4, sugar maple (*Acer saccharum*); 5, red oak (*Quercus borealis*); 6, white oak (*Quercus alba*). The numbers across the top are the compositional index values for the two separate ordinations. They are divided into five lowland (L) and five upland (U) segments, as discussed in the text. Note that the upland ordination is plotted in reverse order from that shown in Figure 9.

forests of willow and cottonwood), and rose through intermediate numbers (representing forests of elm, silver maple, and ash), to high index numbers approaching 3000 for the topographically lowland stands dominated by sugar maple. The compositional index numbers continued to the abutting high numbers, near 3000, (representing mesic sugar maple stands on the upland ordination) and proceeded down through intermediate numbers (representing red oak and white oak forests), to low numbers approaching 300 for the driest stands of black oak and bur oak. The combined ordination is thus a linear presentation of the entire mass of compositional data, arranged along a gradient which is correlated with decreasing amounts of soil moisture.

The behavior of many species becomes much more easily understood when plotted against the combined ordination than when depicted separately on the individual gradients for uplands and lowlands. In Figure 10, the curves for several of the leading tree species are shown plotted against the combined ordination. The curves were made by averaging the importance values for each tree species in all stands in each of five equal segments of the ordinations for both the lowland and the upland series and plotting the averages according to increasing index for lowlands and decreasing index for the uplands, as described above. The proper interpretation of these curves may be aided by a discussion of the curve for sugar maple (Curve 4, Figure 10). This species reaches optimum values in the central region of the ordination which includes the most mesic segments (L5 and U5 on the top of the graph)

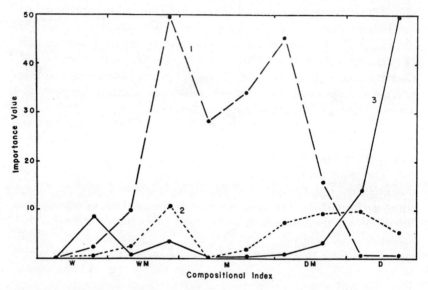

Figure 11.—Bimodal curves for certain tree species along the combined ordination of the southern upland and lowland forests. 1, basswood *(Tilia americana)*; 2, shagbark hickory *(Carya ovata)*; 3, bur oak *(Quercus macrocarpa)*.

of both the upland and lowland series. Stands in this range of the ordination are reasonably similar in appearance and in total composition because of the prevalence of sugar maple (average importance value of 126). They have intermediate levels of basswood (Figure 11), slippery elm, red oak, and American elm and have either no black willow and white oak or very low amounts of these species. On the other hand, the stands in segments L4 and U4 with intermediate levels of sugar maple (average importance value between 30 and 90) do not make such a uniform group. These stands have higher levels of basswood, American elm, and red oak than the central stands in segment L5 and U5 and also showed increased quantities of the other species, but the compositions of the stands toward the wet end (L4) are quite different from those toward the dry end (U4). These differences become greater in segments L3 and U3 in which sugar maple occurs only rarely, while the combinations of species in segments L1 and L2 are almost wholly different from those in segments U2 and U1, none of which have any significant quantity of sugar maple. These curves are all in line with the basic assumption made by ecologists concerning the nature of adaptation of organisms to their environment, namely that the response of each species to any environmental gradient can be represented by a Gaussian amplitude curve with minimum, optimum, and maximum values, or cardinal points. In the present case, the environmental gradient is a complex one of moisture supply which includes soil aeration, soil nutrient supply, and microclimatic features, in addition to actual water content. Each of the species named above appears to have a range along this moisture gradient in which it can grow, and a much narrower range in which it does its best.

In the case of the bimodal curves such as that of basswood and shagbark hickory and the almost U-shaped curve of bur oak, it is apparent that some factor complex other than moisture is active (Figure 11). The troughs of these curves all coincide with the region of high dominance of sugar maple. Such stands have a nearly complete canopy coverage (usually 97 per cent or more) with resultant minimum light intensities on the forest floor (down to 4 foot-candles or less in midsummer). It is probable, therefore, that these bimodal species are responding jointly to water and to light, with a depression in the region of minimum illumination although other competitive factors may be active.

Another highly significant feature of the southern forests is apparent from the curves in Figures 10 and 11. No two of the species have the same or closely similar curves on the basis of amplitude, location, peak, or modal values, nor are there any discrete groups of species present which are separated from other species by gaps along the ordination. This means that the entire arrangement represents a vegetational continuum as was the case in the upland forests alone. Because of this, it is not possible to classify the stands into clearly bounded, uniform groups on the basis of any inherent internal resemblances. A forest dominated by sugar maple and basswood may have red oak as its third most important species, or it may have American elm. The factors responsible for the oak and elm are equal in importance to those responsible for the maple and basswood. To group all stands dominated by the latter two trees into a maple-basswood unit, then, is to place entities with very different characteristics into the same pigeonhole. To say that no inherent means of classification exist does not deny the possibility of classification altogether. The stands can be grouped in any fashion the investigator feels expedient, necessary, or useful for the problem at hand, as long as it is clear that the resultant classification is not an internal property of the stands.

In the case of the P.E.L. studies of the southern forests, the initial twofold division was on a topographical rather than a botanical basis. As can be readily seen from Figure 10, however, such a division is biologically unsound, because a number of important species reach their optimum levels just at the meeting ground of the two topographic types. Since no inherent biological divisions in the ordination are present, it seems best to subdivide the entire series for purposes of discussion by an arbitrary but objective segmentation of the ordination into five unequal units. For this purpose, the two original 2700 unit ordinations are cut into two 1000-unit segments, from 300 to 1299 and 1300 to 2299. The remaining segment is a 1400 unit combination, from 2300 to 3000 on the lowland gradient and down to 2300 on the upland gradient. This increased size of the central segment is justified by the scarcity of stands on either gradient with index values of 2800 or more. Such values are possible only with nearly pure stands of sugar maple and such stands are few or nonexistent in the region. These five segments will be discussed in three groups; the central segment of mesic stands, the two more xeric segments of the uplands, and the two more hydric segments of the lowlands.

CHAPTER 6

Southern forests—mesic

GENERAL DISTRIBUTION OF MESIC FORESTS

The mesic forests of the prairie-forest floristic province are not present in a single contiguous area but rather are scattered quite widely in separate islands (Figure VI-2). The largest areas in the province are in the eastern counties near Lake Michigan (Milwaukee, Ozaukee, Washington, Sheboygan, and Calumet), in an area centering on Richland County and in the St. Croix, Pierce, Dunn county region. Smaller concentrations are to be seen in Grant, Green, and Sauk counties. In Dane County, only a few very small, isolated areas are known, with four of these in small pockets along the northeast shores of the four Madison lakes (Ellarson, 1949). The total amount of the mesic forest type in presettlement times was about 3,432,000 acres, or 9.80 per cent of the total land area of the state.

COMPOSITION AND STRUCTURE

The average importance values for all trees found in the 47 P.E.L. stands (Figure VI-1 and the map at the opening of this chapter) of southern mesic forests are shown in Table VI-1, together with their constancies as determined by the number of stands in which they appeared in the sample of measured trees. Sugar maple (*Acer saccharum*) is in a class by itself. The next most important trees are basswood (*Tilia americana*), beech (*Fagus grandifolia*), slippery elm (*Ulmus rubra*), red oak (*Q. borealis*), and ironwood (*Ostrya virginiana*), while all 19 other species are of relatively minor importance. Notice that beech is in third place with respect to average importance value, although it ranks tenth in constancy. This is a result of its very limited geographical range within the province. Within that range, however, it may achieve high levels of importance, equaling or exceeding sugar maple in this respect. Of the 26 trees found in this forest, eight species reach their optimum development in Wisconsin in the type, including sugar maple, basswood, ironwood, yellowbud hickory (*Carya cordiformis*), and Kentucky coffeetree (*Gymnocladus dioica*).

The numerical structure of a typical mesic forest is shown in Table VI-2. Again sugar maple stands out by itself, with a terrific preponderance over all other species in the three smallest size classes. Slippery elm, ironwood, yellowbud hickory, and basswood are the only other species which are clearly reproducing themselves; butternut (*Juglans cinerea*), black walnut (*J. nigra*), white oak (*Q. alba*), and American elm (*Ulmus americana*) are totally lacking in the seedling classes. The value of 15 small seedlings per acre for red oak was based upon the presence of germinated acorns. The seedlings depend solely upon the stored food within the acorn, and die when all of it is used. Nevertheless, their ability to germinate and persist for even a few years gives them an advantage if an opening appears in the canopy.

Life-history of sugar maple (*Acer saccharum*)

The great importance of sugar maple in the southern mesic forests justifies an examination of its life-history characteristics in some detail, since the unique features of this forest are dependent almost entirely on the behavior of this species. Potentialities for achieving high levels of dominance are based upon several qualities, which include high reproduction rates, great tolerance of shade, and long life. The reproduction rate is evident from the figures in Table VI-2, which show nearly 20,000 seedlings less than 12 inches in height per acre. This measure is regularly at a high level in all the mesic stands, but is subject to quite wide fluctuations from year to year in any one stand; these fluctuations depend upon the size of the seed crops of previous years as well as on the annual vagaries of the weather.

In early spring the flowers of sugar maple are produced in umbels from primordia formed during the previous growing season. Most of the flowers are male and some trees produce only male flowers. Pollination is brought about largely through the aid of bees and other

Table 5

Annual variation in seed yield of sugar maple in a mesic forest in Green County

Year	No. of seeds per acre*	% of seeds with embryos	Year	No. of seeds per acre*	% of seeds with embryos
1947	4,300,000	?	1953	5,270,760	50.0
1948	914,000	15.5	1954	248,290	10.5
1949	423,600	12.2	1955	130,680	13.1
1950	127,080	16.6	1956	4,837,330	31.2
1951	544,000	12.5	1957	40,000	3.1
1952	1,397,880	31.0	1958	78,400	6.0

* Multiply by 2.47 to get number per hectare.

insects, although some pollen may be dislodged by heavy winds. The pollinated flowers produce their characteristic keys or pairs of samaras within a few weeks but these fruits do not ripen until late September or early October. Only one key of a pair is capable of producing a viable embryo, but frequently both keys are empty.

The annual variation in seed yield is very great. Observations have been made in the same forest described in Table VI-2 for a number of years by means of random quadrat sampling. The results are shown in Table 5. Peak yields of four to five million seeds per acre were attained in 1947, 1953, and 1956, while minimum yields of about 130,000 seeds per acre were reached in 1950 and 1955. In addition, a large age-class of seedlings dating from 1944 was observed in 1946, so that high yields seem to be produced every three to five years. The percentage of samaras with viable embryos tended to be directly proportional to the fruit yield. Indeed, in 1953, the sample indicated that 50 per cent of the seeds were filled, which is the top value that is possible.

The seeds may begin germination in the late autumn but emergence of the cotyledons occurs in early spring. The seeds are shed at the same time as leaf-fall so they are partially buried in the new litter. The seedling root may directly penetrate the leaves beneath, but it is frequently deflected laterally and finds its way to the mineral soil by passing around the edges of the lower leaves. Seedlings with the latter habit are especially likely to be killed during the first winter or during dry periods in either the first or second summers.

Germination and mortality studies of the 1953 seed class in the Green County forest showed that 55.7 per cent of the potentially viable seeds germinated in the spring of 1954 to give 1,505,400 seedlings per acre. These numbers declined to 80,420 seedlings per acre by late summer of that year. The first winter caused a further reduction to 45,640 per acre in the spring of 1955. The summer of 1955 was very hot and dry and the seedlings entered the winter under drought conditions. By the spring of 1956, only 14,320 living seedlings per acre were present, or less than one per cent of the number that had initially germinated. These figures were obtained from permanent quadrats and reflect the average conditions in the forest. Actually, however, only the initial germination appeared to be fairly uniform throughout the stand. Thereafter, the survival was correlated with the amount of light available. The seedlings which germinated beneath older seedlings or saplings died much more quickly during the first summer than the seedlings in openings or under large trees without an understory. As a result, the older seedlings became more aggregated each year, with survival greatest where the most light was present and almost nil in the darkest spots. Since the openings were mostly the result of the death of single trees in the canopy, these areas of high survival are usually the size of a single crown, or about 20 to 25 feet in diameter. As the seedlings in such an area develop to sapling size, the shade they produce is sufficiently dense to prevent the successful establishment of new crops of seedlings.

Thus a single age class is likely to take command of any new opening produced in the forest in the year following the first heavy seed crop after the opening was made. This age class remains in control until one or more of its members attains a dominant position in the canopy, after which the class group is no longer distinguishable.

On the basis of the figures given, this means that an initial crop of around 15,000 seedlings in a single opening may be reduced to about 150 during the first three years. The mortality rate thereafter is unknown in detail, but observations on clumps of seedlings from 3 to 6 feet in height indicate a density of about 40 or 50 in an area of the size discussed. These are reduced to no more than one or two individuals by the time the trees mature, but the time intervals involved are unknown.

Gleason (1923) made similar estimates for the maple forests of northern Michigan. He reported that "one may find beneath a mature tree, in any single year, some 4,000 seedlings, 400 two-year seedlings, 40 small plants 2 to 5 feet high, five saplings and one good-sized tree waiting to replace the veteran when it dies."

Individual trees may reach the top of the canopy when they are about 6 inches in diameter at breast height (hereafter abbreviated d.b.h.), which may occur at the age of thirty years under the best conditions (Cheyney, 1942). However, there is no direct proportionality between size and age, since sugar maple is extremely tolerant of shade and is able to withstand suppression for long periods of time. Trees more than forty years old that are less than $\frac{1}{2}$ inch d.b.h. are not uncommon, and more extreme cases of suppression are known. Such suppressed trees are fully capable of resuming normal growth rates if conditions become favorable. Sugar maple is the most shade tolerant tree in Wisconsin, being approached in this character only by beech and hemlock, and this feature goes far to account for its great dominance potential. Frequent years of high seed yield at rates that fully saturate every favorable spot in the forest produce tremendous crops of seedlings which survive well in openings and fairly well in the shade. The shade seedlings persist in a living but scarcely thriving condition for decades and thus greatly increase the chances that an established maple seedling will be available when an accident befalls one of the canopy trees. These potentialities exceed those of any other species, and sugar maple thereby comes to occupy a greater and greater share of the canopy of any forest in which it is found. Once established in the canopy, sugar maple is able to stay because of its great longevity. Trees over four-hundred years old and over 40 inches d.b.h. are known for southern Wisconsin, and trees over two-hundred years old are common.

The great density, size, and age of the sugar maples mean that conditions on the forest floor are largely produced through the actions of this species. The most obvious effects result from the maple leaves which are produced in great numbers and in a complete leaf mosaic, so that practically no direct sunlight penetrates the canopy. In fact, most of the incident radiation has passed through several layers of leaves and is greatly reduced in intensity by the time it reaches the ground. It is also enhanced in its proportion of green light, since both the red and blue ends of the spectrum are absorbed by the chlorophyll in the maple leaves (Coombe, 1957).

Another effect of the leaves is brought about by their high content of basic nutrient elements at the time of leaf shedding. The sugar maple, along with basswood, slippery elm, and black cherry (*Prunus serotina*), is reputed to form an abscission layer across the phloem elements at the base of the petiole early in autumn. At any rate, the nutrients which have been pulled into the leaf by the transpiration stream during summer remain there and are not withdrawn into the trunk as in the oaks and many other species. When the leaves are shed, they still contain high levels of calcium, magnesium, and potassium. The upper layers of the soil thus become enriched in these nutrient elements. The tree must replace these elements in the following year by removal from the subsoil, rather than in large part from stored reserves in the trunk as in the oaks. This combination of events results in a pumping action, with basic nutrients pulled up from the subsoil, moved to the top of the trees during the summer, and deposited on the soil surface during the autumn at an altitude higher than their origin.

The high base content of the maple leaves is correlated with their ease of breakdown by millipedes, fungi, and other soil biota, so that the nutrients are quickly made available to other plants growing on the forest floor. In fact, this enrichment of the soil is second only to the shade cast by the trees in explaining the great influence of maple on the entire ecosystem.

Another behavioral trait that is of importance for an understanding of the role of sugar maple in Wisconsin forests concerns its lack of resistance to fire; the species is perhaps the most sensitive to fire of any of the trees of the state. A thin bark is easily damaged by ground fires, and cambial injury occurs even in trees that show little external charring. Fortunately, a closed-canopy mesic forest is highly resistant to the passage of ground fires because of the high moisture content of the litter in the winter and spring and the frequent total absence of inflammable litter in the summer. The season of hazard is October when the newly fallen leaves frequently have long periods free of rain which allow almost complete dessication of the litter. Fire at this season can result in the complete destruction of practically all trees in the forest. Vegetative sprouting from the bases of fire-killed trees never occurs in mature maples, although small saplings may occasionally send out a vigorous sucker. Such vegetative sprouting is very common in trees, especially in smaller ones, which have been cut but not burned; seedlings and saplings can also regenerate themselves in this way when cut or when browsed by deer or other animals.

Life-histories of other trees

Many of the life-history characteristics of the other trees in the mesic forests resemble those of sugar maple. Insect pollination, for example, is found in basswood, coffeetree, and hackberry (*Celtis occidentalis*), while these trees plus yellowbud hickory, beech, slippery elm, and ironwood possess a relatively high degree of shade tolerance. The nutrient content of basswood leaves at the time of shedding exceeds that of sugar maple. Like the maple, 12 other species or about 50 per cent of the trees have seeds or fruits which are dispersed by the wind. Mammals aid in the dispersal of 11 species, while 3 species are particularly adapted for dispersal by birds.

The shade tolerance of yellowbud hickory is rather remarkable. This species appears to possess an ideal protective mechanism in the form of mimicry. Its nuts closely resemble those of shagbark hickory, but the kernels are bitter rather than sweet and are not eaten by squirrels or other animals. In spite of this, both squirrels and blue jays have been observed carrying the nuts away from the parent trees, and the squirrels have been seen burying them. Apparently the instinct to hoard overcomes the acquired knowledge that the objects being hoarded are worthless. A remarkably high percentage of the buried nuts germinate and produce successful seedlings over a wide range of environmental conditions, from open oak forests to dense maple stands. In the low light conditions of the maple stands, the seedlings persist beyond the period of utilization of stored food in the nut and a high percentage become saplings. As the trees age, however, they appear to become more intolerant, because very few survive to become members of the canopy layer. Dead hickories of 4 to 6 inches d.b.h. are a common feature in most of the southern mesic forests.

Several of the members of the forest utilize vegetative reproduction mechanisms to aid in their competition with the more tolerant maple. Both basswood and hackberry possess the ability to send up sprouts from the roots at the point of junction with the trunk or at some distance from the trunk. Basswood commonly produces a ring of such shoots all around the trunk. When the central tree dies, the shoots develop into mature trees which, in turn, are capable of producing more such shoots. In this case, however, the shoots develop only on the outer periphery of the clump. Eventually, they too will produce mature trees and the perfect fairy rings of large basswoods frequently seen in the mesic forests are the result (Plate 9). Because of this habit, basswood can maintain itself in competition with maple, even though its seedlings rarely become established in the shade of the maples.

Another vegetative mechanism of some interest occurs in the slippery elm. The mature trees will produce stump sprouts only when cut, so that long persistence as in basswood and hackberry is not possible. The seedlings, however, have developed a very unusual ability to send out long rhizomes with erect branches at their tips. Each aerial stem may be 12 to 20 inches high. The plants may eventually form a cluster of 30 or more such stems, very much like the habit of the creeping honeysuckle (*Lonicera prolifera*). In this way, a chance seedling may persist for long periods. When it happens to grow into a microsite with higher light intensities or when death of a canopy tree increases the light, one or more of the branches may rapidly develop into a sapling and eventually a mature tree. So far as is known, this habit is possessed by no other Wisconsin tree.

A number of species which are fairly regular but numerically minor members of the mesic forest are commonly lacking in the smaller size classes, which clearly indicates their inability to reproduce in the existing conditions. Included in the group are the walnuts, oaks, large-toothed aspen (*Populus grandidentata*), black cherry, and white birch (*Betula papyrifera*). A logical extension of such evidence would indicate that they might eventually disappear from the forest. Such a result is probably of rare occurrence, however, because of the ability of the species to utilize gap phase reproduction. This concept, advanced by Watt (1947) and documented in Minnesota by Bray (1956), states that catastrophic destruction of small segments of the canopy by disease, wind, or lightning is certain to occur in any closed forest if a sufficient time period elapses. The gaps thus created possess a very different microclimate from the remainder of the forest, especially in light and moisture conditions; they may have a different soil environment as well, if the catastrophe was caused by wind throw or spot fire. Seedlings of the oaks and other shade-intolerant species can become established in these openings and at least a few may develop to mature trees. The clusters of large aspens frequently seen in mesic forests can be traced beyond doubt to such gap phase origins. On the basis of this gap phase concept, it is clear that the degree of success of the intolerant trees and hence the degree of species mixture to be found in the forest is proportional to the chances for disturbance. Exposed sites on hilltops are more liable to catastrophe than protected places in deep valleys or ravines. Only in the latter does maple have a good chance of eventually asserting total dominance.

Sometimes a widespread destruction of the canopy occurs, too extensive to be covered by the idea of a gap. This may result from a forest fire, a hurricane, or from excessive logging by man. When this happens, the forest may be thrown back to a stage in which relative dominance is captured by the shade-intolerant trees. This is illustrated in a mesic forest in Green County near the Illinois line. A study of age class distributions showed that the sugar maples were the oldest trees in the forest; some of them dated back to the early 1700's, but all ages were present. About 1830, some widespread disturbance occurred, probably utilization of the trees for construction of log cabins by the first settlers on the nearby prairies. In any case, a large crop of red oak and slippery elm became established between 1830 and 1855. In fact, all of the red oaks in the forest date to this interval. White oaks, a single tree of swamp white oak (*Quercus bicolor*), many butternuts, black walnuts, and a few American elms also date back to this time. Thereafter, reclosure of the canopy prevented further regeneration, except for an occasional slippery elm, ironwood, and basswood, and the continuously developing maples.

The stabilizing and purifying influence of the sugar maple in the southern mesic forests is thus occasionally interrupted either locally or generally, with a resultant resurgence of less well-adapted trees. The forest is never static but rather exists in a state of a dynamic equilibrium, compounded of the opposing forces of the super-tolerant sugar maple on the one hand and a particular group of vigorous but less tolerant species, aided by external agents of destruction, on the other. In so far as the term has any discoverable meaning, such a forest may be said to be the closest approach to "climax" of any in Wisconsin.

GROUNDLAYER COMPOSITION

The prevalent groundlayer species of the southern mesic forests are given in Table VI-3. As described in Chapter 4, the prevalent species are determined by arranging all species in order of decreasing per cent presence, and then counting off the topmost names to the number equaling the average species density per stand. The prevalent species which reach their optimum in this forest, in the sense that their presence value is higher here than in any other Wisconsin community, are indicated by asterisks. The remaining modal species of lower presence are also listed in Table VI-3.

Of the 205 groundlayer species of herbs and shrubs found in the stands of the P.E.L. study, over one-half belong to eight families, and the remainder are in 48 other families, which include 24 with a single representative each. The leading families and their species as a percentage of the total number are: *Liliaceae*—9.3 per cent, *Compositae*—8.7 per cent, *Ranunculaceae*—7.8 per cent, *Cyperaceae*—6.8 per cent, *Caprifoliaceae*—5.4 per cent, *Gramineae*—4.9 per cent, *Polypodiaceae*—4.4 per cent, and *Rosaceae*—3.9 per cent. Shrubs and woody vines make up 16.1 per cent of the understory flora. Of these, 81.8 per cent have fruits in the form of berries, while 12.8 per cent of the herbaceous plants have berries. Evergreen plants or plants with green leaves through the winter are relatively rare, and only 11 species or 6.4 per cent of the herbs and shrubs are in this category.

The flowering plants in the groundlayer are predominantly spring bloomers, with 115 species or 60.8 per cent of the group in bloom before June 15. Another 30.2 per cent bloom in the summer between June 15 and August 15, and the remaining 9.0 per cent bloom after August 15; one of them, the witchhazel (*Hamamelis virginiana*) appears as late as November in some years.

Among the spring bloomers is a fairly well-defined group known as the spring ephemerals. These were studied in some detail by Randall (1952) in his investigation of the autecological characteristics of Wisconsin forest herbs. As the name implies, the ephemerals are of short duration, at least as far as their aboveground parts are concerned. They grow very rapidly in early spring, frequently while the last snows are still melting. Both flowers and leaves usually appear together. Full bloom and maximum leaf expansion occur before the trees have expanded their leaf buds. Fruits are ripened quickly, often within three weeks of anthesis. Photosynthesis must occur with great efficiency, since these plants make enough food to complete their life cycle and to provide reserves to last until the following spring in the brief period before the tree canopy develops. By the time the tree leaves are fully expanded in early June, the ephemerals have died back completely with no trace of leaves or fruits to be seen. All of them have some type of underground storage organ, either a corm, a tuber, a bulb, or a fleshy rhizome. Prominent members of this group include the troutlilies (*Erythronium albidum* and *E. americanum*), squirrel-corn (*Dicentra canadensis*), Dutchman's-breeches (*Dicentra cucullaria*) (Plate 3), spring-beauty (*Claytonia virginica*) (Plate 4), toothwort (*Dentaria laciniata*), and false rue-anemone (*Isopyrum biternatum*). Two other species have the same time schedule but have adopted a different method of overwintering. Mermaid-weed (*Floerkea proserpinacoides*) and rough bedstraw (*Galium aparine*) are both very successful winter annuals, which send out their cotyledons in the autumn and develop true leaves and flowers in early spring, after which they die down completely, to pass the summer season in the form of seeds.

Several other modifications of the ephemeral habit are to be seen in other members of the mesic forest understory. Ramp (*Allium tricoccum*) behaves like the typical ephemerals as far as the leaves are concerned, but its flowers are sent up on a naked stalk in midsummer, long after the leaves have disappeared. One of the characteristic orchid members of the mesic forest is the Adam-and-Eve orchid (*Aplectrum hyemale*). This plant has a pair of glutinous, fleshy tubers, from the terminal one of which a single, plaited, oval leaf develops during the first week in September. This leaf remains green during the winter and early spring and presumably carries on photosynthesis whenever temperatures permit. When the tree canopy closes, the orchid leaf dies. The flowers are borne on a naked stem in early summer, appearing about the same time as the ramp. One of the true ephemerals, toothwort, partakes of this autumn habit in some years, sending up a set of leaves in September and October. These die down on the approach of winter and are replaced by more leaves in the spring, this time accompanied by flowers.

The ephemeral habit and its modifications are clearly adaptations of a very complicated sort to the unusual light conditions produced by the dominant trees. The adjustment by a group of herbs to a whole set of growth patterns and to a particular timing cycle produced by another member of the community is to be taken as an indication of a very long period of coexistence. Such evidence forms one of the strongest supports for the concept of Tansley (1949) and other ecologists that the plant community is a quasi-organism, with interrelations between its member species at an organized level nearly comparable with that between the various organs of a single individual plant.

Many other well-known species bloom in the early spring along with the ephemerals but differ because they retain their leaves for part or all of the summer, and commonly ripen their fruits in summer. Included here are bloodroot (*Sanguinaria canadensis*) (Plate 5), Jack-in-the-pulpit (*Arisaema triphyllum*), wild ginger (*Asarum canadense*), cohosh (*Caulophyllum thalictroides*), hepatica (*Hepatica acutiloba*) (Plate 6), woods phlox (*Phlox divaricata* var. *laphamii*), mayapple (*Podophyllum peltatum*), and the trilliums (*Trillium grandiflorum, T. gleasoni,* and *T. recurvatum*).

These species, and the others that bloom in summer or early autumn, are true shade plants, since they can grow or at least persist in the very low light intensities of the summertime mesic forest. Randall (1953) studied certain aspects of the water relations and photosynthesis of these plants. He found that their leaves contained between 7 and 8 micrograms of chlorophyll per milligram of dry matter, as compared to values between 3 and 4 micrograms for plants of dry, open, oak woods. The chlorophyll, however, was much more sensitive to strong light. An ether extract lost over 70 per cent of its original content of chlorophyll after exposure to bright sunlight as contrasted to a less than 40 per cent loss for similar extracts from species growing in the open forests. Still more striking differences were seen between the shade tolerant mesic herbs and the sun plants when the water relations were studied. Randall found that leaves of the shade plants could lose an average of only 3.6 per cent of their turgid water content before showing visible signs of wilting, while the sun plants lost 20.0 per cent before loss of turgidity was evident. Because no growth can take place when plants are wilted, these figures mean that the mesic forest plants must have conditions of uniform moisture availability if they are to thrive, whereas the plants of xeric forests can withstand much wider fluctuations in moisture supply without serious effect.

One of the plants which best illustrates this need for constant moisture conditions is the three-birds orchid (*Triphora trianthophora*). This relatively uncommon species is to be found only in the best-developed and least-disturbed mesic forests. It blooms in late August at a season when most botanists do not visit the maple woods and its small size and general inconspicuousness tend to make it appear less common than it really is. It has very succulent, almost watery stems which arise from tubers of similar texture. It may be devoid of chlorophyll or may have one or two scalelike, green leaves. No visible morphological modifications to prevent water loss are present. A steady supply of moisture is essential for its well-being, a drought period of even one week being sufficient to reduce greatly the flowering population, while a more extensive or a multiple dry period will completely eliminate flowering for the season.

In other phases of his study, Randall (1953) showed that the typical groundlayer plants of the southern mesic forests possess a number of life-history characteristics which differ from those of the xeric forests. Most of the species are of low stature (less than 18 inches or 45 cm tall), are rooted in the rich organic A_1 layer of the soil, and produce flowers from primordial buds which were developed during the previous growing season. A significant number of the species have seeds which are dispersed by animals, including a large proportion with special outgrowths on their seed coats which lead to their transport by ants (Sernander, 1906). Among the last group are wild ginger, Dutchman's-breeches, troutlily, bloodroot, and the trilliums.

With regard to reproduction, the mesic forest herbs vary all the way from the annuals with purely sexual reproduction to species like the troutlily which relies almost exclusively on vegetative propaga-

Table 6
Reproduction rates of herbs in a Green County mesic forest
(In part from Archbald, 1950, and Struik, 1957)

Species	Seeds per flowering plant per year	Approx. no. of seeds per acre per year*	Estimated success (% of seeds which produce seedlings each year)
Allium tricoccum	15	320	1.33
Aplectrum hyemale	200,000	1,279,000	0
Arisaema triphyllum	50	138,000	8.00
Caulophyllum thalictroides	7	220	4.28
Claytonia virginica	48	3,894,000	0.29
Dicentra cucullaria	14	330,000	1.79
Erythronium albidum	2	360	0
Floerkea proserpinacoides	4	73,700	25.00
Galium aparine	26	1,717,100	5.77
Hepatica acutiloba	90	17,800	0.03
Hydrophyllum virginica	18	9,850	0.21
Impatiens pallida	39	4,278,000	2.56
Orchis spectabilis	20,000	ca. 40,000	0
Osmorhiza claytoni	17	1,040	2.20
Podophyllum peltatum	33	2,180	0.11
Polygonatum pubescens	96	59,900	0.52
Sanguinaria canadensis	25	35,000	0.54
Smilacina racemosa	11	57,400	5.64
Trillium gleasoni	52	680	0.04
T. recurvatum	29	138,000	0.35
Uvularia grandiflora	10	6,500	1.33
Viola pubescens	20	10,500	0.15

* Multiply by 2.47 to get number per hectare.

tion. These differences are reflected in certain results obtained by Struik (1957) with regard to the size of the annual seed crop and the percentage success of the seeds for a number of species in a maple woods in Green County. She found that the average number of seeds produced per plant was a relatively fixed quantity, with little variation from year to year. As seen in Table 6, which is based in part on her figures, the yield ranged from 2 seeds per flowering plant for *Erythronium albidum* to 200,000 seeds per flowering plant for *Aplectrum hyemale.* When adjusted to the actual densities of these species per acre, the seed crop is seen to be similarly disparate, but with different species in the extreme positions. The success of the seeds in producing established individuals can be ascertained only in a very rough way, but the estimates in the table are probably of the right order of magnitude. The astonishingly high value of 25 per cent for *Floerkea proserpinacoides* is possibly as high as is ever attained by any plant anywhere. It could only be exceeded by another annual which maintained a constant population level and produced less than 4 seeds per plant.

The seeds of practically all members of the mesic forest understory are dormant when ripe and require a period of exposure to cold moist conditions before they are capable of germinating. Many of them have a complicated type of dormancy, in which the first winter breaks the dormancy of the hypocotyl and allows a root system to develop during the following summer while a second winter is needed to permit the epicotyl to grow into an aerial stem. Germination of all the understory species takes place in the A_1 layer of the soil, usually at its surface. It is probable that the high organic content of this layer actually provides some respirable foods for the developing seedlings. The relatively high populations of saprophytic and semi-saprophytic angiosperms like Indian pipe (*Monotropa uniflora*), squaw-root (*Conopholis americana*), and three-birds orchid testify to this soil richness.

Vegetative propagation methods are well developed and of diverse sorts in the various herb species. Without going into detail about the nature of these methods, it must suffice to say that vegetative propagation enables the successful individual resulting from a single sexually-produced seed to maintain itself, to expand its size, and to migrate into areas remote from its place of origin. Whitford (1951) has devoted considerable attention to the mechanisms of vegetative increase and has developed techniques for estimating the age of single colonies or clones through measurement of their average rate of radial increase per year. These clones consist of individual stems which have arisen through the branching of rhizomes or other underground parts of the original plant. They are, therefore, of identical genetic makeup and frequently may be recognized by peculiar distributions of spots on the leaves as in waterleaf (*Hydrophyllum virginianum*), or troutlily (*Erythronium*), or by characteristic lobing of the leaf as in bloodroot (*Sanguinaria*) and mayapple (*Podophyllum*). Application of these techniques to the mesic forest of Green County described in Table VI-2 showed that the average age of the troutlily colonies was one hundred forty-five years, ranging from forty to three hundred thirteen years. The oldest thus got its start just six years after the discovery of Wisconsin by Nicolet in 1636. In contrast, the estimated average age of the mayapple clones was only forty years. However, it is probable that very old clones of this rapidly expanding species (average rate is 25 cm per year) might have become fragmented and no longer recognizable as to their single origin. Comparable studies of other species indicated that most of them had life spans of twenty to sixty years.

One important result of the vegetative propagation process is that the species possessing the habit tend to be very aggregated in their spatial distribution, with high densities per unit area within the colony but with large areas devoid of the species in spaces between the colonies.

A great amount of study in recent years has been devoted to this question of spatial distribution of plants, since it is of obvious importance both from the standpoint of accurate sampling and from the knowledge it can contribute to an understanding of the dynamic behavior of the species. Much of the work has been done in Wisconsin (Whitford, 1949; Archbald, 1950; Curtis and McIntosh, 1950; Mueggler, 1953; Cottam, Curtis and Catana, 1957) but the local results to date, in common with those of investigators elsewhere (Clapham, 1936; Cole, 1946; Clark and Evans, 1954; Hopkins, 1954; Morisita, 1954), have produced no conclusive answers or methods whereby the degree of aggregation may be measured, except in the simple case where all of the plant individuals under consideration are located in discrete colonies with no isolated single individuals between the colonies.

The southern mesic forests well illustrate that such a simple type of distribution is very rare. The complicated nature of the actual distributions has been reported in detail by Struik (1957). The distribution patterns range all the way from the random or nearly random dispersion of isolated individuals of the trilliums to the nearly discrete and separated colonies of troutlily. The majority of species, however, might be described by saying that they vary, from place to place in the forest, from local areas of high density broken by bare spots of various sizes, to more and more isolated colonies separated by open areas with scattered individuals. In the language of colloidal chemistry, they vary from the continuous phase to the discontinuous phase or from a gel to a sol. It is apparent that no simple method of describing such a situation is possible.

Shrubs and woody vines make up 16.1 per cent of the total ground-layer species but as a group they contribute little to the forest structure because their average presence is only 14 per cent. The vines are the most numerous and widespread, especially woodbine (*Parthenocissus vitacea*), poison ivy (*Rhus radicans*), and bittersweet (*Celastrus scandens*). The last species exists almost entirely in a vegetative state, with very rare instances of fruit production. Among the true shrubs, various gooseberries (*Ribes* sp.) and bladdernut (*Staphylea trifolia*) are the most common, although wahoo (*Euonymus atropurpureus*) may reach high densities in areas with low rabbit populations.

Most other groups of the mesic forest biota have not been studied in detail in Wisconsin. The fungi appear to reach great levels of importance, both in number of species and individuals. The even moisture supply and the rich organic substrata are favorable for various fleshy fungi, especially in the group of *Ascomycetes*. Cup fungi, like *Urnula* and *Peziza*, are common in early spring, and that epicurean's delight, *Morchella*, is present in several varieties and occasionally in high populations. The lichen flora of the tree trunks on the other hand, is very poor with only a single species, *Graphis scripta*, being regularly present in the forest, although *Arthonia radiata* and *Physcia orbicularis* are occasionally found (Hale, 1955). The microfungal flora of the soil is rich, with 36 species reported by Tresner, Backus, and Curtis (1954). Members of the genus *Penicillium* made up over 30 per cent of the total with *P. nigricans, P. admetzi,* and *P. thomii* as characteristic species.

Among the animal members of the biota, only the birds have been studied in Wisconsin. Bond (1957) found a number of species which were characteristic of the mesic forests; they included the cerulean warbler, red-eyed vireo, Acadian flycatcher, hairy woodpecker, ovenbird, red-bellied woodpecker, and pileated woodpecker. The birds of this forest in general were insectivorous. They tended to build their nests high in the canopy or in the sapling layer. Species using shrubs as nesting sites were rare, in keeping with the low importance of shrubs in the type. The apparent low populations of the larger mammals like squirrels and chipmunks may be related to the low proportion of trees and shrubs which produce nutlike seeds.

Regional variation

As seen on the map of the southern mesic forests (Figure VI-2), there are several large blocks of this type of forest plus a number of smaller islands, all separated from each other by prairie, savanna, or oak forest. The largest block, from Kenosha County to Outagamie County, adjoins the northern mesic forest on its northeastern edge.

Several of these isolated regions were sampled by sufficient stands in the P.E.L. study so that relatively detailed comparisons may be made between them. The Green County region, for example, was studied by nine different stands, all in an area 12 by 20 miles, and all on a similar geological substrate. All stands were farmers' wood lots, generally 20 to 40 acres in size. Some impression of the degree of stand to stand variation may be obtained from Table 7 which shows the presence or absence of some of the typical mesic forest herbs in the nine stands, together with their percentage presence and their average frequency. The species in the table vary from the astoundingly prevalent *Galium aparine* (F × P index of 7900) to such rare species as *Triphora trianthophora* (F × P index of 11), recorded for a single stand only.

Using quadrat frequency values for all the herbs in each stand, indices of similarity between stands were obtained which ranged from 48.1 per cent to 71.1 per cent, with an average of 64.4 per cent. A comparable treatment of nine stands in Ozaukee and Washington counties showed an average index of 60.3 per cent, while five stands in Vernon and Richland counties had an average index of 61.2 per cent. When the regions are compared with each other on the basis of average composition, it is found that many species have a high presence in all regions, while others are present only in one or two regions or are much more widespread in one or two. This is seen in Table 8 which shows all of the species which reach a level of presence of 80 per cent or more in at least one region.

The mayapple (*Podophyllum peltatum*), for example, was found in all stands of each region; spring-beauty (*Claytonia virginica*), wood's buttercup (*Ranunculus septentrionalis*), and the red trillium (*Trillium recurvatum*) were found only or predominantly in Green County, while *Brachyelytrum erectum, Cryptotaenia canadensis,* and *Dioscorea villosa* were most abundant in the Richland-Vernon area. Plants with northern affinities, like sarsaparilla (*Aralia nudicaulis*) and bishop's cap (*Mitella diphylla*) were lacking in Green County but plentiful in the other two areas.

On the basis of the presence values for the entire herb flora of the three regions, the indices of similarity were 55.5 per cent between Green County and Ozaukee-Washington, 53.5 per cent between Green and Richland-Vernon, and 61.1 per cent between Ozaukee-Washington and Richland-Vernon; thus they approached the values for the stand to stand variation in any one region. On the basis of this and similar

Table 7
Stand to stand variability—Green County mesic forests

Species	A	B	C	D	E	F	G	H	I	F(Av.)%	P%	F×P
Galium aparine	55	85	95	100	60	100	60	75	80	79	100	7900
Claytonia virginica	30	85	1	100	80	45	90	55	60	61	100	6100
Viola pubescens	35	40	60	45	50	70	40	15	40	44	100	4400
Sanguinaria canadensis	65	20	15	30	50	30	55	40	45	39	100	3900
Trillium recurvatum	35	80		40	15	65	35	50	30	39	89	3471
Erythronium albidum	50	20		80	50		85	35	65	43	78	3354
Arisaema triphyllum	20	5	50	25	20	40	35	25	35	28	100	2800
Dicentra cucullaria		60		10	90	5	85	35	25	34	78	2652
Podophyllum peltatum	25	5	60	15	35	30	25	20	15	26	100	2600
Botrychium virginianum	25	5	45	5	15	25	20	35	5	20	100	2000
Dentaria laciniata	10	5		1	15	30	55		35	17	78	1326
Allium tricoccum		20	20	15	5	25	5	15	10	13	89	1157
Caulophyllum thalictroides		10	5	35	5	10	5	15	5	10	89	890
Uvularia grandiflora	25		60			5		15	20	14	56	784
Phlox divaricata		15		45		20	15			11	56	616
Carex laxiflora		30	5			15	10	5	10	8	56	448
Hepatica acutiloba		20		40	15			10		9	44	396
Isopyrum biternatum				35				30	40	12	33	396
Polemonium reptans	30			1	15			5		6	44	264
Orchis spectabilis	5	5		5			1		5	3	67	201
Trillium gleasoni		5		15		10				4	44	176
Floerkea proserpinacoides	5	20		5	5				5	3	33	99
Panax quinquefolia				5		5			1	1	33	33
Aplectrum hyemale		5								1	22	22
Triphora trianthophora			5							1	11	11

Frequency % in Stand

Table 8
Region to region variability

Presence % of mesic forest herbs—80% or more in one region

Species	Green Co.	Vernon-Richland Co.	Washington-Ozaukee Co.	All of S. Wisconsin
Actaea alba	43	100	72	38
Adiantum pedatum	86	100	43	43
Allium tricoccum	89	80	57	78
Amphicarpa bracteata	0	100	43	35
Aralia nudicaulis	0	80	43	22
Aralia racemosa	0	80	14	30
Arisaema triphyllum	100	100	86	57
Aster shortii	72	100	57	27
Athyrium filix-femina	72	80	29	49
Botrychium virginianum	100	80	57	65
Brachyelytrum erectum	29	100	43	41
Caulophyllum thalictroides	89	100	72	68
Circaea quadrisulcata	72	60	86	65
Claytonia virginica	100	0	29	38
Cryptotaenia canadensis	29	80	14	35
Desmodium glutinosum	14	80	29	30
Dioscorea villosa	0	80	14	14
Galium aparine	100	40	57	70
G. concinnum	86	60	72	54
Geranium maculatum	86	80	100	78
Hepatica acutiloba	44	100	86	62
Hydrophyllum virginianum	100	80	72	60
Menispermum canadense	43	80	29	14
Mitella diphylla	0	100	57	19
Osmorhiza claytoni	86	100	86	89
Podophyllum peltatum	100	100	100	84
Polygonatum pubescens	14	80	86	70
Prenanthes alba	57	80	43	41
Ranunculus abortivus	86	60	57	32
R. septentrionalis	86	0	0	22
Sanguinaria canadensis	100	80	57	65
Sanicula gregaria	57	80	57	65
Smilacina racemosa	100	100	86	76
Smilax ecirrhata	86	80	29	49
S. hispida	0	80	14	11
Solidago ulmifolia	14	100	14	27
Thalictrum dioicum	43	80	86	57
Trillium recurvatum	89	0	0	16
Uvularia grandiflora	56	100	72	54
Viola cucullata	86	40	72	65
V. pubescens	100	60	72	65

evidence for other regions, it is concluded that the southern mesic forests, as delimited in the P.E.L. study, are reasonably uniform, with no geographical trends in variation which are of much greater magnitude than the stand to stand variability in any one local region.

ENVIRONMENT

Microclimate

The main factor of the environment which distinguishes the mesic forests from the other forests of the region is the low light intensity on the forest floor in summer. The tree leaves begin to develop about the first week in May and produce an essentially continuous canopy by the first of June. The shading is not absolutely unbroken, even in a completely undisturbed forest; a few openings are always present. These allow the direct penetration of sunlight in small patches known as sun flecks (Plate 7). In several stands, it was found that these flecks had an average area of slightly less than 2 square feet. They occupied slightly more than 1 per cent of the total surface at any given moment, but they were in constant movement as the inclination of the sun changed; hence a much larger fraction of the surface was exposed to high light intensities for brief periods of time. As a result, the measurement of the average quantity of light reaching the forest floor is a difficult task. In general, the intensity at noon on a clear summer day ranges from 4 to 40 foot-candles in the shade and from 2,000 to 6,000 foot-candles in the sun flecks. It would perhaps be possible to integrate these values mathematically and arrive at an over-all average, but such a result would be biologically meaningless, since even brief periods of exposure to high light intensities may have far different effects on growth substance concentrations than much longer periods at fictitious average levels.

The best that can be said is that the southern mesic forests are very dark and the evidence indicates that the amount of light is less than that required by most plants to manufacture sufficient food for continuous well being. As a result, the forest floor during the summer is very poor in species with active aerial organs (Plate 7). It is probable that most of those species which are present are existing in part at the expense of stored food reserves (Lundegårdh, 1931).

One interesting response to light is frequently seen in mesic forests in which selective logging has been practiced so that large openings have been made in the canopy. The yellow jewelweed (*Impatiens pallida*) regularly forms an almost pure stand under such openings. This succulent and tender annual is very sensitive to light and is markedly reduced in height at diminished intensities. The colonies thus take on the characteristics of an integrating light meter, with the tallest plants in the center of the colony and shorter and shorter plants toward the edges. They produce contoured mounds which reflect the chance peculiarities in shape of the canopy opening with surprising accuracy. The wood nettle (*Laportea canadensis*) also demonstrates this phenomenon at times.

The dense layering of leaves in the forest canopy influences other features of the microclimate in addition to light. Many of these are correlated with the protective action of the canopy against wind movement and air turbulence. Actual wind velocity is regularly less than one-tenth of that measured in the open, outside of the forest. Humidity readings in various Wisconsin forests in midsummer showed values ranging from 10 to 32 per cent higher in the forest than in the open on bright, clear days. No continuous records of evaporation have been made here, but Aikman and Smelser (1938) in Iowa, and Wolfe et al. (1949) in Ohio have shown that the rate in summer is only one-tenth to one-third of that in the open. The lack of air turbulence allows the development of a stratified layer of air, which acquires markedly different properties near the ground surface from those in the canopy region, especially in water vapor and carbon dioxide content.

One of the most significant features of the temperature regime in the mesic forests is concerned with the abnormally high temperatures which are reached in the leaf litter during clear days in early spring preceeding the opening of the leaf buds on the trees (Plate 8). The mat of leaf litter allows the direct penetration of light through its upper layers. This light is absorbed by the lower leaves and converted to heat, which is prevented from escaping by the excellent insulating qualities of the multilayered mat. The temperatures therefore rise to levels of 120°F. or 130°F. These brief periods of high heat appear to be effective in breaking the dormancy of both the flowers and leaves of many of the ephemeral species in the groundlayer.

The insulating qualities of the leaf litter are also influential during the winter. When a blanket of snow is present, the added insulation may prevent all frost penetration into the soil, or allow a previously frozen soil to thaw in midwinter. This complete lack of freezing temperatures in the soil, however, is less common in the southern mesic forests than in comparable types in the north, due to the irregular occurrence of a continuous snow cover.

A summary of the macroclimate is given in Table VI-4.

Soils

The southern mesic forests may develop on a wide variety of soils, which include those developed from glacial till and those formed on loess. The soil series include Miami and Dodge in the glaciated areas and Dubuque and Fayette in the Driftless Area. The typical position is about mid-catena, with good but not excessive internal drainage, and with a deep layer of material fine enough to allow root penetration. Most of the forests under discussion occur on calcareous substrata but this is true of all communities of the region because of geological reasons and has no special significance for the mesic forests. Fosberg (1949), in an extensive study of the soils under maple-basswood forests in Green, Dane, and Jefferson counties, found a fairly uniform profile in most stands. In those classed as Gray-Brown Forest soils, the A_1 layer was neutral or slightly alkaline (pH 7.0 to 7.5) and had a very high content of exchangeable calcium (18 to 32 milli-equivalents per 100 grams). Nitrogen (0.34-0.48 per cent) and exchangeable potassium (135-150 p.p.m.) were also high. The leached A_2 layer had the lowest calcium, magnesium, potassium, and phosphorus contents of any layer, although the acidity was usually greatest in the B layer (pH 4.8 to 5.2). All nutrients tended to increase in the deeper layers of parent materials, although they did not reach the concentrations shown in the surface A_1 layer.

The very ancient maple forests of southern Green County occur on soils sometimes classed as Brown Forest soils. These were developed on a thin layer of loess over weathered glacial till of Farmdale age. Free carbonates in the form of limestone pebbles are to be found at depths of 18 inches. The profile has been studied by Fosberg (1949), Steinbrenner (1951), and Pierce (1951). The black A_1 layer, enriched with humus, extends to a depth of 7 to 9 inches, gradually decreasing in blackness with depth. A typical, highly leached A_2 layer is absent, with the A_1 layer grading directly into the B layer. The soils had been mapped in the early reconnaissance soil surveys as prairie soils, in spite of the presence on them of maple trees four hundred years old. Pierce arrived at the conclusion that the soils were originally prairie soils and that the organic matter was a relic from former prairie occupation. Actually, the existing evidence indicates that the reverse is probably true; instead the prairie soils owe their present characteristics to a former occupation of the site by mature mesic forest (see Chapter 14). In any case, the soils in these Green County forests are very rich, with a total pore space by volume of 72.5 per cent, a water-stable aggregate (over 0.5 mm. in size) content of 73.8 per cent, and a very high infiltration rate. Steinbrenner studied the soil moisture changes in the top 48 inches of the profile by means of buried gypsum blocks. No evidence of lack of water was shown by the mid-profile layers at any time, although the top 4 inches showed a considerable decrease in moisture from mid-October to mid-November.

Youngberg (1951) studied the soils under maple-basswood forests in Vernon County. He noted a higher level of soil fertility under the mesic forest than under an adjacent mixed oak stand on the same soil type:

> The most significant improvement brought about by the presence of the maple-basswood type is the increase in cation-exchange capacity of the soil. The rise in exchange capacity is accompanied by a higher content of replaceable bases. Great increases were also revealed in the contents of total nitrogen, available phosphorus, and available potassium. Thus, the analytical data strikingly illustrate the "nutrient-pumping" ability of sugar maple and basswood.

This pumping action also influences the content of lesser nutrients in the surface soils. Spectrographic analyses of the ash of the A_o layer of Green County mesic forest soils showed 145 p.p.m. of boron, 113 p.p.m. of copper, 167 p.p.m. of lead, and 76 p.p.m. of zirconium, while the content of radioactive substances as shown by alpha counts from the ash were at the level of 3.27 counts per hour per square centimeter, as compared with 1.11 counts per hour per square centimeter in the subsoil (Curtis and Dix, 1956b).

GEOGRAPHICAL RELATIONS

Studies of mesic hardwood forests in neighboring states show that the composition is very similar over a wide area. Daubenmire (1936) investigated the "Big Woods" of Minnesota. All of the tree species he found are also present in the list of Wisconsin trees (Table VI-1), while of the 23 most abundant groundlayer species, 17 are on the list of prevalent Wisconsin species (Table VI-3), and the remainder are represented at lower levels of presence on the total flora list. The average importance values of the trees in two of the Minnesota stands, Minnetonka and Northfield, show a 59.3 per cent similarity to the average of all 47 Wisconsin stands in Table VI-1, by the $2\ w/a+b$ test.

The leading dominants in a mesic forest in central Iowa studied by Kucera (1952) were basswood, red oak, black maple (*Acer saccharum* var. *nigrum*), and ironwood, with lesser amounts of white ash, slippery elm, American elm, and shagbark hickory. J. J. Jones (1952) reported on the composition of a mesic forest in Union County, Indiana. Of the 19 tree species over 1 inch d.b.h., all but 4 are on the Wisconsin list. Included in the four species were the tuliptree (*Liriodendron tulipifera*), whose range does not extend into Wisconsin, the flowering dogwood (*Cornus florida*), the red mulberry (*Morus rubra*), and the sour gum (*Nyssa sylvatica*), which barely reach Wisconsin. In spite of these floristic differences, the index of similarity is 56.9 per cent. Kucera and McDermott (1955) studied mesic stands in central Missouri with a compositional index over 2500. Only 3 tree species of a total of 16 were found which were absent from the Wisconsin list. These three were *Quercus muhlenbergii* Engelm., *Cornus florida,* and *Sassafras albidum,* all at the extreme edge of their range in Wisconsin. The index of similarity between the Missouri stands and the average Wisconsin forest was 67.3 per cent. Such a high value of the index would rarely be exceeded by any single Wisconsin stand in comparison with the average. Of the 43 herbs in the Missouri forests, only 11 are not found in the Wisconsin stands.

Other stands in Indiana and Michigan are summarized in Table 9 with those just discussed. Within such a limited summary, several features of significance are apparent. First of all, the stands show a relatively great statistical similarity, but this similarity is achieved by different species in different regions. Secondly, trends in compositional change are apparent along both east-west and north-south topoclines. Thus, basswood is much more prominent in the states west of Wisconsin than in those to the east, and beech shows the reverse trend. Similarly, stands to the south contain more species not represented in the Wisconsin forests. Prominent among these are *Liriodendron tulipifera, Aesculus glabra,* and *Nyssa sylvatica* in the canopy layer, and *Cornus florida, Sassafras albidum,* and *Asimina triloba* in the intermediate layer.

As is to be expected, stands in regions still more remote from Wisconsin usually differ considerably in composition, although in some areas a rather high degree of similarity exists. Thus, in New Jersey, an upland mesic forest studied by Niering (1953) in the High Point State Park had 13 tree species of which only two, *Betula lenta* and *Betula lutea,* are missing from the Wisconsin lists. On the basis of cover and frequency values, the most important species in the New Jersey forest were, in order, sugar maple, basswood, ironwood, red oak, beech, and slippery elm, which are also the leading six species in Wisconsin although not in the same order. Buell and Wistendahl (1955) examined a mesic forest on the lowland terrace of the Raritan River of New Jersey. All nine tree species they found are present in the Wisconsin mesic forest. The leading species in terms of impor-

Table 9
Comparison with forests in other midwestern states

Importance values of major dominants

Species	Wis.	Minn.	N. Ind.	Mo.	S. Ind.	Mich.
Acer saccharum	126.0	103.8	90.9	102.0	105.9	68.3
Tilia americana	34.1	94.3	4.9	61.0	9.4	4.4
Fagus grandifolia	30.3	....	19.3	....	40.6	115.0
Ulmus rubra	25.5	12.0	11.8	46.0	7.2	....
Quercus borealis	21.2	30.0	8.5	10.0	2.9	9.0
Ostrya virginiana	15.8	....	11.6	....	1.0	22.0
Fraxinus americana	7.0	....	30.3	24.0	16.2	....
Quercus alba	5.7	....	....	6.0	....	....
Carya cordiformis	5.3	1.6	8.1	4.0	0.9	....
Juglans cinerea	5.2	....	....	....	....	....
Ulmus americana	5.0	25.3	....	4.0	0.9	27.3
Prunus serotina	1.9	....	26.0	....	1.5	8.8
Gymnocladus dioica	1.8	....	....	6.0	....	....
Carya ovata	1.8	....	....	4.0	....	....
Celtis occidentalis	1.6	....	3.1	6.0	0.6	....
Juglans nigra	0.7	....	16.8	....	2.4	....
Acer rubrum	0.6	....	4.2	....	0.6	17.3
Fraxinus pennsylvanica	0.1	6.2	....	7.0	....	....
Number of additional sp. not on Wisconsin list	....	0	5	3	4	1
Σ I. V. of additional sp.	....	0	38.5	21.0	36.1	23.9
% similarity to Wis.	....	59.3	56.9	67.3	56.1	45.1

Wis. P. E. L. study
Minn. 2 stands, Daubenmire (1936)
N. Ind. 1 stand, Jones (1952)
Mo. 2 stands, Kucera and McDermott (1955)
S. Ind. 1 stand, Potzger and Friesner (1940)
Mich. 1 stand, Cain (1935)

tance values were sugar maple, white oak, red oak, basswood, and yellowbud hickory, all in the top ten in Wisconsin.

Studies in which data on density, frequency, and dominance are given are few in most other states. In the absence of an adequate basis for exact comparisons, recourse must be had to species lists, abundance tables, or occasionally to canopy coverage values. E. Lucy Braun has compiled much information of this sort in her book on the *Deciduous Forests of Eastern North America* (1950). From her summaries it is apparent that the mesic forests of Wisconsin, of the Midwest in general, and in view of the New Jersey results, possibly of the entire glacial boundary region, are closely related to the mesic forests of the coves and flat valleys of both the southern Appalachian and the Ozark regions. These southern forests are richer in woody species than those of the northern, sometimes by a factor of three times or more, but with a few exceptions, most of the additional species are rare or occur only in the intermediate layers. The mesic forests of the Cumberlands and the Great Smokies differ in that the northern basswood (*Tilia americana*) is replaced by another species, *Tilia heterophylla*, and the buckeye (*Aesculus octandra*) assumes a position of major dominance. However, the leading dominants over much of the region are sugar maple, beech, red oak, white oak, and white ash.

The forests of the Cumberland Mountains of Kentucky were studied in great detail by Braun. She recognized a community similar to the Wisconsin stands in which 21 species attained a position in the upper canopy. On the basis of simple density, the most important species were sugar maple, 19 to 36 per cent, basswood (*Tilia* sp.) 11 to 23 per cent, and buckeye (*Aesculus* sp.) 13 to 25 per cent. Altogether, eight species were present which also occur in the mesic forests of Wisconsin, but the sum of density of the differing species, 61 per cent, exceeded the same value for the species in common, 39 per cent. A summary of the composition of all of the mesic forests of the Cumberland Mountains (Braun—Table 1) lists 33 tree species which achieve canopy dominance. Of these, only 13 are present in the mesic forest of southern Wisconsin. The species found in all of the Cumberland stands listed (100 per cent presence) are sugar maple, basswood, red oak, white ash, butternut, and black walnut, plus tuliptree, chestnut (*Castanea dentata*), magnolia (*Magnolia* sp.), sour gum, and chestnut oak (*Quercus montana*). The differing species accounted for 52 per cent of the total density.

The mesic forests of ravine slopes in Mammoth Cave National Park, Kentucky, had 14 tree species in their canopy layer, according to Braun. Of these, 8 are present in Wisconsin mesic forests, with beech and maple the leading dominants. On the basis of tree counts, the non-Wisconsin species had a total density of 23 per cent, while species present in Wisconsin made up the remainder or 77 per cent. The mesic forests of the Ozark mountains of southern Missouri and Arkansas were also studied by Braun. Of the 17 species attaining canopy size, 9 are found in Wisconsin, and the sums of density were exactly the same as in the Mammoth Cave forest, 23 per cent *vs.* 77 per cent.

Many of the mesic stands studied in the southern states were immature, because a considerable portion of the canopy dominance was still held by such shade intolerant species as white oak, tuliptree, chinquapin oak, and chestnut which were not represented in the reproduction layers. When this fact is taken into consideration, it is evident that a high degree of similarity exists between the mature mesic forests over a very large portion of the United States, from New Jersey to Minnesota, and from the Carolinas to Arkansas. The richest forests are to be found in the southeast, with a gradual decrease in the number of species to the north and west. Accompanying this change in floristic richness is a corresponding shift in the combination of leading dominants, with sugar maple imparting a unifying character as it shares dominance successively with southern basswood and buckeye, with beech, and finally with northern basswood. The mesic forests of southern Wisconsin occupy an intermediate position on this compositional continuum, with maple and northern basswood in the leading roles as in the states to the west, but with a rich assortment of accessory species more like the region to the south and east. The groundlayer species show a similar degree of uniformity.

As will be seen in later chapters, this relatively great floristic homogeneity of the American mesic hardwood forests is a rare phenomenon not duplicated by any other major community. It has been explained on the basis of a very long geological history. The evolution of great shade tolerance is apparently a difficult step, involving fundamental physiological properties. Only a few plants have been able to achieve this particular ability, but over the ages a number of complementary mechanisms, like the ephemeral habit, have appeared in otherwise unrelated species. The severe light conditions imposed by the trees have held down competition from the more numerous intolerant species. In effect, the whole complement of internal microenvironmental factors has acted to screen out any but the relatively limited group of adapted species, and these have come to occupy suitable sites over a wide geographical area and have produced a community with a very high level of integration and floristic uniformity. This alikeness of the most mesic members of all the deciduous forest types is largely responsible for the concept of Clements and other early American plant ecologists that the plant formation may be described and delimited solely in terms of its most mesophytic or climax community, and that all other communities of the region may be related in one way or another to this climax type. If the relations be considered as possible trends rather than as positive eventualities, then the concept certainly receives much support from the available evidence from the mesic forest.

Interestingly enough, the maple-basswood forests of Europe show practically no direct floristic relationship with those of Wisconsin, in the sense of species-in-common. On the other hand, generic relationships are very high. Thus Oberdorfer (1957), in the maple-basswood forests of south Germany, reported only a single species (*Milium effusum*) out of a total of 167 which is also found in Wisconsin, but 45 or 37 per cent of the total genera were the same. These included *Acer, Carpinus, Fagus, Fraxinus, Quercus, Tilia,* and *Ulmus* among the trees; *Cornus, Ribes,* and *Viburnum* among the shrubs; and *Actaea, Asarum, Carex, Cardamine, Clematis, Galium, Geum, Impatiens, Orchis, Polygonatum,* and *Viola* among the herbs.

UTILIZATION AND CURRENT MANAGEMENT

The major impact of man on the southern mesic forests has been one of destruction through land clearing for agricultural croplands. In the stands that have thus far escaped this action, several types of utilization are practised that greatly influence composition and structure of the forest. The most widespread of these activities, selective logging, has had a beneficial or constructive influence. This is due to differences in the relative merits of the various species to be used as lumber on the farm. The oaks, both red and white, the hickory, walnuts, and to some extent the elms are used for fence posts, gate boards, wagon boxes, and shed sidings. Demand for such purposes is slow but steady, so that the farmer-owner tends to remove a tree or two from his wood lot every year, but he rarely needs quantities sufficient to cause a general opening of the canopy. As seen above, the species wanted are all in the group that shows little or no reproduction. As a result, the continuous selection, or high-grading, tends to produce a forest composed of greater and greater proportions of sugar maple, basswood, ironwood, and the other shade tolerant trees. The application of this process over a period of more than a century in many parts of southern Wisconsin has given rise to a considerable acreage of maple forest on lands originally dominated by oaks.

The sugar maple, itself, has been of high value for such specialized products as flooring and furniture but the supplies have not been sufficient to justify the local existence of the necessary factories and transportation costs to distant factories have been prohibitive. In the period since World War II, however, the demand has overcome the distance barrier, and high prices for stumpage have greatly increased the harvest of the maple, especially for bowling pins and other high-value products.

Some use is made of the mesic forests for the production of maple sugar. Formerly, such activity was widespread and was practised in relatively small wood lots. In recent years, the trend has been toward larger operations which need extensive acreage and, as a result, the production of sugar is now centered in the northern mesic forests. The process of sugaring apparently had little influence on the ecological conditions of the forest, except in so far as it tended to protect the forest against other destructive uses.

The mesic forest has little value as pasture, since the browse production by shrubs is very low and the yield of grass is practically nil. Many wood lots are open to cattle from adjacent permanent pastures because of the cooling effect of the shade on hot summer days rather than because of any potential forage value, but on the whole, the mesic forest is less subject to grazing than any other forest type of the region. In wood lots which are converted to pasture, there is an almost complete destruction of the understory. One interesting exception is the spring-beauty (*Claytonia virginica*). This plant is able to resist damage and actually to thrive under the new conditions. The plant, a typical ephemeral, blooms in early spring and usually dies down before the cows are turned out to pasture. The increased light in the partially opened stands increases its vigor and floriferousness to the point where such pastured wood lots appear from a distance to be covered by a blanket of pink snow when the blooming season is at its peak.

The spring-beauty and the other showy wild flowers of the mesic forests were familiar to the early settlers in Wisconsin, because they had seen them in the forests of New England and Pennsylvania. Many of them were also known to the immigrants from Europe, in so far as related species are common in the woods of Germany and other north European countries. As a result, the woods flowers gained and held a position of esteem never attained by the more beautiful but unknown plants of the prairie. To this day, the first appearance in the early spring of trillium, Jack-in-the-pulpit, violet, and Dutchman's breeches brings joy to the school child and hope to his parents. In the P.E.L. studies, a small but heartening percentage of the best mesic forests were found, upon inquiry, to be preserved because of the love of the owner or his wife for the spring display of wild flowers to be found there. Whether this interest will be maintained in an era of a "farm is a factory" philosophy, consolidated schools reached by school buses, and education by television is a moot question.

CHAPTER 7

Southern forests—xeric

GENERAL DISTRIBUTION

The xeric forests of southern Wisconsin are here considered to include all of the closed canopy forests on the uplands of the area southwest of the tension zone which are included within the range of 300 to 2300 on the compositional index described in Chapter 5. For purposes of certain discussions, this group will be divided into two segments, the dry forests from 300 to 1300 and the dry-mesic forests from 1300 to 2300, but this division is purely arbitrary. The xeric forests as a whole are predominately oak forests. They are located on well-drained sites on either sandy and porous flat lands, on south and west slopes of hills, or on thin soils on hilltops and ridges. As pointed out by Chamberlin (1877), they frequently adjoin savannas on the dry side and mesic forests on the more protected side, with gradual transitional stages on boundaries of both.

The total acreage of oak forest at the time of settlement was 1,386,700 acres or 3.96 per cent of the land area of the state (Figure VII-2). This was only about 20 per cent of the acreage in oak savanna. No large continuous tracts of closed oak forest, comparable to the massive blocks of mesic forest, were present. The greatest concentration of the type was in the west central counties, toward the northern boundary of the Driftless Area, especially in Monroe, Dunn, and Buffalo counties. Only a single small island was present in Dane County (Ellarson, 1949) and some other southern counties were completely devoid of stands of this type.

COMPOSITION AND STRUCTURE

In the 127 xeric stands in the P.E.L. study, which are shown in the figure reproduced at the opening of this chapter, 29 species of trees were found. These are shown in Table VII-1, which gives average importance values and per cent constancies for each of the two segments of the compositional gradient. There are seven species of oak on the list, including the three most important trees. Hickories, which are prominent components of the so-called oak-hickory forest in the southern states, are of minor significance in Wisconsin. From the standpoint of both importance value and constancy, white (*Quercus alba*), red (*Q. borealis*), and black oak (*Q. velutina*), are far in the lead; no other species are even close. Just as sugar maple (*Acer saccharum*) controlled the mesic forests through its overwhelming dominance, so the oaks control the xeric forests. The other members of the forest are adjusted to the environment as modified and influenced by the oaks. The dry segment is characterized by the prominence of black oak, black cherry (*Prunus serotina*), bur oak (*Q. macrocarpa*), Hill's oak (*Q. ellipsoidalis*), chinquapin oak (*Q. muhlenbergii* Engelm.), trembling aspen (*Populus tremuloides*), and boxelder (*Acer negundo*); the dry-mesic segment has red oak, large-toothed aspen (*Populus grandidentata*), and red maple (*Acer rubrum*), as characteristic members, plus a number of invaders from the mesic forest, including significant quantities of basswood (*Tilia americana*), sugar maple, slippery elm (*Ulmus rubra*), white ash (*Fraxinus americana*), and ironwood (*Ostrya virginiana*). White oak, shagbark hickory (*Carya ovata*), black walnut (*Juglans nigra*), and green ash (*Fraxinus pennsylvanica* var. *subintegerrima*) are represented about equally well in both segments of the xeric forest.

The structure of a typical initial oak forest is well shown in

Table VII-2, which is taken from Cottam's detailed study (1949) of a forest near Verona in Dane County (Plate 11). The compositional index of this stand is 1006 so it is near the top of the dry segment. The stand possessed a total density of 123 trees per acre with a dominance of 105 square feet per acre. The canopy coverage was 83 per cent and this allowed sufficient light to reach the forest floor so that reproduction of the dominant trees, white and black oak, could occur. Notice, however, that although the bur oaks are among the largest trees in the forest, they are not reproducing and that black oak has far less reproductive material than white oak. This is in line with the relative shade tolerances of these three species. Red oak, the most tolerant of the major oaks, is not represented in this sample.

Table VII-2 also shows the structure of a typical intermediate, or dry-mesic, oak forest, based on data collected in a stand in Columbia County. The reproduction here is almost completely derived from the mesic forest species which are present, with practically no replacement of the dominant intermediate species. The importance of red oak in the larger size classes is typical of the behavior of this species. In many stands it attains an even greater preponderance; in one stand on record it contributed 91.4 per cent of the total dominance, at the very high rate of 106 square feet per acre (Plate 12).

LIFE-HISTORY OF THE OAKS

In many respects, the various species of oak which reach dominance in the dry southern forests resemble each other in their life-history characteristics, with small but important differences in certain features of ecological significance. The species are very closely related and frequently are difficult to identify satisfactorily in the field. The two main groups, white oak and black oak, are perfectly distinct, but within each group the taxa are all connected by long series of intergrades. These are probably the result of large-scale introgression from long continued hybridization and backcrossing, with the result that certain gene pools associated with particular morphological characters have taken on definite geographical ranges. Within these ranges they may be associated with different combinations of other characters, each with its own range. Thus a particular type of acorn may be found on all of the bur oaks in a given area, but part of the population will have one combination of leaf, twig, and bark characters, and another part will have a different set of these characters. Since the basic taxonomy of the oaks is so poorly understood, it seems appropriate here to give brief descriptions of the species as interpreted by the P.E.L. study in Wisconsin, with the understanding that the descriptions refer to the modal types of greatest frequency and that many intergrades exist.

The most xeric species in the black oak group is Hill's oak (*Quercus ellipsoidalis*), frequently referred to as scrub oak, but sometimes called northern pin oak. This tree has small, ellipsoidal acorns, usually with prominent, dark, longitudinal stripes. The leaves are hard, shiny, and very deeply lobed, often to within 5 or 6 mm. of the midrib in both sun and shade. The lower branches persist for long periods as dead stubs and give a ragged appearance to the trunks. Hill's oak is predominately a tree of dry, sandy places, both sand plains and sandstone hills; it reaches extensive pure populations only on such sites. It ranges throughout the state, and is frequently associated with jack pine in the north. The trunks are very susceptible to damage by ground fires, but the base of the tree at the junction of roots and trunk has great powers of regeneration, so that a circle of stump sprouts appears after a fire. Recurrent fires merely induce further sprout production, reducing the tree to a bush; whence the name scrub oak. The species is the most intolerant of all Wisconsin oaks and cannot reproduce in the shade of any other tree species.

The black oak (*Quercus velutina*) is found on better soils with more available moisture than Hill's oak. It has oval acorns in deep cups with loosely appressed scales. The leaves are thinner than Hill's oak, shiny, and deeply lobed in the sun but less so in the shade. The bark is black and is broken into small, equidiametric patches on mature trees. Its range in Wisconsin is sharply delimited to the area south of the tension zone. Black oak is fairly intolerant to shade, but is able to grow beneath both Hill's oak and bur oak. It reacts to fire as does Hill's oak and can produce the same type of bushy scrub.

The most nearly mesic of the Wisconsin oak species is the red oak (*Quercus borealis*). This species has very large acorns with straight sides which are produced in flat, saucer-shaped cups. The leaves are large, thin, dull on the surface, and shallowly lobed in both sun and shade. The bark on mature trunks is arranged in long, continuous vertical strips which are gray on their surfaces but black between the strips. Like Hill's oak, the red oak is found throughout Wisconsin. It reaches greatest average dominance in the western counties bordering the Mississippi but may be locally prominent elsewhere. Throughout the state, it does best on deep, well-drained soils, rich in nutrients, and well supplied with moisture. It is the most tolerant of all oaks and can replace all other oak species. The trunks are very susceptible to fire damage but regeneration is quite poor, so that maintenance of burned trees by stump sprouts is rare or absent.

Among the white oaks, the most xeric species is the chinquapin oak (*Quercus muhlenbergii* Engelm.), sometimes called yellow oak. The acorns are ovoid in outline, in thin cups which cover $\frac{1}{3}$ to $\frac{1}{2}$ of the acorn. The leaves have slender petioles and oblong blades which are merely toothed rather than deeply lobed. The bark is gray and flaky. In Wisconsin it has a very restricted range in the southernmost tier of counties, with best development in Grant County along the Mississippi. There it grows on hot, dry hillsides exposed to the southwest. It is very intolerant, and appears to equal bur oak in this respect. Insufficient information is available to assess accurately its reaction to fire.

The bur oak (*Quercus macrocarpa*) is represented by at least two races, neither one of which has the large acorns responsible for the specific name *macrocarpa* as found in the race which occurs in southern Illinois, Missouri, and southward. Rather, the Wisconsin forms have either intermediate acorns about 12 mm. in diameter in the most common type or very small acorns scarcely 9 mm. in diameter in the race in the northwestern counties. Both forms have deep acorn cups with the heavy fringe of coarse hairs which are implied in the name bur oak. Both forms have thick leaves which are widest near the tip and have a single pair of deep indentations toward the base which impart a fiddle-shaped outline to the leaf. The upper surface is dark green and shiny while the lower surface is whitish or very light green. The twigs, which are very coarse and often irregularly bent, are covered with thick corky ridges in many individuals. The bark on mature trunks is very thick, black, and deeply furrowed. The ecological amplitude of the bur is very great, as it is found on acid sands, deep silt loams, and thin rocky calcareous soils. On the moisture scale, it occurs from moist peat lands bordering sedge meadows, through mesic uplands, to the driest rocky cliffs, hill crests, and glacial moraines. Throughout this entire soil range it remains very intolerant of shade, being only slightly less extreme than Hill's oak in this respect. It is an important tree south of the tension zone only, although scattered occurrences of it in the north are known. It is a fire tree par excellence, since it not only has the stump sprouting ability of the scrub black oaks but also has a very high resistance to initial damage by fire, due to its very thick, fire-resistant bark.

The most tolerant member of the white oak group is the white oak (*Quercus alba*) itself. This tree has oval acorns of medium size in deep cups with tightly appressed scales. Its leaves are extremely variable in size and shape but usually they have two or more equal pairs of indentations, which may be very deep or very shallow. The bark is light gray and scaly textured, usually as a result of the almost universal occurrence of a fungus (*Aleurodiscus*) which inhabits the bark and causes it to flake off continuously, leaving a nearly smooth surface. The ecological amplitude is only slightly less extreme than that of the bur oak, as it is found in nearly the same range of sites except for the very wettest and the very driest. The geographical range is also similar, with importance attained only in the region south of the tension zone. It is more tolerant of shade than any other oak except the red; hence it can replace all others with the exception of the red oak. White oak is intermediate in fire resistance and is more susceptible to damage than bur oak yet less susceptible than red, but has rather poor powers of regeneration.

The last member of the oak genus in Wisconsin, the swamp white oak (*Quercus bicolor*), is more properly considered a member of the lowland forests of river bottoms and is found on upland sites only rarely. It has the distinction of being the only Wisconsin oak to produce acorns on long peduncles, often 3.5 to 5 cm in length. All other oaks produce sessile acorns or have very short peduncles. The swamp white oak has shallowly and evenly lobed thick leaves which are very dark green and shiny on their upper surfaces and white on their lower surfaces from a dense covering of hairs. The bark is gray and flaky, but somewhat more furrowed than the white oak. The bark on the twigs frequently peels in long, dark papery layers. The tree does best on poorly drained soils along rivers or in other swampy situations. Its response to shade is about the same as that of the true white oak or possibly it is a little more intolerant. Swamp white oak is resistant to ground fires and consequently forms a savanna tree like the bur oak which often grows with it. When burned, it can regenerate from the stump. Its range in Wisconsin is restricted to the major river valleys and the streams entering Lake Michigan in the area south of the tension zone, except for a more or less isolated population in the Wolf River valley.

All of the oaks flower in May, the bur and black oak in the first week, and the red and white in the second or third week in southern Wisconsin. The flowers are of two types, the staminate form in long catkins and the pistillate in reduced clusters. The pollen is produced in great abundance and is wind-borne. In the white oak group, the fruits or acorns develop in the same year as the flowers, while in the black oak group, two years are required for fruit maturity. Both groups show great fluctuation in the size of the annual acorn crops with the red oak the least variable in this respect. Part of the fluctuation is due to damage to the staminate flowers caused by late spring frosts, but since this effect is delayed in expression for two years in the black oaks, there is usually a crop from either a black or a white oak in a particular year. In general, there is a period of three to five years between heavy crops in one species. Sometimes there is a resonant coincidence in the separate cycles so that very few acorns are produced by any species. This is shown in Table 10, which gives the number of

Table 10

Annual variation in size of acorn crop in a black oak-white oak (*Q. velutina-Q. alba*) forest in University of Wisconsin Arboretum

Year	No. of acorns per acre*	Year	No. of acorns per acre*
1949	157,000	1954	93,800
1950	23,000	1955	119,600
1951	40,000	1956	10,700
1952	69,000	1957	170,600
1953	63,000	1958	37,700

* Multiply by 2.47 to get number per hectare.

acorns produced per acre for a ten-year period in a mixed forest of white oak and black oak in the University of Wisconsin Aboretum in Madison. In 1956, there was an almost total failure of the acorn crop, which resulted in severe effects on the squirrel populations. Ordinarily, the squirrels and chipmunks consume a very large portion of the crop and the only time significant quantities of acorns are available for germination is the years of bumper crops when the supply exceeds the demands of these rodents. Other large predators include the introduced pheasant and the deer, but perhaps the most important agent of destruction is the acorn weevil, which may infest as much as 90 per cent of the crop in some years. Such infested acorns are a favorite food of blackbirds, who descend on the oak forests in huge mixed flocks in the autumn.

Germination takes place in the spring in the black oaks, but the whites usually put out their hypocotyl in the fall and may have a well-developed root system by spring. In wet years, it is not unusual to find the white oak acorns beginning germination while still on the tree, in a feeble attempt at vivipary. The seedlings survive partially at the expense of the stored food in the acorn for the first one or two years. During this time and for the next several years they are very susceptible to attack by mice and rabbits, but all species are capable of putting forth new shoots from the plentiful supply of adventitious buds above the root collar. A period of high rabbit populations is thus able to keep down a whole series of annual crops of seedlings to a condition of tiny one-year growths. When the rabbit population crashes, as it does periodically, the entire batch of seedlings may get away to the sapling stage, thus producing an apparently even-aged stand from a group of plants of very different actual ages (Leopold, 1949). Recurrent fires, followed by temporary or chance periods of fire protection, will produce the same effect.

All oaks grow slowly in comparison to most hardwoods, but their rates are greatly influenced by the nature of the site. Thus bur oaks have been measured which were only 4 inches in diameter at one hundred years of age, while others on deep rich soils attained 30 inches at the same age. All species are potentially long-lived, but the bur oak is the one most frequently seen in the older age classes, with trees three hundred to four hundred years old not uncommon.

Some of the Wisconsin oaks show their relationships to the live oaks of more southern regions because they drop only a portion of their leaves in the autumn, with some or most remaining on the tree over winter in a dead and brown condition. This habit is best developed in Hill's and black oak, less well in white oak, and not at all in bur and red oak. The load of dry and highly combustible leaves greatly increases the chances of destructive crown fires in the stands of Hill's and black oak.

Among the other tree species of the xeric forests, the hickories, walnuts, and beeches have seeds which are dispersed primarily by mammals; a total of 12 species or 41 per cent have this habit. The cherry and hackberry are distributed by birds; thus more than half of the species are dependent upon animals for invasion into new territory. The remainder have seeds adapted for wind dispersal.

The black cherry (*Prunus serotina*) plays an unusual role in the xeric forests. As indicated in Table VII-2, it is frequently present in great numbers in the seedling and sapling classes, sometimes to the point of total dominance in the latter group, but it is rare as a canopy tree. The young cherries have a relatively high degree of shade tolerance but as they get older they become more intolerant and are unable to withstand long periods of suppression. When proper openings in the canopy occur, they can mature into trees and may make majestic, straight-trunked forest giants 30 to 36 inches in diameter on good sites. The shagbark hickory is similar; it is regularly present in good numbers in the younger size classes, particularly in the initial stands, but only rarely does it achieve the large size and age of which it is capable.

According to the P.E.L. studies, the black walnut (*Juglans nigra*) reaches its state-wide optimum in the dry-mesic stands, but the relentless search for these trees for their valuable lumber in the past has so reduced their populations that few positive statements concerning their true ecological behavior can be made. From the available information, it would appear that the habitat requirements of black walnut are similar to those of red oak, where the best growth is on deep, rich, permanently moist soils as in ravines, along the base of north slopes, and on river terraces. The early settlers commonly used the black walnut as an indicator of potentially good agricultural soils; hence the typical sites on less extreme topographies may have been long since destroyed.

A number of the species are restricted to the area south of the tension zone and may be considered indicators of the type. Besides the five oaks with this range mentioned previously, shagbark hickory, black walnut, boxelder, and hackberry are found only in the south, while slippery elm, yellowbud hickory (*Carya cordiformis*), and butternut (*Juglans cinerea*) are of importance only in the south although they are present as scattered individuals in the north.

GROUNDLAYER COMPOSITION

The prevalent groundlayer species of the southern xeric forests are given in Tables VII-3 and VII-5 for the dry and dry-mesic groups separately. The other modal species are also listed in these tables.

A floristic summary of the groundlayer plants of the initial and intermediate groups indicates that, in both cases, nine families include more than one-half of the total species, but the families concerned are slightly different in each group, with the *Primulaceae* of the dry stands replaced by the ferns in the dry-mesic type. The *Compositae* comprise 15.1 per cent of the total species in the dry stands, but this figure drops to 10.0 per cent in the dry-mesic stands and to 8.7 per cent in the mesic stands. The *Leguminosae* show the same trend, while the *Liliaceae* behave in reverse fashion. Other families in the group of nine are the *Gramineae, Ranunculaceae, Rosaceae, Umbelliferae,* and

Caprifoliaceae. The number of species of both herbs and shrubs which have berries for fruits are lower in the two xeric groups than in the mesic forest, and the percentage of plants with sticktights and evergreen leaves remains essentially constant in all groups.

On the basis of the prevalent species in Tables VII-3 and VII-5, the xeric forests show a great decrease in the number of spring bloomers, with corresponding increase in plants which bloom in summer and autumn, as compared with the mesic forests. Such characteristic oakwoods plants as hog peanut (*Amphicarpa bracteata*), enchanter's nightshade (*Circaea quadrisulcata*), wild spikenard (*Aralia racemosa*), ticktrefoil (*Desmodium glutinosum*), lopseed (*Phryma leptostachya*), and bottle-brush grass (*Hystrix patula*) bloom in midsummer. These and the other late bloomers also differ from typical mesic forest plants because they are tall, well over 18 inches (45 cm) in height on the average. They have well-developed leaf systems which carry on active photosynthesis during the entire growing season. The great majority of the xeric forest understory plants produce flowers from flower buds developed during the current season rather than during the previous year. The main exception is found in the spring flowering shrubs, like the dogwoods and the hazelnut.

According to Randall's summary of the life-history characteristics of Wisconsin forest plants (1952), the xeric southern forest plants have the following features at levels statistically higher than the mesic forest plants: tap roots are common; root systems in general penetrate to considerable depths and tend to be below the humus-rich A_1 layer; vegetative reproduction by stolons is more common than by rhizomes, although adventitious buds on roots provide another frequent means of propagation; the leaves are smaller and thicker, and have higher osmotic pressures (14 atmospheres in extreme xeric stands compared to 9 atmospheres in mesic stands); the chlorophyll content of the leaves is low; and finally, the plants do not show external evidence of wilting until they have lost a considerable amount of water (up to 20 per cent of their turgid content).

Very little information is available on seed reproduction and size of seed crop. The majority of the seeds are dispersed by animals and wind. The seed crop per plant varies from very low levels of six or eight in the windflower (*Anemone quinquefolia*) to the high thousands in the orchids (Curtis, 1954a), like rattlesnake plantain (*Goodyera pubescens*), and yellow ladyslipper (*Cypripedium pubescens* Willd.), and in the pink shinleaf (*Pyrola elliptica*). As in the mesic forests, yields per acre are dependent on total density as well as individual seed load;

species like lopseed (*Phryma leptostachya*), smooth bedstraw (*Galium concinnum*), wild geranium (*Geranium maculatum*), and the like probably produce the greatest total yields although precise figures are lacking.

Most of the seeds produced are dormant but nearly all respond to the stratification effects of a single winter and do not show the complicated epicotyl dormancy found in many mesic forest plants. Annuals are inconspicuous and none ever reach the density levels shown by false mermaid-weed (*Floerkea proserpinacoides*), rough bedstraw (*Galium aparine*), or touch-me-not (*Impatiens pallida*) in the mesic forests.

Many of the xeric forest herbs have effective means of vegetative reproduction, but in spite of this, the presence of dense, tightly aggregated clones is not as common as in the mesic forests. Such rhizomatous species as the false Solomon's seal (*Smilacina racemosa*), the yellow ladyslipper (*Cypripedium pubescens* Willd.), and the true Solomon's seal (*Polygonatum pubescens*) frequently grow for many years without producing side branches. Unbranched rhizomes over twenty years old have been found in the ladyslipper. The rattlesnake plantain (*Goodyera pubescens*) and the windflower (*Anemone quinquefolia*) produce tight clones but these remain small in size. There are a few typical oak plants which do produce prominent clones; these include creeping honeysuckle (*Lonicera prolifera*), interrupted fern (*Osmunda claytoniana*), and pink shinleaf (*Pyrola elliptica*). The most frequent instances of extreme aggregation, however, are found in those species which will achieve their optimum in the succeeding mesic forests and which are recent invaders of the oak forest. Outstanding instances of such scattered but very dense colonies are to be seen in mayapple (*Podophyllum peltatum*), waterleaf (*Hydrophyllum virginianum*), bloodroot (*Sanguinaria canadensis*), and bellwort (*Uvularia grandiflora*).

Another species sometimes producing very distinct clones is the heart-leaved aster (*Aster macrophyllus*), which is actually much more common in the northern forests and is found in only about one-third of the southern xeric stands. This aster apparently produces a very active antibiotic substance (Curtis and Cottam, 1950) since its colonies have no other species within their confines, except for an occasional individual of the lopseed. Such instances of antibiosis are not as common in Wisconsin forest vegetation as in the prairies or other open communities. The opposite type of direct chemical interaction, that of the production of beneficial or pro-biotic substances, is much more difficult to demonstrate. One method that has been employed gives evidence that certain pairs of species or even groups of species do occur together more often than would be expected by chance. In every case where this has been shown, one or more of the species involved has been a nitrogen-fixing legume, so the possibility exists that the mutual groups are benefiting from the extra nitrogen provided by the legume. This possibility has not yet received experimental verification, however.

Cottam (1949) studied the occurrence of such groups in a xeric forest by an ingenious use of punched-card records of occurrence and by appropriate machine calculations. Twelve species were found which grew together more frequently than expected by chance. The most significant group was associated with the tick-trefoil (*Desmodium glutinosum*) and included: woodbine (*Parthenocissus vitacea*), hog peanut (*Amphicarpa bracteata*), enchanter's nightshade (*Circaea quadrisulcata*), smooth bedstraw (*Galium concinnum*), wild geranium (*Geranium maculatum*), and lopseed (*Phryma leptostachya*).

The most conspicuous feature of the groundlayer of the southern xeric forests is the great abundance of shrubs. The shrub layer is frequently so thick that it impedes ready passage, especially when it contains a large element of thorny species like the blackberries (*Rubus*), gooseberries (*Ribes*), and prickly ash (*Zanthoxylum*). This prevalence of shrubs is in great contrast to their virtual lack in the best mesic forests, as is shown by the sum of presence for shrub species on the prevalent species lists, where shrubs total 507 in the dry forests, 622 in the dry-mesic forests, and only 46 in the mesic forests. A total of 48 species of shrubs and woody vines were found in all of the P.E.L. xeric stands, both dry and dry-mesic. The two shrubs mentioned by Chamberlin (1877) as characteristic of the oak forest, gray dogwood (*Cornus racemosa*) and hazelnut (*Corylus americana*), are the most widespread and most numerous on the basis of the modern study. Members of the genera *Rubus* and *Ribes* are also very common, though no one species equals the dogwood or hazelnut. The shrub densities may reach levels of 21,000 per acre or more (Cottam, 1949) of which nearly 19,000 are due to gray dogwood, hazelnut, and blackberry.

The shrubs exert a profound influence on dynamics of the oak forest, since their canopy intercepts a high fraction of the meager amount of light penetrating the tree canopy. Intensities on the forest floor, where tree germination takes place, thus are too low for the intolerant species of the canopy, which might be able to survive in their own shade if the shrubs were not present. On occasion, the shade cast by the shrubs may also cause the disappearance of the typical oak forest herbs. This is particularly the case when witchhazel (*Hamamelis vir-*

giniana) forms a solid understory in a rich red oak forest in the intermediate group. The groundlayer in such cases becomes enriched in typical mesic forest species like *Erythronium, Hepatica,* and *Dicentra,* in great contrast to comparable stands without the witchhazel.

Among the other members of the oak forest biota, the lichens, soil fungi, and birds have received special study in Wisconsin forests. In an exhaustive investigation of the lichen communities occupying tree trunks in the southern forests, Hale (1955) was able to show that the species were definitely associated with the total forest composition but, superimposed on this, there was a strong element of host specificity. *Parmelia rudecta,* for example, was most common in the dry stands, less abundant in the dry-mesic stands, and lacking in the mesic stands; but within this over-all gradient, the lichen was restricted to trees of the black oak group. Thus it was present on black oak and red oak, but was missing from bur and white. Other species showed the reverse pattern on the oaks while still others were either indifferent to the kind of oak or were missing from all kinds. Among the more common lichens found in the xeric forests by Hale were species in the genera *Allarthonia, Buellia, Candelaria, Leconora, Lepraria, Parmelia, Physcia, Ramalina,* and *Rinodina.*

The microfungi in the soil (Tresner, Backus, and Curtis, 1954) were more numerous in the xeric oak forests than in the mesic forests at all seasons of the year. They were especially numerous in the winter and spring, when numbers from 650,000 to 750,000 per gram of dry soil were found. Members of the *Mucorales* were prominent, making up over 20 per cent of the total species in the dry segment. *Mucor angulisporus* and *M. ramannianus* were typical members of this group. Other species which were restricted to the oak forests and which reached high frequency levels were *Penicillium janthinellum, Oospora sulphurea, Absidia cylindrospora,* and various species of *Mortierella, Mesobotrys,* and *Cephalosporium.* A few species were present which were equally abundant in the xeric and the mesic forests. Among them were *Penicillium simplicissimum, Cladosporium herbarum, Myrothecium verrucaria,* and several species of *Trichoderma.* The total number of species of microfungi per stand varied from 30 in the low index dry stands to 70 in the dry-mesic stands dominated by red oak. Kitzke (1949) found that the interesting group of slime molds known as the *Acrasiales* were much more numerous in the oak woods of Dane County than in the mesic forests. Members of the genera *Polysphondylium* and *Dictyostelium* reached frequencies as high as 80 per cent in the initial stands.

The bird populations were studied in detail by Bond (1957). He found that birds which nested in shrubs were most numerous in the xeric forests and that plant foods rather than insects were most important in the diet of the birds of this forest. There was a considerable shift in species composition from the stands of the dry segment to those of the intermediate or dry-mesic segment. In the dry stands, the most important species belonged to a colorful group including the scarlet tanager, black-capped chickadee, downy woodpecker, rose-breasted grosbeak, cardinal, blue jay, Baltimore oriole, red-eyed towhee, and red-headed woodpecker. The birds of the intermediate forests on the whole were less brightly colored than the last. They included the wood thrush, least flycatcher, redstart, blue-gray gnatcatcher, yellow-throated vireo, ruby-throated humming bird, and the veery. There were 19 species of birds per stand on the average in both the dry group and the dry-mesic group, but the total density increased slightly from 302 birds per 100 acres in the dry to 336 birds per 100 acres in the intermediate. These figures compare with 16 species and 331 birds in the mesic forests.

No extensive investigations of the mammal populations of Wisconsin oak forests have been made, but repeated sampling in a few stands in the University of Wisconsin Arboretum by J. T. Emlen of the Zoology Department, and his co-workers, and scattered studies elsewhere clearly indicate a much richer combination of species than that found in the mesic forests. Among the most prominent mammals are the gray, flying, and fox squirrels and the chipmunk. The mice and other very small mammals are much more numerous than these, but also much less conspicuous. The cottontail rabbit, the woodchuck, the raccoon, the opossum, the red fox, and the white-tailed deer are other important members of the mammal community.

SUCCESSION

In contrast to the mesic forests, the southern xeric forests are relatively unstable. The dominance exerted by any one oak species is likely to be only one generation in duration, since other species with greater shade tolerance will replace it. This successional replacement due to differing tolerances is considerably influenced by the topography of the site and it is most rapid on flat lands, where the light on the forest floor is entirely in the control of the tree canopy. On steep hillsides, a solid canopy is rarely formed, so direct penetration of light from the side is always possible. Under such conditions, even the most intolerant of trees will occasionally receive enough light for a certain amount of successful reproduction and thus remain a continuing component of the community. The topography is also important in its influence on the moisture conditions. A southwest hill slope is considerably hotter and drier than a northeast slope, due to the prevailing southwest winds in summer. The buildup of soil organic matter and the consequent retention of soil water is thus greatly retarded, with the result that the more xeric species, like chinquapin, Hill's, or bur oak, are able to remain in control for long periods of time. On some of the most severe sites, the moisture conditions may prevent the growth of closed forest and may allow only savanna or prairie to grow.

On flat or gently rolling lands, the cessation of reproduction by the various oaks is associated with a certain degree of closing of the canopy. When measured as percentage of the sky covered by leafy branches as seen from below, it appears that each species has its own critical level or threshold. Numerous studies in southern Wisconsin have shown this to be 85 per cent for white oak, 78 to 80 per cent for black oak, and less than 75 per cent for bur oak. The red oak threshold has not been clearly determined, but it appears to be about 88 to 90 per cent or possibly a little higher. Within the range of values, if seed sources are present, any given oak species may be replaced by one with a higher threshold. It does not follow that this regularly or even frequently happens, however, since the moisture and light conditions in forests with a canopy over 80 per cent are also suitable for the establishment of the typical mesic forest trees, especially slippery elm, basswood, and ironwood. The best that can be said is that no oak will maintain itself in its own shade or ever be replaced by another species of lower shade tolerance in the absence of catastrophe. It may or may not be replaced by an oak species of higher tolerance. The most common successional replacements are shifts from bur to black oak and from black to white oak. Only a very few instances are known of a replacement of white by red oak. The stands of very high red oak dominance have had a different type of origin, as will be seen later.

Another kind of instability in the xeric forests is seen in the frequent occurrence of local openings in the canopy. These may be due to spot lightning fires, to wind throw by tornadic winds, or to attacks by the oak wilt fungus (*Chalara quercina*). This disease is widespread and apparently native. Its actions were first described in Wisconsin by Warder (1881), who reported the extensive dying of black oaks but not bur oaks near Sun Prairie. The disease occurred in both pastured and unpastured groves; the first attack was in one corner or small spot. Then the trouble seemed to extend in all directions, and a few trees died out every year. More recent research by Riker, Kuntz, and co-workers at the University of Wisconsin has shown that an initial infection causes the death not only of the infected tree but also of a gradually increasing circle of neighboring trees by means of spore transmission through root grafts. Trees of the black oak group are much more susceptible to the disease than are the white oaks. In nearly pure stands of black oak, these circles of destruction may reach several acres in size. In forests with an admixture of other species, the root graft transmission system sooner or later runs into barriers of non-susceptible hosts and the disease is stopped. The affected trees show a rapid wilting of the leaves, which is followed by loss of leaves, and eventually by sloughing of the bark. The naked stubs may remain standing for several years.

The gaps thus produced in the forest show no soil disturbance and they continue to receive the protection against wind movement afforded by the surrounding trees. The main change is a greatly increased light intensity and a greater range in temperature fluctuation. Two species, the black cherry and the shagbark hickory, benefit greatly by these new conditions. Saplings of both, which are usually present before the oak wilt attack begins, quickly fill in the gap with a canopy of their own. The frequently observed discrete patches of cherry or hickory to be seen in the xeric forests throughout southern Wisconsin probably have originated in this way. Some of them are even-aged stands ninety years old or more. This temporary and small scale replacement of the dominant oaks by less vigorous species of lower tolerance is an almost perfect example of the gap phase concept of Watt (1947). No conceivable examination of the existing environmental conditions within such a filled gap could possibly explain its existence, since it is due entirely to some chance happening in the remote past.

Another gap phase tree of importance in the xeric forests is the large-toothed aspen. This species requires soil disturbance as well as increased light; hence it is found in those gaps created by lightning fires or other spot fires. The ashes on the burned soil surface offer favorable conditions for germination; the area may be filled by a solid stand of seedlings if suitable seed trees are present anywhere in the remote vicinity.

Discrete patches of greatly different species combinations similar to gap phases may be produced by very different means. Larsen (1953) studied a forest in Dane County in which red maple (*Acer rubrum*) was invading a stand of oaks. The initial red maple trees were apparently chance introductions of single seedlings distributed at random throughout the forest during some period of disturbance in the past, since the oldest red maples were only slightly younger than the dominant oaks. No reproduction took place, however, until the oaks had matured and produced a closed canopy. Then the red maples began to seed in, in tight colonies surrounding each original tree. These colonies were very dense and the shade cast by their canopy of saplings and young trees was sufficiently dark to wipe out the typical oak groundlayer of shrubs and light-demanding herbs. The soil beneath the red maples was thus nearly bare since no typical mesic forest herbs were present in the vicinity. The red maple colonies are continuing to spread, destroying the oak community as they grow, much as a cancer destroys its host. For many years to come, this forest will present a non-homogeneous appearance, until the maple colonies coalesce or are augmented by new colonies and begin to be invaded by more tolerant members of the mesic forest. This replacement of oak by red maple has been observed in several stands in Wisconsin but it is by no means common. In each case, the original forest was dominated by red oak.

The normal transition from xeric forest to mesic forest may also proceed in this island-like manner when the initial invader is sugar maple. McIntosh (1957) reported such a case in a Green County forest. The original stand was dominated by white and black oak with an island of red oak containing a few maples and basswoods. Due to complete fire protection afforded the stand in the last 50 years, the mesic trees began to spread out, basswood going first and farthest, followed by an almost solid wall of young sugar maples. As in the red maple stands, the shade from the maples brought about the death of the typical oak forest understory. McIntosh made observations on the vitality (Braun-Blanquet, 1951) of the groundlayer plants, and found that the first effect was on reproduction, with flower and fruit production greatly reduced. As the period of exposure to low light lengthened, the oak plants gradually died out altogether, although some persisted for decades in a weak, entirely vegetative condition. The few mesic forest understory species were enabled to reach unusual combinations of dominance because of the lack of competition. Thus the heart of the maple island had high densities of blue cohosh (*Caulophyllum thalictroides*), bellwort (*Uvularia grandiflora*), and Gleason's trillium (*Trillium gleasoni*), which rarely attain such proportions in a typical mesic forest.

Not all existing stands of xeric forest were derived from more pioneer forests by progressive succession based on shade tolerance, soil improvement, and generally increased mesophytism. The type of reversion found in gap-phases can also exist on a large scale when a catastrophe destroys a large area of forest. It has been shown in Chapter 6 how red oak and slippery elm invaded a mesic forest in Green

County 130 years ago as a result of disturbance. Various lines of evidence indicate that most of the existing stands dominated by red oak had a similar origin because complete canopy destruction by fire permitted a retrogressive replacement of the dominant maples and basswoods by large-toothed aspen, and the subsequent development of a stand very rich in red oak. There are also good indications that many of the red oak forests were derived by retrogression from relic stands of northern rather than southern hardwoods. This is particularly true in the Driftless Area, where the red oak stands are frequently found on the same kind of northeast facing slopes as are the remaining northern forest relics. The main evidences for this type of origin are in the soils and the understory plants. A typical podzol profile, with a strongly leached, ashy-gray A_2 layer, is present in such stands but is lacking in stands of comparable development derived by ordinary succession. The groundlayer of these northern-related stands contains a rather high representation of species which are typical of the northern pine-hardwoods but are missing from other southern stands. Included among them are pipsissewa (*Chimaphila umbellata*), round shinleaf (*Pyrola rotundifolia*), bunchberry (*Cornus canadensis*), ground pine (*Lycopodium lucidulum*), and partridgeberry (*Mitchella repens*). It is not certain that all such stands were derived by retrogression from northern hardwoods. A few may have been the result of ordinary progressive succession from relic stands of white and red pine (*Pinus strobus, P. resinosa*), whose passing was accelerated by the utilization pressures of the early settlers. Relic stands of pure pine, pine-hardwoods, and hemlock-hardwoods are known to exist throughout the region today. Many of the red oak stands may simply represent former relic conifer stands which have disappeared only recently, many since the coming of white man.

Thus the xeric forests of southern Wisconsin are seen to be a series of rapidly changing species combinations whose local complexity is the result of progressive and retrogressive processes induced by the biological characteristics of the dominant species and by the repeated interference of outside agents of destruction. An over-all pattern is imprinted on the group by the behavior syndromes of the dominant oaks so that an orderly continuum of average conditions is readily discernible. Lesser species possess adaptations enabling them to fill special niches caused by the temporary or local relinquishment of dominance by the oaks. The entire complement of dominants, gap trees, and subsidiary species forms a broad community with a much lower level of integration than that shown by the mesic forest.

ENVIRONMENT

Microclimate

The great range in species combinations found in the xeric forests is accompanied by a similar range in microclimatic conditions. The canopy coverages usually range from 55 to 80 per cent in the initial segment, and from 70 to 90 per cent in the intermediate segment. Midsummer light intensities vary accordingly, from 1500 to 4500 foot-candles in the initial stands down to levels below 100 foot-candles in the most nearly mesic of the intermediate stands. The sparse branching of the oaks contrasts with the complete leaf mosaic of the maples so that considerable light penetrates the canopy even when it is nearly complete.

Various studies in the southern oak forests have shown that humidity and evaporation conditions differ considerably from those recorded in the open. Cottam (1949) found that two to three times as much water was lost from evaporimeters in exposed locations as was lost from those in protected places in the forest. In the University of Wisconsin Arboretum, comparative studies over a number of years in a white oak stand and an adjacent prairie during midsummer showed that the relative humidity at 12 inches above the ground at 2 P.M. average 60.8 per cent in the forest and 49.9 per cent in the grassland. The air temperatures at the same time and place were 74.5°F. and 81.2°F. respectively, while soil temperatures at 1 inch depth were 69.2°F. and 77.5°F. A comparison of all microclimate studies shows that the xeric forests are intermediate between the mesic forests and the open, and most factors more nearly resemble those of the mesic forest.

Average values for certain features of the macroclimate are given in Tables VII-4 and VII-6.

Soils

The xeric forests tend to occupy the upper portions of the various soil catenas of the region, although the lower slopes frequently have red oak or other intermediate stands. The most important soil associations are the Dodge-McHenry-Elba (which includes the Miami), the Fox-Rodman, and the Fayette-Dubuque, according to the classification scheme of Hole and Lee (1955). The first two soil associations are in the glaciated area and the last is in the Driftless Area. Together, the three soil associations cover over seven million acres, or 21.3 per cent of the land area of the state. The vegetal cover of these soils is not restricted to the xeric forests but includes some of the mesic forest and large areas which originally were in savanna as well. The zonal soils under oaks are in the Gray-Brown podzolic soils group and belong to the special type called grood soils by Wilde *et al.* (1949) to indicate their close historical relationship with grasslands. They vary from dark-colored loams resembling prairie soils in the stands that resemble savannas the most, to highly leached light-colored soils in the most nearly mesic stands, particularly those derived from northern forest relics. The Fox-Rodman association includes poorly developed profiles on sand and nearly skeletal soils on the coarse gravels of glacial moraines.

Analyses of soils in the P.E.L. studies show a greater acidity and lower nutrient content than comparable figures for mesic forest soils. The acidity is apparently the result of organic compounds, especially tannic acids, produced by the oak leaves. The ash content of leaves of various species of oaks varies from 4.5 to 7 per cent, in contrast to levels of 10 to 12 per cent in sugar maple and basswood. Of this amount of ash, only about $\frac{1}{3}$ is in the form of calcium and magnesium oxides, compared to $\frac{2}{3}$ in the mesic trees. The low lime content is associated with low palatability by millipedes and other soil fauna and may be partially responsible for the low rate of decay induced by other soil biota. In any case, the leaves of Hill's, bur, black, and white oaks tend to accumulate on the ground and to form thick layers of duff or mulch in the A_o layer, frequently 4 to 7 inches deep. Preliminary observations indicated that complete disintegration of a given crop of leaves may be delayed for 3 to 5 years. Such a thick blanket of slowly decaying leaves gives rise to mor humus, which is contrasted to the mull humus that is produced when the leaves decay in the year following their deposition and leave a bare surface of the A_1 layer by late summer. It is obvious that a mor humus may have markedly different effects on germination of tree and herb seeds than a mull humus, but the exact details have not been studied. There is a good correlation between the nature of the groundlayer and the type of humus. Typical mesic herbs, especially the spring ephemerals, are largely restricted to mull, and typical summer-blooming xeric herbs are associated with mor humus. The mor type is characteristic of dry and early dry-mesic stands. In the red oak stands, both types are found; some stands have the deepest layer of duff of any xeric forest and others show complete decay of the leaves within one year. A thorough microbiological investigation of these stands might shed much light on the nature of the transition from mor to mull and would greatly increase our understanding of the successional process.

GEOGRAPHICAL RELATIONS

Many studies have been made of the xeric upland forests in other states. A summary of the tree composition of a number of the stands shows that great similarities exist between the Wisconsin forests and those of Michigan, Minnesota, northwestern Indiana, northern Illinois, Iowa, and Nebraska, with the majority of the communities containing only trees which are also found on the Wisconsin lists. Some trends are apparent in proportions, with areas to the south usually having a better representation of hickory and those to the west having more American elm. The oak forests of Michigan were shown to follow a continuum pattern remarkably similar to that in Wisconsin by Parmalee (1953).

In more distant areas, considerable shifts occur in the floristic makeup of the xeric forests, particularly in the pioneer stands on the driest sites. On a percentage basis, trees not on the Wisconsin lists constitute an increasing proportion of the canopy (either as total importance value or total density), with values of 31 to 46 per cent in three stands in New Jersey studied by Niering (1953), 31 per cent in Kansas (Hale, 1955), 37 per cent in the Mammoth Cave region of Kentucky (Braun, 1950), 55 per cent in Virginia (Braun), and 73 per cent in Oklahoma (Braun). These forests show a relatively high degree of similarity from a gross appearance or physiognomic viewpoint. The floristic differences result from the generally restricted ranges and varied ecological behaviors of the many species of oak. The differences are due largely to a shift in the dominant species of this genus, with chestnut oak (*Quercus montana*), post oak (*Quercus stellata*), scarlet oak (*Quercus coccinea*), black jack oak (*Quercus marilandica*), and southern red oak (*Quercus shumardii*) being important replacing species. Other additions include several species of hickory, tuliptree (*Liriodendron*), chestnut (*Castanea*), sour gum (*Nyssa*), and short-leaf pine (*Pinus echinata*).

The oak-hickory association of the eastern deciduous forest as recognized by Braun (1950) and many other ecologists is usually thought to center on the Ozark Mountains and to have spread out from that *refugium* in postglacial times into the regions to the north. Many of the herbaceous species of the Wisconsin xeric forests have geographical ranges which include the Ozark region, so the hypothesis apparently has a certain amount of supporting evidence. However, an examination of the species concerned shows that many of them are represented in Wisconsin by morphological subspecies or by ecological strains or ecotypes which are quite different from the typical forms now found in the Ozarks. As mentioned earlier, the bur oak of Wisconsin has a completely different kind of acorn from that found in the Ozarks. The red oak is a different subspecies; the snakeroot (*Veronicastrum virginicum*) has flowers of a different color; the white gentian (*Gentiana flavida*) has differences in number and size of flowers and great differences in growth habit; and the downy phlox (*Phlox pilosa*) has a different blooming season and growth habit, as shown by transplant experiments. One of the major Wisconsin dominants, *Quercus ellipsoidalis*, is lacking in the Ozarks but has a range which centers on the Driftless Area of Wisconsin. Although these differences are individually of minor importance, taken together they indicate that the Wisconsin forest has been separated from the Ozarks for a long period of time, although it may possibly have originated there in the distant past. The simplest explanation for its present composition would call for a Pleistocene *refugium* in the Driftless Area, from which the forest has spread out in postglacial Wisconsin time, to be enriched by some additions from the Ozarks, some from the eastern Appalachian region, and by certain residual species from the northern conifer-hardwood forests.

The rapid change in composition of the oak forests with distance in the United States is supported by the very low similarities between the American and the European oak forests. Oberdorfer (1957) reported on the species compositions of a number of types of oak forest in South Germany. Only 2 of the 120 total species of the anemone-oak type were in common with Wisconsin and these were the very widely ranging *Fragaria vesca* and *Campanula rotundifolia*. The genera in common included such important groups as *Anemone, Aster, Cornus, Galium, Geranium, Lonicera, Prunus, Ribes,* and *Solidago*.

UTILIZATION

On an area basis, the xeric forests make up the great bulk of the farmers' wood lots of southern Wisconsin. The total acreage at present is roughly 1,950,000 acres, or about 40 per cent greater than in presettlement time, according to the *Land Economic Inventory* of the Wisconsin Department of Agriculture. This is mostly in blocks of 40 acres or less in the glaciated region, but some stands of 500 to 1000 acres or more are to be found in the hill country of the Driftless Area. The small wood lots are maintained for their potential yield of wood products. An overwhelming majority of the farmers believe that maximum gains are to be had from a multiple use program which calls for double duty by the wood lots as both pasture and wood factory. Over 85 per cent of the wood lots are grazed in most counties as a result of this belief. Conclusive proof that such double use is financially unsound, as shown by the investigations of Ahlgren *et al.* (1946), has been without noticeable effect. The roots of this traditional behavior pattern are deep and inextricably interwoven with questions of small farm size, of quadrangular land survey patterns, and of lack of proper incentives.

The ecological effects of grazing in the xeric forests are definite and easy to observe. Grazing is almost universally accompanied by the practice of burning the woods, in the correct belief that the annual or biennial removal of leaf litter by a controlled ground fire will increase the amount of grass in the groundlayer. The observed changes in grazed forests are as much the result of burning as of grazing and their individual actions cannot be separated with certainty. The most obvious effects are exerted on the groundlayer. Any invading seedlings of mesic forest trees are the first to disappear, due to their great sensitivity to fire, but these are quickly followed by the xeric tree reproduction, which succumbs to the trampling actions of the cattle. The shrub layer disappears next, along with the taller members of the herb stratum. Native grasses, sedges, and other branching plants with growing points near the ground increase in relative density. These are augmented by the invasion of a number of exotic weeds of similar habits, including chickweed, dandelion, and bluegrass (see Chapter 23). The repeated ground fires sooner or later open fire scars on the bases of the tree trunks; these increase in size by action of decay fungi as well as by further burning. The total lack of reproduction means that no replacements are available for the mature trees which now die at an accelerated rate or which are harvested on a slow, selective basis. The canopy coverage thus decreases with time, bringing about a general increase in the xeric nature of the microenvironment. The final result, necessarily, is the gradual change from forest to savanna, with a few fire-tolerant trees scattered throughout a bluegrass pasture.

Soil changes accompanying this floristic degradation are largely the result of compaction by the hooves of the grazing cattle. Infiltration rates are reduced by a factor of 40 to 100 fold, so that surface-runoff is greatly increased and internal soil storage of rain water greatly reduced. Erosion losses may be phenomenally stimulated. Soil changes are perhaps more serious than botanical changes, since they are not readily reversible. As a result, recovery of the forest when protected from grazing is very slow. A lightly pastured dry-mesic forest of red oak in the University of Wisconsin Arboretum in Madison was given complete protection in 1932. No visible changes in the groundlayer became apparent until 1945 and complete recovery had not occurred by 1957 after 25 years of protection. Other cases are known where even the initial recovery has been delayed for over 25 years in very dry stands and may not occur in the foreseeable future.

Many of the xeric forest stands which have escaped grazing are used for timber harvest purposes. In the smaller wood lots, this harvest is usually the slow removal of one or a few trees per year, and depends upon temporary farm needs. As in the case of similar operations in the mesic forests, this tends to hasten successional changes and may result in a complete type conversion, from oak forest to maple-basswood or more usually to maple-slippery elm if a small amount of grazing has been permitted. In the larger stands, commercial harvesting for tie-cuts, flooring stock, or barrel bolts may lead to a removal of 70 or 80 per cent of the canopy trees or sometimes to clear-cutting in a forest of mature, even-aged trees dominated by a very high percentage of red oak. The logging operators profess to cut only overmature trees, but their definition of an overmature tree frequently is any trunk that does not bend when pressed by a chain saw.

Any cutting operation which opens more than 50 per cent of the canopy results in a resurgence of intolerant herbs, shrubs, and tree seedlings which may have been present in the groundlayer. The blackberry and the hazelnut are the most prominent of these liberated species and they frequently make impenetrable tangles in the cutover forests. The oaks are all capable of sending out stump-sprouts; these grow with great vigor and usually produce a new canopy within 15 to 20 years. If fires occur, there may be an admixture of large-toothed and trembling aspens, but the fires do not interfere with the oak sprouts. Second-growth forests in southern Wisconsin are readily recognizable because of the great prevalence of trees with two, three, or more equal-sized trunks which originated as sprouts. This coppice growth tends to be subject to heart rot unless the original trees were cut very close to the ground. The back-breaking effort required to do this with a two-man crosscut saw means that most of the existing stands are highly defective. Perhaps the advent of the chain saw will improve the situation in the future.

CHAPTER 8

Southern forests—lowland

GENERAL DISTRIBUTION

The moist forests of southern Wisconsin are found on two different types of physiographic sites—river valleys and lake plains. The valley forests are commonly known as bottomland or floodplain forests, and the lake border types are usually called hardwood swamps. They are similar because the soil moisture supply is in excess of that falling as rain. They differ principally in the amount of soil disturbance present; the bottomlands receive frequent additions of silt from spring floodwaters. They differ in the variation in water supply; the bottomlands again show the greatest fluctuation. This varies from actual submergence during flood times to nearly xeric conditions during midsummer low-water stages. Because of their more nearly constant supply of water, the hardwood swamps or lacustrine forests tend to have a higher content of organic matter in the soil that sometimes approaches peat in nature.

The floodplain forests are present along all of the major rivers of the area. The largest stands are on the Mississippi, Wisconsin, Rock, Fox, Wolf, Black, Chippewa, Sugar, and Pecatonica rivers (Figure VIII-2). The streams entering Lake Michigan usually have only a minor development of this type of forest because of their rapid fall. The lacustrine forests are found around the larger existing lakes and also on extinct glacial lakes, especially in Dane, Jefferson, Waukesha, and Dodge counties and the area bordering the Fox River Valley lakes. A few areas of hardwood swamp are to be found in the Lake Michigan border counties on poorly drained clay uplands that resulted from Valders glaciation.

The total area now in the southern lowland forests is approximately 290,000 acres, according to the *Wisconsin Land Economic Inventory*. This acreage represents an unusually large fraction of the area so occupied at the time of the original land survey a century ago, when about 420,000 acres were covered. The high degree of preservation of this type of forest is a result of the low agricultural value of most of the wet lands. Because of their low altitudinal position, they cannot be drained, and their high probability of being flooded further reduces their usability as cropland. This does not mean that they are totally undisturbed, in as much as they are used for grazing perhaps more than the upland forests.

COMPOSITION AND STRUCTURE

The stands studied are shown in the figure reproduced at the opening of this chapter. The average composition of the tree layer as found in the P.E.L. study is shown in Table VIII-1 for the wet and wet-mesic segments of the vegetational gradient previously discussed. The figures are based in large part upon the studies of Dietz (1950) and Ware (1955). Of the 37 species listed, 21 were found in the initial or wet segment and 36 in the intermediate or wet-mesic segment. The lowland forests thus have more species of trees than any other community in Wisconsin. In large part this is due to a number of species of southern derivation which have entered the state along the river valleys and are not found elsewhere. These include the smooth buckeye (*Aesculus glabra*), the river birch (*Betula nigra*), the honey locust (*Gleditsia triacanthos*), and the sycamore (*Platanus occidentalis*).

The average values in Table VIII-1 tend to obscure the fact that several different combinations of species are included within the wet segment. On pioneer sites along sand bars, mud flats, and other open places of recent soil disturbance near the water's edge, the usual forest is dominated by black willow (*Salix nigra*) and cottonwood (*Populus deltoides*). On open sites near the upland edge of the wet ground, river birch or swamp white oak (*Quercus bicolor*) are the usual dominants. As both of these types mature, they are invaded by silver maple (*Acer saccharinum*) and American elm (*Ulmus americana*), thus accounting for the high values attained by these species in the averages.

In his very thorough analysis of the southern lowland forests, Ware (1955) noted the overwhelming dominance of silver maple and American elm, with one or the other of these species being the leading dominant in 61 per cent of the stands studied. In combination with green ash (*Fraxinus pennsylvanica* var. *subintegerrima*), these species clearly form the most widespread and important community group in the southern moist forest. In the wet-mesic segment, red oak (*Q. borealis*) and basswood (*Tilia americana*) tend to form another incipient community node, although it is of minor areal importance. The rather low levels of both average importance value and constancy for the leading dominants in comparison to upland stands is a reflection of the greater competition for space offered by the greater number of companion species. In general, the greater the number of species adapted for life in a particular forest, the lower will be the relative contribution of any one of them.

The structure of typical moist forests is shown in Table VIII-2. The initial stand of willow and cottonwood in Dane County is a lacustrine type resulting from the artificial raising of the lake level by a dam on the outlet stream. The intermediate stand is on the Sugar River bottoms in Rock County and is especially notable for the large size of its dominant trees. The low densities of seedling reproduction in both of these stands are characteristic of the lowland forests in general, as are the abruptly lowered values in the sapling class. Conditions for germination are frequently poor, due to flooding. The seedlings that do succeed in occasional good years are likely to be destroyed by subsequent floods or periods of continued submergence so that the over-all densities in the lower size classes are much less than those found on upland sites. It is to be noticed that silver maple and American elm are invading the initial willow-cottonwood stand, while in the wet-mesic stand where these species are members of the canopy they are maintaining a good rate of self-regeneration in spite of the entrance of basswood and red oak.

The moist forests, particularly the riverine stands, tend to have a

Plate 11.—Interior of southern dry forest, dominated by white and black oaks (*Quercus alba* and *Q. velutina*). Stewart's Woods, Dane County.

Plate 12.—Interior of southern dry-mesic forest, dominated by large red oaks. Most of understory is witchhazel (*Hamamelis virginiana*). Decatur Lake Woods, Green County.

Plate 13.—Open-grown oak, surrounded by straight, forest-grown trees. New Observatory Scientific Area, Dane County.

Plate 14.—Southern lowland forest in early spring of a dry year, showing the multiple trunks on the silver maples (*Acer saccharinum*). Avon, Rock County.

Plate 15.—Southern lowland forest along the Wisconsin River, still flooded in late May. Wyalusing Scientific Area, Grant County.

Plate 16.—Same forest as in Plate 14. Photograph made in late August.

Plate 17.—Interior of northern mesic forest, dominated by hemlock (*Tsuga canadensis*). Note the low light and the bare condition of the forest floor. Plum Lake Scientific Area, Vilas County.

Plate 18.—Mature white pine–red pine (*Pinus strobus, P. resinosa*) forest. U.W. Trout Lake Biological Station, Vilas County.

very high total basal area per acre, with an average of 14,200 square inches (22.6 sq. m. per ha.). This is due to the large size of the trees rather than to high densities. Indeed, the typical mature river forest of silver maple and American elm has only about 85 trees per acre, which is less than usual stocking for upland sites. Silver maple and cottonwood reach very large sizes; individual trees between 17 and 20 feet in circumference are not unusual. Swamp white oak also attains tremendous girth. One tree in the Sugar River bottoms is over 15 feet in circumference. The canopy spread of these large trees tends to be proportionally large, so that a complete canopy coverage is maintained in spite of the low average densities. The well-known umbrella shape of the typical, planted, roadside elm is maintained to a large degree by the elm that grows in forest stands. The interlacing of the branches of adjacent trees occurs only near the top of the canopy, and conveys a strong resemblance to the arched and vaulted ceiling of a cathedral. Some impression of this effect is given by Plate 14, which also shows the multiple-stemmed nature of many of the trees. This last characteristic is typical of most of the species of the moist forest and is thought to be the result of damage to the trunk bases by ice floes and other flood-carried debris during the spring period of submergence.

Life-histories of dominants

As a group, the typical lowland forest trees are less well known from the standpoint of ecological behavior than are the mesic or xeric forest species. The leading dominants, silver maple, American elm, green ash, black willow, cottonwood, river birch, and swamp white oak, bloom in the spring, are wind-pollinated, and all but the oak have seeds or fruits which are wind disseminated. All grow rapidly and may attain very great size. All may produce stump sprouts and all tend to produce a many-branched trunk but none has a well-developed ability to grow from root sprouts or to form dense vegetative clumps. As mature trees, all but the willow have a moderate or high degree of resistance to ground fires. Destruction of the living bark on as much as one-half to two-thirds of the circumference of the trunk base is insufficient to kill well-established trees of green ash, cottonwood, silver maple, river birch, and swamp white oak.

No studies have been made in Wisconsin of the annual variation in seed production of the lowland forest trees, but scattered observations would indicate a high and relatively even production for such species as the American elm, silver maple, cottonwood, willow, and green ash. The yield by both willow and cottonwood is more than ample to fully saturate the ground for a considerable distance from the seed trees. In the University of Wisconsin Arboretum at Madison, road-building operations in 1936 exposed a smoothly graded surface of wet, marl-rich soil away from a willow-cottonwood forest through an adjacent marsh. The final grading coincided with the peak period of seed dispersal of both the willow and cottonwood. The surface became covered with a solid carpet of seedlings of these species for a distance of 1200 feet from the forest edge. Counts made in 1937 revealed an average density of 1,560,000 seedlings per acre, of which two-thirds were willows. This value dropped to 3720 per acre by 1947 and to 414 per acre by 1957. The proportion of cottonwoods had increased to 51 per cent. This is in accord with the findings of Barclay (1924) who showed that cottonwood seedlings form a taproot while the roots of willow tend to spread laterally. Because of the formation of taproots survival is greater for cottonwood if there is any drop in the water table.

The question of introgression between related species in the tree flora of the lowland forests received considerable attention from Ware (1955). Instances in his study where hybridization was found or where puzzling forms not readily referable to known taxa were observed were widely distributed through the flora. Thus the American elm of the Wisconsin River floodplains showed the rough upper leaf surfaces and the dark cork cambium layers usually associated with slippery elm. The ashes proved to be very troublesome in the field. Ware concluded that the lowland areas of eastern Wisconsin mostly harbored an introgressive cline of green ash-white ash, while the Wisconsin River Valley showed a complete gradation from green ash to red ash. Perhaps the most spectacular example of hybridization was that between silver maple and red maple (*Acer rubrum*), where many populations, particularly in the Wolf River Basin, were completely intermediate between these two supposedly distinct species. Other putative hybrids noted by Ware included cottonwood-balsam poplar (*Poplus balsamifera*) in the Green Bay area and river birch-white birch (*Betula papyrifera*) hybrids in the Wisconsin River Valley. He also reported the swamp white oak-bur oak hybrid which was studied later in detail by Bray (1955).

The great apparent importance of hybridization in the riverine forests is related to the linear nature of the habitat, where areas of suitable environment are confined to narrow strips along the rivers which pass through upland sites of highly varied nature and totally different vegetations. Genetic variation induced by either mutation or hybrid introgression has a much greater chance of becoming fixed in relatively pure local populations under such linear circumstances than under uniform upland conditions where outcrossing can occur in all directions (Wright, 1952). Ware was of the opinion that this great genetic diversity was causally related to the high degree of vegetative vigor and environmental adaptability displayed by the lowland forest trees. In his view, it is no accident that the trees most commonly employed for planting along city streets are lowland trees rather than typical upland species present in the region traversed by the streets. The almost exclusive prominence of American elm, silver maple, green ash, and honey locust on the streets of southern Wisconsin towns and of boxelder in their waste grounds lends ample support to this hypothesis.

GROUNDLAYER

The prevalent groundlayer species of the two segments of the lowland forest are given in Tables VIII-3 and VIII-5. The average species densities of these two groups are considerably lower than the equivalent values for the southern xeric forests. The floristic homogeneities as shown by the ratio of prevalent species sum of presence to total species sum of presence are also very low, lower than for any other major community in Wisconsin.

Floristic analyses of the total flora reveal that the same seven families include 50 per cent of the total species in both segments of the forest, although their ranking is different. The main change in family representation, as compared to the upland stands, is the prominent role of the sedge and mint families and the increased importance of the nettle and carrot families. The five leading families in each segment are listed in Tables VIII-4 and VIII-6.

The low values of average presence for the prevalent species of the wet stands (Table VIII-3) are an indication of the great variation to be found from stand to stand. In large part, this is due to frequency of spring flooding in the riverine stands. The photograph of the forest in Plate 15 was taken on May 21, at a time when the understory flora of the adjacent uplands was at its peak of development and bloom, yet no herbaceous plants are visible on the bottomland floor which was still inundated over most of its area. In some years this condition persists through the month of June, so that spring blooming perennials are at a distinct disadvantage. Full development of the groundlayer in these floodplain forests is frequently delayed until mid-August (Plate 16). Considerable changes in the composition of the dominant groundlayer community take place from year to year, apparently in response to the varying abilities of the different species to recover from the effects of submersion. Thus a particular stand may be covered by a dense layer of *Laportea canadensis* in one year and by a mixture of *Boehmeria cylindrica, Pilea pumila,* and *Impatiens biflora* in another. Other species which show great annual fluctuations are *Amphicarpa bracteata* and *Leersia virginica.*

In the intermediate or wet-mesic stands dominated by American elm and green ash, the periods of submergence are neither so frequent nor so prolonged, which allows the development of a fairly rich spring flora. Among the typical but not necessarily widespread spring plants are the green dragon (*Arisaema dracontium*), a close relative of the Jack-in-the-pulpit, toothwort (*Dentaria laciniata*), woods phlox (*Phlox divaricata*), Virginia bluebells (*Mertensia virginica*), and a group of species in the *Umbelliferae* including *Cryptotaenia canadensis, Osmorhiza claytoni, Sanicula gregaria,* and *Chaerophyllum procumbens.* By all odds, the most characteristic family of the wet-mesic moist forests is the sedge family, with 37 species in the genus *Carex* alone. The relative sum of presence of these sedges (10.1 per cent) is not exceeded even by the usually prominent composites (8.8 per cent).

Very little concrete knowledge is available concerning specific life-history details of the lowland groundlayer species, so little can be said about rates of seed production, germination requirements, or methods of vegetative propagation. Aggregation is usually very pronounced, often associated with minor elevational differences in the soil surface. The grasses and sedges tend to form dense colonies by vegetative division, as do the two nettles (*Laportea canadensis* and *Urtica dioica*).

Both floodplain and lacustrine forests have an unusually high content of lianas, including both woody and herbaceous forms. Apparently the high content of soil moisture presents few barriers to the vine life-form with its extensive canopy supplied by conducting stems of small diameter. Lianas have the highest presence values of any of the understory species, with values of 80 per cent for poison ivy, woodbine, and grape. Other vines of frequent occurrence are moonseed (*Menispermum canadensis*), hog peanut (*Amphicarpa bracteata*), parasitic dodder (*Cuscuta gronovii*), wild yam (*Dioscorea villosa*), clematis (*Clematis virginiana*), groundnut (*Apios tuberosa*), wild cucumber (*Echinocystis lobata*), carrionflower (*Smilax herbacea*), and bittersweet (*Celastrus scandens*). The total sum of presence for all vines is 444 and 484 for the wet and wet-mesic segments, respectively, or 15.1 per cent and 12.3 per cent of the total sum of presence. These figures compare with 6.5 per cent, 10.4 per cent, and 10.3 per cent for the mesic, dry-mesic, and dry stands of the southern forest.

Detailed comparative studies of the liana component have been made in typical stands of dry-mesic, mesic, and wet-mesic forest in Green County, Wisconsin. The number of trees bearing one or more lianas per tree was much greater in the wet stands (6 to 7 per tree as compared to 1 or 2 in the other types). Most of the lianas in the dry and mesic stands were in poor vigor and were confined to the basal portion of their host trees. In contrast, the lianas of the lowlands tended to climb to the tops of the trees, over 60 per cent of the total number of individuals reaching the tops, as compared to less than 1 per cent in the dry stands. Estimates of the contribution made by the leaves of the lianas to the total forest canopy showed that over 25 per cent of the canopy of the lowlands was contributed by the lianas, in contrast to 0.1 per cent in the dry forests.

Poison ivy is a typical example of a species that clambers well into the tops of even the tallest bottomland trees where it sends out vigorous, rigid branches to some distance from the host trunk. On occasion, more than one vine will grow up the same tree and the branching of the vine stems and their extensive development of clinging roots may kill the host tree by strangulation. Their leafy branches develop a canopy nearly equal to that of the original host, thus producing a pseudo-poison ivy tree that persists until the host decays. This behavior is a close approach to the strangling habit of such tropical lianas as the strangling fig and the rose vine and serves to heighten the subtropical atmosphere often conveyed by the bottomland forests.

The importance of lianas contrasts strongly with the relatively insignificant role of the true shrubs. On the basis of sum of presence, shrubs contributed only 9.1 per cent of the total in the wet stands and 9.0 per cent in the wet-mesic stands. These low values approach the mesic forest value of 7.5 per cent but are far below the values of 13.6 per cent and 16.2 per cent for the dry-mesic and dry stands. Among the shrubs which are most characteristic of the moist forest are wahoo (*Euonymus atropurpureus*), buttonbush (*Cephalanthus occidentalis*), and silky dogwood (*Cornus purpusi*).

The other members of the biota have received little attention. Fleshy fungi are often prominent, especially species of morels (*Morchella*) in the spring. Various gill fungi grow abundantly in wet years; the oyster mushroom (*Pleurotus ulmarius*), which grows on dead trunks of both elms and maples, is a typical example. Among the animals, a rather distinctive group of birds of southern affinities is associated with this forest, particularly along the larger rivers. Such species as the prothonotary warbler, the cardinal, and the tufted titmouse are common residents. The woodpeckers as a group find the bottomland forests a favorable habitat; perhaps this is due to the regular presence of flood-killed trees. Notable for their comparatively high population levels are the red-bellied, red-headed, and pileated woodpeckers, and the flicker. Among the lesser animals, the insects make their presence known through their ubiquitous and frequently overwhelming representative, the mosquito.

SUCCESSION

Ordinarily, forest succession is motivated by the autogenic changes induced in the environment by the first occupants of the site in terms of decreased light intensity, improved soil nutrients, and water supplies. The varying shade tolerances of the component tree species allow an orderly replacement of one by another to take place, with the more shade tolerant generally requiring richer and more mesic soil conditions than the less tolerant. In the floodplain forests, only a limited autogenic change is possible. Light is decreased by the tree canopy but soil improvement is rarely possible. According to Ware (1955):

. . . the effect of floodplain vegetation on the excessive soil moisture is negligible. Though many tons of water in a stand are lost by transpiration during a growing season, the forest is just as likely to be flooded by the river whether transpiration occurs or not. The effect of organic matter on the floor of a floodplain forest is also negligible. In other words, significant improvement of moisture conditions by the vegetation is trivial in comparison to the soil moisture improvement possible in upland forests. The lowland forest must await external or physiographic improvement of soil moisture conditions before other trees can enter.

Thus we find that the theoretical succession implied by the changing composition shown in the graphs in Chapter 5 is actually a difficult process that is rarely accomplished in its entirety.

The pioneer stages of the lowland forests occur on two quite different sites. Toward the river's edge, on wet, newly deposited banks of sand or mud, conditions are favorable for the establishment of willow and cottonwood, as has been described earlier. As these mature, they tend to be replaced by silver maple and American elm in the absence of further disturbance by the river. Toward the upper or landward side of the floodplain, soil conditions are more stable, although not necessarily drier. In the lower portions of their valleys, many rivers are in an aggrading stage, in which they deposit coarse materials in the area immediately adjacent to the channel. This deposit builds up a natural levee which gradually raises the river above the level of the plain which is some distance away. Seepage through the levee plus runoff from the surrounding uplands may keep this distant plain in a very wet or permanently submerged condition, in contrast to the fairly dry surface conditions on the levee itself. The wet area is frequently occupied by a sedge meadow, a cattail marsh, or other semi-aquatic non-forest vegetation. Fires, sweeping from the uplands during dry seasons, burn across these open communities and eat away at the floodplain forest proper. The newly exposed ground thus produced is initially reinvaded by two characteristic species, either the river birch or the swamp white oak, although not usually both at the same time. In the absence of further fires, the forests of these species, plus such associated trees as bur and white oak, are also invaded by silver maple and American elm.

The linear nature of the river system and the fact that its flood periods are simultaneous throughout the length of the river tend to impress a uniformity on its forests, which includes even the catastrophic events in its history. Consequently it is not exceptional to find long stretches covered with a very similar combination of species, either the two pioneer types, mixtures of them and the intermediate species, or the pure intermediate type of silver maple, American elm, and green ash. Beyond this point, however, further successional changes are of rare occurrence. As seen in Table VIII-2, the major dominants of the wet-mesic type are capable of reproducing themselves beneath their own canopy, and forming an all-aged, semi-stable community. Successful invasion by the mesic species is prevented by the unfavorable soil moisture conditions and may be delayed for hundreds or thousands of years. Such invasion can occur only when the river has cut through a hard rock dam or other barrier down stream and subsequently becomes recessed to a new and lower terrace in the old floodplain. The improved internal drainage which follows in the forest remaining on the old terrace allows a mesic forest to develop. Only a very few examples are known in Wisconsin of this mesic, potentially climax forest on floodplains. These conditions, of course, do not obtain in the lacustrine areas and terminal mesic forests are therefore more common in such situations.

ENVIRONMENT

Climate

No studies of the internal microclimate have been conducted in the moist forests of Wisconsin and very few have been made in other states. The information that is available comes from the Weather Bureau records of towns located within the floodplain of the major rivers. Lone Rock, for example, is on the second terrace of the Wisconsin

River, and is immediately adjacent to the bottomland forest. Temperature records at Lone Rock show a greater annual fluctuation than do nearby stations on the uplands, with an absolute range from −53°F. to 107°F. (−47.2°C. to 41.7°C.). Cold air drainage is responsible for the very low winter minima. The high summer temperatures combine with a long frost-free season, a high humidity, and a plentiful soil moisture supply to provide excellent growing conditions. These semitropical summer conditions are found along the Mississippi and Wisconsin rivers and in the lowland trough containing the Rock and the Fox-Wolf systems. The lacustrine swamps have less extreme conditions with much lower summer temperatures, particularly those in the Lake Michigan counties. Average macroclimatic figures are included in Table VIII-4.

Soils

The alluvial soils in the riverine forests show little or no development of typical profile layering. In the willow-cottonwood stands, the parent materials are in alternating bands or lenses of sand, silt, clay, or gravel, with little or no humus enrichment of the top layers. In the wet-mesic stands, a distinct A_1 layer may be present, but frequently such layers are buried by new flood sediments before they achieve any great depths. In recent years, the flood deposits have been rich in nutrients resulting from artificial fertilization of the agricultural fields on the watersheds. The radioactivity of these materials is also higher than formerly, due to the high uranium content of many of the phosphate fertilizers in common use.

The extreme heterogeneity of the soil materials results in great differences from place to place within a single stand. However, an average of all the analyses of the A_1 layer in the P.E.L. study shows the general level of plant nutrients to be high, with 3790 p.p.m. of calcium, 348 p.p.m. of magnesium, 42 p.p.m. of phosphorus, 75 p.p.m. of potassium, 8 p.p.m. of nitrate, and 12 p.p.m. of ammonia. The median pH was 6.5. Ware (1955) showed that calcium, magnesium, and nitrate nitrogen increased along the compositional gradient, whereas phosphorus and ammonia nitrogen decreased. Potassium, as usual, showed no definite trend.

A more important soil character than either the texture or nutrient content is the presence of a gley layer (Wilde, 1940). This zone is sticky, often highly dispersed, and shows a blue or green mottling because of the presence of reduced iron. The layer may be near the surface or at some depth below it. All of the stands in the P.E.L. study had a gley layer within 3 feet of the surface, thereby indicating the poor conditions for internal drainage, since the layer is produced only when the soil is regularly saturated.

GEOGRAPHICAL RELATIONS

The most obvious relations of the Wisconsin lowland forests are with those of the river valleys in states to the south, although the lacustrine types also show resemblances to the forests of the poorly drained till plains of the Ohio River valley. An examination of forest composition along the major rivers of Indiana, Illinois, and Iowa reveals that stands of very similar makeup are found in Iowa (Aikman and Gilly, 1948) where 19 of the 20 tree species recorded are also found in the Wisconsin forest. In Illinois, a number of studies reveal that stands resembling those of Wisconsin can be found only in the northern portion (Telford, 1926). From the mouth of the Illinois river southward, the bottomland forests show a greatly increased complexity, with 20 to 45 per cent of their total species not growing in Wisconsin. Prominent in these additional members are: pin oak (*Q. palustris*), pecan (*Carya pecan*), persimmon (*Diospyros virginiana*), sour gum (*Nyssa sylvatica*), and sweet gum (*Liquidambar styraciflua*). In Indiana, Lee's study (1945) of the White River system showed that 35 per cent of his 40 species were not in the Wisconsin stands. In both Illinois and Indiana, however, the leading dominants in most stands were the same as those attaining great importance in Wisconsin, with silver maple and American elm often in the top positions.

Farther south, along the Mississippi River in Missouri and southward, the bottomland forests become still richer, with very high numbers of tree species. The Wisconsin stands are clearly on the periphery of a widespread type of great age and complexity.

The hardwood swamps of Ohio (Sampson, 1930) show a fair resemblance to the lacustrine stands of eastern Wisconsin, with about three-fourths of the species in common. Farther east, along the Atlantic seaboard, few studies of the moist hardwood forests have been recorded. A recent study by Buell and Wistendahl (1955) of the floodplain forests of the Raritan River in New Jersey revealed them to be fairly similar to the Wisconsin forests. All of the trees in one stand on a terrace were also present in Wisconsin. However, the index of similarity based on importance values was only 26.4 per cent, which was due principally to the absence of silver maple and green ash in the Jersey forest and the prominence of white ash (*Fraxinus americana*) in that forest. The changes in proportions and numbers of dominant species with distance are thus very great in the floodplain community and are in distinct contrast to the situation in the mesic forests where a widespread uniformity was apparent.

Lowland forests in natural condition are apparently quite rare in Europe. They have been studied in France by Tchou (1948), in Germany by Passarge (1956) and Oberdorfer (1957), and in Austria by Wendelberger and Zelinka (1952), among others. Oberdorfer found 107 species in the ash-elm community on the upper Rhine. Of these, 13 were in common with the Wisconsin stands, including a number of species of *Equisetum* and such weedy forms as *Humulus lupulus, Lysimachia nummularia, Solanum dulcamara,* and *Urtica dioica.* Among the 46 genera in common between the two regions were *Salix, Populus, Ulmus, Fraxinus, Betula, Acer,* and *Tilia* among the trees; *Clematis, Cornus, Crataegus, Euonymus, Ribes,* and *Vitis* in the shrubs; and *Allium, Asarum, Campanula, Carex, Impatiens, Ranunculus,* and *Viola* in the herbs.

UTILIZATION

As indicated earlier, a relatively large fraction of the original acreage of lowland forest remains uncut. This is not the result of any conscious attempt at conservation but, as is so often the case, it is merely because better use for the land could not be found. Extensive grazing is practiced, particularly when the forest adjoins a meadow on the landward side. One noticeable effect of this grazing on the Mississippi and Wisconsin drainages is the great increase in river birch. Apparently this species is unpalatable to cattle and is aided in germination by the soil disturbances created by the animals.

Moderate selective harvesting of usable timbers appears to simulate the natural death of trees from flood injuries so that forests utilized in this way do not differ in apparent composition from totally uncut forests. In recent years a considerable demand has developed for sound river birch logs for use in the birch veneer industry, but the species is in little danger of decimation, because most of the trees are defective from old fire and flood injuries.

Neither of the two main dominants, silver maple or American elm, are of great value for either commercial or farm use as lumber. In some areas, notably along the central Wisconsin River, the silver maple is used for the production of maple sugar.

Part 3

Northern forests

CHAPTER 9

Northern forests—general

GENERAL DISTRIBUTION AND COMPOSITION

The northern floristic province, north of the tension zone, contains a wide variety of vegetational types, both forest and non-forest. The main provincial area and a number of southern outliers are shown on the map at the opening of this chapter. The forests are typically characterized by the presence of coniferous trees, which include pine, spruce, hemlock, fir, cedar, and tamarack, although a large hardwood element is also present. These forests occur on a wide range of topographic sites, from very wet to very dry and from thin rocky soils to deep loams and clays. Accurate figures on the original acreage of these forests are difficult to obtain, but an estimate of 16,900,000 acres, or 48.3 per cent of the land surface, appears reasonable. As in the south, the major problem in determining areas lies in the difficulty of separating the closed forest from the savanna-like pine barrens or scrub lands which were of widespread occurrence. In contrast to the southern forests, a large proportion of the total acreage is still covered by forest although the current species compositions are frequently very different from the original.

Early descriptions of the northern forests date back to writings of the French explorers in the 17th century (Perrot, 1667; Hennepin, 1698) but accounts which can be tied accurately to known locations were first produced by geologists and boundary surveyors in the early part of the 19th century (Schoolcraft, 1834; Cram, 1840; Owen, 1848). Thus Owen (1852) in describing the land near Lake Court Oreille, in Sawyer County, said, "There the country becomes more open; the dense pine forest gives place to a more stunted growth of evergreens and aspens. . . . for four or five miles before reaching the lake . . . a few stunted and half-decayed pines were the only trees visible." Again, concerning the Namekagon River, "The prevailing growths are pine and birch, usually of small size; the undergrowth is chiefly ferns. The land is level and the woods open. . . . the waters of the Namekagon are not as highly colored as those of the principal branches of the Chippewa but they are warmer and less palatable. The Indians who inhabit its banks are wont, before drinking it, to mix it with maple sugar." While traveling the St. Croix River, Owen noted that the ridges near the Kettle River "are clothed with 'pine openings.' Four or five miles above the mouth of Kettle River is a beautiful stretch of bottomland, with picturesque groves of white oak and aspen, above the reach of high water."

Norwood (in Owen 1852) described the country between the Bois Brule and Montreal rivers (near Lake Superior) in 1847 as follows, "the ridges support a dense growth of both hard and soft woods, while the marshy valleys and low grounds are covered with tamarack, spruce, and hemlock. So dense is the forest over most of this region, that it is difficult to see from the top of one range to another." Norwood noted that the shores of lakes in the northwestern lake district differed greatly in appearance.

Some of them are surrounded by gentle grassy slopes, with occasional trees scattered along them; while others are bordered by extensive marshes, often overrun by the cranberry plant [*Vaccinium oxycoccus*]; and again, the shores are rather abrupt, with a dense, dark forest skirting and overhanging the margin. The shores of many of them are chosen as sites for villages by the Indians, who show their taste by selecting the most beautiful and picturesque, in sections where the soil is of a quality suitable for gardens.

On a journey in 1847 from the present site of the city of Superior south to Lake St. Croix, Norwood reported a number of "first-rate pines" about 15 miles south of Superior, although most of the "lands in this neighborhood are covered with an excellent growth of sugar maple, with white birch, ash, and linden [*Tilia*]." About 25 miles south, a section of country was observed of which "a good deal of it is prairie, covered with whortleberry bushes [*Vaccinium* sp.] and strawberry vines; while, in the low grounds, hazel abounds." The remainder was covered with "small pine, birch, and shrubby oak." Near Lake Pokegama, "the ridges bear a dense growth of birch, linden, boxelder [?] and fir, with large pines," while toward the St. Croix, the "principal timber is sugar maple, poplar, oak, ash, walnut, elm, hornbeam [*Carpinus*], and some birch and pine." Near Reed Creek, however, "small prairies began to appear; the high intervening prairie land supported a thin growth of oak timber."

In the autumn of 1847, Norwood made a trip from La Pointe, on the Apostle Islands, overland to the headwaters of the Wisconsin River, and thence down that stream to Prairie du Chien. He described the vegetation along the route, not only the vegetation seen from a canoe but also the vegetal covering that he saw from all of the ridges along the way, which he climbed in pursuit of his work as geologist. In general, the flat or gently rolling uplands were sparsely timbered, while the heavy forests were restricted to the deeper valleys. Thus, on the way from the Montreal River to Lac du Flambeau he reported:

the hills are covered with a growth of small timber, mostly pine, with some sugar maple, oak, and a few aspens, while the valleys support a good growth of sugar maple, with undergrowth of the same. . . . the trail runs over a sand barren, with the exception of the last half mile, which passes through one of the worst tamarack swamps I have ever seen. A few stunted pines, with occasional patches of coarse grass, is the only vegetation supported on the high ground.

The portage from the Manitowish River to Trout Lake in Vilas County "passes for some distance over a sandy plain supporting a few scattering pines." Trout Lake itself "is surrounded by drift hills, from twenty-five to forty feet high, supporting a sparse growth of small pines and birch."

On the passage down the Wisconsin River, Norwood noted that the country around the present site of Eagle River was "covered with pine, fir, and spruce, with a few aspens, and small birch. The low grounds, which frequently intervene between the river and the high banks, support elm, and, where very low, tamarack in abundance."

In Lincoln County, "a narrow strip of small pines line the banks of the river at intervals; but, as you recede into the country, there are few trees of any size to be seen. Clumps of very small birch and pine are scattered over it." Near Grandfather Falls, the "ridges are densely timbered with hard and soft woods." Just north of Stevens Point, the west side of the river supported "a small growth of oak, elm, and aspen, while east of the river a beautiful undulating prairie extends as far as the eye can reach." In the central Wisconsin area, north of Wisconsin Rapids "the country is a rolling sand plain, with a few pine bushes and dwarf oaks scattered over it." The view from the top of Petenwell Rock revealed a "level or gently undulating plain, dotted here and there with groves of small oak and pine."

A number of descriptions of the northern forest were made by geologists who studied various local regions in connection with the preparation of the *Geology of Wisconsin* (1877). In 1876, L. C. Wooster described the plant communities of the lower St. Croix district (centering on St. Croix and Dunn counties) with considerable care, appending the first, detailed, colored map of the vegetation of any region in Wisconsin (in *Geology of Wisconsin,* Vol. 4, 1882). Much of the country was covered by prairie and oak forest, but a portion of the northern area had a mixed hardwood-conifer forest with patches of pine, in which "the bodies of White Pine are frequently skirted by Red Pine, (*Pinus resinosa*) wrongly called Norway Pine, with a tree termed Black or Jack Pine (*Pinus banksiana*) in the drier situations." The mixed forest contained white pine with such hardwoods as sugar and red maple, basswood, slippery and American elms, white ash, ironwood, butternut, and yellowbud hickory.

F. L. King made a survey of the Upper Flambeau river valley in 1877 and found, that in the lake region towards the headwaters of the river, the trees were small, stunted, and scattering, and in the lower valley (at the present site of the Flambeau River State Forest) they were large, vigorous, and dense. In the former region the forest was mostly red and jack pine, while in the latter, white pine, hemlock, and yellow birch were the dominant trees, with patches of spruce and balsam in suitable places. "Black alder and kinnickinnic [*Cornus stolonifera*] fringe the streams, and upon some of the flats, small black ash, elm, and soft maple grow." King made the following observation on the behavior of yellow birch:

It is interesting to note how long the trunk of a large white pine will withstand disintegration when it falls in a damp shaded forest. On the North Fork [of the Flambeau River], there was seen a yellow birch sixteen inches in diameter, growing with its roots astride a white pine log more than two feet in diameter and 50 feet long, which still retains its outline sharply. Of course the seed of the birch must have fallen upon the pine log after it had become moss-clad and had decayed sufficiently to furnish nourishment and support for the young birch until its roots could penetrate the earth.

The upper valley of the Wisconsin River was described by Chamberlin (1877). He reported various combinations of trees on the uplands. These included red pine-jack pine, white pine-white birch-trembling aspen, white pine-hemlock-yellow birch-American elm-basswood, and sugar maple-red maple-red oak-yellow birch-balsam fir. The swamps were occupied by white cedar, tamarack, or black spruce.

Chamberlin recognized seven major forest types in his description of the vegetation of northeastern Wisconsin. They were the Hardwood and Conifer Group, the Pine Group, the Tamarac Group, the Arborvitae Group, the Spruce Group, the Black Ash Group, and the Yellow Birch Group. In addition, he described a Comprehensive Group of both northern and southern derivation which "consists of a commingling of nearly all the arboreal species of the Oak, Maple, Beech, and Conifer types." This heterogeneous type was actually located within the tension zone, as in Outagamie and Brown counties. Chamberlin reported, "Clusters may be selected that are representatives of nearly all the other classes, but in general the species are curiously mingled."

The Hardwood and Conifer Group was considered to be a modification of the Maple and Beech Group (Chapter 5) and was characterized by the lack of representatives of the Oak Group, by the decreased importance or absence of ironwood, black walnut, and butternut, and by the presence of white pine, red pine, hemlock, balsam fir, spruce, and white cedar. The underbrush included witchhazel, mountain maple (*Acer spicatum*), and many berry-producing shrubs. The Pine Group included all those lands dominated by pine, with white pine being given first importance. The Tamarac, Arborvitae (*Thuja occidentalis*), and Spruce Groups were thought to be closely related communities occupying swamps, where "the *Ericaceae* form the chief undergrowth, and the Sphagnoid mosses carpet the peaty bottom, forming a well-marked flora. From the habit of these three paludal conifers of mingling, it is sometimes difficult to classify a given swamp. . . . A not infrequent arrangement consists of a predominance of the Arbor-Vitae around the borders of the swamp, and of the Tamarac toward the center."

In the Black Ash Group, the black ash is the predominating plant, with the black alder, as "a subordinate and quite constant associate." The black ash and black alder characterize the group. Arborvitae and witchhazel are sometimes present. The Yellow Birch Group was thought to be "not a well defined group." It was confined largely to Door County and was characterized by the abundance of yellow birch, associated with hemlock, maple, and beech. "Under the dense shadow of these," several species of ground pine (*Lycopodium* sp.), yew (*Taxus canadensis*), thimbleberry (*Rubus parviflorus*), shinleaf (*Pyrola* sp.), and baneberry (*Actaea* sp.) were present.

These rather extensive quotations are of value not only in showing the widespread occurrence of rather similar groups of species but also in emphasizing that by no means all of northern Wisconsin was covered by a dense carpet of virgin forest. In fact, the frequently held picture of a forest primeval, "with murmuring pines and hemlocks" was actually of rather infrequent occurrence. Barrens, parklike areas of scattered and stunted trees, and stretches of thin forest of small pine, oak, and aspen were mixed with widely spaced patches or larger tracts of mature forest.

Following Chamberlin, no other account of the vegetation of the region based upon actual floristic combinations was produced until the 1930's. However, reports by foresters were made which included information on the quantities of merchantable timber by species and by counties or other local units of area. The most complete of these was the report on "Forestry Conditions of Northern Wisconsin," written by Filibert Roth in 1898. Many other observations by various writers were recorded which gave information on the effect and extent of forest fires, on the origin of "barrens," and on other special problems.

The most complete record of northern forest composition is embodied in the notebooks of the government land surveyors who covered the area in the 1850's and early 1860's. The relatively late arrival of the surveyors means that a period of thirty or more years had elapsed between the defeat and virtual disappearance of the inhabiting Indian tribes and the date of the survey, so that conditions described by the surveyors were not the true pre-European conditions. Only limited amounts of lumbering had taken place, nearly all of which were confined to the immediate vicinity of the larger rivers. Great stretches of uplands between the rivers were literally devoid of any human influence, except for an occasional fur trapper, and the vegetation reflected this inadvertent protection. Over large areas the surveyors recorded stands of pine which were only 3, 4, or 5 inches in diameter that had obviously sprung up in the immediate past.

With these considerations in mind, valuable information can be learned about the northern forests from the survey records. Fassett (1944), Stearns (1949), and Ward (1956) used the records in comparing the original and present day vegetations of local areas, while Bourdo (1956) has made detailed comparisons of past and present composition in portions of the adjacent Upper Peninsula of Michigan.

From the studies mentioned above, from detailed analyses of joint occurrences of species pairs at witness corners in comparison to random expectation, and from the general accounts of the earlier travelers, it is possible to construct with fair accuracy a picture of the major forest types of northern Wisconsin and to understand something as to the behavior of the individual tree species. The wet lands contained either conifer swamps, dominated by tamarack, black spruce, and white cedar, or hardwood swamps with black ash and yellow birch. The dry lands were dominated by pine, with jack and red pine on the lighter sands and white pine on the sandy loams. The heavier soils were typically covered by mixed conifer-hardwoods, with white pine, hemlock, balsam fir, and white spruce as the conifers, and sugar maple, basswood, yellow birch, beech, American elm, red oak, and ironwood as the deciduous species. Some boreal forest stands of spruce and fir were found in the coldest localities, and combinations of hardwoods typical of the southern forests were present near the tension zone.

All modern studies have served to confirm this general account, but they have also contributed more detailed information of value for an understanding of the dynamics of succession and of the nature of community integration. The first quantitative ecological investigation of the northern forest was made by Eggler (1938) in Washburn County. He described a series of stands dominated by sugar maple and basswood, that usually had some white pine present, and presented figures for density, frequency, and dominance for all of the trees and density and frequency for the understory species. The most interesting of his results concerned the history of the forest as deduced from extensive age determinations made from tree cores that were procured with an increment borer. In the same period, Kittredge (1938) studied the aspen community of parts of Minnesota and Wisconsin. He presented quantitative analyses which led to a description of characteristic groups of understory species typical of the major forest types of the region and which he used to determine the historical origin of the various aspen stands.

P.E.L. STUDY OF NORTHERN FORESTS

The first study of forests in the northern floristic province by members of the University of Wisconsin Plant Ecology Laboratory was made by F. W. Stearns (1949, 1950, 1951), who investigated various aspects of the upland forests in Forest and Vilas counties. This was followed by a study of a stand in Taylor County by Anderson (1948) and by an extensive survey of the upland types by Brown and Curtis (1952) which was based in part on field work directed by Grant Cottam. Phytosociological and autecological investigations of the understory plants of the same forests were conducted by M. L. Gilbert (1953) and W. E. Randall (1952), while W. L. Culberson (1955a) studied the lichen and moss communities of the tree trunks.

The lowland forests were examined by E. M. Christensen (1954) and by J. J. Jones (1955), the latter concentrating on the conifer bogs. The special roles of beech and hemlock and the factors limiting their range were studied by R. T. Ward (1956) and H. A. Goder (1955). The soil microfungi of the upland stands were investigated by M. Christensen (1956) as part of the mycological-ecological program conducted jointly with M. P. Backus. The distribution of radioactivity in the various forest types was reported on by Curtis and Dix (1956a, 1956b). Much of the material in this and the following three chapters is based on the combined results of the aforenamed workers.

Stands chosen for detailed investigation were selected according to objective criteria of size, degree of disturbance, and land-form type, with a basic division into upland and lowland stands as in the southern studies. Quantitative measurements of the trees were obtained in 146 stands of upland forest and 121 stands of the lowland group. In most of these, data on quadrat frequency were obtained for the herbs and shrubs, as well as information on soil type, soil nutrients, and light intensity.

ORDINATION OF THE FORESTS

Two separate ordinations were constructed, one for the uplands by Brown and Curtis (1952) and one for the lowlands by Christensen, Clausen, and Curtis (1959). These were arrived at by using methods similar to those described for the southern forests (Chapter 5), and by using actual joint occurrences of species as a guide in the erection of a behavioral classification of the trees. As in the south, sugar maple was clearly the most shade tolerant member of the mesic forests and the species most successful in successional replacement of all others on suitable sites. The opposite extremes of the adaptational series were occupied by jack pine on the dry side and by tamarack on the wet side; all other species were intermediate in their behavior.

The adaptation numbers for the trees are presented in Table 11. As can be seen, many species occur in both the uplands and the lowlands, although they may not maintain the same relative position with respect to sugar maple in the two sites.

Table 11

Adaptation numbers for trees of northern forests

Species	Upland	Lowland	Species	Upland	Lowland
Abies balsamea	7.0	4.5	P. resinosa	3.0	3.5
Acer rubrum	6.0	6.0	P. strobus	5.0	4.0
A. saccharum	10.0	10.0	Populus balsamifera	...	4.5
A. spicatum	6.0	6.0	P. grandidentata	2.0	5.0
Betula lutea	8.0	7.5	P. tremuloides	2.0	3.5
B. papyrifera	5.0	5.0	Prunus pensylvanica	3.0	4.0
Carpinus caroliniana	7.0	8.0	P. serotina	4.0	6.0
Carya cordiformis	8.0	8.0	Quercus alba	4.0	...
Fagus grandifolia	9.5	9.5	Q. borealis	6.0	7.0
Fraxinus americana	8.0	7.0	Q. ellipsoidalis	2.0	...
F. nigra	6.0	5.0	Q. macrocarpa	2.0	...
Juglans cinerea	5.0	8.0	Sorbus americana	...	4.0
Larix laricina	...	1.0	Tilia americana	8.0	8.5
Ostrya virginiana	9.0	9.5	Thuja occidentalis	7.0	4.0
Picea glauca	6.0	4.0	Tsuga canadensis	8.0	8.5
P. mariana	...	2.0	Ulmus americana	8.0	5.5
Pinus banksiana	1.0	3.0	U. thomasi	7.0	7.0

Compositional indices were calculated as before by weighting the measured importance value of each species in each stand by its adaptation number and summing to obtain a stand index. Separate ordinations were constructed for the upland and the lowland series, and the behavior of each species plotted in the same manner as that shown for red oak in Figure 8. It was found that the compositions of the most mesic stands on the two ordinations were almost identical. Therefore, it appeared that the ordinations might be combined into a single linear gradient, representing the full moisture regime from very wet, through mesic, to very dry, as had been done in the study of southern forests (Chapter 5). The combination was achieved as before by reversing the value of the upland compositional index so that the high numbers were on the left instead of the right, and then putting this high value end of the graphs against the high end of the lowland series.

When the importance values of individual tree species averaged by five equal segments in each half are plotted against this combined ordination, curves are obtained which show the same range in behavior as was demonstrated for the southern trees (Figures 12 and 13). Two species, tamarack (*Larix laricina*) and black spruce (*Picea mariana*), have maximum values at the left or wet end of the gradient and decrease rapidly to the right, while another pair of species, jack pine (*Pinus banksiana*) and Hill's oak (*Q. ellipsoidalis*), show the reverse pattern. Jack pine, however, also has a small rise near the left end. Several species show peak importance values near the mesic center of the gradient, including sugar maple (*Acer saccharum*), beech (*Fagus grandifolia*), hemlock (*Tsuga canadensis*), ironwood (*Ostrya virginiana*), basswood (*Tilia americana*), and white ash (*Fraxinus americana*). In the intermediate region on the moist side, white cedar (*Thuja occidentalis*), balsam fir (*Abies balsamea*), and yellow birch (*Betula*

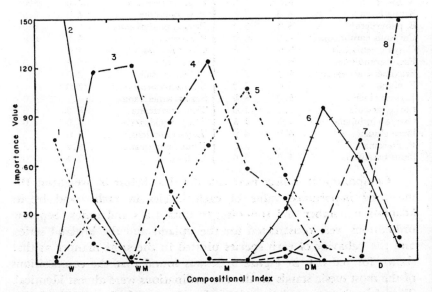

Figure 12.—Behavior of major tree species on the combined ordination of northern upland and lowland forests. 1, tamarack (*Larix laricina*); 2, black spruce (*Picea mariana*); 3, white cedar (*Thuja occidentalis*); 4, hemlock (*Tsuga canadensis*); 5, sugar maple (*Acer saccharum*); 6, white pine (*Pinus strobus*); 7, red pine (*Pinus resinosa*); 8, jack pine (*Pinus banksiana*).

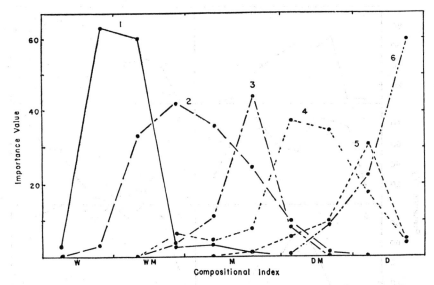

Figure 13.—Behavior of lesser tree species on the combined ordination of northern forests. 1, balsam fir *(Abies balsamea)*; 2, yellow birch *(Betula lutea)*; 3, beech *(Fagus grandifolia)*; 4, red oak *(Quercus borealis)*; 5, large-toothed aspen *(Populus grandidentata)*; 6, Hill's oak *(Quercus ellipsoidalis)*.

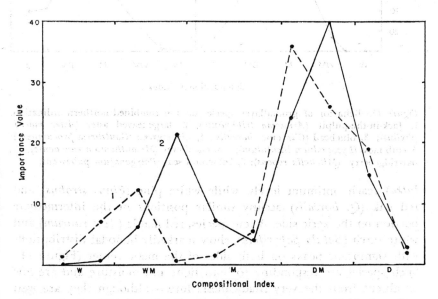

Figure 14.—Bimodal curves for two lesser tree species on the combined northern ordination. 1, white birch *(Betula papyrifera)*; 2, red maple *(Acer rubrum)*.

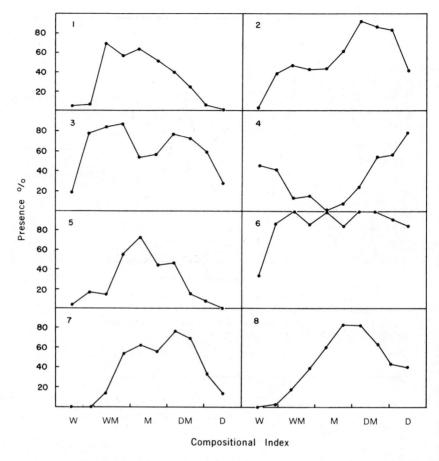

Figure 15.—Behavior of groundlayer species on the combined northern ordination. 1, Jack-in-the-pulpit (*Arisaema triphyllum*); 2, large-leaved aster (*Aster macrophyllus*); 3, bluebead (*Clintonia borealis*); 4, wintergreen (*Gaultheria procumbens*); 5, club moss (*Lycopodium lucidulum*); 6, mayflower (*Maianthemum canadense*); 7, partridgeberry (*Mitchella repens*); 8, Solomon's seal (*Polygonatum pubescens*).

lutea) reach optimum levels, while white pine (*Pinus strobus*) and red oak (*Q. borealis*) occupy similar positions in the intermediate portion on the xeric side. A few species, red maple (*Acer rubrum*) and white birch (*Betula papyrifera*), show markedly bimodal distributions with significant peaks on both sides of the mesic center (Figure 14). Such species are responding to both light and moisture and are low or absent from the very shady mesic forests although they are well adapted to the soils and moisture conditions of this segment, as is shown by their behavior in secondary successions.

The herbs and shrubs (Figure 15) show almost exactly similar behavior patterns, with some species showing optimum response at every locus along the gradient and with a number showing bimodal distributions of various intensities, from nearly smooth curves with a slight central depression, like partridgeberry (*Mitchella repens*), through prominent mesic recessions, like bluebead (*Clintonia borealis*) to actual U-shaped curves as in wintergreen (*Gaultheria procumbens*). Some herbs, like Canada mayflower (*Maianthemum canadense*), have a much broader amplitude of tolerance than any tree, and occur at high levels throughout the full moisture gradient. It is obvious that plants of the latter type have no indicator value with respect to either sociological or environmental conditions, although some have been suggested for this purpose by both foresters and soil scientists.

For purposes of detailed discussion, the combined northern forest ordination will be divided into the same five objective segments employed in the southern forest; namely four 1000 unit composition index groups (300-1299 and 1300-2299 on both upland and lowland), and a 1400 unit group, which will include the mesic stands (2300-3000-2300). These will be discussed in three chapters on mesic, xeric, and moist forests, as before. It should perhaps be re-emphasized that this is purely a method of convenience and does not represent any natural, inherent biological division into discrete communities.

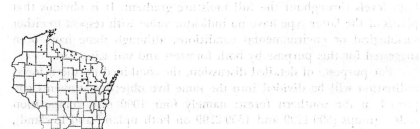

CHAPTER 10

Northern forest—mesic

GENERAL DISTRIBUTION

The mesic forests of the northern floristic province are widely distributed through the entire area north of the tension zone. There were at the time of settlement large contiguous areas of this type in the counties along Lake Michigan north of Sheboygan and in the central portion from Langlade through Marathon, Price, and Sawyer to Barron and Polk counties (Figure X-2). The total acreage was 11,740,750 acres or 33.54 per cent of the total land area, by far the largest community in Wisconsin. Throughout the region, the forest displays a certain similarity but its composition is not as uniform as in the southern mesic forests. This is due to the fact that the ranges of several of the important dominants do not coincide with the range of the type. The beech (*Fagus grandifolia*) is restricted to the counties along Lake Michigan and Green Bay, the hemlock (*Tsuga canadensis*) extends as far as western Sawyer and Bayfield counties, while the yellow birch (*Betula lutea*) rapidly loses importance in the far western areas approaching the St. Croix river. The concentric nature of these ranges means that the mesic forests are most complex in the northeastern corner of the state and become simpler toward the west and southwest. In fact, the stands near the tension zone in Washburn and adjacent counties are very similar to the southern mesic forests. They differ mainly because white pine is an important member of the community and because the herb and shrub species are like those to the northeast.

A few isolated relic stands of northern mesic forest are to be found on steep north-facing hills and cliffs in the Driftless Area of southern Wisconsin (McIntosh, 1950a). These are usually dominated by hemlock, with sugar maple (*Acer saccharum*), mountain maple (*Acer spicatum*), and yellow birch in association. Such relics are known in Iowa, Grant, and Richland counties. They also are found in a very narrow strip along Lake Michigan as far south as northern Milwaukee County (Bruncken, 1902; Whitford and Salamun, 1954).

COMPOSITION AND STRUCTURE

The average composition of the 70 mesic stands (see figure reproduced at chapter opening) in the P.E.L. study is shown in Table X-1. As in the south, the leading dominant is sugar maple, but hemlock is a much stronger second contender than basswood (*Tilia americana*) was in the southern forest. In both hemlock and beech, the figures shown are the averages for all stands, and include those stands located beyond their range limits. Within their range, both species may attain values equaling or exceeding those of sugar maple. Both yellow birch and basswood are of widespread occurrence, as shown by their high constancy values, but their average importance is not very great. Several other tree species occur in one-quarter or more of the stands, including ironwood (*Ostrya virginiana*), American elm (*Ulmus americana*), red oak (*Q. borealis*), red maple (*Acer rubrum*), white birch (*Betula papyrifera*), white ash (*Fraxinus americana*), and balsam fir (*Abies balsamea*).

The structures of two typical mesic stands are given in Table X-2. They are similar because sugar maple shows an overwhelming preponderance in the seedling classes and is well represented in the smaller tree classes. Both have white pine as their largest trees, but the densities are low, and no individuals are less than 10 inches d.b.h.

The major differences between the two stands are in the amounts of smaller reproduction of the lesser trees, with ironwood and white ash leading in Door County and yellow birch and hemlock in Vilas County. This pattern is frequent in hemlock, where the number of stands with no visible reproduction whatsoever exceeds the stands with active reproduction. Beech also varies from conditions of no replacement to situations where its saplings outnumber those of sugar maple.

LIFE-HISTORIES OF DOMINANTS

Yellow birch (*Betula lutea*)

Yellow birch actually reaches a position of slightly higher numerical importance in the wet-mesic forests on the lowlands than it does in the mesic forest, but its high constancy and fairly high importance value in the mesic stands justifies its discussion here. It is perhaps the most characteristic member of the hardwood forests of northern Wisconsin, since only in Wisconsin does it achieve such a position of prominence. The genetic status of the species is not clear. Practicing lumbermen claim that they recognize two types (by their bark and other unspecified criteria) which provide logs of different quality when cut. No detailed study has been made of the species, but many individuals have been seen which closely approach the sweet birch (*Betula lenta*) of the East. Clear cases of hybridization with white birch have been noted, and hybrids of yellow birch and bog birch (*Betula glandulifera*) have been reported for Wisconsin. It is possible that the preference for wet soils shown by many of the Wisconsin populations has been derived by introgression with the bog species, but no concrete evidence is available.

The flowers of yellow birch are produced in catkins which open in spring at the time of expansion of the leaves. The winged seeds ripen in September and October and are slowly liberated as the cones gradually disintegrate. Many of the seeds which fall during the winter are transported for long distances by a wind-powered skidding action over the glazed surface of the snow. Germination takes place in the spring on moist, mineral soil surfaces if these are available. Most of the seeds which fall within the mesic forest land on the leaf litter above the mor humus layer, where they are incapable of germinating. Those which chance to fall on the bark of exposed logs or stumps germinate in great numbers, regardless of the state of decay of the substrate, but only the logs with a moderate blanket of mosses are likely to produce successful seedlings. As King (Chapter 9) observed in 1882, some of the seedlings which begin on the top of large logs put forth roots which reach the soil on each side of the log. As the birch matures, these roots increase greatly and form a prop root system which may finally become the sole support of the trunk after the host log has decayed away. Such trees closely resemble the so-called mountain mangroves of the tropical cloud forests in appearance. Occasionally, several birches will get their start on the same log, with a resulting linearity in arrangement of the mature trees that is surprisingly artificial in appearance.

The yellow birch is intermediate in its shade tolerance and requires an opening in the canopy for successful establishment, but this opening need not be extensive in area. Once established, the tree is potentially long-lived and may reach an age of three hundred years and a height of 90 to 100 feet, with a diameter of 4 feet. It is very sensitive to fire, but possesses a fair ability to regenerate from root sprouts.

Hemlock (*Tsuga canadensis*)

The hemlock is a true member of the mesic forest, but, like the yellow birch, it does best on the more moist soils and is more likely to be dominant in valleys, at the base of slopes, or on areas of poor internal drainage than it is on the well-drained uplands or hill crests. Considerable information is available concerning the ecological behavior of hemlock in Wisconsin as a result of the P.E.L. studies of Goder (1955). He was primarily interested in the nature of the factors which were active in limiting the range of the species. Of great interest were his investigations of the conditions under which reproduction takes place.

Hemlock seeds are shed from cones on the tree in the autumn and from fallen cones on the ground which open and close in response to moisture changes in spring and summer. Germination occurs in spring after a winter's exposure to cold, damp conditions and takes place on any moist substrate, although the highest percentage germination is found on acid, humus-rich soils. However, Goder found that successful germination, in the sense of production of seedlings which survive to become saplings, was much more narrowly restricted for it was confined to certain rather specialized micro-habitats. By far the best survival occurred on old stumps of pine and hemlock, with an average of 44.9 saplings per 100 square meters of these surfaces. The next best site was the side of a tip-up mound. These characteristic, gravelike mounds of earth result from the raising of a quantity of surface soil enmeshed in the roots of a tree which has been tipped over by the wind. As the roots decay, the soil slumps to form a hillock at the base of the decumbent trunk. The area from which the soil came persists as a depression or pit. Goder found that 91 per cent of the saplings in the stands he studied were growing on such tip-up mounds. However, since the mounds averaged 6.9 square meters in area and numbered about 320 per hectare, the total mound surface was much greater than that of the stumps, so that the average rate was only 6.4 saplings per 100 square meters. The down-trunks of the wind-thrown trees were also favorable, with 3.2 saplings per 100 square meters. The undisturbed soil surfaces of the intervening areas between the mounds had only 0.8 per cent of the total saplings, or 0.02 per 100 square meters. The pits, which are frequently inundated for long periods in the spring, were the worst habitat, with only 0.004 saplings per 100 square meters.

In a detailed examination of seedling production on logs, Goder was able to show good correlations between the state of decay of the wood, the completeness of the moss covering, and the number of hemlock seedlings. No seedlings were found on freshly fallen trees, but the numbers increased in proportion to the increase in moss cover, to a maximum of 1600 seedlings per 100 square meters on logs with a 90 per cent covering. Thirteen species of mosses were recorded on such logs, including *Hylocomnium proliferum, Dicranum rugosum, Mnium cuspidatum,* and *Thuidium* species. A few herbaceous plants were also present on the optimum logs, including Canada mayflower (*Maianthemum canadense*), starflower (*Trientalis borealis*), and woodsorrel (*Oxalis acetosella*).

Best germination of hemlock seedlings occurred on hemlock or pine logs. Large numbers of yellow birch seedlings were present along with the hemlock, but best germination of the birch took place on logs of sugar maple or the birch itself, with a maximum rate of 355 per square meter in one stand.

The hemlock shows a remarkable ability to withstand suppression. It is not as tolerant of shade as sugar maple, in the sense that it requires more light for active growth, but it is able to persist in very dense shade for very long periods in an essentially quiescent state. Many so-called saplings have been observed which were less than 2 inches d.b.h. but which were more than one hundred years old. Examination of the stumps of mature trees frequently reveals a stunted core or heart of this type, surrounded by rings of normal width resulting from release of the suppressed sapling by death of surrounding trees. Stearns (1951) reported many trees which had experienced more than one period of such suppression before they finally reached a position of dominance in the forest canopy. Estimations of tree ages from their diameters are always subject to large error, but few species are as difficult to estimate as hemlock, in which deviations of several hundred years from the correct age are not unusual.

The needles of hemlock are very acid and contribute to a high acidity in the surface layers of the soil. Extreme podzolization occurs in pure hemlock stands, with a deep, highly leached, nearly pure-white A_2 layer and frequently with a cemented B layer. The susceptibility to wind throw and heart rot so apparent in older hemlocks has been ascribed in part to the severe condition of soil degradation brought about by the trees. One of the interesting aspects of the hemlock litter is the high content of calcium in some stands. In spite of acidities which range from pH 3.5 to pH 4.5 in the A_1 layer, the available calcium may attain levels of 0.5 to 0.7 per cent. These values are exceeded only by the dry lime prairies of southern Wisconsin.

Hemlock seedlings and saplings are very palatable to deer, so that effective reproduction may be totally lacking in areas where deer populations are high. A browsed hemlock seedling has very little potential for regrowth in contrast to the equally favored sugar maple seedlings, which resprout vigorously and may persist for ten years or more with annual pruning. This differential response throws the balance more sharply in favor of the maple than its slightly higher shade tolerance would warrant and will serve to hasten the approach of pure maple forests if the deer are allowed to remain uncontrolled.

Beech (*Fagus grandifolia*)

Beech is taxonomically related to the oaks and resembles them in many of its life-history characters. The seeds, the beechnuts, are produced in great numbers only at infrequent intervals. Schorger (1949) found evidence from historical records for an average period of five years between heavy crops in the period from 1853 to 1893. No detailed recent records are available but it is known that some of the modern heavy crops are made up mostly of empty seeds. Matthews (1955) reported that the European beech had high yields every 5 years on the average over a 30 year period. The high years were correlated with both high July temperatures in the preceding year and the absence of late spring frosts in the bearing year. Germination occurs in the spring on moist soils rich in humus and lime. The cotyledons become leaflike and do not contain enough stored food to carry the seedling for more than one year. Consequently, seedling success is low in mor humus of hardwood origin, because of failure of the seedlings to reach the mineral soil.

In most stands, reproduction from root sprouts is more common than from seedlings and it may be the exclusive method in some cases. The roots of beech are superficial and extend for considerable distances from the trunk. At any point where a root approaches the soil surface, an erect leafy shoot may be produced from an adventitious bud. These shoots may develop into saplings and eventually into trees if a suitable opening occurs in the canopy. The exact conditions responsible for such vegetative reproduction are not well understood and should be investigated further (Ward, 1956).

The beech is very shade tolerant and closely approaches sugar maple in this respect. There is occasional evidence that beech saplings can withstand suppression even better than maple, but this is due to continued nourishment of root sprouts by the mother tree.

Like the closely related oaks, the beech is not a nutrient pumper, because its leaves are strongly acid and low in bases. The net result is a trend towards soil podzolization which often is masked by the high percentage of sugar maple normally associated with the beech. The beech has a complicated genetic history which is imperfectly understood. According to Camp (in Whittaker, 1954) there were three ancestral types of beech, the white, the red, and the gray, which differed in morphology and in ecological requirements. Migrations associated with Pleistocene climatic changes permitted introgressions between all three of these types. The Wisconsin form at present seems to be mostly derived from the red beech but it has significant amounts of characters from both of the other types.

GROUNDLAYER

The prevalent shrubs and herbs of the northern mesic forest are listed in Table X-3, together with the additional modal species. Altogether, 227 species were found in 66 stands of the P.E.L. study, with an average species density of 33. The three most widespread species, Canada mayflower (*Maianthemum canadense*), Solomon's seal (*Polygonatum pubescens*), and twisted-stalk (*Streptopus roseus*), are members of the lily family which has the highest total sum of presence. Of the 58 families represented in the total flora, these nine contribute one-half of the species—*Liliaceae* 8.4 per cent; *Cyperaceae* 7.5 per cent; *Rosaceae* 5.7 per cent; *Compositae* 5.7 per cent; *Ranunculaceae* 4.8

per cent; *Caprifoliaceae* 4.8 per cent; *Polypodiaceae* 4.4 per cent; *Orchidaceae* 4.4 per cent; and *Gramineae* 3.5 per cent. Of the prevalent species, 70 per cent bloom before June 15, exactly equaling the figure for the southern mesic forests. Less than 5 per cent bloom after August 15.

The shrub and woody vine content of the northern mesic forests is much lower than that found in either the wet or dry forests, but it is nearly double that of the southern mesic forests. The increase is due largely to the presence of prostrate, creeping, evergreen vines like partridgeberry (*Mitchella repens*), bunchberry (*Cornus canadensis*), and twinflower (*Linnaea borealis*), which are lacking in the south. The evergreen Canadian yew (*Taxus canadensis*) is another characteristic shrub of the northern forests which is now much more rare than formerly, due to heavy browsing pressure of deer since the herds began to build up in the 1930's. This plant is preferred above all others by the deer and is sought out and utilized by them to the point of virtual extinction in many counties. Density values for the taller mesic shrubs are available for the three stands studied in detail by Stearns (1951). There were 1439 shrubs per acre on the average. The most numerous was the beaked hazelnut (*Corylus cornuta*) at 588 per acre, followed by leatherwood (*Dirca palustris*) at 355, swamp currant (*Ribes lacustre*) at 220, and arrow-wood (*Viburnum acerifolium*) at 85.

One group of plants is particularly noticeable in the northern terminal stands, although its members are neither of high presence nor high density. This is the group of chlorophyll-free seed plants, both saprophytic and parasitic. Included are the two Indian pipes (*Monotropa* sp.), the coral root orchids (three species of *Corallorhiza*), beech drops (*Epifagus virginiana*), and squaw-root (*Conopholis americana*). The saprophytes are undoubtedly being fed by the rich organic matter in the humus layers of the soil and are further benefited by the very constant moisture supplies present.

The total density of the understory vegetation in the northern mesic forests is relatively low on the average, due to the nearly total lack of plants beneath dense groves or clusters of hemlock (Plate 17). The forests dominated by beech also have a low density, particularly in the summer, frequently with less than one individual of any species per square meter, although the early spring flora of some stands approaches that of the maple woods. This sparseness of the groundlayer in beech and hemlock forests is frequently attributed to the very low light intensity in such stands, but the recent work of Winter and Bublitz (1953) on the toxic effects of water extracts of European beech leaves may indicate that an antibiotic reaction is in part responsible for the absence of many herbaceous species. Similar reactions were reported for the litter of various European conifers by Grümmer (1955) but no detailed investigations of the American hemlock have been reported.

The spring ephemerals are of considerable interest, since they involve many of the same genera as found in the southern mesic forests but are represented by different species. This phenomenon of the existence of closely related species pairs in closely related plant com-

Table 12

Presence of species pairs in northern and southern mesic forests

Northern species	N.	S.	Southern species	N.	S.
Circaea alpina	33%	5%	Circaea quadrisulcata	23%	65%
Claytonia caroliniana	14	0	Claytonia virginica	12	38
Corylus cornuta	41	5	Corylus americana	5	8
Dentaria diphylla	5	0	Dentaria laciniata	2	22
Dicentra canadensis	14	3	Dicentra cucullaria	14	22
Erythronium americanum	9	5	Erythronium albidum	11	24
Hepatica americana	33	11	Hepatica acutiloba	29	62
Mitella nuda	26	0	Mitella diphylla	35	20
Oryzopsis asperifolia	52	3	Oryzopsis racemosa	5	8
Panax trifolium	16	3	Panax quinquefolium	5	19
Sambucus pubens	44	8	Sambucus canadensis	3	19
Trillium cernuum	9	5	Trillium gleasoni	5	16
Uvularia sessilifolia	39	0	Uvularia grandiflora	47	54

munities has been interpreted as evidence for the independent existence of the two communities over long geological time periods. In the present case, this would mean that the northern mesic forests of Wisconsin, in spite of their great physiognomic resemblance to the southern mesic forests, are actually of different origin and of different history. Their separation for study purposes thus has biological support beyond that afforded by the tension zone and its consequences.

The species pairs involved are either wholly exclusive to their respective communities or else show great quantitative differences in the two forest types (Table 12). Thus *Dentaria diphylla* is restricted to the north while *Dentaria laciniata* is essentially confined to the south; *Claytonia caroliniana* is restricted to the north whereas *Claytonia virginica* is far more abundant in the south; *Dicentra canadensis, Hepatica americana,* and *Erythronium americanum* are most numerous in the north and their counterparts, *Dicentra cucullaria, Hepatica acutiloba*, and *Erythronium albidum* reach highest levels of importance in the south; and the same situation obtains in *Panax trifolium* and *Circaea alpina* as compared to *Panax quinquefolium* and *Circaea quadrisulcata*. In the miterworts, *Mitella nuda* is confined to the north, while *Mitella diphylla* is found in both northern and southern forests. This is the case also with the bellworts, where *Uvularia sessilifolia* is in the north and *U. grandiflora* is in both, but more numerous in the south. Among the grasses, *Oryzopsis asperifolia* of the north is replaced by *O. racemosa* in the south, while the shrubs have such species pairs as *Sambucus pubens* (north) versus *S. canadensis* (south) and *Corylus cornuta* (north) versus *C. americana* (south and north) although the last is not a common member of the mesic forests in either area.

The geographic distribution of mesic stands containing a large representation of the spring ephemerals is of some interest, since it appears that their range is not coextensive with the type as judged by the dominant trees. Special investigations were conducted in early spring when these species were in full bloom by Margaret Gilbert in 1952 and by the author in 1957. Stands were noted with at least six species from the following list: ramp (*Allium tricoccum*), Jack-in-the-pulpit (*Arisaema triphyllum*), wild ginger (*Asarum canadense*), spring-beauty (*Claytonia caroliniana* or *Claytonia virginica*), toothwort (*Dentaria diphylla* or *Dentaria laciniata*), squirrel-corn (*Dicentra canadensis*), Dutchman's breeches (*Dicentra cucullaria*), troutlily (*Erythronium albidum* or *Erythronium americanum*), hepatica (*Hepatica acutiloba* or *Hepatica americana*), waterleaf (*Hydrophyllum virginianum*), false rue anemone (*Isopyrum biternatum*), miterwort (*Mitella diphylla*), woods phlox (*Phlox divaricata*), bloodroot (*Sanguinaria canadensis*), large trillium (*Trillium grandiflorum*), and large bellwort (*Uvularia grandiflora*). The stands with these species form a relatively continuous band along the southern portion of the mesic forest region, and are lacking in northern Door County, in most of Marinette County, in the central region of Vilas, Oneida, Price, and Iron counties, and along the entire western portion of the type. The stands along the extreme southern border, which were in or near the tension zone, frequently had both the southern and the northern representative of the paired species discussed earlier, while the more northerly stands usually had only the northern representatives.

Many of the stands contained 12 to 14 species of the list of 20. All of them were distinctly different in gross appearance from the other mesic stands which lacked the ephemerals, conveying a much greater impression of richness and luxuriance. The ground was literally covered with the leaves and flowers of the spring herbs, frequently with a multiple layer of several species, arranged one above the other, with spring-beauty and troutlily on the bottom, squirrel corn and toothwort above them, and trillium, phlox, and bellwort projecting through to a top layer. A single footprint might trample as many as 15 or 20 individuals of 3 or 4 species. In contrast, the mesic forests lacking this group are extremely bare, with large areas devoid of any herbs whatsoever. Most of the ground cover in such sites is composed of tree seedlings, with sugar maple predominating. The majority of the herbs which are present are of boreal conifer forest affiliation, and include such species as bluebead (*Clintonia borealis*), mayflower (*Maianthemum canadense*), and starflower (*Trientalis borealis*).

These very great differences in the groundlayer are not reflected in the dominant trees, as all of the stands, both with and without ephemerals, are composed of sugar maple, yellow birch, hemlock, basswood, ironwood, American elm, and red oak. It is true that many of the ephemeral areas had a relatively pure stand of hardwoods with small amounts of hemlock, but at least one of the stands had better than 50 per cent dominance by the latter species. This case seems to be the best example so far discovered in Wisconsin of the unions or layer societies described by Lippmaa (1939) for the hardwood forests of Esthonia and by Daubenmire (1952) for the coniferous forests of Idaho and eastern Washington.

In general, the life-history characteristics of the northern mesic herbs are the same as those of their southern counterparts (Randall, 1952). Unfortunately, no quantitative data are available concerning reproduction rates but a general survey would indicate that more species producing large numbers of very small seeds are present in the north, especially in the orchid and heath families. The potential aggregation of the species, due to innate tendencies to form vegetative colonies is about the same as in the south, but the actual degree of aggregation is greatly intensified by the antibiotic effects of the hemlocks and beeches, and by the unevenness of the terrain caused by the numerous tip-up mounds. The net result is a very patchy distribution pattern, with homogeneity achieved only in areas one-quarter acre or larger in size.

OTHER BIOTA

The thick layer of mor humus in many of the mesic stands provides a very favorable habitat for the larger fleshy fungi as well as the saprophytic seed plants. In wet years, the profusion of sporophores exceeds that to be found in any other Wisconsin plant community. Prominent are members of the genera *Russula, Amanita, Polystictus, Boletus, Hydnum, Thelephora, Clavaria, Phalloidea, Auricularia, Mitrulla, Spathularia,* and *Helvella*. The soil microfungal flora is also very rich, with *Spicaria elegans, Sporotrichum roseum, Trichoderma album,* and species of *Acrostalagmus, Penicillium,* and *Phialophora* as typical members, according to the studies of Martha Christensen (1956). The slime molds are also very numerous, especially *Fuligo, Stemonitis,* and *Cribraria* (Greene, 1932).

The epiphytic flora on the tree trunks is much better developed than in comparable southern forests. Culberson (1955a) found among the lichens that *Graphis scripta* was best developed in the mesic forest as Hale had found in the south, but he also found a number of other species which reached their optimum development in the northern mesic forests. These included *Cladonia coniocraea, Parmelia aurulenta, Physcia grisea, Physcia orbicularis, Physcia tribacoides,* and *Rinodina halei*. Among the mosses and liverworts, *Dicranum montanum, Frullania eborocensis, Lindbergia brachyptera, Neckera pennata, Platygyrium repens,* and *Porella platyphylloidea* were most prominent on the aerial trunks, while *Bazzania trilobata* and *Chamberlainia acuminata* were additional species common on the trunk bases. As in the southern forests, a large element of host specificity was present, with noticeable differences between the conifers and the hardwoods.

ENVIRONMENT

Microclimate

The outstanding feature of the microclimate is the low light intensity on the forest floor. The beech and the hemlock both cast as heavy a shade as the sugar maple and of course the hemlock is effective on a year-round basis. As a result, the average light intensity is about the same in midsummer in the north as compared to the south but is lower when averaged over the entire growing season. This decrease in intensity is compensated for in part by the longer days in midsummer, but no figures of actual total radiation received are available for examination. Measurements of incident light with a photometer in many stands were made by R. T. Brown and R. T. Ward in the P.E.L. studies. Many readings of less than 10 foot-candles were obtained, but the modal value was between 60 and 75 foot-candles.

Soils

As indicated in the general discussion, the northern mesic forests are frequently found on deep loam soils of non-extreme topographies but they are by no means confined to such sites. One of the finest beech-maple-hemlock forests in Wisconsin formerly grew on a large stabilized sand dune in Door County until it was removed for a real estate development a few years ago. Many examples of mature maple-beech, maple-birch, or mixed maple-basswood-birch-hemlock forests were found in the P.E.L. study on Vilas Sands and other sandy or sandy-loam soils. Similar variations were reported by Westveld (1933) in Michigan. The potentially high climatic moisture supply, combined with the soil improving qualities of the maple, birch, and basswood, allows the development of a humus-rich topsoil, regardless of the coarse structure of the parent material. This topsoil is fully equivalent to an original glacial loam in both nutrient and moisture supplying powers.

The soil profile under mature northern mesic forests is typically a Gray-Brown Podzolic type, with well-developed A_1 and A_2 layers. In younger stands which still contain a high percentage of hemlock or have recently supported that species, and in stands at the far north on the younger Valders Drift the soil may be a true Podzol, with a thick mat of partially decomposed mulch, a thin or nearly absent A_1 layer, and a deep, light-colored A_2 layer. It is commonly believed that these differences are due solely to the activities of the hemlock, but the cause and effect relation is almost certainly not as simple as that. In any case, the differentiation of the soils into mor and mull humus types is usually quite distinct. This was noticed as long ago as 1860 by Increase Lapham on a trip to the Penokee Range near Lake Superior:

The soil on the Range is much deeper and of better quality than that on the lower grounds, and sustains a different growth of timber. . . . Vegetable mould or soil appears to accumulate here, as in other places; while over a large share of the Lake Superior country there seems to be some cause operating which is unfavorable to such accumulation. What this cause can be I am unable to decide; but the fact is quite evident.

The passage of a century has brought us little closer to an adequate explanation, in spite of the obvious silvicultural and ecological importance of the phenomenon.

One of the strongest correlations between vegetation and humus type is to be seen in the relation of the spring ephemerals to mull humus. All of the stands mentioned earlier with a high content of these early flowering herbs had a definite mull layer correlated with the complete breakdown of a given year's leaf litter by the end of the following summer. This was true even in the stand with over 50 per cent dominance by hemlock. On the other hand, the apparently similar stands which lacked the ephemerals invariably had a mor humus, with a deep layer of partially undecomposed and matted leaves and needles, often 3 or 4 inches in depth. Such a layer would obviously be unfavorable for the growth and development of most of the spring herbs. The factors responsible for the presence of one or the other of these humus types in a particular stand are not clearly understood, but the geographical distribution of mull humus primarily along the southern-half of the northern forest region lends some weight to the suggestion that postglacial migrations of southern hardwoods (Chapter 22) may be causally related to the current pattern.

The dual nature of the soil profiles in places with and without hemlocks is naturally reflected in their acidity and nutrient content. The average of the P.E.L. soil analyses in 13 stands dominated by hemlock showed a pH of 4.9, available calcium 3140 p.p.m., and available potassium 84 p.p.m. in the A_1 layer. The water-retaining capacity of that layer was 305 per cent. In contrast, 19 stands dominated by maple or beech had a pH of 5.7, a calcium content of 4100 p.p.m., potassium 98 p.p.m., and a water-retaining capacity of 194 per cent. The alpha radioactivity of the A_0 layer of hemlock soils was very high, at 9.14 counts per hour (c.p.h.) under standard conditions (Curtis and Dix, 1956b). This compares with 6.77 c.p.h. in the non-hemlock northern stands and 2.63 c.p.h. in the southern mesic forests. The great difference may be due to specific chelating properties of the hemlock organic matter, since similar high values (9.87 c.p.h.) were found under hemlocks in the Great Smoky Mountains of Tennessee.

STABILITY OF THE MESIC FOREST

The northern mesic forests exhibit a high degree of compositional stability. Sugar maple is the strongest potential dominant and has the ability to replace both hemlock (Brown and Curtis, 1952) and beech (Ward, 1954), its two most shade-tolerant competitors. The other major members of the mesic forests, yellow birch, basswood, American elm, and white ash can maintain themselves only in openings in the forest created by the death of a group of canopy trees. Such gaps are frequent as a result of wind throw and lightning fires. Larger openings created by major blowdowns of hurricane origin may be filled by balsam fir, red oak, or white pine, particularly if a fire starts in the down wood. On the other hand, such major wind catastrophes may sometimes increase the dominance of maple in the absence of fire, because the ever-present crop of maple seedlings has a distinct advantage over any species which must seed into the gap. Some of the stands in the P.E.L. study have an index over 2900 on the compositional gradient, with up to 98 per cent of their total basal area due to sugar maple. Such stands resulted from catastrophic destruction by hurricanes in the past. No comparable consociation of maple was ever found in the south.

GEOGRAPHICAL RELATIONS

The resemblance of the northern Wisconsin mesic forests to those of neighboring states is not as great as that of the southern mesic stands, largely because of the concentric ranges of the major species and the resultant east-west gradient in floristic complexity. Thus the terminal forests of northern Minnesota (Grant, 1934) contain no trees not found in Wisconsin but they have far fewer species. The major missing trees are hemlock and beech, and the yellow birch is of greatly reduced importance. In Michigan, on the other hand, 83 per cent of the species listed by Westveld (1933) in his study of 43 stands in the Upper Peninsula are in common with the Wisconsin list, and three additional species are present. The only other extensive study is that of Dansereau (1943) in Quebec. In the 180 stands examined by him, 63 per cent of the total species found are also present in the Wisconsin flora. There are 15 additional species. However, in both of these studies, the species-in-common are more important than the extra species, in terms of per cent presence, so that their sums of presence as a per cent of the total sum is 96.4 per cent in the Michigan stands and 90.3 per cent in the Quebec stands. Differences in proportionate representation of the various species are not great, with index of similarity values of 81.0 per cent for Michigan and 72.2 per cent for Quebec.

Toward the south, differences become much greater. Thus in Pennsylvania (Hough and Forbes, 1943) the species-in-common are 72.8 per cent of the total, in Ohio (Williams, 1936) they are 68.4 per cent, and in North Carolina (Oosting and Bordeau, 1955) they are only

40.0 per cent. An examination of these and many other studies of single stands reveals certain trends in composition on a NW. to SE. gradient. At the edge of the range of the type in Minnesota, basswood achieves levels of importance not found elsewhere. In Wisconsin, the species complement is larger and is marked by the great relative importance of yellow birch. From eastern Wisconsin eastward and southward, hemlock and beech become more nearly equal to sugar maple, while toward the southeastern limits sugar maple decreases in relative importance to be replaced by a mesic ecotype of red maple, as in Pennsylvania, or by a combination of mesic species which includes magnolia (*Magnolia* sp.), buckeye (*Aesculus* sp.), silver bell (*Halesia monticola*), and flowering dogwood (*Cornus florida*), as in North Carolina. In some instances in the region, sugar maple retains its position as the terminal member of the forest, as Dix (1957) has shown for protected forests in Washington, D.C. Along the northern border, from northern Minnesota to Quebec and New England, balsam fir, and either white spruce (*Picea glauca*), or red spruce (*P. rubra*) are important constituents of the type. Thus it can be seen that although Wisconsin is near the edge of the range of the forest, it contains an adequate representation of all of the important species of the type, at least in the northeastern portion of the state.

Some attempts have been made to divide this northern conifer-hardwood forest into sub-units or "associations" on the basis of the relative order of the leading dominants. Braun (1950), for example, recognizes a beech-maple forest and a maple-beech forest, among others. A similar approach to the classification of the eucalyptus forests of South Australia, using the two most important dominants in the order of their importance, was reported by Crocker and Wood (1947). When such a classification scheme was applied to the P.E.L. stands, 15 different combinations based on the two most important species were found. Sugar maple was in first or second position in 12 of the combinations, thus showing its great dominance potential. Beech and hemlock were present in four combinations each, and yellow birch was present in three. The birch, however, was never the leading dominant in any stand. When this procedure was applied to the first three leading dominants, the number of groups was greatly increased, while there were no two stands which had the same combination of the first four leading dominants. Thus any attempt at subdivision based on the dominant trees can lead only to the individualistic concept of Gleason (1939) which says that no two stands in either place or time are ever alike.

The Wisconsin forests, especially those containing beech, show much greater relationships to the mesic conifer-hardwood forests of Eurasia (Rübel, 1932; Suzuki, 1953) than was the case with the southern mesic hardwood. Oberdorfer (1957) reported on the composition of the melic grass-beech community of the Black Forest of Germany. Of the 107 total species, 11 were in common with the Wisconsin stands, these including *Athyrium filix-femina, Dryopteris austriaca, Milium effusum,* and *Oxalis acetosella.* There were 44 genera in common, with *Abies, Acer, Carpinus, Fagus, Fraxinus, Quercus, Tilia,* and *Ulmus* among the trees, and *Anemone, Carex, Circaea, Dentaria, Lonicera, Sambucus, Taxus,* and *Viburnum* in the understory.

UTILIZATION AND CURRENT MANAGEMENT

The northern mesic forests have had an interesting history of utilization in which changes in relative economic value of the various species have been of great importance. Prior to 1900, the big demand was for white pine. Hardwood stands with only 2 or 3 pine trees per acre were highly profitable, since these few trees were likely to be forest giants from 3 to 6 feet or more in diameter. Such an amazingly thorough "high-grading" took place, that it is exceedingly difficult today to find a mesic forest with the original pines still intact. Rather, mile after mile of otherwise virgin forest reveals scattered white pine stumps as evidence of the former activity. In some areas, hemlock was harvested for its bark, which was used in the tanning industry, and the logs were left on the ground to decay. The hardwoods were used only when a furniture factory or other consumer was present in the immediate vicinity. These logs could not be floated down the streams and long distance transportation by overland means was too expensive in relation to the worth of the wood. By World War I, however, the demand for yellow birch for propeller stock and veneer, and for the other quality hardwoods had begun in earnest. In those days, and for a long time thereafter, clear cutting was the usual method of harvest. This was dictated by economics, the tax laws, and fear of forest fires. Following the initiation of an adequate system of fire protection in the 1930's, changes in the tax structure, and improvement in the economics through the "unearned increment" of increasing value of stumpage, some holdings, both private and governmental, began to be harvested on a selective logging basis. In this system, the cuttings are limited to trees of certain size and form, and are made in such numbers that they improve the quality of the stand. The Goodman Lumber Company lands in Marinette and Florence counties and the Menominee Indian Reservation in Shawano County are good examples where this system is practiced.

When properly conducted, the selective logging procedure introduces little more disturbance in the stand than that caused by natural death from wind throw and disease, since excessive canopy openings are avoided and soil disturbances kept to a minimum. One of the difficulties encountered results from the dominating qualities of sugar maple. Gaps created by logging tend to be filled by maple, so that the stand loses in yellow birch, basswood, and hemlock. A "waste not, want not" philosophy results in a scarcity of down logs, which, as we have seen, are favorable sites for the regeneration of birch and hemlock. The complete solution to this problem will no doubt be based upon a more complete understanding of the autecology of the component species, as studied under natural conditions.

It was with this view in mind that a large portion of the Scientific Area in the Flambeau River State Forest was ordered by the Conservation Commission to be allowed to remain in an untouched state, with no cleanup following wind throw and no harvest of so-called overmature trees. Such forests can provide the necessary information on the methods whereby each species maintains itself under natural conditions, with the expectation that such methods can be duplicated or improved upon in the commercially harvested forests.

The mesic northern forests are used to a considerable extent for the production of maple sugar. The usual deep snow cover in the area means that the proper combination of warm days and cool nights necessary for good flow of sap will occur in early spring, and the large contiguous acreages which are available assure the large-scale operations needed for efficiency.

CHAPTER 11

Northern forests—xeric

GENERAL DISTRIBUTION

The two xeric segments of the northern forest include the pine forests of the state. They are found in all sections of the area north and east of the tension zone but are particularly numerous on the outwash sands and the sandy glacial lake beds of Marinette County, Vilas and Oneida counties, Burnett, Washburn, and Bayfield counties and in the central area centering on Adams, Juneau, Wood, and Jackson counties. Elsewhere they are of less extensive occurrence but may be seen in limited development on the banks of most of the lakes (Plate 58) and streams of the entire area. The total area shown on the map in Figure XI-2 is 2,269,400 acres or 6.48 per cent of the land area. In addition, numerous relic stands of pine forest occur in southern Wisconsin on the sandstone cliffs of the Driftless Area and the region of Farmdale glaciation.

In eastern Wisconsin, east of the Wisconsin River, the transition from the xeric oak forests of the south to the xeric pine forests of the north is sharp and amounts to a virtual replacement of one by the other at a very narrow tension zone. To the west, however, in an area from Clark County south through Monroe County to Richland County and northwest to Buffalo County, the ecotone is much more gradual, with mixed forests of pine and oak, or with patches of pure pine on rocky or sandy sites, and pure oaks on the deeper soils. The pine here is largely one species, the white pine (*Pinus strobus*), but its presence is sufficient to impart a distinctly northern appearance to the landscape. A few groundlayer species share this range with the white pine, but the majority of the northern herbs and shrubs stop at a tension line considerably farther north.

STRUCTURE AND COMPOSITION

In the figure at the opening of this chapter the stands studied are shown. Thirty-four species of trees were found in the P.E.L. studies of the northern xeric forests (Table XI-1). When the two segments are considered separately, it is seen that the three species of pine, jack (*Pinus banksiana*), red (*P. resinosa*), and white (*P. strobus*), are the most important members of the initial or dry segment, whereas the intermediate or dry-mesic forests have white pine, red maple (*Acer rubrum*), and red oak (*Quercus borealis*) as their three most important trees. The dry forests have considerable quantities of Hill's Oak (*Q. ellipsoidalis*), trembling aspen (*Populus tremuloides*), large-toothed aspen (*P. grandidentata*), and white oak (*Q. alba*), while the associated species of significance in the dry-mesic segment are white birch (*Betula papyrifera*), sugar maple (*Acer saccharum*), hemlock (*Tsuga canadensis*), and red pine. The remaining tree species are mostly of rare or sporadic occurrence.

The pines may occur in nearly pure stands of a single species, or in mixtures of two or all three (Plate 19). They may also be present in nearly any proportion in mixed forests with oaks, aspens, or maples. Consequently, it is difficult to adequately portray the structure of typical stands with a few limited examples. Those given in Table XI-2 were chosen merely to give some idea as to the behavior of some of the more important species.

The forests dominated by jack pine may attain very high densities (up to 308 trees per acre in some of the P.E.L. stands) but the total basal area is usually quite low due to the small size of the trees, with a maximum of only 95.5 square feet per acre in a stand in Marinette County. The red and white pine forests, however, may produce enormous values for total dominance (Plate 18), with one of the current stands in Sawyer County showing 282.3 square feet per acre, of which 221.0 feet were contributed by white pine and 10.2 feet by red pine. The growth form of the pines, with their predominating central trunk, great height, small taper, and small branch fraction, means that their basal areas represent a much greater percentage of potential lumber than would an equivalent area of hardwoods (Plate 20). Since the P.E.L. studies had an ecological rather than a forestry approach, no measurements of actual volume of wood were made. Available records from the early lumbering camps, however, indicate amazing yields from certain areas. Thus, one forty in Bayfield County which was cut in 1921 was reported to have yielded an average of 70,000 board-feet per acre (Bordner, 1942). Roth (1898) reported that many single acres could be selected which would yield 100,000 board-feet, although 25,000 was considered a good cut while the average yield was less than half the latter amount. The very few virgin pine forests remaining in Wisconsin are usually one hundred years or less in age and date back only to the period of settlement. Most of the big pines cut in the heyday of the lumbering business were about four hundred years old and stemmed from widespread catastrophes in the 1400's. The occasional giants of 7 to 10 feet d.b.h. reported by the surveyors must have been still older. Results from modern studies, therefore, cannot give a true picture of the actual magnitude and majesty of a mature pine forest at its optimum and this should be kept in mind in the interpretation of contemporary findings.

LIFE-HISTORIES OF DOMINANTS

White pine (*Pinus strobus*)

If there is one tree which can be used to exemplify the northern forests, it is the white pine. It is the only species which is present in appreciable quantities in all segments of the full moisture gradient from wet bogs to xeric sand plains. It is the largest and the longest lived tree of the region. When the vacation-bound traveler from the hot and steamy cities of the south sees his first white pine, he knows that he is entering the "north woods." This innate and almost totally unrecognized ecological skill at judging climate by vegetation is probably a vestigial holdover from the days when man was more intimately connected to his landscape than he now believes himself to be. In any case, the judgment is a sound one, since white pine is truly a sign of pleasant summer days and cool nights. Its corollary, deep snow and bitter winter cold, is less well appreciated.

Like the oaks, the pines are similar to each other in most of their life-history characteristics; there are just enough differences in environmental requirements to bring about significant ecological separations. All of the pines are wind-pollinated; all produce wind-disseminated, winged seeds; all germinate best on exposed mineral soil; all grow rapidly; none is tolerant of shade; and all exert similar influences on the soil.

Of the three species in Wisconsin, white pine is the most exacting with respect to the moisture and nutrient supply required for optimum development. It grows best on deep loams or sandy loams which have an appreciable quantity of available bases, but it can survive as stunted specimens on any soil in the area or, indeed, on no soil at all, as it does in the cracks on sandstone cliffs. The seeds of white pine are shed in autumn in large numbers after a two-year ripening period, with intervals of 3 to 5 years between bumper crops. Germination occurs in the spring on moist surfaces which need to be moderately rich in organic matter for good survival. Invasion of open sites can occur but the presence of a nurse crop of aspen, birch, or other pioneer species promotes the best regeneration. The young trees grow rapidly and may reach 8 to 10 inches d.b.h. at forty years, with a height of 60 feet or more. Ultimate heights over 200 feet are common, and ages of five hundred years or more are easily possible.

The statement is frequently seen in forestry publications that white pine avoids all types of organic soils. This is definitely not true in Wisconsin, where 37 per cent of the stands of bog conifers in the P.E.L. study which were dominated by tamarack or black spruce had white pine as an associate. Several stands in the north are known where white pine is the leading dominant in swamp forests located on peat beds which are more than 10 feet thick. The old tamarack swamps in stream valleys of the Driftless Area are frequently invaded by white pine. However, no stands were observed in which white pine was more than a minor and accidental component in forests on mineral soils which were subject to frequent or prolonged inundation by flood waters. It is not known whether the trees of organic swamps represent distinct ecotypes or not.

Red pine and jack pine (*Pinus resinosa and Pinus banksiana*)
Red pine closely resembles white pine in most details of its life-history. It differs largely in that it is less shade tolerant, can germinate and grow well on drier sites with a lower nutrient content, and is more

resistant to ground fires and other basal injuries. This better adaptation to xeric conditions is correlated with a greater restriction to the areas of sandy soil within its range. Although red pine may become a very old and a very large tree by ordinary standards, it usually does not equal white pine in these respects.

Jack pine possesses a number of special attributes which serve to differentiate its behavior from that of the other two pines. The most significant is its habit of producing serotinal cones, many or most of which do not open normally so that their seeds may be shed. Rather, they remain on the tree in a closed condition for many years and some actually may become buried in the wood of the trunk as the new growth rings engulf the cones. Excessive heat and dryness are required to open these delayed cones and these conditions are best provided by a forest fire, which may completely consume the branches but which only scorches the cones, and causes their scales to open and scatter the seeds on the burned ground. This burned surface is a very favorable seedbed for jack pine and extremely dense stands of seedlings may result from such a sequence of events.

Eyre and Le Barron (1944) reported that mature stands of jack pines may have a reserve supply of seeds in unopened cones of as many as two million seeds per acre. Some seeds are liberated from a few of the cones while the trees are still living, but these ranged only from 2700 to 10,500 per acre, according to these investigators. Growth of the seedlings is very rapid and sexual maturity comes early, so that the new stand may be producing more cones within five to ten years. Repeated fires at intervals greater than this will serve to perpetuate jack pine on the site, while more frequent fires may cause the area to degenerate to a pine barren with only a few scattered trees.

The jack pine is usually a small tree but its form and ultimate size appear to vary from place to place in Wisconsin. Thus in the central sand plains and in the disjunct stands around Muscoda on the lower Wisconsin River, the tree approaches a bush in form, with persistent lower branches which become partially ascendant and increase in diameter at a rate only slightly less than does that of the central trunk. Farther north, this bushiness is less extreme, with many stands in Vilas County and in the northwestern sand plains made up of tall, straight-trunked trees 80 to 100 feet tall. North and west of Lake Superior, the tree reaches still better proportions and approaches red pine in aspect. How much of this is due to introgression from the hybrid swarms with lodgepole pine as found in Alberta is not known. The change in form with lower latitude in Wisconsin, at least, appears to have a genetic, ecotypic basis and is not due simply to the effects of warmer temperatures. Artificial plantations in the south made with trees of northern provenance in the University of Wisconsin Arboretum at Madison appear to be attaining their hereditary shape.

Associated hardwoods

The three most characteristic hardwoods to be found in the pine forest regions are red maple (*Acer rubum*), trembling aspen (*Populus tremuloides*), and white birch (*Betula papyrifera*). The first of these is responsible for the brilliant displays of autumnal coloration which are the glory of the north woods in September. It is a tree of very wide environmental tolerance which reaches highest levels of importance in the dry-mesic stands with a secondary peak in the wet-mesic moist forests; the bimodality thus reflects its intermediate degree of shade tolerance. The red maple responds well to various types of disturbance and may seed in heavily on burned-over areas. However, some internal lack almost invariably prevents it from achieving a position of true dominance in any of the local forest types. Susceptibility to disease may be an important factor, since the species rarely attains either great size or great age.

The aspen, or popple as it usually is called in the north, is primarily a pioneer invader following forest fires. The seeds, which are produced in great abundance, are very light in weight, and are carried great distances by the wind. Optimum germination occurs on moist loam soils which have a surface enrichment of nutrients provided by ashes. Growth of the seedlings is fantastically fast, often exceeding three feet per year for the first decade. After that, the height growth slows down appreciably as the crowding effects begin to take hold. However, since the saplings are extremely intolerant of shade, minor differences in size from tree to tree are quickly magnified as some trees take over dominance and the remainder become suppressed and soon die. Several stands in Wisconsin have been studied which showed a decrease in stems per acre from an initial seedling value of 10,000 or 12,000 down to 800 or 900 after 10 years, and to 300 to 350 after 20 years. By 30 years, the canopy of the aspen stands begins to open up. This opening is hastened by the susceptibility of the trees to the aspen canker (*Hypoxylon pruinatum*) which attacks the trunks and causes them to snap off in heavy winds. In the normal course of events, an understory of less pioneer species, such as white pine, red maple, red oak, or American elm will have become established beneath the aspens by the time this degradation sets in, so that the initial stand is replaced after one generation (Plate 21). In the absence of suitable seed trees for this succession, the aspen can maintain itself by root sprouts which are readily produced and which serve to regenerate the stand after cutting or fire.

White birch resembles trembling aspen in many respects and can replace it or grow along with it in similar secondary successions. In general, it requires slightly more mesic conditions and is far more likely to be found as a gap-phase tree in small openings in the forest than the aspen, not only in xeric stands but mesic stands as well. It frequently reaches greater sizes and ages than the aspen, especially along Lake Superior and Lake Michigan.

GROUNDLAYER

The dry segment of the northern pine forests in the P.E.L. study had a total of 239 species of herbs and shrubs whereas the dry-mesic segment was slightly richer with 252 species. The prevalent species in these two segments are listed in Tables XI-3 and XI-5. As can be seen, there are great similarities between the two lists, which have a very high index of similarity (See Chapter 4, p. 83) of 67.6 per cent, thus indicating the essential uniformity of the pine forests as a whole. The other species in these forests are likewise very similar in the two segments. The important species which differ significantly between the two types include *Corylus americana* (66 per cent in dry *vs.* 25 per cent in dry-mesic), *Melampyrum lineare* (37 per cent *vs.* 8 per cent), *Actaea alba* (0 per cent *vs.* 48 per cent), *Brachyeletrum erectum* (5 per cent *vs.* 38 per cent), *Dryopteris austriaca* (13 per cent *vs.* 55 per cent), *Mitchella repens* (26 per cent *vs.* 75 per cent), and *Prenanthes alba* (5 per cent *vs.* 50 per cent).

The number of species which bloom in the summer is greatly increased as compared with the northern mesic forests. Thus, considering the prevalent species only, 47.2 per cent of those in the dry stands bloom between June 15 and August 15, compared to only 26.7 per cent in the mesic forests. The plants blooming in the autumn are also increased, from none in the mesic forests to 6.5 per cent in the dry-mesic forests. A floristic summary by families reveals that nine families contain one-half of the total species in each segment. The first five in each segment are the same—*Compositae, Liliaceae, Ranunculaceae, Rosaceae,* and *Ericaceae.* The *Violaceae, Caprifoliaceae,* and *Gramineae* are among the first nine in each, although not in the same order. The orchids in the dry-mesic segment are replaced by the ferns in the dry segment.

Herbs or shrubs with evergreen leaves are important in the xeric

Plate 19.—Young stand of mixed jack pine (*Pinus banksiana*) and red pine (*P. resinosa*). Buckatabon Lake, Vilas County.

Plate 20.—Mature stand of red pine, with red maple (*Acer rubrum*) and red oak (*Quercus borealis*) understory. Finnerud Forest near Hazelhurst, Oneida County.

Plate 21.—Second-growth stand of aspen (*Populus tremuloides*), white birch (*Betula papyrifera*), and red oak (*Quercus borealis*). Northern Highlands State Forest, Vilas County.

Plate 22.—Continuous canopy of groundlayer in pine forest, mostly of bunchberry (*Cornus canadensis*) and club moss (*Lycopodium complanatum*).

forests, with more than 15 per cent of the total species so equipped in each segment of the compositional gradient. Of these chamaephytes (Raunkiaer, 1934), the greatest number are members of the *Ericaceae,* and include such common plants as pipsissewa (*Chimaphila umbellata*), trailing arbutus (*Epigaea repens*), and wintergreen (*Gaultheria procumbens*). The true shrub content of the xeric forest understory is very high, with a sum of presence in the dry stands which is 29.7 per cent of the total sum. The comparable value in the southern xeric forests is only 16.2 per cent. In contrast, the liana component, which includes both woody and non-woody species, is only 8.4 per cent compared to 10.3 per cent in the southern forest.

No detailed studies of aggregation in the northern forests have been reported. A few exploratory investigations plus general observations indicate that the pine forests resemble their oak counterparts in the south in having a low or medium degree of aggregation on the average. Many of the ericads spread by runners or by creeping rhizomes and tend to form small colonies or clumps, but these are usually so well distributed throughout the forest that they impart a certain amount of randomness to the distribution. The club mosses (*Lycopodium* sp.) behave in a very similar manner. Perhaps the most extreme cases of aggregation are to be seen in the jack pine forests, where bunchberry (*Cornus canadensis*) frequently forms dense clones with an almost complete leaf mosaic (Plate 22). Bearberry (*Arctostaphylos uva-ursi*) and sweetfern (*Myrica asplenifolia*) also tend to grow in tight colonies, usually associated with openings in the canopy.

The mature red and white pine stands often show an intimate mixture of species, which may occur in relatively great numbers on small areas. The average number of species per quarter mil-acre quadrat in the pine stands of the P.E.L. study was 6.2. This compares to a value of only 3.6 per quadrat in the northern mesic forests.

OTHER BIOTA

The lichens of the pine forests, as reported by Culberson (1955a) showed considerable evidence of host specificity, with some differences in the flora of the several pine species and an almost completely different flora on the aspens in the same stands. Thus *Alectoria nidulifera, Baridia chlorococca, Cetraria ciliaris, Evernia mesomorpha,* and *Parmelia sulcata* were found on all pines but reached peak frequencies on jack pine, while *Cetraria atlantica, Lecanora subfusca, Lepraria aeruginosa,* and *Parmelia caperata* were similar but attained optimum levels on white pine. In contrast, all of these species except the *Lecanora* were rare or absent on aspens, which showed high populations of such species as *Caloplaca aurantiaca, Physcia aipolia, Physcia ciliata,* and *Xanthoria polycarpa.*

The soil microfungi, as studied by Martha Christensen (1956) showed some relationships with the oak forests of the south. Important species were *Penicillium janthinellum, P. melinii, Pullularia pullulans,* and *Mortierella ramanniana.* The *Mucoraceae* as a group reached highest importance in the initial pine stands.

The red squirrel is perhaps the most characteristic mammal of the pine forest fauna. It can be seen, and heard, working over the cones in almost every stand in the north. Other typical rodents are the red-backed mouse and the lesser chipmunk. Among the birds, the Canada jay, the red-breasted nuthatch, and many warblers find conditions in the pines to their liking. Detailed studies similar to those of Bond in the southern forests have not yet been made for the northern forests in Wisconsin.

ENVIRONMENT
Microclimate

The light relations of the pine forests have been studied in detail by Shirley (1932) in Minnesota. Measurements of canopy coverage and of incident light in the Wisconsin P.E.L. investigations have given similar results. In general, the light intensity on the forest floor decreases as shade tolerance of the dominant species increases, with maximum light in the jack pine stands and minimum readings in the white pine-hardwood intermediate stands.

Atmospheric moisture conditions were studied by Van Ardsel (1954) in his investigation of the microclimatic requirements of the white pine blister rust. Perhaps the most significant finding was the high frequency of periods of atmospheric saturation of 24 hours duration in the northern pine stands. Such periods were rare or absent in most of the pine relics in the Driftless Area, and accounted for the absence of blister rust there. These saturation periods are necessary for the success of the disease organism.

Various measurements of rainfall interception have been made in Wisconsin pine stands and they show that only 76 to 83 per cent of the total available rain penetrates to the ground beneath the canopy. Interception of snow during the winter is no doubt far greater by comparison but no measurements have been reported.

The "dryness" of the xeric northern forests is to be considered as a relative condition only. The southern xeric forests, the savannas, and the dry prairies are actually much drier. The slightly lower total precipitation in the north (Tables XI-4 and XI-6) is more than balanced by the lower potential evaporation there. The dry conditions in the pine forests are largely the result of the excellent internal drainage and low storage capacity of the soil and the high light intensity.

Soils

The major soil series represented in the xeric forests are the Cloquet, Hiawatha, Omega, and Vilas. All are Podzol or podzolic soils. The surface layer is of the acid, mor humus type, that has resulted from the accumulation of several years' production of pine needles. The A_1 layer is thin or nearly absent, while the leached A_2 layer is generally well developed. The soils of some stands now supporting jack pine are anomalous because they are nearly neutral in reaction and show several inches of incorporated organic matter in the topsoil. These unusual conditions result from a former grass cover during a pine-barrens stage when fires were more frequent than they have been in recent years. Still other jack pine stands show a different history because they have almost no profile development but are on nearly pure sand with a thin layer of needles on the surface. These either developed on recent aeolian deposits or on soils which have had their entire organic fraction destroyed by repeated heavy fires. Jack pine is one of the few Wisconsin species which can successfully invade an area of the latter type and there produce a forest which begins the process of rebuilding the soil.

SUCCESSION

The general nature of the upland successions in the northern forest has been pointed out by many investigators (Kittredge, 1934; Graham, 1941), and is clearly evident from the results of the current study (Chapter 9). The major changes are of an autogenic nature and result in a successional replacement of intolerant, pioneer species by others which are more tolerant. The theoretical order of the dominants in this replacement is shown by their adaptation values (Table 11). The order for the major species is jack pine-red pine-white pine-hemlock-beech-sugar maple. As in the southern oak forests, however, this order is not necessarily or even usually followed in a complete series. Rather, any given species may be replaced by any other higher on the list; the chances of this higher species being one which is close to the first is much greater than for one which is remote. Thus, jack pine is more likely to give way to red pine or white pine than it is to hemlock or sugar maple. Furthermore, any given stage, short of the terminal forest, can be produced by catastrophic retrogression as well as by progressive succession.

The initial stages in the normal successions usually begin on land which has been devegetated by forest fires. If these lands are sandy, poor in nutrients, and subject to drought, the initial invaders are likely to be jack pine or Hills' oak, or both. On heavier soils, the first trees are trembling aspen, white birch, or pin cherry, although direct invasion by white pine, red maple, or even yellow birch is possible if the soils are suitable and nearby seed sources are present. Under typical conditions the jack pine forests are invaded by red pine, white pine, or red oak before they reach maturity at sixty to one hundred years, while the aspen and birch stands act as short-lived shelter crops for white pine, red maple, hemlock, or sometimes for sugar maple. Both very dry and more mesic sites, therefore, have the opportunity of developing a stand of white pine or one with a large white pine component. Further successional changes in such a white pine stand are slow because of the great size and age of the white pines. However, a nearly closed forest of mesic species is usually to be found beneath the high pine canopy within 200 years. This forest may have a high component of hemlock or it may be made up largely of such hardwoods as yellow birch, basswood, beech, and sugar maple. The stand near Peavy Falls in the Menominee Indian Reservation is an excellent example of this successional stage, where a forest of pines 100 to 130 feet tall towers over a solid understory of hemlock, sugar maple, and beech that is 60 to 80 feet tall. Barring catastrophe, the white pines gradually die out and probably would be entirely gone within 600 years under natural conditions, leaving a climax mesic forest which would gradually increase in sugar maple dominance. Uninterrupted successions of this sort must have been very rare in presettlement times, since the time period of 800 to 1000 years necessary from initial stand to final forest is so long that the chances of having no catastrophe are practically nil.

The two major disasters which are most likely to affect the forest are fire and wind throw. Fire is most frequent in the initial pine forests because the low moisture and high light of such sites frequently lead to conditions of high combustibility. Following the entrance of the hardwoods, chances of fire decrease due to the more even moisture supply on the forest floor. Wind throw, however, may affect any stage in the succession although even here the mesic maple forests are less liable to total destruction.

Severe crown fires during very dry periods may cause the death of all trees in the stand. Such conflagrations were less frequent in the past than ground fires which burned the shrub layer and the needle mat. Many instances of retrogression were a result of such ground fires. Thus, initial stands of pine which have been protected from fire for 100 years or more may show a successional replacement of most of the original jack pines by red pines, with only a few relic jacks in the stand. Severe or repeated ground fires at such a stage may destroy most of the red pines and the jacks, but they may stimulate a new crop of jack pines from serotinal cones. Many stands are known where this has happened, where a scattered overstory of tall red pines exists above a nearly pure stand of jack pines of varying states of maturity. Jack pine may thus remain in control of a given site almost indefinitely, so long as forest fires occur with a periodicity of 10 to 200 years. The droughty sand soils, which are the optimum for this species almost inevitably are visited by natural or man-made fires within this interval, so that such soils continue to support a pine forest with a varying jack pine component. Similar ground fires may cause a retrogression from white pine to red pine or from white pine-hemlock-red maple to white pine, but the nature of the undergrowth in most of the intermediate and mesic forests would discourage such fires. Furthermore, the soil amelioration which has accompanied the successions may make conditions more suitable for aspen or birch than for pine, so that infrequent ground or crown fires are more likely to be followed by these species than by the pines.

The relatively high degree of similarity of the herb and shrub layers in the dry and dry-mesic segments of the xeric forests can be explained largely on the basis of a repeated series of succession and retrogression cycles, from jack pine to white pine or beyond, and back again. Evidence from fossil pollen in the bogs of the region (Chapter 22) indicates that some areas have maintained an important pine component for at least 6000 to 8000 years, which would allow for at least 20 or 25 cycles of 300 years each. Thus the understory would be purified to a rather uniform flora of fire-tolerant species which would show only minor shifts due to light and nutrient fluctuations.

Frequently, fire and wind destruction went hand-in-hand. Areas of blown-down trees presented so great a fire hazard that inevitably they were visited by fire. The prevalence of violent windstorms in the entire northern forest has been discussed in detail by Stearns (1949), who relied on the records of the land surveyors and other early travelers for much of his evidence. The storms vary from tornadoes, which destroy everything within their long but narrow path, to hurricanes which topple only a part of the stand but do this over a very wide area. The findings of Stearns indicate that both tornadoes and hurricanes were prevalent in the 1860's and 1870's, with an apparent area of concentration centered on Sawyer County. Records thereafter are scanty, but evidence from the forests themselves would imply that such storms were less frequent until the 1940's and 1950's, when a series of tremendous winds of hurricane intensity caused severe damage in the few remaining virgin stands in the Flambeau River Valley and in Vilas and Forest counties.

Some of the early descriptions of windfalls are impressive. Hoyt described the vegetation of Clark County in 1860:

There are also in these townships a great many windfalls, some of which are miles in length. Here can be seen thousands of acres of trees that have been blown down or broken off, the stubs of which are still standing. The fire runs through these windfalls every year or two, and kills all the vegetation, leaving nothing but the blackened timber that lies upon the ground, or the stubs of those that were broken off.

The geologist Irving (1880) had this to say about a windfall in the Lake Superior region:

The great windfall of September, 1872, so far as my knowledge goes, is as much as forty miles in length, though it is reported to have a greater length than this. It crosses the Chippewa river ... with a width of about one and one-half miles. When last traversed in 1877, this windfall had been partly burned over, but was, for the most part, made more impenetrable than ever on account of the new and dense growth of bushes and small trees. In crossing it on the Penokee Range, it was found necessary to cut the way with an axe for nearly a mile and a half.

The amount of energy liberated by such a storm in knocking down all the trees on an area of at least 60 square miles can scarcely be comprehended. The total influence of these wind throws on the structure of the forest as a whole can be partially appreciated by the fact that "a single storm on July 11, 1936, blew down 30,000,000 board-feet of timber within the Menominee Indian Reservation" (Stearns, 1949). As another instance, 1,500,000 board-feet, or 27 per cent of the standing timber in the 1-square mile block of virgin hemlock hardwoods in the Flambeau River Forest Scientific Area, were blown down in the storms of October 1949, July 1951, and June 1952.

Stearns concluded from his study of the northern forests of Wisconsin, "Windfall alone or with the other agents of mortality, fire, drought, glaze storms, insect or fungus infestation and senescence, keeps the forest in a constant state of flux which results in a continuous variation in composition, both in space and time, within the limits of the species present." The climax, as a potential steady state at the end point of a succession, is a goal towards which the forest tends but to which it rarely attains.

NATURE OF THE SOUTHERN PINE RELICS

As mentioned earlier, relic stands of pine are to be found on rocky cliffs in many places in the Driftless Area (Plate 55) and in a few places on the area of Farmdale glaciation. A number of these have been studied in detail by McIntosh (1950a). Of the 22 stands he investigated, 10 are on northeast facing cliffs, 8 on northwest cliffs, 4 on southwest, and none on the southeast. All are on sites with a slope of at least 40° and all are on sandstone rocks of Cambrian or Ordovician age. The most common site is a waterworn cliff, with the undercutting stream still present at or near the cliff base. The back slopes are covered with prairie or oak savanna. Considerable light usually penetrates these stands from a lateral direction even though the trees are close together. As a result, seedlings and saplings of the pines are often present. In a few cases, the pine relics contain an admixture of hemlock, yellow birch, mountain maple, and sugar maple. White pine is the most important species. It occurred in all but three of McIntosh's stands. Rarely, red pine is the sole pine present, but it is usually mixed with white pine. One area, in the new Governor Dodge State Park in Iowa County, has white, red, and jack pines in mixture. This stand has been set aside as a scientific study area by the State Board for Preservation of Scientific Areas.

The herbs and shrubs found in the pine relics by McIntosh show a close resemblance to those of the northern pine forests, with 28 of the prevalent species in Table XI-5 appearing in the pine relics. However, only a few of these species occurred in any one relic, with a slightly different complement in each other stand. The average size of the relics is small, from $\frac{1}{4}$ to 4 or 5 acres, but the fragmentary representation per stand cannot be due entirely to this factor. Rather, the isolation of the stands results in a situation where any chance destruction of a single species in a given stand cannot be repaired easily by migration from adjacent areas, as would be the case in the contiguous forests of the north.

The fossil pollen record (Chapter 22) indicates quite clearly that these relics have persisted in the Driftless Area since at least the time of the Cary ice 12,000 years ago. They may well have been present for much longer periods, most of that time isolated from the major areas of the forest they represent. Partial evidence for this is seen in the fact that the hemlocks of the relic stands are ecotypically different from those of the north (Goder, 1955) and in the probable existence of ecotypes in white pine, beaked hazel (*Corylus cornuta*), and starflower (*Trientalis borealis*), based upon general observations of growth form.

Although these relic stands are considerably removed from the area of their climatic optimum, there is no indication that they are in a retreating or moribund condition. On the contrary, when the upper slopes above them are protected from fire or grazing, the pines show great vigor in establishing themselves beneath and around the open-grown bur oaks (*Q. macrocarpa*) that were originally present. In a few instances, the pines have produced healthy closed stands on the uplands which have overtopped the bur oaks and caused their death by shading.

Another example of current vigor is to be seen in the jack pine stands on the Wisconsin River terraces near Muscoda in Grant County. These stands now occupy many square miles of land that were prairie at the time the first settlers arrived in the 1830's. Injudicious plowing of these sand flats was quickly followed by wind movement of the sand and the production of sand dunes. The jack pines spread from a few isolated remnants to the new dune areas, where such spread continues, as greater and greater areas become engulfed by the moving sands. The flora beneath these Wisconsin River jack pine stands is very poor in northern species and is made up mainly of prairie and sand barrens plants. It is of interest that Culberson (1955b) found the lichen flora of the jack pine trunks to be similarly depauperate, with only three species per tree as compared to ten in northern counties. Many of the characteristic jack pine lichens in the north, such as *Evernia mesomorpha* and *Alectoria nidulifera*, were lacking altogether in the Muscoda stands.

GEOGRAPHICAL RELATIONS

In spite of the very great economic importance of the northern pine forests of the United States and Canada and of their former great areal extent, there have been very few phytosociological studies reported for the type. As a result, it is amazingly difficult to make comparisons between the forests of Wisconsin and those of other areas on any meaningful quantitative basis. In contrast to the mesic hardwood forests which had their optimum expression in states to the south and east of Wisconsin, the northern pine forests appear to reach their best development in the Upper Great Lakes region, in Michigan, Wisconsin, and Minnesota. Extensive areas in New York and the New England states formerly supported this type also, but they were harvested so long ago that little record of their exact composition remains.

Shirley (1932) studied a red pine forest in the Chippewa National Forest of north central Minnesota. All 9 of the tree species in the stand are found on the Wisconsin list (Table XI-1), and all of the 53 shrubs and herbs except the doubtful *Prunus susquehanae* and the introduced *Vicia cracca* were found in the Wisconsin stands. In a study of the forests of the Adirondack region of New York, Heimburger (1934) reported on 23 stands dominated by pines or with a high representation of pines. All of the 28 trees he found are on the Wisconsin lists except for striped maple (*Acer pensylvanicum*), tamarack (*Larix laricina*), black spruce (*Picea mariana*), and red spruce (*Picea rubra*). The index of similarity was 59.7 per cent, with the greatest differences in the higher amounts of balsam fir (*Abies balsamea*), white birch, white spruce (*Picea glauca*), and black spruce in New York. Of the 124 herbs and shrubs reported by Heimburger, 84 per cent are on the Wisconsin list. The leading species and their presence percentages were bracken fern (*Pteridium aquilinum*)—100 per cent, Canada mayflower (*Maianthemum canadense*)—96 per cent, wintergreen (*Gaultheria procumbens*)—83 per cent, the two blueberries (*Vaccinium* sp.)—78 per cent, bunchberry (*Cornus canadensis*)—74 per cent, mountain-rice (*Oryzopsis asperifolia*)—70 per cent, bush-honeysuckle (*Diervilla lonicera*)—65 per cent, witherod (*Viburnum cassinoides*)—65 per cent, and Pennsylvania sedge (*Carex pensylvanica*)—61 per cent. All but the witherod, which reaches the edge of its range here, are also of great importance in the Wisconsin forests.

A study of the few available species lists from white pine forests in other areas reveals the same situation which obtained in the forest types discussed previously, namely that the floristic resemblances became less as the distance from Wisconsin increases. These differences involve a slight increase in prairie flora toward the northwest, an abrupt increase in circumpolar boreal forest species toward the north, and a gradual increase in Appalachian elements toward the south and east. Prominent in the latter group are the larger shrub members of the Ericaceae, including mountain laurel (*Kalmia latifolia*), the rhododendrons (*Rhododendron* sp.), and the blueberries, bilberries, and whortleberries (all *Vaccinium* sp.). In spite of these differences, there remains a considerable element of similarity of the northern xeric forests at least throughout the region of overlap of the ranges of the three pine species.

This similarity may have been in part responsible for the delimitation of the area as a separate plant formation by Weaver and Clements (1928). These authors, supported by their theoretical and totally unwarranted assertions regarding the impossibility of having both conifers and hardwoods in the same formation, created a special plant formation, the so-called "Lake Forest" to cover the vegetation of the upper Great Lakes and the St. Lawrence valley. Because conifers and hardwoods cannot grow together, according to their a priori reasoning, the actual forests of mixed hemlock and hardwoods in the region were totally disregarded. Instead, it was avowed that the climax forest was dominated by white pine, red pine, and hemlock. This rather amazing conclusion was reached without benefit of a single quantitative study in the region and is an indication of the pitfalls that may be met when attempts are made to place vegetation into a preconceived framework without supporting quantitative evidence.

The Eurasian pine forests show considerable similarity to the Wisconsin type in general structure and appearance (Yoshioka, 1949; Grosser, 1956). A typical example is the moss-pine community as studied in South Germany by Oberdorfer (1957). Of the 124 species reported, 11 per cent are also present in Wisconsin, including *Campanula rotundifolia, Chimaphila umbellata, Fragaria vesca, Monotropa hypopithys, Pyrola secunda,* and *Pteridium aquilinum.* Of the 38 genera in common, many are represented by closely related species in such typical groups as *Aster, Aquilegia, Betula, Galium, Hieracium, Lycopodium, Melampyrum, Polygala, Populus,* and *Vaccinium.* The main differences seem to be in the lower representation of ericads and composites in the European forests.

UTILIZATION AND CURRENT MANAGEMENT

The pine forests of the north were the scene of one of the most thorough exploitations of a single natural resource ever witnessed in Wisconsin. Beginning in the early 1800's, the white pine was sought out and harvested on a scale that staggers the imagination. By 1899 pine logs were sent down the rivers of Wisconsin at the annual rate of 3.4 billion board-feet. Logging railroads were pushed into every known stand of pine, while the hardwood forests were scavenged for their one or two pine trees per acre to an unbelievable point of completeness. By 1932 the harvest was over, with a total cut of only 0.3 billion board-feet which was the lowest since 1850. In this entire cycle, it is estimated that 103.4 billion board-feet of pine were cut in the state. This lumber, at current prices, would be worth over 10 billion dollars.

The logs were floated to mills in St. Louis and Chicago as well as to many cities in Wisconsin where they were sawed into lumber that found its way into the houses and factories of the rapidly expanding population centers throughout the Middlewest. One major factor in the expansion was the cheap and abundant supply of housing materials from the pine forests of the Lake States. For many years it has been the fashion to defame the early timber cutters as ruthless lumber barons who plundered the country for greedy personal profit, but it is well to remember that the consumers of the lumber were primarily responsible for the slaughter of the pines.

The prevailing philosophy during the early years of the harvesting cycle was that "the plow follows the axe." The lumbermen believed that they were aiding the farmers who would follow them by getting rid of the trees. Consequently, they were not particularly careful about fires which frequently started in the slashings left after the logs were removed. These fires served to clean the ground even better than the cutting itself. The impression that farms would replace forests was strengthened by the apparent success of local settlers who provided milk, meat, and vegetables for the workers in the nearby lumber mills. It was realized gradually, when the mills began closing, that pine lands and farm lands were non-compatible. Lack of a ready market quickly caused the abandonment of many of the farms in the area. This process continued, with serious socio-governmental consequences brought about by widespread tax delinquency and a decrease in populations. This is not the place to describe the problems of the "cutover area" in detail. Suffice it to say that the few remaining settlers, scattered thinly throughout a huge area of deforested and burned-over land, were not the best possible supporters of a sound conservation program of forest restoration. Their activities led to a continuation of the fires and to a continual harvest of the last remaining remnants of usable timber. The desolation of much of the pine area in the 1920's and early 1930's is difficult to describe to anyone who did not see it. In many places the entire landscape as far as the eye could see supported not a single tree more than a few inches in diameter. Only the gaunt stumps of the former pines, frequently with their root systems fully exposed as a result of the consumption of the topsoil by fire, remained to indicate that the area was once a forest rather than a perpetual barren.

The inauguration of an adequate system of fire control in the 1930's and the resettlement of isolated farmers in accord with the zoning laws adopted in the same decade, led to far-reaching vegetational consequences. Within 20 years, the great bulk of the area became covered with a pioneer stand of aspen, birch, pin cherry (*Prunus pensylvanica*), and now, in less than 30 years, significant quantities of pine, balsam, and other more valuable species are beginning to appear beneath the pioneers. This rejuvenation of the northern forests is best seen in the winter, when the aspens are leafless and when the conifers stand out clearly against the white snow cover. If present land use practices are continued, there is every reason to believe that the north woods of Wisconsin will return to levels of productivity equaling or exceeding those originally present over a century ago.

Some impression of the composition of the new forest can be obtained from the results of a transect 118 miles long from Vilas County to Ashland County as recorded by the *Wisconsin Land Economic Inventory* (Bordner, 1942). A complete tally of all the trees on this sample of record-breaking length revealed the following relative densities; aspen–16.7 per cent, sugar maple–15.2 per cent, white birch–12.1 per cent, balsam fir–10.0 per cent, hemlock–8.0 per cent, red maple–8.0 per cent, all pines–0.9 per cent, and eleven other species–29.1 per cent. Much of the land was originally in mesic hardwoods or conifer swamps but a considerable portion was in pine land. The very low representation of pines in the second growth is due to the complete harvest of these species before the fires, and a consequent lack of seed sources. The new forest will thus be of very different composition than the original stand but it will not necessarily be inferior in commercial value. Our utilization of forest products has changed greatly in the past one hundred years and can be expected to change fully as much during the next one hundred before the current crop of trees becomes mature.

CHAPTER 12

Northern forests—lowland

NATURE AND GENERAL DISTRIBUTION

The swamp forests of the northern lowlands have the distinction of being the first vegetation type to be described for Wisconsin. Nicolas Perrot, nearly three hundred years ago in 1667, gave a general account of the swamps, with remarks on their extent and composition. Two centuries later, T. C. Chamberlin in 1877 described them more fully and showed their relations to the glacial topography of the region, as shown in the quotations in Chapter 9. The two segments of the compositional gradient, the wet and the wet-mesic communities, include the tamarack-black spruce bog forests, the white cedar-balsam fir conifer swamps, and the black ash-yellow birch-hemlock hardwood swamps. They are found largely on two topographic types—lake beds and river floodplains. The pitted glacial outwash of north central Wisconsin has numerous depressions whose bottoms are below the surface of the water table. These may be small kettles or potholes in regions of recessional moraines, or larger, more extensive basins in areas of ground moraine or outwash. Many of these are still occupied by open-water lakes, especially in and near Vilas County and in the northwestern portion of the state. Others are partially or completely filled by the action of encroaching vegetation and consist of a spongy, organic substrate called peat, which is formed by the dead but undecomposed remains of mosses, sedges, or woody plants. These peat beds may be covered either by forest or by non-forest vegetation. The latter is usually either open sphagnum bog or sedge meadow, both of which will be described in later chapters.

The forests may occur also on the lowlands which border the rivers and streams of the region. True floodplains, in which the annual or seasonal rise of the water floods the bordering lands and deposits a fresh layer of mineral sediments, are relatively rare in northern Wisconsin. More commonly, the streams meander sluggishly through a valley of varying width or are rapidly flowing with "white water" on numerous cascades and rocky rapids. The latter type commonly has very little streamside vegetation of the lowland kind. The meandering streams, on the other hand, are frequently dammed by beaver or by wind-thrown logs with the resultant formation of ponds and flowages. These ponds accumulate organic sediments or are filled by vegetational encroachment in a manner similar to the glacial lakes. They, also, may be occupied either by open or forest vegetation. The former type, called a "beaver meadow," is one of the characteristic features of the northern landscape. The typical tree of the forested streamside is white cedar although black ash is also very common.

Because of the rather specialized topographic requirements of the northern lowland forests, they tend to be distributed as small discrete bodies and rarely cover any extensive contiguous areas (Figure XII-2). The total area on the map is 2,241,400 acres or 6.40 per cent of the land surface. Certain large or locally well-known swamps include those in the bed of former Glacial Lake Wisconsin in Juneau and surrounding counties, the Cedarburg Swamp in Ozaukee County, the Sheboygan and Manitowoc Swamps in counties of the same names, and the New London Swamp in Waupaca County. Perhaps the greatest concentrations of swamps are to be found in southern Iron County, northern Forest County, Price County and Oconto County.

A number of conifer swamps are found in the prairie-forest province south of the tension zone. These are most numerous east of the Rock River in the glaciated area but a few occur in the Driftless Area on streams that were dammed by outwash from the melting glaciers.

These southern swamps are largely covered by tamarack forest with a groundlayer of herbs and shrubs typical of the region north of the zone. For various reasons these southern tamarack swamps are frequently considered to be relic outliers of the northern forest. They will be so treated in the present chapter since their floristic affinities are obviously with the northern lowlands or even the boreal forest of spruce and fir, but it is recognized that their relic status is by no means definite or simple.

The major P.E.L. studies of the northern lowland forests were made by Christensen (1954) and Jones (1955—see Clausen, 1957). Much of the material in the present chapter is based on these works, supplemented by environmental and other studies by Habeck (1956) and the author. The stands studied are shown in the figure reproduced at the opening of this chapter.

COMPOSITION AND STRUCTURE

The two lowland segments of the northern forests contained 27 tree species (Table XII-1). Only eleven of these attain importance values in excess of 10.0. The wet segment is dominated by black spruce (*Picea mariana*) and tamarack (*Larix laricina*), (Plate 23), with white cedar (*Thuja occidentalis*), balsam fir (*Abies balsamea*), and jack pine (*Pinus banksiana*) as important associates. The cedar and the fir are the leading dominants in the wet-mesic segment (Plate 24), with hemlock (*Tsuga canadensis*), yellow birch (*Betula lutea*), black ash (*Fraxinus nigra*), and American elm (*Ulmus americana*) of secondary importance. Sugar maple (*Acer saccharum*) and red maple (*A. rubrum*) achieve a prominent position in some swamps. All other tree species are so rare as to be called accidental.

The initial stands on the wettest sites show all admixtures from pure tamarack to almost pure black spruce, but the spruce stands tend to be on the firmer peat of more completely filled lake basins, while the tamarack is more frequently found in areas where an open bog mat is still advancing over open water. South of the tension zone, black spruce is absent and tamarack fills both niches, where it is found around open bogs and in extensive forests on well-solidified peat or even on partially mineralized muck, as in the Jefferson County swamps.

The pines are frequent associates of the tamarack and spruce in the far north. Ordinarily, white (*Pinus strobus*) and red pine (*P. resinosa*) do not reach a position of leading dominance under swamp conditions, although white pine may approach this condition in certain areas in Vilas County. Jack pine, on the other hand, may achieve as high as

100 per cent dominance in certain bogs. This may seem surprising in view of the position of jack pine as a typical member of the driest sand barrens in the region, but in the boreal forests of northern Ontario, jack pine is most typically found on poorly drained flats in company with black spruce, Labrador tea (*Ledum groenlandicum*), sphagnum, and pitcher-plant (*Sarracenia purpurea*). The jack pine swamp forests of Wisconsin can be regarded as outliers of this more widespread Canadian community.

The initial lowland forests vary in spacing from savanna-like muskegs with widely scattered trees or clumps of trees on the open bog borders to gradually more dense forests with 60 to 70 per cent cover. In the latter, the tamaracks and black spruces commonly grow in patches with an almost solid canopy, interspersed with open glades carpeted by bog ericads but with no mature trees present. In tamarack swamps south of the tension zone, the marked aggregation is not so evident, with large areas being covered by a fairly uniform dispersion of even-aged trees. Even here, however, the cover rarely exceeds 75 per cent.

In the intermediate or wet-mesic forests, the cover increases to 80 per cent or 85 per cent but the aggregation noticeable in the initial type is accentuated in those stands containing white cedar. This species regularly grows in dense clusters with no other trees present. Within the cluster, the canopy coverage may reach 100 per cent with a very great foliage density. Perhaps no other situation in Wisconsin forests can match these cedar thickets in light interception, with photometer values at the forest floor frequently dropping below 4 foot-candles on a bright summer day. The two other major dominants in the intermediate swamp, balsam fir and black ash, tend to grow in a random manner resulting from the chance establishment of individual seedlings but a truly random spatial dispersion is rarely or never attained in the forests with white cedar, since the cedar clumps effectively prevent the establishment of other species.

Some impression of the internal arrangement of the northern swamp forests can be obtained from the structural figures for two typical stands given in Table XII-2. The number of seedlings and saplings is much lower in the tamarack-black spruce swamp than in the white cedar-balsam fir forest. In fact, the latter stands approach the upland maple forests in numbers of small seedlings per acre, with values over 20,000 commonly attained. Of some interest is the surprisingly high figure for red oak seedlings less than 1 foot high in the cedar swamp. This is not an isolated instance but rather represents a frequent condition. Similar seedlings are often found in tamarack-spruce swamps and sometimes out on the open bog surface. Apparently the acorns are carried from the nearby uplands and either dropped or intentionally buried by some undetected bird or mammal. The seedlings survive for two or more years on the food reserves of the acorn but die when this supply is exhausted.

The average size of the mature trees in the two stands in Table XII-2 is not far from the average size for all of the stands studied. However, this average is far below the ultimate size which the species are capable of attaining. When the Federal land surveyors were laying out the Fourth Principal Meridian through central Wisconsin in 1830 they reported several swamps in which the tamaracks were 3 feet or more in diameter. No records of black spruce of similar size are available, but white cedar may reach this size in swamp environments. The largest cedars, however, are those of the upland forests, particularly along Lakes Michigan and Superior. Unfortunately, almost no information is available as to the maximum age reached by either the swamp conifers or the swamp hardwoods, but numerous specimens have been seen which were between one hundred fifty and two hundred years. It seems probable that ages in excess of three hundred years are attainable, but trees of this antiquity are very rare in the type today.

Because of the lack of fully mature stands it is not possible to give values for the maximum height that may be reached by the swamp forests. Present day stands of all types do not often exceed 60 feet in height and many are between 40 and 50 feet. One outstanding characteristic of the conifer swamps, both tamarack-black spruce and white cedar-balsam fir, is the very uniform level of the tips of the trees. When viewed from an adjacent hill, the surface of an extensive swamp has a sheared appearance which approaches that of a well-kept hedge. Closer examination reveals a few trees whose terminal leaders extend 6 inches to 1 foot higher than the general level but these tips are almost invariably dead. Apparently those trees with extra vigor are cut back to size by wind action akin to physiological tip-burning. The general level of the forest rises only as the average mass of trees elongate, thus providing mutual protection.

LIFE-HISTORIES OF MAJOR DOMINANTS

Tamarack (*Larix laricina*)

This species has the distinction of being the only deciduous conifer native to Wisconsin. Late in September or early in October the needles turn a golden-yellow and shortly thereafter they are shed from the trees. The relative proportions of tamarack and black spruce in the initial lowland stands are most readily appreciated in this season of autumnal coloration, for the deep green of the spruce is accentuated by the bright yellow of the tamaracks.

The trees begin to leaf out in early spring before the ground has thawed. The needles develop slowly, usually requiring four to six weeks for full maturity. The cones flower during the early stages of this development period, while the elongation of the stem does not begin until its close (Cook, 1941). The rate of growth is very dependent upon the substrate conditions. Trees on the open bog border make extremely small diameter increments. On sites with more compacted peat, or better drainage, or both, growth is more rapid. Maximum rates under optimum conditions exceed those for any other swamp conifer. Bruncken (1902) reported an average diameter growth of 0.12 inches per year for trees in a seventy year old stand in Waukesha County with a maximum rate of 0.26 inches per year.

The root system of the tamarack is very shallow but very wide. Depths rarely exceed 1.5 feet, but the spread of the roots commonly is greater than the height of the tree. Adjustment of the root system to the increasing depth of the moss layer on the forest floor is brought about by the production of new roots from the main stem which are above the earlier ones. This capacity for adventitious root production may also lead to a vegetative reproduction of the trees through the development of roots on lateral branches which become buried in the ever deepening moss layer (Bray, 1921).

Tamarack has the rather unusual ability among conifers of producing root sprouts, sometimes at a distance of 25 to 30 feet from the mother tree (Lewis, Dowding, and Moss, 1928). Such root sprouts can become established under conditions definitely unfavorable to true seedlings.

Seeds ripen in the autumn, with 90 per cent shed from the cones by mid-October (Duncan, 1954). The seeds are small and produced in great numbers. Duncan estimated that mature stands might yield as many as 5,000,000 germinable seeds per acre in a year of heavy production. Such good seed crops are borne at intervals of four to six years in Wisconsin. The seeds are dormant and will not germinate until after exposure to the cold of winter. Germination occurs on moss carpets, exposed peat, or mineral soil, if any is present. The main requirement is a constant moisture supply, but standing water over the surface is usually fatal to the seedlings (Duncan, 1954). The tamarack is extremely intolerant of shade and will not grow in its own shade; it succeeds only when full overhead light is present.

As is frequently the case with trees which grow in pure stands, tamarack is subject to epizootic attacks of insects. The main offender is the larch sawfly (*Pristophora ericksonii*). This defoliater caused a serious loss in Wisconsin tamarack swamps between 1900 and 1910 and has become serious on the new generation since about 1950. The extensive swamps in Waukesha and Jefferson counties were largely destroyed during the earlier outbreak, since the normal tendency for regeneration from seedlings was not allowed to express itself by the farmer-owners, who converted the land to pasture or fields wherever possible. The tamarack casebearer (*Coleophora laricella*) has been a serious pest in recent years but it has not caused the death of the trees in most cases.

The bark of the tamarack is very thin and quite fire-sensitive. Even a moderately severe ground fire will cause the death of an entire stand. Fortunately, the habitat of the trees is usually wet enough to discourage ground fires.

Black spruce (*Picea mariana*)

The black spruce is one of the spire-formed conifers, with a tall, narrow crown, tapering to a sharp point. Its evergreen needles are dark green, sometimes with a bluish cast. Ordinarily a small tree, it rarely exceeds 80 feet in height and is usually seen in the range of 40 to 60 feet. Such trees are 8 to 12 inches in diameter and are within thirty years either way of being one hundred years old. Like the tamarack, black spruce is highly variable in growth rate. Open bogs of loose peat may support a stand of scrub trees less than 6 feet high which are eighty years old or more, while trees on good sites may reach 60 feet at the same age (Le Barron, 1948).

Exact age determinations are difficult to make on black spruce trees, because the habit of producing new roots above the old in response to growth of the moss layer is even more pronounced than in the case of tamarack. A given tree may have grown only a few feet in its first decades and then have begun to grow rapidly in response to changing conditions in the swamp. As the ground surface builds up, this portion of the trunk with its retarded core becomes effectively buried and is not available for the usual age determinations made at stump height.

Production of clumps of trees by the layering of branches is very common in black spruce. Many investigators believe that more reproduction occurs by this vegetative means than by seed, but there has been little quantitative work to support or refute the contention. From the standpoint of interaction with other members of the community, the frequent production of a dense circle of little trees around a mother tree growing in open situations is of considerable importance. The heavy needle cover of the young trees creates a dark and protected micro-environment within the clump which effectively eliminates the light-demanding bog ericads and other understory species originally present and allows herbaceous species typical of the mature forest to become established. *Moneses uniflora, Habenaria obtusata,* and other species usually found in cedar-fir swamps thus gain a foothold in the bog forest and contribute to the high degree of aggregation of its groundlayer.

The root system of the black spruce is shallow and wide-spreading, although not as wide as the tamarack under ordinary conditions. Wind throw by uprooting is a relatively common cause of loss in the spruce, especially when growing on shallow peats.

The cones flower in spring, usually in April. They ripen in September or October but do not open completely. Rather, a few open throughout the year and in the succeeding one or two years, so that a rather steady fall of seeds is assured. One-half or more of the seeds in the cones are still present at the end of the first year. Le Barron (1948) in his study of black spruce in Minnesota reported monthly yields of 20 to 40 thousand seeds per acre with total yearly variations over a five-year period only from 198,000 to 439,000 per acre. This slow shedding of the seeds resembles that of the serotinal cones of the jack pine. The response to fire is similar also. The cones are largely localized in the extreme tip of the tree, so that fires which kill the trees rarely do more than scorch the cones, thus hastening their opening.

Great acreages of black spruce forest in Ontario have originated as even-aged stands from such fire liberated seeds. Many of the more extensive swamps of nearly pure spruce in Wisconsin appear to have had a similar origin. Beginning with a mixed, closed stand of tamarack and black spruce, a series of catastrophic destructions of succeeding generations by fire would serve to gradually eliminate the fire-susceptible tamarack and promote the spruce, through this behavior of its cones.

The exposed peat surface that results from a fire is a favorable medium for seed germination. In the absence of fire, seeds germinate and become established best on rotten wood and on certain kinds of slow-growing sphagnum hummocks. Many first-year seedlings are seen in all kinds of sphagnum but these usually are killed by a smothering action of the moss. Regardless of substrate, successful establishment of spruce seedlings requires abundant light, as the species is only slightly less intolerant than tamarack.

One of the unusual parasites attacking black spruce, in addition to the usual butt rots and other fungi, is the dwarf mistletoe (*Arceuthobium pusillum*). This is the only Wisconsin representative of the family producing the Christmas plant of osculatory fame. The dwarf mistletoe plants are only 6 to 12 mm. tall but they may occur in great numbers on the young spruce twigs where they cause an upset in growth-substance relations which leads to the production of witches'-broom and a general stunting of the host plant.

White cedar (*Thuja occidentalis*)

The wet-mesic forests of the northern lowland are dominated by white cedar. This tree has a number of common names, among which arborvitae is the most interesting. It is presumed to have been named by French voyageurs who found that the Indians used the leaves as a preventative against scurvy.

The tree is usually seen in small or medium sizes, although individuals 70 to 80 feet tall and 5 to 6 feet in diameter are known. There is an island in Cedarburg bog in Ozaukee County which contains a number of white cedar stumps all over 4 feet in diameter. The growth rate varies tremendously with the substrate. In most swamp soils, growth is slow although it is better on woody peat than on raw sphagnum peat. White cedar is by no means confined to swamps in Wisconsin. It occurs in extensive forests on the clay banks of Lake Michigan where it achieves maximum growth rates. The tree also can be found on rocky cliffs throughout its range, the root system apparently being well adapted for securing water and nutrients from cracks in the rocks. It occurs on the dry limestone cliffs of the Niagara escarpment in Door County and also on the acidic trap rock of Copper Falls in Iron County and on other rock outcrops near Lake Superior. It is possible that the limestone cliff forms are a different ecotype than are the typical swamp forms.

White cedar flowers in April or May. Its cones are ripened and begin to open from early September near Milwaukee to mid-October near Lake Superior. Like the other conifers, the cedar produces heavy seed crops at intervals of about five years, but the intervening years always produce a fair to medium crop. No figures are available on seed yield per acre.

The seeds germinate readily on a variety of moist substrates. Seedlings become successfully established, however, on a much more limited range of surfaces. Several investigators (Nelson, 1951; Habeck, 1956) have shown that 75 to 85 per cent of the seedlings are found on rotten logs and stumps, although these surfaces occupy only about 16 per cent of the forest floor. The actual density was found to be about 8.6 seedlings per square meter. In contrast, all other types of substrate produce only about 1.1 seedlings per square meter. Sphagnum moss is relatively favorable for germination but tree litter and sedges are poor. On occasion, excellent germination takes place following fire on swamp soils. Apparently the main requirement for successful establishment is a constant supply of moisture. Even brief periods of drought lead to excessive mortality in the seedlings.

White cedar exhibits the same type of vegetative reproduction as that shown by tamarack and black spruce; namely, the production of adventitious roots from lower branches which become embedded in the deepening moss blanket. In addition, cedar has a pronounced ability to continue growth on trees that are tipped by wind. In spite of an extensively developed root system, cedar is very prone to tipping, so much so that perfectly straight trees are a rarity in most swamps (Plate 24). The tipped trees usually go down slowly over a period of many years, so that gradual curves, pistol butts, and other peculiar shapes of the lower trunk are commonly seen. When the tree is finally uprooted, the main trunk comes in contact with the peat surface. The branches which happen to be in a vertical or nearly vertical position at this time continue growing, usually at an increased rate relative to that when they were aerial branches. As the moss blanket smothers the old log, roots are formed at the base of the sapling-like branches and they take on independent existences. The straight rows of even-aged cedars so frequently seen in intermediate swamps are the result of just such a tipping process. When the new trees mature they are capable of repeating the process, so that the actual age of a given tree in terms of years since seed germination is almost impossible to ascertain. This habit is perhaps an even better reason for the name arborvitae than are its antiscorbutic properties.

Black ash (*Fraxinus nigra*)

The black ash is the only important hardwood species which is characteristic of the northern lowlands. All other hardwoods are also found in other communities, usually in greater amounts although yellow birch reaches a numerical peak here. The ash is a medium-sized tree, 45 to 70 feet high and mostly less than 2 feet in diameter. The rather coarse branches grow in an upright fashion and form a narrow crown in the swamp forests.

The tree flowers in May and ripens its fruits in September. Many of the winged fruits remain on the tree during the winter and are gradually shed during that period. When the snow surface is slightly glazed, the winter seeds may be blown for long distances, after the manner of ice boats. Black ash, together with blue ash (*Fraxinus quadrangulata*), and others in the same section of the genus, has rather unusual requirements for germination. The embryo is immature when the seeds are shed; it requires a period of warm temperatures under moist conditions to become fully developed. The fully ripened embryo is still dormant and must be exposed to a period of cold moist conditions before the dormancy is broken. Seed germination in nature, therefore, takes place in the second spring after fruit production. The best germination occurs on consolidated peat or mineral soils with a high organic content. Seedlings rarely reach high densities in terms of numbers per square foot.

Vegetative reproduction has not been observed to be of importance in black ash, although the species has good abilities to sprout from cut branches or sapling stumps. Heavy utilization by deer in some swamps produces a much branched, bushy growth, but one year's protection from attack is frequently sufficient to enable the tree to send out an erect branch which is tall enough to escape further browsing injury. The annual length increment on some of these sucker shoots frequently exceeds 5 feet.

GROUNDLAYER

The prevalent groundlayer species of the wet and wet-mesic segments are listed in Tables XII-3 and XII-5, together with the modal species of lower presence. The total flora of the wet stands included 193 species, while the wet-mesic had 228. The index of similarity between these two segments is 50 per cent, which is considerably lower than the corresponding figure between the pine forests of the dry and dry-mesic segments. In fact, the wet forests are more closely related in a floristic sense to the open bogs than they are to the wet-mesic forests. Among the groundlayer species which have significantly higher presence percentages in the wet than in the wet-mesic stands and which may be used as differential indicators of the type are the ericaceous bog shrubs, like Labrador tea and bog Rosemary, the cotton grasses, and the three-leaved Solomon's seal. All of these, however, are equally or more common in the open bogs.

With regard to phenology, a relatively high percentage of the northern swamp plants produces their flowers in the early part of the season. Considering the two segments together, 64 per cent of the prevalent understory species bloom before June 15, 45 per cent between June 15 and August 15, and 9 per cent after August 15. All ericaceous shrubs, most other shrubs, and many herbaceous plants bloom in the spring. A number of the orchids, especially *Habenaria* and *Listera*, bloom in summer, as do *Pyrola, Moneses,* and *Monotropa* of the Ericads. Members of the mint and composite families are about the only plants to bloom in late summer or early fall, especially the genera *Scutellaria* and *Aster*. Most of the groundlayer species have small and inconspicuous flowers, so that the floral display is usually restrained. Perhaps the most colorful scene is exhibited in the initial tamarack-black spruce swamps when the bog shrubs are in bloom in late May, especially the bog laurel (*Kalmia polifolia*), bog Rosemary (*Andromeda glaucophylla*), and Labrador tea (*Ledum groenlandicum*).

A floristic analysis of the understory plants of the two forest segments reveals that nine families are of highest importance, with the top three being the *Cyperaceae* (13.5 per cent), *Compositae* (6.7 per cent), and *Ericaceae* (6.7 per cent) in the wet forests and the *Compositae* (9.2 per cent), *Liliaceae* (6.1 per cent), and the *Cyperaceae* (5.7 per cent) in the wet-mesic stands. The grasses, orchids, and ferns are included in the group of nine families containing 50 per cent of the species in both segments. Altogether, 12 species of orchids were found in the P.E.L. study, although a number of other species are known to occur in conifer swamps but they are so rare that they were not found in this study. Included in this group are: *Calypso bulbosa*, reported to be quite common in white cedar swamps by Dr. and Mrs. C. W. Finnerud of Minocqua, but missed by most investigators because of its very early blooming season; *Cypripedium arietinum*, one of the rarest American orchids; *Habenaria dilatata,* which is locally abundant but very spotty in occurrence; and *Malaxis monophyllos,* which is exceedingly inconspicuous. Altogether, Fuller (1933) reported 25 species of orchids as sometimes occurring in northern lowland forests.

Very little is known about the detailed life-histories of the typical plants of the wet and wet-mesic forests. One characteristic of the ground flora as a whole is the richness in plants with evergreen leaves Accurate statistics are not possible because of lack of knowledge about the evergreenness of sedges and other plants in the sedge family. If this group is excluded, however, 12.3 per cent of the remaining species are evergreen or have green leaves through the winter.

The two most characteristic families, *Ericaceae* and *Orchidaceae,* are very similar in many phases of their life cycles. Both have very small seeds, frequently consisting only of an undifferentiated embryo suspended in an airfilled sac or testa. Germination under ordinary conditions is dependent upon the intervention of a soil fungus (Curtis, 1939) which provides the carbohydrates and perhaps other substances needed for development. Growth is very slow, with periods of 12 to 16 years from germination to maturity common in the ladyslippers of the genus *Cypripedium* (Curtis, 1943). Both families have complicated mechanisms for the transfer of pollen from anther to stigma through the aid of insects. In the orchids, the roots are infected with endotrophic mycorrhizal fungi, while in the ericads, the entire plant is so infected. Most of the above characteristics are usually thought to be detrimental to survival; it is not easy to see how they adapt the plants to the swamp environment. Perhaps the main factor is concerned with the need for high organic content in the soil during the young stages. The peaty soil of the northern swamps certainly provides this factor in abundance.

For the most part, the ground flora is very shallowly rooted; many of the roots are in the still-living top layer of mosses and sedges. Some of the species, however, have rather special requirements which influence their root habit. Fuller (1933), in his book *Orchids of Wisconsin*, illustrates an interesting situation in *Cypripedium reginae*. This is by far the most beautiful orchid in Wisconsin and is occasionally found in abundance in cedar forests. In all such cases, Fuller found that the plants were rooted in an underlayer of neutral or alkaline marl or clay although the surface layers may have been composed of sphagnum moss or peat which was highly acid.

The shrub portion of the groundlayer in this forest is rich in species which produce berries. Nearly three-quarters of the shrub species found in the Wisconsin studies had berries or other fleshy fruits. In addition, 22 or 9.5 per cent of the herbs had berries. Such plants as the cranberries, blueberries, snowberry, bearberry, and wintergreen in the ericads, the viburnums, honeysuckles, hollies, dogwoods, gooseberries, and dewberries are typical examples. They perhaps help to explain why the northern lowland swamp is a favorite habitat for the black bear.

One of the berried shrubs worthy of special mention is the poison sumac (*Rhus vernix*). This is a tall bush with colorful leaves and attractive clusters of white berries and is very commonly found around the borders of conifer swamps, in the moat between the swamp and the upland, especially in the southern portion of the state. Either it produces more of the irritant toxic oil than common poison ivy or gives it off more freely, since many people who are reasonably immune to the ivy are susceptible to the sumac. Actually, many of the better southern bogs owe their botanical excellence to this protective barrier of poisonous shrubs, since even their owners are hesitant about passing through the pale.

While there are many characteristic flowering plants, it is the non-seed plants which are truly the hallmark of the conifer swamp. Various species of sphagnum moss carpet the forest floor in the initial stages. As dense cover develops, the sphagnums give way to feather mosses of the genera *Hypnum, Thuidium, Aulocomnium,* and to such forms as *Climacium, Bryum,* and other clump producers. The leafy liverworts, too, are abundant and often conspicuous, especially *Bazzania trilobata.* The trunks of the trees are hosts to many corticolous bryophytes, both mosses and liverworts, but their special glory is the almost solid blanket of lichens. The total number of lichen species is perhaps not greatly in excess of that of other forest types, but the individual species reach density levels not found in any other plant community in Wisconsin. The ferns are represented by numerous species, their total being exceeded only by that of the rock cliff communities.

The herbs of the northern lowland forest do not commonly produce large clumps by vegetative clone formation. There are a few prominent exceptions, such as cinnamon fern (*Osmunda cinnamomea*) and certain of the Lycopodiums, which do produce dense clumps. The lack of innate tendencies toward aggregation is more than counterbalanced, however, by a secondary aggregation induced by the presence of clumps of black spruce and thickets of cedar. Openings caused by wind throw frequently involve a group of neighboring trees and the clear spaces which result also tend to increase the unevenness of distribution of the groundlayer. All in all, the lesser plants are about as non-random as those in other forest types.

One of the interesting features of the northern lowland forest is its use as wintering range by the white-tailed deer. Under severe winter conditions the deer tend to gather together into tight social groups for mutual protection and probably for mutual compaction of deep snows, in areas termed "deer yards." These yards must offer both food and cover if they are to permit significant survival of the animals using them. White cedar is the single tree offering maximum amounts of these factors. Christensen (1954) found that 76 per cent of the lowland yarding areas that he investigated were conifer swamps dominated by this species.

Plate 23.—Northern wet forest, dominated by tamarack (*Larix laricina*) and black spruce (*Picea mariana*). Northern Highlands State Forest, Vilas County.

Plate 24.—Interior of northern wet-mesic forest, showing tipped trees of white cedar (*Thuja occidentalis*). Trout Lake Cedar Swamp Scientific Area, Vilas County.

Plate 25.—Bog mat growing over Devil's Track Lake, Vilas County. Photograph made in 1940.

Plate 26.—Repeat photograph from same spot, made in 1957. Note great change in height of conifers on the uplands, but slow advance of the mat over the water.

STABILITY AND SUCCESSION

The successional development from open lake to conifer bog is a familiar one that has been described by numerous investigators. One of the first complete accounts for the bogs of this region is that of Transeau (1905) in Michigan. The process in Wisconsin is essentially the same as that reported by Transeau. A characteristic feature of many of the small, non-outlet lakes of northern Wisconsin is a band of vegetation that floats out over the water around the shore of the lake bed. This is commonly arranged in a concentric or zonal pattern, with the open bog community toward the center, having sedges or sedgelike plants (especially *Carex lasiocarpa*) immediately at the water's edge. These are backed by an area of low vegetation made up of a mat of sphagnum moss that has a covering of leatherleaf (*Chamaedaphne calyculata*), bog Rosemary (*Andromeda glaucophylla*), cotton grasses (*Eriophorum* sp.), scattered individuals of pitcher-plant (*Sarracenia purpurea*), and sundew (*Drosera rotundifolia*). Farther back there is a ring of scattered saplings of tamarack and perhaps black spruce. Still closer to the original shore, the closed forest of tamarack and black spruce begins. The mat is supported by the buoyancy of the roots and rhizomes of the sedges and cotton grasses, but the increasing weight of the sphagnum and shrub mat gradually pushes the base down until it touches the solid bottom of the lake bed. Continued growth of the surface layers adds more material and compresses the lower layers until they become semisolid peat. Usually this partial solidification begins before the trees invade the mat. In any case, it is well developed by the time a closed forest is produced.

It is possible to determine the rate of advance of this concentric series of communities by aging the trees along radial lines from the open water to the solid shore. Several such determinations in Wisconsin have shown that the mat and the forest advance at only about 30 cm per year. Plates 25 and 26 are from Devil's Track Lake in Vilas County in 1940 and again in 1957. They show the slowness of the encroachment very well, especially in the narrows between the two sections of the lake.

Certain adverse factors may serve to slow the growth of the mat. Windstorms of great violence sometimes break large portions of the mat loose. In rare instances, these loosened portions will contain some of the young tamaracks and spruces. They then form the floating islands which are reported from time to time in the local press and are so disconcerting to the cottage owner who wakes up some morning to discover his newly built pier surrounded by a young forest.

Most of the initial forests of tamarack and black spruce described above were formed originally by this classical process of primary plant succession. In many instances, an open water pool of some size is still present in the center of the forest. In other cases, the process has gone to completion and the entire basin is covered with forest. The trees that are visible today in these last instances, however, are not the same trees which initially invaded the open mat since the forest is highly subject to catastrophes of various kinds, especially wind throw, fire, and changing water level. There is a kind of internal succession going on constantly from the wind-tipping of individual trees. The openings created by these downfalls allow sufficient light to enter so that both tamarack and black spruce seedlings can become re-established. Occasional storms of greater intensity may open up large areas for recolonization. On those basins which have an outlet, flooding by beaver dams is an important cause for regression. The mature trees are drowned and the succession is set back to the bog shrub stage.

Fire is an important cause of swamp destruction, particularly on extensive swamps which allow a full sweep of the wind. During drought periods the trees, most other woody plants, and sometimes the upper layers of the peat as well, are destroyed. Return of normal water levels may cause the development of a particular type of vegetation called a sedge bog. This is made up of a surface layer of very loose sphagnum with a relatively even cover of mixed leatherleaf and sedges. A good example of such a community is the Powell Marsh, an extensive tract lying on the border of Iron and Vilas counties just north of the Lac du Flambeau Indian Reservation. Analysis of the peat here showed charcoal fragments in layers all the way from the top to the underlying sand.

When the bog basin is partially drained by the down-cutting of the outlet stream, then the surface layer of the peat begins to oxidize and turn to muck, a black, amorphous, almost completely organic material. Such oxidized peat is a favorable place for the growth of alder (*Alnus rugosa*) and alder thickets commonly form, especially following fires (see Chapter 17). One of the common invaders of such alder thickets is the white cedar. It appears that many of our cedar forests have originated in this way. However, the cedar may seed in directly on the burned peat without the intervention of alder. The dense, even-aged thickets of cedar so common in Marinette and other northeastern counties are probably of this type.

In the absence of fire or water-level changes, the initial forests are probably invaded by cedar only very slowly and only after the peat has had opportunity to become consolidated and firm. The vegetative reproduction of cedar is important in this connection, since the few successful cedar seedlings which become established on logs in the tamarack-black spruce swamps are then enabled to spread out and engulf the original trees by vegetative means (Clausen, 1957). Once entrenched, the cedar is able to prevent further regeneration of the intolerant tamarack or black spruce within its dense clumps. By this gradual means, the forest becomes dominated by white cedar, sometimes almost to the total exclusion of all other species. In most instances, however, balsam fir, which is extremely tolerant of shade as a sapling, is present in admixture with the cedar.

Such intermediate forests, like the earlier ones, are capable of perpetuating themselves for very long periods. Wind throw of the cedars only results in rapid regeneration by the branches, so that few new species have the opportunity to invade. Further succession is probably dependent upon gradual changes in the peat, changes which lead to a decreased water-holding capacity and a general trend toward mesic conditions. Black ash is one of the first trees to benefit from these changes and forests dominated by ash represent the next stage in a normal succession. The breaking of the continuous evergreen cover by the invasion of the deciduous ash leads to considerable changes in the understory, particularly in the amount and kinds of shrubs. Other deciduous trees follow quickly, including the red maple and yellow birch. This transitional stage is probably of very short duration, which may help to explain why concrete examples are so difficult to find in the field.

ENVIRONMENT

The major environmental feature which distinguishes the northern lowland forest from the other forest types of Wisconsin is the prominence of an organic or peaty substrate. All other forests have a mineral soil. As a generic term, peat refers to the partially decomposed plant remains which form stratified layers in old lake basins or other wet grounds. As seen above, peat may originate from a floating mat, but this is not always the case. Lowlands along rivers or around beaver ponds may be invaded directly by sedges, whose subterranean parts may give rise to substantial layers of peat, and other permanently moist basins may also produce peat beds in the absence of open water. The underlying sediments in any of these cases may be of sand, clay, or marl. Their chemical nature has little if anything to do with the kind of peat that eventually may form above them.

The classification of kinds of peat is a complicated subject (Dachnowski, 1912; Huels, 1915). Oversimplifying the situation, it can be said that four main types are common in the northern swamps of Wisconsin—sedge peat, sphagnum peat, hypnum peat, and woody peat, with their origins being indicated by their names. Pure beds of any one type are very rare, except for certain extensive areas of sedge peat. Usually there is an alternation of kinds and frequently an intermixture within one layer. The alternations are a clear indication of past retrogressions in the successional history, and the frequent occurrence of layers of charcoal between the peat types is a clue to the cause of the reversals. In general, the peats toward the surface of a bog tend to be loose and spongy. At greater depths, the peat is more consolidated and has a lower water-holding capacity, although most peats from any depth will hold over six times their own dry weight of water.

The peat substrates of the wet forests are briefly characterized as being highly acid, very wet, and very low in oxygen. The reducing conditions are of sufficient intensity to cause the precipitation of "bog iron" and "bog manganese," often in the form of large masses. Undetermined elements which emit alpha radioactivity are also present in high concentration relative to the usual amounts in ordinary soils (Curtis and Dix, 1956a). The anaerobic conditions are largely responsible for the lack of decay in the peat, as the most active micro-organisms are unable to function in such an environment. The water-retaining capacities averaged 670 per cent in the P.E.L. samples.

The soils of wet-mesic forests are less extreme in their acidity, have much lower water contents (with water-retaining capacities usually less than 300 per cent on a dry weight basis), and have moderate levels of oxygen. As a result, the surface layers of peat become oxidized to muck. This is frequently accompanied by an appreciable mineralization brought about by long continued overwash of mineral salts leached from the surrounding uplands.

The climatic conditions of the northern swamps are greatly influenced by their topographic positions (Tables XII-4 and XII-6). Lying as they do in undrained basins or other low grounds, the swamps receive cold air drainage, especially in the summer when the uplands cool off rapidly by radiation. The cool air finds its way to the low areas and very frequently reduces their temperature to below the dew point. Under such conditions a dense night fog develops which bathes the swamp plants in a saturated atmosphere for periods of eight to ten hours. Considerable condensation on to the twigs and leaves takes place. The abundance of lichens and bryophytes is due largely to this nocturnal condition of high moisture.

In the open bogs and the sparsely wooded muskegs, the night temperatures may drop to below freezing levels on almost any clear night during the summer when suitable air masses are present. As a result, only the most frost-hardy plants are able to grow. Many of the same plants are present in the forest, of course, but it is doubtful if they are subjected to frost during the growing season. Unfortunately, no records of the seasonal, internal microclimate of northern lowland forests are available. A series of single readings, made on a number of days in the summer of 1935, indicated the midday air temperatures in a white cedar swamp in Vilas County to average 67.6°F., in contrast to 59.3°F. at the surface of the peat substrate and 79.1°F. in the open on the surrounding uplands. These figures, however, fail to indicate another important feature of the microclimate of the conifer swamps. This is the great shortening of the growing season caused by the insulating qualities of the sphagnum moss on the forest floor and the consequent retardation of the melting of the ice in the spring. It is not uncommon to find ice at a depth of less than 1 foot in mid-June in the northern counties and ice has been found in the first week in July on several occasions. The tops of the plants may be subject to high temperatures at the same time the roots are bathed in ice water. The bog ericads such as Labrador tea (*Ledum*), bog Rosemary (*Andromeda*), and bog laurel (*Kalmia*) are adapted to such conditions and may even burst into flower while the ground is still solidly frozen, but most ordinary plants would be unable to survive such severe stress. The xeromorphic characters of many of the swamp plants have been interpreted as an adaptation to these unusual conditions.

The same insulating qualities of the sphagnum which hold the ice in spring also prevent its rapid formation in the fall, so that the ground is commonly unfrozen by the time the snow blanket comes. Deep penetration of frost frequently does not occur until February.

The general influence of the forest community on its microclimate is thus a moderating one. Summers tend to be shorter, cooler, and more humid than the surrounding lands, while winters are warmer and of more even temperature.

GEOGRAPHICAL RELATIONS

The conifer swamp has always been considered an exceptionally distinct plant community, with great similarity throughout its range.

As Transeau pointed out in 1903, this type reaches best development in the glaciated portions of the continent. At any given region within this large area, the bogs are usually sharply delimited from the surrounding vegetation because of their peculiar topographic requirements for poorly drained basins. Furthermore, a given swamp is likely to be very similar to another swamp in the immediate vicinity. If one examines swamps which are more remote, however, then gradual differences begin to appear, with the differences roughly proportional to the distance. This change over distance is shown in Table 13, which is based upon published accounts of the floristic composition of conifer swamps from the mountains of Maryland to the northwest corner of Alberta,

Table 13

Geographical relations of Wisconsin conifer swamps

Location of study	% of flora in common with Wisconsin stands	Approximate distance from Wisconsin (miles)
N. Minnesota (Conway, 1949)	96	200
N. Michigan (Gates, 1942)	83	220
N. Illinois (Waterman, 1923)	91	250
S. Michigan (Transeau, 1905)	63	330
N. New York (Bray, 1921)	65	770
Maryland (Shreve. 1910)	58	830
NW. Alberta (Moss, 1953b)	46	1600

which is adjacent to the Northwest Territory. These studies varied considerably in scope and intensity and resulted in an incomplete listing of total flora in many cases. Thus, the most meaningful comparisons are those made on a relative basis. In this case, the species listed by each author which are also present in the northern lowland forests of Wisconsin are expressed as a percentage of the entire complement of species. As can be seen, the greatest resemblances are between Wisconsin and the immediately surrounding states, with values above 80 per cent for Minnesota, Illinois, and northern Michigan. As the distance increases, the similarity decreases. Both Maryland and northwest Alberta have the same number of species in common with Wisconsin, but these two remote areas have no species in common with each other except for *Larix laricina* and *Picea mariana*. Both also have some species of *Sphagnum*.

The evidence indicates that the apparent unity of the conifer swamp is really a physiognomic unity, imparted by the two trees, tamarack and black spruce. The presence of one or more varieties of sphagnum and the general similarity of certain unusual edaphic features also help to bind the community together. In spite of this, there is a complete shift in composition of the understory from one extreme to the other of the geographical range in North America. Wisconsin swamps, being in the center of this range, may be considered to be typical of the entire group, since they show partial or high similarities to all.

UTILIZATION

The lowland conifer swamps in the southern part of the state have been used differently than those in the northern. The great majority of the southern areas have been totally destroyed by conversion to agricultural crops through the use of extensive drainage systems. In the late 1800's, many of the tamarack swamps were cut over and converted to mowing meadows or pastures, while some of those most easily drained were changed to cropland for corn and forage. The drainage boom culminating in the 1920's produced expensive and extensive drainage schemes, with complicated, semipolitical district organizations. Many of the schemes failed for lack of suitable crops adapted to the exceptional soils and the microclimate of the swamp basins. The major exception was in the area adjacent to Milwaukee, Racine, and Kenosha where cabbage, celery, and other market vegetables were produced. This intensive use was extended in later years to other counties. In the period following World War II, the drainage program has been greatly expanded. Onions and potatoes have been major crops, but increasing areas in Jefferson and Marquette counties have been devoted to peppermint and head lettuce.

The peat in the swamp basins rapidly becomes oxidized to muck following drainage. Continued cultivation causes progressive oxidation and these losses are augmented by losses due to severe wind erosion. As a result, the years of potential agricultural productivity are definitely limited. It is probable that the high water-retaining capacity of the original peat and its consequent ability to store water and temper the flow of streams would have been of greater value to the entire economy of the state than any temporary returns gained by a few owners at the price of destruction of the reservoirs.

In the north there were a few attempts at swamp drainage also, but the generally more severe climatic conditions prevented any successful agricultural utilization. The main destructive agent in the north was fire but the excellent fire protection system in effect since

1930 has nearly stopped further damage from this source. Currently, the northern lowland forests are used largely as sources of wood products. White cedar is the main species, with a relatively constant demand for fence posts, fence pickets, and other uses where its superior decay resistance is of value. Black spruce is used as a source of paper pulp in areas where extensive stands are present. Both the cedar and spruce are selectively cut, with little or no attention to the silvicultural needs of these or other species. In general, the northern swamps are not managed, but are merely tolerated and exploited for what they will produce.

A highly specialized use of the conifer swamp is made by a certain highway commissioner in one of the northern counties. On his daily rounds to inspect construction and repair jobs, he frequently becomes very thirsty. To allay this thirst, he draws on his practical knowledge of ecology and microclimatology by caching a case of bottled beer in the sphagnum mat of a number of widely dispersed tamarack swamps. In this way he can be sure of a drink of cold beer within easy distance of any job. Should this practice become widespread, future paleobotanists examining the peat for fossil objects may receive some surprises.

CHAPTER 13

Boreal forest

NATURE AND GENERAL DISTRIBUTION

The boreal forest is actually a circumboreal formation, with very close similarities between the Eurasian and the North American communities. The forest in this hemisphere is largely Canadian in its distribution, with lateral extensions southward on the Coast Ranges and the Rockies in the West and on the Appalachians in the East. There is also a limited development around the Upper Great Lakes, mostly in northern Minnesota and Michigan. Its occurrence in Wisconsin was very restricted in presettlement times (672,300 acres, or 1.92 per cent of land area), when the best development was along the shore of Lake Superior and near the tip of the Door County Peninsula in Lake Michigan (Figure XIII-2). These stands were noted by several of the early geologist-explorers, one of whom was Whittlesey (1852) who said:

In this region [Penokee Range], we have a state of things in regard to soil, the reverse of what is usual in mountainous countries. The best soil is on the mountain ridges and slopes. The low grounds are generally swampy, and covered with thickets of tamerack, birch, white cedar, balsam, and spruce, and occasionally pine. At the foot of the mountain ranges, where the slope graduates into the low lands, there is an abundance of hemlock, and the ground is covered with moss, and a shrub called ground hemlock [*Taxus canadensis*], to the continual annoyance of the traveller. Higher up the slopes, and on the summit, the prevailing timber is sugar maple, of a strong, heavy growth, a few yellow birch and pine trees interspersed.

He discussed the interrelations of climate and soil along the Lake Superior shore as follows:

It is a universal rule, that the immediate coast of the Lake, and, in general, the deep gulfs through which some of the rivers and streams flow, produce a tangled forest of cedar, spruce, balsam, and birch, much more forbidding than it is a few miles back from the Lake. The moist atmosphere next the water, and the increased circulation and force of the winds, together constitute a local climate, which is favourable to those hardy evergreens, and to the birch. It may be said that between the summits of the mountain ranges and the level of the Lake, there are three climates, indicated by the changes in the growing timber. Where the soil is good, the highest portions produce sugar maple, black oak, and white pine. Towards the base of the most elevated ridges, hemlock begins to flourish, which graduates into cedar, balsam, and spruce, on the swampy portions adjacent. On the red clay plains, corresponding in level with the swampy portions, spruce, dwarf pines, balsam, aspen, and birch, spring up very thick, the result of a peculiar and tolerably good soil. This thick wood, extending over a large tract nearly on a level, serves to check the winds, and protect both animals and the soil, in some measure, from the severity of winter. But I do not instance this peculiarity as a climatic result. It belongs rather to the soil, and the physical characteristics of the flat clay region.

Sweet (1880) reported a similar mixture of trees for Douglas County:

Along the lake shore in Douglas county and extending into Bayfield county, and reaching back as far as the Copper Range, there is a very dense growth of small trees consisting of about equal numbers of poplar, birch, cedar and balsam. These trees are ordinarily less than a foot in diameter, and from twenty to eighty feet in height. In this area, large white pines are frequently found, as are also the red or Norway pines, and sometimes elm, basswood, spruce and tamarac. The underbrush and vast numbers of fallen trunks of trees, make traveling in this area extremely arduous. On the Copper Range, and on the high ridges to the east and south of it, the prevailing growth is sugar maple, black oak, birch, poplar and yellow pine [*Pinus resinosa*]. These trees are generally less than eighteen inches in diameter and are not nearly so close together as those farther north, just mentioned. Scattered among them occur cedar, tamarac, balsam and white pine trees.

The presence of tamarack in the upland forests was noted by Owen (1852):

Tamerack (*Larix Americana*) and cedar are the prevalent growths on the head waters of the St. Croix. Of the former, there are two species, or at least varieties; one, which frequents the wet, swampy ground; and the other, the low ridges. The leaves of the former are of a grayish green colour, and have the appearance of moss, at a distance, and the tree has a rugged look. The latter has a dark green colour, runs up into a pointed summit, and its whole outline is much more formal. I am not certain whether this is the *Larix microcarpa* of Lambert. On the southeast is a ridge similar to the one crossed on the portage from the Brule to the St. Croix, supporting a growth . . . of pine. This timber, however, is soon replaced by tamerack. From the appearance of these ridges, they would probably afford good ground for a road up the valley of the St. Croix.

Practically no remaining examples of the original boreal forest exist in Wisconsin, except for a few small remnants in the Red Cliff Indian Reservation which are rapidly being cut. However, extensive secondary stands exist along Lake Superior and also inland in Vilas, Iron, Ashland, and Bayfield counties. Most of these stands are on heavy soils which are frequently quite wet. A number of the better stands are found in deep drainage ravines in the Superior red clays or in pockets and kettle holes north of the Valders Moraine.

Due to the scarcity of stands for study in Wisconsin, the P.E.L. investigation included the adjacent portions of Minnesota, Michigan, and Ontario. This study was conducted by P. F. Maycock (1956, 1957) and included data from 110 stands, 49 of which were in Wisconsin and are shown in the figure reproduced at the opening of this chapter. The only other P.E.L. study of the boreal forest is that by Jones and Zicker (1955), who reported on the structure of a stand on the Rahr Reserve in Ontanagon County, Michigan. A number of investigations have been reported for other nearby stands; these include Cooper's study of Isle Royale (1913) and the papers by Buell and Niering (1957) on the region around Lake Itasca. The present discussion is based largely on the P.E.L. data for the Wisconsin stands, but comparisons with the broader study are included, primarily to show the community of optimum presence of many species found in Wisconsin.

STRUCTURE AND COMPOSITION OF TREE LAYER

The average presence and importance values for the trees in the Wisconsin stands are listed in Table XIII-1. The stands selected for study were required to have either balsam fir (*Abies balsamea*) or white spruce (*Picea glauca*) as a member of the dominant canopy layer. As it turned out, both of these species were present in every stand, but the average importance of the fir far exceeded that of the spruce. White pine (*Pinus strobus*), white cedar (*Thuja occidentalis*), and white birch (*Betula papyrifera*) all had importance values slightly higher than did white spruce. Three species of aspen (*Populus*) and three of maple (*Acer*) were other important members of the community.

The borderline or ecotonal nature of the Wisconsin boreal forest is well shown by the trees. Only 9 of the 32 total species are conifers, and they contribute only 35.7 per cent of the total sum of constancy. The importance of the hardwoods might appear to justify the classification of this aggregate group as a boreal-hardwood forest, but many of the stands are closer to the normal boreal type, because conifers are almost the sole dominants. This is shown by the average figures for importance value, where the conifers represent 63.3 per cent of the total sum of I.V.

The boreal forest in Wisconsin is present in three partially distinct types—the first, the old stands of relatively pure conifers with balsam fir and white spruce as the major dominants, associated with large quantities of white pine, red pine (*Pinus resinosa*), or white cedar as found especially along Lake Superior (Plate 27); the second, the mixed conifer-hardwood stands, particularly on inland mesic sites, with the shade-tolerant hardwoods gradually replacing the spruce and fir; and the third, the young stands of dense balsam fir and white spruce under an aging and decadent canopy of trembling aspen or white birch (Plates 28 and 29), as found throughout the range. The majority of the hardwood species in Table XIII-1 are found in the second type. The hardwoods associated with the first type are white birch, mountain-ash (*Sorbus americana*), red maple (*Acer rubrum*), and mountain maple (*A. spicatum*), while the third type usually has only white birch or one of the aspens or poplars, with balsam poplar (*Populus balsamifera*) occasionally reaching significant levels of importance. Obviously, the last two types do not represent distinct and stable entities but reflect successional recovery from recent disturbances.

Plate 27.—Mature stand of white spruce (*Picea glauca*) and balsam fir (*Abies balsamea*). Ridges Sanctuary and Scientific Area, Door County.

Plate 28.—Young stand of spruce and fir under mature aspens. Northern Highland State Forest, Vilas County.

Plate 29.—Intermediate stand of spruce (*Picea glauca*) and fir (*Abies balsamea*) which has developed under a former cover of aspen. High Lake Scientific Area, Vilas County.

Plate 30.—Young stand of spruce and fir under maturing jack pine (*Pinus banksiana*). Vilas County.

The life-histories of the two major coniferous dominants—white spruce (*Picea glauca*) and balsam fir (*Abies balsamea*)—are comparable in many ways. Both trees have a similar life-form, with a very slender, spirelike crown, and a very slowly tapering trunk. The white spruce regularly reaches larger sizes than does the fir, with individuals more than 110 feet tall and over 28 inches d.b.h. on record. The largest fir found was 24 inches d.b.h. but this was highly exceptional, because most of the large trees of this species are less than half this size. Both species flower in late May and ripen their cones in September. The white spruce sheds its seeds very rapidly, with practically all gone before winter. Heavy seed crops occur at four to six year intervals in both species, but some seeds are produced every year. A very thorough study was made by Place (1955) of the influence of seedbed conditions on the regeneration of spruce and fir in New Brunswick. He found that light intensities greater than 10 per cent of those in the open were necessary for appreciable growth of the seedlings. The larger seeds of the fir gave larger seedlings with a deeper taproot than the spruce; hence the fir was better able to establish itself on thicker layers of litter (up to 3 inches thick). Thin layers of living moss provided a very favorable substrate, although germination was also good on moist mineral soils. The major deterrent to successful establishment was an extended drought period during July or August of the first year.

These establishment characteristics seem to be true of the Wisconsin strains also. They permit both species to pioneer into open sites but indicate a more favorable result when an overstory of sparsely-canopied trees such as aspen or white birch are present. The balsam fir is subject to destruction by epidemics of the spruce budworm (*Archips fumiferana*), especially when the tree is present in high densities over large areas, as it is in parts of Canada. In Wisconsin, both fir and spruce are subject to attack, usually without serious consequences. A number of heart rot fungi seem to be the main agents of destruction of the fir. They attack trees of medium size which have suffered from ice-breakage or other injury. Both species are shallow rooted, with the fir slightly more subject to wind throw than the spruce. Both are also quite shade tolerant, but the fir seems to be able to withstand suppression better than the spruce. It is very widely distributed throughout the northern mesic hardwood forests where it attains relatively high densities in the sapling class but rarely enters the canopy layer.

The major hardwood associates of the spruce and fir are white birch (*Betula papyrifera*) and trembling aspen (*Populus tremuloides*),

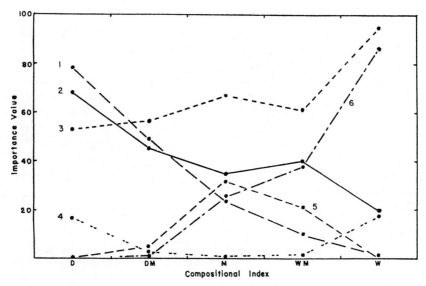

Figure 16.—Behavior of major trees in the boreal forests of the Great Lakes region, plotted on moisture gradient from data of Maycock (1957). 1, white pine (*Pinus strobus*); 2, white spruce (*Picea glauca*); 3, balsam fir (*Abies balsamea*); 4, black spruce (*Picea mariana*); 5, sugar maple (*Acer saccharum*); 6, white cedar (*Thuja occidentalis*).

both of which also are widely distributed in the pine-hardwood forests to the south. A less common but much more characteristic species is the mountain-ash (*Sorbus americana*). The clusters of bright red berries on this species are a conspicuous sight in the boreal forest in autumn and early winter. The fruits are particularly favored by such birds as the grosbeaks and usually are consumed before midwinter. Ordinarily, the mountain-ash is a small tree 6 to 10 inches d.b.h., but trees up to 2 feet d.b.h. have been reported from the Chequamegon Bay area on Lake Superior.

In Maycock's study of the Great Lakes boreal forest, the stands were arranged along a moisture gradient, from wet through mesic to dry, both by subjectively classifying the sites and by an ordination technique based on degree of similarity which used the $2w/a+b$ test. The results were highly correlated in the two methods. The behaviors of the major trees along this gradient are shown in Figure 16. Spruce and fir are seen to differ as mirror images of each other, with spruce favoring dry sites and fir favoring wet sites. Two trees, black spruce and jack pine, have bimodal curves, with peaks in the two extreme conditions. A considerable number of herbs and shrubs, including *Gaultheria hispidula, G. procumbens, Ledum groenlandicum, Polygala paucifolia, Rubus pubescens, Vaccinium angustifolium,* and *V. myrtilloides* also showed this bimodal pattern. Similar distributions were noted in the northern pine-hardwood forest (Chapter 9), but they were neither so numerous nor so distinct.

The Wisconsin stands were mostly located along the central half of the entire compositional gradient, with a few wet stands, but almost none at the dry extreme.

GROUNDLAYER

The prevalent species of the Wisconsin stands are shown in Table XIII-2 with the modal species of low presence in the same stands. The modal species were chosen on the usual basis, namely that their presence values in this community were higher than in any other in the state. Actually, many other species reached optimum levels of presence in one of the moisture gradient segments of the extended study of the community in the upper Great Lakes region. These are shown in Table XIII-3 for their value in an understanding of the true position of many species of the northern forests in general. Some of these species might not have been classified as truly modal if comparable studies of the pine forests and conifer swamps of the other regions were available, but their generally high levels of presence in the boreal stands indicates their high importance there in any case.

The essential unity of the boreal forest in Wisconsin is shown by the index of homogeneity of 57.7 per cent and by the fact that 43.1 per cent of the prevalent species were also modal. Another indication of the unity of this forest in general is shown by the exceptionally high number of ubiquitous species. Altogether, there were 15 species which had a presence value of 40 per cent or more in each of the five moisture segments of the extended P.E.L. study; these included one species, *Maianthemum canadense,* which was present in every single stand of the 110 examined. The others, with their minimum presence values are: *Aralia nudicaulis* (93 per cent), *Aster macrophyllus* (57 per cent), *Clintonia borealis* (73 per cent), *Coptis trifolia* (40 per cent), *Cornus canadensis* (88 per cent), *Corylus cornuta* (50 per cent), *Dryopteris austriaca* (46 per cent), *Galium triflorum* (64 per cent), *Lonicera canadensis* (46 per cent), *Pyrola secunda* (46 per cent), *Rubus pubescens* (69 per cent), *Streptopus roseus* (46 per cent), and *Trientalis borealis* (75 per cent). These species had an average presence of 85.2 per cent in the Wisconsin stands, and eight of them were in excess of 90 per cent.

A floristic analysis of the Wisconsin stands showed a total of 296 understory species in 61 families. There was no single family with a clear predominance. Rather, there were ten families which contributed 50 per cent of the species. These top ten were: *Compositae* (9.8 per cent), *Cyperaceae* (5.7 per cent), *Liliaceae* (5.7 per cent), *Ranunculaceae* (5.7 per cent), *Gramineae* (4.7 per cent), *Polypodiaceae* (4.4 per cent), *Caprifoliaceae* (4.4 per cent), *Orchidaceae* (4.1 per cent), *Rosaceae* (4.1 per cent), and *Saxifragaceae* (3.7 per cent). Other families of importance are the *Ericaceae* with 10 species, *Violaceae* with 9 species, and *Lycopodiaceae* with 5 species. The species are about evenly divided between spring and summer bloomers, with only 3.1 per cent of the prevalent species sum of presence contributed by autumn flowering species. The shrubs are well represented, with 56 species in the total flora list, 80 per cent of which have fleshy fruits. Shrubs contribute 30 per cent of the sum of presence of the prevalent species. Evergreen plants, both shrubs and herbs, are also important, with 32 species on the total list, exclusive of sedges.

The other biota of the boreal forest have received almost no attention in Wisconsin. It is known that the woodland caribou and the moose were present in this type in the early 1800's in Douglas and Bayfield counties. Among the birds, the spruce grouse is perhaps most characteristic, but many warblers, the white-throated sparrow, the crossbill, and other smaller species are also numerous in the spruce-fir stands. Fungi are abundant as are epiphytic and terrestrial lichens, mosses, and liverworts.

ENVIRONMENT

Microclimate

No studies of the microclimate within the boreal forests of Wisconsin have been reported to date, except for P.E.L. records of light intensity. Actually, the macroclimate of the regions supporting optimum growth of spruce and fir can be considered as a type of microclimate, since it is unusual and has a highly restricted geographic distribution (Table XIII-4). The winds from Lake Superior greatly influence the climate in a narrow band near the shore and are felt to a lesser extent all the way up to the fall line on the high ridges south of the shore. This is indicated in the following quotation from Sweet (1880):

The climate of the Lake Superior country is somewhat rigorous during the winter, and quite cool during the greater part of summer. In the vicinity of the lake shore, however, it is by no means so subject to sudden and great extremes of temperature as the inland district, several degrees to the south. For at least fifteen or twenty miles south of the shore, the temperature is considerably modified by the proximity of the vast body of water in Lake Superior. The cold winds of winter, in their long passage across the lake, which never freezes, are very sensibly warmed, and to some extent transfer their warmth to the country over which they pass. During the warm days of summer a cool breeze usually springs up from the lake towards evening, and the nights are always cool. Thus the great lake, although, of course, to a much less extent, has an oceanic effect in equalizing the temperature in its vicinity.

This was realized earlier by Owen (1852) who said: "The temperature of the head waters of the Brule is from twelve to fifteen degrees cooler than that of the St. Croix. This difference of temperature was found generally to hold good between the streams flowing into Lake Superior and the tributaries of the Mississippi."

A similar influence of the on-shore winds is even more strikingly evident in the Door County Peninsula, where Bailey's Harbor on the Lake Michigan shore frequently has a midday temperature of 68°F. to 72°F. when Ephraim, only nine miles away on the Green Bay shore has a temperature of 90°F. to 95°F. These lake breezes serve to keep the summer temperatures and evaporation rates low. In the winter, the effect on temperature is reversed, with the shore areas markedly warmer than those at inland stations. Winter precipitation is high, with 4 to 6 feet of snow frequently on the ground. The spring season is slow in arriving, due to the refrigerative effect of the lakes. Sweet (1880) gives some indication of this in the following passage:

Winter sets in about the last of October, and continues cold and usually clear until the first of April, in ordinary seasons. In 1873 we had freezing nights in the early part of September, on the Copper Range, and a heavy snow storm upon the third of October. The snow usually lies from three to four feet deep in the forest all winter. In 1877, the snow was nearly gone on the 10th of April, and all gone by the first of May, although ice was found in the swamps two or three weeks later. Maple buds were considerably expanded April 25th, and a week later the first flower (*Ranunculus*) was noticed in the barrens. It may be stated, however, in this connection, that the spring of 1877, in this district, was about a month earlier than usual.

The boreal forest climate, as represented on its extreme borders in Wisconsin, is thus seen to be characterized by a cool, relatively equable temperature, a short growing season, abundant available moisture during the growing season, and deep snows in the winter.

SOILS

The soils under the boreal forest are mostly in the Gray Wooded Group, although some true Podzols are present. They include the Ontanagon, Pickford, and Ewen series along Lake Superior, the Longrie along Lake Michigan, and the Gogebic and Wakefield series in the north central counties. They vary in texture from clay to sandy loam, usually with a gley layer within 36 inches. The clay soil of the Lake Superior region has been ably described by Sweet (1880):

It is the predominant, and almost exclusive soil on the north slope of the district. It is usually covered with a very slight coating of humus or vegetable mould, which is frequently so light that a slight scraping with the foot exposes the surface of the red clay, which, interstratified with layers of fine sand, is many feet in depth. The physical properties of this clay, so far as I have observed, are the same, whether taken from the surface, or several feet below it. Where it is unmixed with sand, it is very tenacious, and is exceedingly retentive of moisture, making a "heavy, cold" soil. As a general rule, it contains too little sand, in the Lake Superior country, to give it the most desirable drainage and "lightness" for agricultural purposes in this latitude. It however appears to be an excellent grass soil. In the eastern part of the district it is heavily timbered with hemlock and pine, and in the western portion with scrub pines, poplar, birch, balsam and spruce.

Analyses of the A_1 layers of soils in the P.E.L. series of Wisconsin stands showed a median pH of 5.1 with a range from 4.2 to 6.0. The average available amount of calcium was 2780 p.p.m., magnesium 150 p.p.m., potassium 150 p.p.m., phosphorus 70 p.p.m., and ammonia 18 p.p.m.

The water-retaining capacities of the soils in the extended study of the Great Lakes region are highly correlated with the compositional gradient of the stands, varying from an average of 165 per cent in the dry stands to 370 per cent in the wet stands. In spite of this great change in moisture supply from one end of the gradient to the other, a number of species occur with equal abundance at the two ends, as mentioned earlier. Rowe (1956) found an almost exactly similar situation in the boreal forests of Saskatchewan. The exact nature of the compensating factors present which allow this bimodal behavior are unknown, but they probably include light, acidity, and nitrogen supply. In any case, the phenomenon is an expression of a basic ecological principle which states that plants become more sensitive to differences in soil moisture as temperatures increase. Within Wisconsin, for example, we find a number of species occurring throughout the entire moisture gradient here in the boreal forest stands, fewer such species (although largely the same ones) with that behavior in the northern pine-hardwoods, and only very few that range from willow-cottonwood to bur oak-black oak in the southern counties. Farther south, in the Mississippi valley, virtually no species are found in common between the cypress-tupelo (*Taxodium-Nyssa*) swamps and the adjacent hilltop forests of blackjack oak (*Q. marilandica*) and post oak (*Q. stellata*) as in Missouri and Arkansas. In the tropics, there may be several sets of species present between the two extreme conditions of moisture. The same principle can be seen at work on an altitudinal basis in mountainous regions. Thus, in Whittaker's (1956) nomograms of species behavior in the Great Smoky Mountains, a considerable shift in species composition occurred with decreasing moisture at low elevations, while at high elevations, the spruce-fir forest was found on nearly the full range of the moisture gradient.

SUCCESSION AND STABILITY

Little observational evidence exists concerning the stability of the nearly pure, mature conifer stands in Wisconsin. In the major range of the boreal forest in Canada, it is apparent that catastrophe is of frequent occurrence, caused either by fire, wind, or the spruce budworm. As a result, there is a tendency for entire stands to be destroyed and replaced simultaneously, with resultant even-age forests over wide areas. In Ontario, there may be a direct replacement of spruce-fir forest by spruce-fir forest, with no intervening stage. More frequently, white birch and trembling aspen form an intermediate and short-lived cover crop, particularly after fire. On both the dry sand soils and the wet or poorly drained flatlands, there may be an initial stage of black spruce and jack pine. Both of these species can readily replace themselves after fire, so the type tends to be semi-permanent in areas subject to frequent fire. Examples of spruce-fir invasion of old jack pine stands can also be found in Wisconsin, as in the photo in Plate 30, taken in Vilas County.

In Wisconsin and adjacent portions of Upper Michigan, young stands of boreal forest, very heavily dominated by balsam fir, are found extensively on lands formerly occupied by mesic northern hardwoods. Judging by the relic appearance of very large spruces (and more rarely firs) in the remaining mature hardwood stands of the region, it seems highly probable that these young boreal stands will similarly give way to hardwoods in the future. At present, however, this is only a conjecture, since the existing stands show no positive evidence of such an invasion. It is possible that a slight shift in climatic conditions might allow the spruces and firs to retain dominance for long periods. The only support for such a hypothesis comes from the fossil pollen profiles of the region, many of which show an increased importance of these species in their upper layers (Chapter 22). How much of this increase is due to postsettlement changes and how much to changing climate is difficult to determine.

Balsam fir is an extremely vigorous species in its young stages, and can be found on a variety of sites where, no doubt, it will not be present in the final mature forests. It is especially common as an invader of logged-over or burned cedar swamps and other areas of high soil moisture. It also is one of the few species which successfully invade the bracken-grasslands (Plate 42).

One of the difficult problems in the study of community dynamics of northern Wisconsin is the assessment of the relation between conifer swamps and terminal mesic forests. In Chapter 12, it was indicated that the succession may go from white cedar-balsam fir-black ash swamp through yellow birch and hemlock, and finally to sugar maple or beech, if a sufficient improvement in soil drainage occurs. It appears equally probable that the conifer swamp may give rise to a terminal forest of white spruce and balsam fir if the climatic conditions are suitable. This is shown by the high index of similarity of the swamp community and the boreal forests of wet or wet-mesic sites. In the P.E.L. studies, there was no objective way whereby conifer swamps could be separated into those destined for hardwoods and those destined for boreal forest; hence no differentiation was possible. It may be that certain compositional differences were present, but they were not obvious in the field. Actually, the tamarack swamps south of the tension zone, which eventually evolve into wet-mesic southern hardwoods, showed much higher similarities with the boreal forest than they did with the southern forest in which they were present. It is possible to consider all of the conifer swamps of Wisconsin as wet-ground stages of the boreal forest, with the central and southern swamps isolated by postglacial climatic changes and only those in the extreme north in a position to develop into spruce-fir forest by successional processes. Such a view is useful in certain respects, but should not be held too rigorously, because the supporting evidence available is inadequate for firm judgments.

GEOGRAPHICAL RELATIONS

A rather large proportion of the boreal forest flora is circumboreal in distribution, while many other species are transcontinental in this hemisphere. As a result, there is a rather high degree of similarity between the Wisconsin forests and those of other regions. Thus, on Isle Royale, Cooper (1913) reported 25 species, of which 88 per cent were on the Wisconsin list. In Ontario, Maycock (1956) studied five stands in the Algonquin Park region. Of the 85 understory species found there, 84 per cent were on the Wisconsin list. On Cape Breton Island in the maritime province of Nova Scotia, Collins (1951) found 51 species of which 80 per cent were in common with Wisconsin. In the Ontario stands, the major trees were white spruce, white pine, balsam fir, trembling aspen, and white birch. Comparing the average importance values of all trees with those for Wisconsin by the $2w/a+b$ test, an index of similarity of 48.5 per cent is obtained. In the Cape Breton study, the only important tree species were balsam fir, white spruce, white birch, and mountain-ash, and the index of similarity was therefore very low, largely because of the low importance of both pines and hardwoods.

Farther south, Oosting and Billings (1951) presented comparative studies of boreal forest stands in the White Mountains of New Hampshire and in the Smoky Mountains of Tennessee and North Carolina. These forests differ from the Wisconsin stands in that red spruce (*Picea rubra*) replaces white spruce in both regions, and Fraser fir (*Abies fraseri*) replaces the very closely related balsam fir in the Smoky Mountain region. Otherwise, the general aspect of these two mountain types is quite similar to that seen in Wisconsin. The White Mountain stands had 77 per cent of their species in common, while the Smoky Mountain stands had only 35 per cent in common. However, 29 of 32 genera were the same between Wisconsin and the Smokies, and were often represented by very similar but taxonomically discrete species in the latter region.

The greater resemblance of the northern stands is supported by the work of Heimburger (1934) in the Adirondack forests of New York, a region where both white spruce and red spruce are present. In his so-called *Oxalis-Hylocomnium* type, for example, a total of 42 herbs and shrubs was found, of which 83 per cent are in common with Wisconsin.

Towards the northwest and west, a high level of physiognomic similarity is maintained all the way to the Cascade Mountains by a

Table 14
Geographic relations of boreal forest

Locations of study	% of flora in common with Wisconsin stands	Approximate distance from NW. Wisconsin (miles)
Isle Royale (Cooper, 1913)	88	160
Algonquin Park, Ontario (Maycock, 1956)	84	580
Adirondack Mountains, New York (Heimburger, 1934)	83	910
Cape Breton Island, N. S. (Collins, 1951)	80	1500
White Mountains, New Hampshire (Oosting and Billings, 1951)	77	1000
NW. Alberta (Moss, 1953)	59	1500
Smoky Mountains, Tennessee (Oosting and Billings, 1951)	35	920
Idaho (Daubenmire, 1952)	21	1240
Rocky Mountains, Wyoming (Oosting and Reed, 1952)	14	830

rather uniform generic dominance in the tree layer, but the actual species composition changes quite rapidly. Thus, Rowe (1956) listed 89 species for the mesic boreal forests of central Saskatchewan, of which 89 per cent were in common, Moss (1953b) found 107 species in the boreal white spruce forests in Alberta, of which 59 per cent were in common, and Oosting and Reed (1952) reported 43 understory species for the sub-alpine boreal forests of Wyoming, of which only 14 per cent were in common. In the first case, white spruce, balsam fir, jack pine, black spruce, white birch, balsam poplar, and trembling aspen made up the canopy; while in Alberta, white spruce, trembling aspen, balsam poplar, and white birch were important tree dominants; and in the Wyoming forests, the coniferous element was represented by Englemann spruce (*Picea engelmannii*) and sub-alpine fir (*Abies lasiocarpa*). In northern Idaho and adjacent Washington, Daubenmire (1952) found the last two conifers dominant in sub-alpine regions, with three different groundlayer unions beneath them. Of the 28 species in these unions, only 21 per cent were in common with Wisconsin, but 23 of 28 genera were the same, with slightly different species, which include such important groups as *Mitella, Trillium, Coptis, Anemone, Clintonia, Lonicera, Viola, Vaccinium,* and *Luzula*.

These geographic relations of decreasing floristic similarity in the boreal forest with increasing distance are summarized in Table 14.

Forests dominated by spruce and fir have been studied extensively in Europe (Sukachev, 1928; Poliakov, 1950; Aichinger, 1952). They show a great structural similarity to the American types, with many of the same species in both areas and a great number of genera with related species. Thus Oberdorfer (1957) in South Germany listed 109 species in his spruce-fir community, of which 66 were in common, including *Athyrium filix-femina, Dryopteris phegopteris, Goodyera repens, Monotropa hypopithys, Oxalis acetosella,* and *Pyrola secunda.* More than one-half of the 101 genera listed were present in both areas, among them such typical groups as *Abies, Picea, Acer, Pinus,* and *Sorbus* among the trees; *Lonicera, Rubus, Ribes, Sambucus* and *Vaccinium* in the shrubs; and *Anemone, Aquilegia, Corallorhiza, Listera, Maianthemum, Melampyrum, Petasites,* and *Viola* in the herbs.

UTILIZATION AND MANAGEMENT

The boreal forest in Wisconsin is of such limited extent that it is not of great economic importance. Young stands of balsam fir are important sources of Christmas trees, while the older second-growth stands of both spruce and fir are used as sources of pulp. Considerable efforts have been devoted to the reforestation of certain heavy soil areas with white spruce, again for pulp supplies, but almost no plantings of balsam fir have been made. This type cannot be considered to be of importance in the over-all forestry picture in Wisconsin.

Part 4

Grasslands

CHAPTER 14

Prairie

NATURE AND GENERAL DISTRIBUTION

The prairies of Wisconsin include some of the most interesting though least widely known of any of our plant communities. The word prairie is of French origin and means meadow. It was applied to the open, grass-covered, treeless landscapes of middle America by the early French explorers, since they had no other word for such a community. When the English settlers arrived, they adopted the name prairie, since they, too, had no exact term for these huge grasslands.

Wisconsin lies on the northeastern boundary of the American prairies. Within the state, prairies are located southwest of the tension zone, in the triangular area extending from Racine to Grant to Polk counties. They occupy the greatest area in the southwestern corner and gradually become smaller and more scattered as the tension zone is approached. Many of the larger and more important prairies received distinctive names by which their important place in the agricultural history of the area is remembered. Among them were: the Barnes Prairie in Racine County; the Walworth and Rock prairies, in the counties of the same names; the Arlington Prairie in Dane and Columbia counties; the Sauk Prairie in Sauk County; the Military Ridge Prairie, from Verona nearly to the Mississippi; the Star Prairie in St. Croix County; and many others. The location of the major prairies is shown in Figure XIV-4. The area shown on the map is 2,101,400 acres, or 6.00 per cent of the total land area.

The prairie is a plant community dominated by grasses rather than by trees as in a forest. Growing with the grasses are many other species of non-grassy herbs which are known by the collective name "forbs." Many woody shrubs are present in the prairie as well, and, under certain circumstances, tree seedlings may also be found. The prairies frequently grade imperceptibly through an oak savanna to a denser oak forest, although on occasion there may be a rather abrupt boundary between grassland and forest, particularly at rivers or at places of rapid topographic change. On moist soils, the prairies blend into marshlands dominated by sedges rather than by grasses. Since these transitional boundaries are present, it is necessary to establish the criteria by which a prairie may be recognized. For purposes of the present discussion, a prairie is defined as an open area covered by low-growing plants, dominated by grasslike species of which at least one-half are true grasses, and with less than one mature tree per acre. Similar communities with more than one-half of their dominants in the sedge group are called sedge meadows, while areas with more than one tree per acre but with less than one-half of the total area covered by the tree canopy are called savannas. As a further restriction, areas meeting the above criteria are called prairies only if they are located south of the tension zone. This last distinction separates the bracken-grasslands of the northern floristic province (see Chapter 15) from the prairies of the south, at least for study purposes.

Early descriptions of the prairie of both botanical and literary interest were made by many of the early travelers. The discrepancy between the lack of thoroughness in their treatment of the forests and the fullness of their accounts of the grasslands attests not only to their wonder at seeing a new phenomenon but also to the inherent beauty and grandeur of the virgin prairie. The first mention of the Wisconsin prairies is by Father Dablon in 1670 (*Jesuit Relations*, 1899), who described the area around the upper Fox River:

... in every direction, prairies only, as far as the eye can reach. ... All this prairie country extending ... more than three leagues [*ca.* 10 miles] in every direction ... affords ample subsistence to the elk not infrequently encountered in herds of four or five hundred each. These, by their abundance, furnish adequate provision for whole villages, which therefore are not obliged to scatter by families during the hunting season, as the case with the savages elsewhere. In these rich pasture lands are also found buffalo. ...

Another early account is by Father Hennepin in 1683 (Shea, 1853), who described the area near the Mississippi River: "... between the mountains and the river there are large prairies, where you often see herds of elk browsing. ... Beyond these mountains you discover vast plains." At the site of the present city of Prairie du Chien there was "a beautiful, even grassy meadow of six miles in length and one or two broad."

Among other early descriptions of the prairie are those of J. F. St. Cosme in his "Letter to the bishop" (in Shea, 1861), written in 1699 about the Little Fox River country in Racine and Kenosha counties and the account of the same general region as seen in 1721 by P. F. de Charlevoix (1761). The latter contains additional references to the high populations of grazing animals, as in this passage:

Nothing to be seen in this course but immense prairies, interspersed with small copses of wood, which seem to have been planted by hand; the grass is so very high that a man is lost amongst it, but paths are everywhere to be found as well trodden as they could have been in the best peopled countries, though nothing passes that way excepting buffaloes, and from time to time some herds of deer and a few roe-buck.

Later accounts of Guigna (1728, in Shea, 1861) and Carver (1781) mention the occurrence of prairie on steep hillsides. Carver, for example, described the country along the lower Wisconsin River thus: "For miles nothing was to be seen but lesser mountains, which appeared at a distance like haycocks, they being free from trees ... and only a few groves of hickory and stunted oaks covered some of the vallies."

The great expanse of the prairies is indicated in the report of the "March of the 5th Regiment, in June 1819, from Green Bay to Prairie du Chien, by an officer of that Regiment," (Anon., 1819), where the country around Portage is described as "prairie interspersed with large clumps of oaks, looking like islands, studding an immense sea of verdure." Again, Keating (1824) traveled from Chicago to Prairie du Chien "... at first through thin woods, which gradually disappeared, their place being supplied by an extensive and apparently boundless prairie, which occupied us a whole day in crossing it." The beauty of the prairie is described by Smith (1837) in this account of the view from the top of Belmont Mound, in La Fayette County:

The view from this mound . . . beggars all description. An ocean of prairie surrounds the spectator whose vision is not limited to less than thirty or forty miles. This great sea of verdure is interspersed with delightfully varying undulations, like the vast waves of the ocean, and every here and there, sinking in the hollows or cresting the swells, appears spots of trees, as if planted by the hand of art for the purpose of ornamenting this naturally splendid scene.

Again, Owen in 1848 said, "On the summit levels spreads the wide prairie, decked with flowers of the gayest hue; its long and undulating waves stretching away till sky and meadow mingle in the distant horizon." These allusions to the ocean and its waves are almost universal in the early accounts. Perhaps the most vivid is that by Lieutenant D. Ruggles (1835), who described the prairies around Fort Winnebago in Columbia County:

In some instances, prairies are found stretching for miles around, without a tree or shrub, so level as scarcely to present a single undulation; in others, those called "rolling prairies," appears in undulation upon undulation, as far as the eye can reach presenting a view of peculiar sublimity, especially to the beholder for the first time. It seems when in verdure, a real troubled ocean, wave upon wave, rolls before you, ever varying, ever swelling; even the breezes play around to heighten the illusion; so that here at near two thousand miles from the ocean, we have a fac-simile of sublimity, which no miniature imitation can approach.

The prairies have received more attention than any other plant community in the state. The *Bibliography of Wisconsin Vegetation* (Greene and Curtis, 1955) lists 65 titles dealing with the prairie. The first detailed floristic description of a Wisconsin prairie was by Lueders (1895) in his account of the vegetation of the Township of Prairie du Sac in Sauk County. Gates (1912) gave complete species lists for the extremely interesting prairies on the Lake Michigan shore south of Kenosha. The present distributions of remnant colonies of prairie plants were mapped by Anthoney (1937) for Rock County and by Gould (1937) for Dane County.

The first quantitative, phytosociological investigation of Wisconsin grasslands was made in 1948 by Phoebe Green (1950), who examined the mesic prairies of Rock County by the quadrat method. Other P.E.L. community studies of prairies include those of Partch

(1949), Curtis and Greene (1949), Wagner (1951), Anderson (1954), and Curtis (1955). Related studies dealing with soil fungi were made by Orpurt (1955); with effects of fire by Curtis and Partch (1948, 1950), Dix and Butler (1954), and Archbald (1953); with life-histories of prairie plants by Curtis and Cottam (1950), Curtis and Greene (1953), and Butler (1954); with effect of grazing by Dix (1953); with establishment of prairie plants by Robocker (1951), Archbald (1954), Miller (1954), and Greene and Curtis (1950); and with the role of legumes in community dynamics by Bard (1957). The following discussion is based largely on these P.E.L. studies.

GENERAL COMPOSITION OF PRAIRIE

The first attempts at understanding the behavior of prairie species were based upon the apparent differences in species composition of prairies on different topographic sites (Curtis and Greene, 1949). In a study of 65 remnant prairies throughout the prairie forest border, it was noted that certain species were most abundant on dry, thin, soil prairies on limestone ridges and steep hillsides, others reached their maximum development on mesic sites with deep loam soils and good internal drainage, while the inundated flatlands and poorly drained prairies supported optimum growth of a different combination of species. With this soil catena differentiation as a background, it was found possible to select groups of indicator species, each member of which reached its optimum within a certain narrow range of soil moisture conditions as represented by segments of the catena. For this purpose, the moisture gradient was divided into five units; wet, wet-mesic, mesic, mesic-dry, and dry. Ten species were chosen which attained clear-cut peaks of optimum growth in the sites representing each of these moisture conditions. Species with very broad moisture tolerances which occurred abundantly in all or most of the prairies and very rare species whose behavior could not be assessed with certainty were not used in the selection. The resulting set of fifty species, in five groups of ten each (Table 15) thus included those species with an intermediate range of tolerance (Curtis, 1955). Similar use of indicator groups was made by Raabe (1949), Ellenberg (1952), and others.

The indicator groups were used to characterize a particular stand of prairie by determining the number of species in each group present in that prairie, expressing those numbers as a percentage of the total number of indicators present, and then weighting the values by multi-

Table 15
Indicator species for five segments of prairie gradient

Wet prairie species
　Aster novae-angliae
　Calamagrostis canadensis
　Hypoxis hirsuta
　Oxypolis rigidior
　Pycnanthemum virginianum
　Solidago gigantea
　Spartina pectinata
　Thalictrum dasycarpum
　Veronicastrum virginicum
　Zizia aurea

Wet-mesic species x 2
　Cicuta maculata
　Desmodium canadense
　Dodecatheon meadia
　Fragaria virginiana
　Galium boreale
　Helianthus grosseserratus
　Heuchera richardsonii
　Lathyrus venosus
　Phlox pilosa
　Rudbeckia hirta

Mesic species x 3
　Aster laevis
　Ceanothus americanus
　Cirsium discolor
　Desmodium illinoense
　Eryngium yuccifolium

Mesic species (cont.)
　Helianthus laetiflorus
　Liatris aspera x
　Panicum leibergii x
　Ratibida pinnata
　Solidago missouriensis

Dry-mesic species x 4
　Anemone cylindrica x
　Asclepias verticillata
　Helianthus occidentalis x
　Linum sulcatum
　Panicum oligosanthes
　Petalostemum candidum
　Potentilla arguta
　Scutellaria leonardi
　Sporobolus heterolepis
　Stipa spartea

Dry species x 5
　Andropogon scoparius x
　Anemone patens
　Arenaria stricta
　Artemisia caudata
　Aster ptarmicoides
　Aster sericeus
　Bouteloua curtipendula
　Panicum perlongum x
　Petalostemum purpureum
　Solidago nemoralis x

plying the wet indicators by one, the wet-mesic by two, and so on. A summation of the weighted values thus produced an index number which indicated at a glance the composition of a given stand in terms of its content of species adapted to a particular level of soil moisture. Stands containing only those indicators which peak in wet areas have an index of 100 whereas stands whose only indicators are characteristic of the driest sites have an index of 500. Any mixture of intermediate species results in an intermediate number, with the mesic prairies centering around 300. When this procedure was applied to all of the 181 stands in the P.E.L. study, a range of from 100 to 485 was obtained. By plotting the compositional index on the abscissa and the measured importance of individual species on the ordinate, it is possible to represent graphically the behavior of the species in relation to this moisture-based gradient. The graphs in Figures 17 and 18 show such curves for the dominant prairie grasses and for selected forbs. As can be seen, these curves are very similar to those given in Chapters

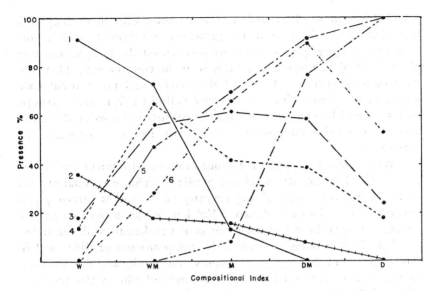

Figure 17.—Behavior of major prairie grasses along compositional gradient based on moisture indicator species. 1, bluejoint grass *(Calamagrostis canadensis)*; 2, prairie muhly grass *(Muhlenbergia racemosa)*; 3, Leiberg's panic *(Panicum leibergii)*; 4, wild rye *(Elymus canadensis)*; 5, little bluestem grass *(Andropogon scoparius)*; 6, needle-grass *(Stipa spartea)*; 7, side-oats grama grass *(Bouteloua curtipendula)*.

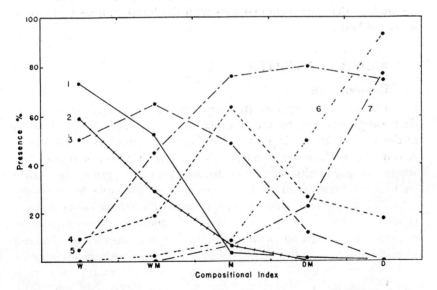

Figure 18.—Behavior of selected prairie forbs along compositional gradient. 1, *Oxypolis rigidior*; 2, *Aster simplex*; 3, *Desmodium canadense*; 4, *Desmodium illinoense*; 5, *Aster ericoides*; 6, *Anemone patens*; 7, *Verbena stricta*.

5 and 9 for the southern and northern forests. Like them, they indicate that the prairies form a vegetational continuum, with a continually changing species composition based on the unique responses of the individual members to changes in the environment. There are no groups of species with identical behavior and no clear-cut combinations of species which are sharply delimited from other groups, such as would be present if the prairies consisted of a series of discrete and recognizable communities, each with its own environmental optima.

When certain features of the measurable environment are plotted against the floristic compositional gradient, definite correlations are noted. Thus, the water-holding capacity of the soil is inversely proportional to the index as Partch (1949) had found earlier in a limited sample, whereas the calcium content increases directly with the index.

For the following discussion, the compositional gradient will be arbitrarily divided into three segments, wet, mesic, and dry, as it was in the forests. The wet includes the ranges of 100 to 180 and from 180 to 260 units, and involves 53 stands; the mesic from 260 to 340 with 45 stands; and the dry from 340 to 420, and from 420 to 500 with 83 stands. The locations of the stands in the three segments are shown in Figures XIV-1 to XIV-3, while all stands are indicated on the map at the opening of this chapter. In some cases, all five segments will be used, so that direct comparisons with the forests can be made upon an equal basis.

XERIC PRAIRIES

Composition

The prevalent species (the top species, as ranked in order of decreasing presence, to the number of the average species density) of the dry and dry-mesic prairies are given in Tables XIV-1 and XIV-3. A number of these are restricted to the xeric prairies or reach overwhelming predominance there. Included in this group are fairly well-known plants such as the pasque flower *(Anemone patens)*, silky aster *(Aster sericeus)*, side-oats grama grass *(Bouteloua curtipendula)*, yellow puccoon *(Lithospermum incisum)*, and false boneset *(Kuhnia eupatorioides)*, together with such less familiar species as *Acerates viridiflora, Arenaria stricta, Castilleja sessiliflora, Panicum perlongum,* and *Physalis longifolia*. The prevalent lists also contain such wide-ranging species as leadplant *(Amorpha canescens)*, big and little bluestem grasses *(Andropogon gerardi* and *A. scoparius)*, bergamot

(*Monarda fistulosa*), and common milkweed (*Asclepias syriaca*), many of which are not solely confined to prairies but are found in many other plant communities as well.

Some of the less abundant modal species indicated in the tables have been collected but rarely in Wisconsin, such as *Orobanche fasciculata*; others are restricted to the extreme northwestern edge of the prairie region, *Artemisia frigida, Astragalus crassicarpus, Liatris punctata* but may be common there; and still others are widely distributed but are never abundant in any prairie, *Acerates lanuginosa, Cirsium hillii, Lactuca ludoviciana*.

Anderson (1954), in a thorough study of the high lime prairies of Wisconsin, made both frequency and presence determinations in a large number of xeric stands, that were mostly in the dry segment of the compositional gradient discussed above. He compiled a master list of species which he arranged on the basis of a factor obtained by multiplying their average frequency (in one square meter quadrats) by their presence (Chapter 4). Of a total list of 223 species, 10 had factors over 2000, 8 were between 1000 and 2000, 13 were between 500 and 1000, 25 were between 100 and 500, and the remaining 167 were below 100, including 90 with factors less than 1. Thus the extreme importance of a relatively few species stands in great contrast to the relative infrequency of the great majority of the total flora. The ten leading species, with their factors, are as follows: *Andropogon scoparius*–7130, *Bouteloua curtipendula*–6240, *Andropogon gerardi*–4779, *Aster sericeus*–4100, *Euphorbia corollata*–3360, *Amorpha canescens*–3117, *Solidago nemoralis*–3072, *Panicum perlongum*–3064, *Petalostemum purpureum*–2949, *Sporobolus heterolepis*–2103. The sum of the factors for these ten species (59.8 per cent of the total) is more than that of the remaining 213 species combined.

A floristic analysis of the xeric prairie flora reveals that one-half of the species are members of only five families. In the dry prairies, these are distributed with 27.5 per cent in the *Compositae*, 13.7 per cent in the *Gramineae*, 5.3 per cent in the *Leguminosae*, 4.6 per cent in the *Rosaceae*, and 4.6 per cent in the *Asclepiadaceae*, with the values in the dry-mesic segment essentially the same and in the same order. The tremendous numbers of Composites in these and all other prairies would suggest that, on floristic grounds, the prairie should be called a "daisyland" instead of a "grassland." However, the grasses more than make up in numbers of individuals what they lack in numbers of species. On the basis of Anderson's frequency-presence factor, the sum of the factor for grasses is 27,060 compared to 20,302 for the

Composites and 19,379 for all other families combined; figures for densities would favor the grasses still more.

In both the northern and southern forests, a large proportion of the flora blooms in the spring, with up to 70 per cent in flower before June 15 in the mesic stands. In the prairies, a very different situation exists, with only about one-third of the species flowering in spring. The greatest number bloom during the summer, but one-quarter or more of the total flower after August 15, which is in contrast to insignificant numbers of autumn bloomers in the forest. Butler (1954) studied the phenology of the prairie plants along with life-form and other life-history characteristics and noted an interesting correlation between flowering date and average height of the species. Thus, in the dry-mesic and dry segments of the compositional gradient, he found that the species which bloomed in May averaged 13 cm in height; those in June, 23 cm; those in July, 32 cm; those in August, 41 cm; and those in September, 46 cm. The same progression was seen in all other prairie types, although the levels attained were greater with increasing soil moisture, and reached a maximum of 91 cm in the September plants of the wet prairies. The average level of the leaf canopy of the entire community paralleled these flower heights but was 3 to 6 cm lower at any given season.

Among the earliest species to bloom on the xeric prairies are the pasque flower (*Anemone patens*, Plate 33), pussy toes (*Antennaria neglecta*), prairie buttercup (*Ranunculus fascicularis*), prairie smoke (*Geum triflorum*), and rock cress (*Arabis lyrata*); all have basal rosettes of leaves from whose centers arise the flower stalks. The lack of deep snow on the exposed lime prairies and their frequent inclination to the south and west allows for maximum insolation and rapid warming during the early spring days. Thus the very first flowers to appear in Wisconsin, always excepting the preternaturally early skunk cabbage (*Symplocarpus foetidus*), are those on the xeric prairies, with the pasque flower in undisputed first place.

By the first of June, a number of other species reach their peak of flowering, including such beautiful and horticulturally promising plants as downy phlox (*Phlox pilosa*), blue-eyed grass (*Sisyrinchium campestre*), birdfoot violet (*Viola pedata*), the puccoons (*Lithospermum* sp.), and shooting star (*Dodecatheon meadia*). Both the phlox and the shooting star attain greater distribution in less xeric prairies but they never appear to such good advantage as when they stud the steep hillsides with a cloud of pink and white. The shooting star stirred the imagination of Linnaeus, who called it Dodecatheon or twelve gods, and locally it has called forth many other poetic names, which include prairie sparks, diamond sparks, star shower, and bird-bill.

Summer sees a host of species in bloom, largely in the legume and composite families, but the xeric prairies suffer by comparison with other prairies which reach an intensity of floral display unequaled in any Wisconsin community. By autumn, the xeric prairies again contend for the lead in beauty, since their display of asters, goldenrods, and gentians are not hidden by the sea of tall grasses as frequently happens in more mesic areas. All of the asters are colorful, but the most characteristic species of the dry prairies, the silky or mouse-eared aster (*Aster sericeus*), is also the most striking in appearance, with very large individual flower heads of deep amethyst-violet set off against a background of silver-green, shining leaves. Many of the xeric prairies support large populations of the downy gentian (*Gentiana puberula*), which is by all odds the most beautiful member of this famed genus in Wisconsin and which at its best compares favorably with the species from the high Himalayas that are so prized by rock gardeners. The downy gentian may produce a cluster of 15 to 20 stems, each of which supports five or six wide-open, chalice-like flowers, unrivaled in the clarity and intensity of their deep-blue color. Another gentian, *Gentiana quinquefolia,* is frequently found on the dry prairies. It has a very interesting capacity to fluctuate between wide population limits. This is a biennial whose exact requirements for germination and establishment remain unknown. It has varied from no discoverable plants and certainly no blooming plants on a five-acre prairie in Green County to a conservatively estimated 200,000 plants within two years, only to drop again to a very few individuals in some subsequent years, and none in others. What the period between peaks may be is not known at present.

In Anderson's study of the dry lime prairies, a number of tests revealed a rather high degree of homogeneity between the various stands. The greatest differences observed were those between the prairies along the Mississippi River, and the prairies in Dane County and other more eastern stations, but these differences were of a minor nature and reflected slight changes in importance of a few species along an east-west gradient. Species which did best toward the east included *Arenaria stricta, Asclepias verticillata,* and *Geum triflorum,* while those with an optimum along the Mississippi included *Bouteloua hirsuta, Aster oblongifolius,* and *Scutellaria leonardi.*

One of the interesting geographic variations in the xeric prairies concerns the distribution of *Muhlenbergia cuspidata,* a plains grass which may reach high levels of dominance in some of the driest Wisconsin prairies. Anderson found that this species was absent from all of the prairies in the glaciated region, although it occurred throughout the Driftless Area, sometimes in abundance on prairies only a few miles from the drift border of both the Cary and Farmdale sheets. The pomme de prairie *(Psoralea esculenta)* showed a similar distribution. De Selm (1953) reported the same relationship for the *Muhlenbergia* on the glaciated and non-glaciated areas of Ohio. This obvious connection between prairie composition and glacial history is one of the evidences used by Anderson to support his contention that the dry prairies persisted in the Driftless Area throughout the Pleistocene period of glaciation.

The initial study by Curtis and Greene (1949) separated a group of prairies on the basis of the sandy nature of their soils. Later observations indicated that these sand prairies had a very low degree of homogeneity, with few if any species restricted to them. Rather, the flora on a given sand prairie closely resembled that on heavier soils of the same moisture content. Thus, the wet sands in old lake beds had species like the other wet prairies which were free of sand. Nevertheless, by the very nature of sand deposits, the majority of sand prairies are on the xeric side of the gradient. Few of them are as dry as the driest high lime prairies with very thin soils on steep slopes; hence they tend to appear in the dry-mesic segment of the compositional gradient. Among the plants in Tables XIV-1 and XIV-3 which seem to grow best on sandy substrates are pasque flower *(Anemone patens),* birdfoot violet *(Viola pedata), Aster linariifolius, Callirhoe triangulata, Cirsium hillii,* Venus's looking-glass *(Specularia perfoliata), Cassia fasciculata, Penstemon grandiflorus,* and sand cherry *(Prunus pumila);* but most of these are of minor importance in the over-all prairie group.

Other Biota

The only other members of the xeric prairie biota that have been studied in detail in Wisconsin are the soil fungi. In an investigation of the soil microflora of the entire prairie gradient, Orpurt and Curtis (1957) were able to show that a number of soil fungi, including *Penicillium lilacinum, Aspergillus terreus, Aspergillus fumigatus, Penicillium funiculosum, Spicaria violacea,* and *Tilaclidium* sp. reached optimum frequencies in the xeric prairies. Among the other cryptogams, a number of ground lichens are present, as well as the xerophytic liverwort *(Reboulia hemisphaerica).*

A rather distinct group of birds finds the conditions on the xeric prairies very favorable, including such ground nesting species as the prairie horned lark, the grasshopper sparrow, and the western meadowlark. The longspurs, pipits, and snow buntings are seen in winter or in spring migrations. Among the mammals, the thirteen-lined spermophile and the Franklin's gopher are normal inhabitants of the dry prairies, apparently because of the low total height of the vegetation. In the northwestern part of the state the pocket gopher is abundant on the intermediate prairies, particularly those with sandy soils. The spotted skunk has been reported from prairies along the Mississippi River.

Structure

No density determinations have been made in the xeric prairies of Wisconsin, so that no exact information on the numbers of individuals present can be given as they were for the dominants in the forest studies. This lack is due largely to the great difficulties attendant upon such a determination. In a forest there is usually no question as to the status of a single tree and it can be counted and measured as a discrete unit in company with the other trees. In a grassland, however, the problem as to what constitutes a countable or measurable unit becomes paramount. Is a clump of grass to be counted as one plant, comparable to a tree; or is each stem to be compared to each tree trunk? In the latter case, should flowering stems only be counted, or should vegetative stems or tillers be counted as well? These questions perhaps could be settled on an arbitrary basis, with preference going to the method which equates a clump of grass with a single tree, but all methods fail when a sod-forming grass such as side-oats grama is encountered, since here no clumps are discernible, the production of flower stems varies greatly from year to year, and the number of vegetative shoots varies even more widely from season to season in a single year. In the face of these practicably insurmountable difficulties, the best analytic method is a delimitation of simple frequency, a determination of the percentage of small quadrat samples which contain the species, irrespective of the number of individuals present in any one sample. Abundant experience has shown that species behavior diagrams based upon such frequency information are identical to those based upon density values, in species where the latter are obtainable and the assumption is here made that equal information is also given in those cases where density is not available. The same line of reasoning applies to the relationship between presence and frequency since many investigations have shown that the behavior of a species is the same when based upon presence or absence in a stand as it is when based upon presence or absence within sub-unit samples of that stand (Gilbert and Curtis, 1953). This concordance of frequency and presence values is the main reason why Anderson's factor based on the product of the two gives such clear-cut separations of the important species.

In the analysis of frequency data for any community, it is usually found that rare species of low frequency greatly outnumber the common species of high frequency and the xeric prairies are no exception. In the dry-mesic stands (Table XIV-3), for example, only four species occurred in more than one-half of the total quadrats, 11 species in more than one-quarter, and 28 species in more than one-tenth of the total. The three leading species were *Andropogon scoparius* (84 per cent), *Bouteloua curtipendula* (69 per cent), and *Andropogon gerardi* (61 per cent). These figures can be regarded as the chances of finding the species at any random spot in the prairie. Since they were based on quadrats of one square meter, they mean that the chances of finding *Andropogon scoparius* within 50 cm of the investigator's right foot at any place he happened to stop in any one of the prairies would be 84 in 100 on the average, though they might exceed this in some prairies and be less in others. In any case, such a high average frequency attests to the ubiquity of this grass in the xeric prairies. Since the second most common species, side-oats grama, is another grass of medium stature which, like the little bluestem, rarely exceeds 60 cm in height, it is possible to obtain a mental picture of not only the species composition but also the mass appearance of the prairie from these frequency figures.

Another aspect of the physiognomy of the xeric prairies involves the spatial distribution of the component species. In order to reach such high average frequencies as 84 per cent, little bluestem must be uniformly distributed throughout the entire area of the stand, but another species with 10 or 20 per cent frequency may convey a vastly different impression depending upon whether it occurs as scattered individuals over the whole area or as one or a few dense clumps or colonies in localized spots. This latter case of aggregation has been discussed for the understory species of the various forests, but the same phenomenon seems of more importance in the prairie, since there are no trees, saplings, or shrubs to obscure the vision; a tight colony of some Composite in full bloom may be a very conspicuous sight even though it is located at some distance from the observer.

The lack of information on density means that the usual methods for measuring degree of aggregation are unavailable. To circumvent this difficulty, Anderson (1954) devised a measure of aggregation based upon the ratio of frequency as determined by large (1 square meter) quadrats to frequency as determined by 16 small quadrats located within each large quadrat. The method is based upon the fact that when one of the large quadrats falls within a clump of an aggregated species, all or most of the smaller quadrats will contain the species, hence the two frequency values will be similar and their ratio to each other small, while if the species occurs as randomly scattered individuals then each hit by a large quadrat is likely to have only one of the small quadrats occupied, and the small quadrat frequency will be much smaller than that from the large quadrat. The range is from 1 for highest aggregation to 16 for most random. Upon application of this method, Anderson found that the dry prairie plants showed the full range in possible behavior. The most aggregated species were those like *Asclepias verticillata, Helianthus occidentalis, H. laetiflorus, Geum triflorum,* and *Commandra richardsiana,* all with ratios of 3.0 or less, and all with a strongly developed rhizome method of vegetative propagation. On the other hand, the more nearly random species, with ratios of 9.0 or more, included annuals like *Ambrosia artemisiifolia* and *Erigeron strigosus,* biennials like *Lobelia spicata,* and single-stemmed perennials with no apparent means of vegetative spread like *Psoralea esculenta* and *Anemone patens.* The aggregated species greatly exceeded in number those which approached randomness and gave the prairie an over-all appearance of patchwork. The clumps of one species were widely separated from one another by spaces containing similar clumps of the other species, and the whole was tied together and unified by the ever-present matrix of the ubiquitous grasses and the general occurrence of the scattered random species.

Environment

The xeric prairies may occur on topographies varying from flat to precipitous, but the majority of the present remnants are on steep hillsides, usually sloping towards the southwest (Plates 31 and 32). The angle of slope is so great as to make passage over the area a hazardous undertaking and this danger has prevented the utilization of many of the areas for pasturage and thus has insured preservation of their vegetation. The runoff of rain water is extreme on these slopes, with continual surface movement of soil particles. In consequence, the soil blanket is very thin, except where dissolution pockets occur in the underlying limestone, and may average less than 4 inches in depth over extensive areas. On the flat uplands, as in southwestern Dane County, wind erosion accomplishes the same end as water erosion, and the soil blanket remains similarly undeveloped (Plate 1). Profile structure is poorly differentiated, the entire soil layer from surface to parent rock material frequently being a uniformly black mixture of organic matter, loess, and silica particles. Depending upon the nature of the bedrock, there also may be a surface accumulation of chert nodules or iron concretions mixed with the finer soil particles. Both water-holding properties and nutrient supplies in the true soil are very good, but the thinness of the layers means that total quantities of both water and nutrients are severely limited. There is a great uniformity from prairie to prairie in many of the chemical properties of the soils, particularly in pH and calcium content. Dry prairies more acid than pH 8.0 were very rare in the P.E.L. study, while calcium values of 5000 p.p.m. or more were the rule.

The lower water storage capacity of the soil combined with the strong insolation of the southwest slope and the high wind velocities because of exposure would seem to indicate a degree of dessication in excess of the xerophytic adaptations of the flora. In other words, the dry prairie vegetation, even though it is reduced in size and total productivity in comparison to other Wisconsin communities, still appears to be too luxuriant for the site. The probable explanation is that significant quantities of water are obtained from condensation, rather than from precipitation. During the hottest and driest seasons of the year, it is possible to walk through one of these hillside prairies in the early morning and get one's shoes thoroughly wet from the heavy dew on the grass. The very exposure which leads to desiccation by midday also increases the net radiation of energy during the night, bringing temperatures below the dew point. The rich organic content of the soil aids in the process by acting as a sponge to hold the condensate and by insulating the surface layers from the stored heat in the rocks below. Accurate measurements of the actual proportion of the total plant needs for water produced in this way would be of great interest (Stone, 1957).

MESIC PRAIRIES

Composition

The mesic prairies must have presented a scene of unequaled splendor to the first settlers but they were not regarded with high favor by these early viewers, largely because of the belief that any land which did not grow trees must be of low agricultural value. Small-scale trials quickly disproved this idea, with bumper yields of all crops that they

Plate 31.—Dry prairie on steep hillside overlooking the Mississippi River. Brady's Bluff Prairie Scientific Area, Trempealeau County.

Plate 32.—Dry prairie landscape in Green County.

Plate 33.—Pasque flower (*Anemone patens*) in dry prairie.

Plate 34.—Compass plant (*Silphium laciniatum*) in mesic prairie.

tried. General settlement of the prairies, however, was dependent upon the introduction of the steel moldboard plow because the prairie sod was so tough that it defied the efforts of a wooden plow. Once the means for subduing the prairie were at hand, these areas commanded top prices and were quickly placed under complete cultivation. This utilization was so complete that huge areas of such famous prairies as the Rock and the Arlington now are totally devoid of any remnants whatsoever of the original community. Because of this nearly total occupancy, the actual composition of the mesic prairie is a difficult matter to determine at this late date. During more than a decade of the P.E.L. investigations, intensive search turned up only 45 stands of this prairie type, mostly on old cemeteries, on a few railroad rights of way (Plate 66), and very rarely on undisturbed portions of some lands being held for gravel or mining rights rather than for agriculture. Judging by the rate of falloff in new discoveries, it is estimated that these 45 stands represent two-thirds or three-fourths of all that are in existence. Even this pitifully small fraction has been reduced during the course of the study and the remainder is in great danger of similar destruction.

The prevalent species of the mesic prairies are listed in Table XIV-5 along with the modal species. Several other species are known which once grew in this type, such as *Asclepias meadii* and *Lespedeza leptostachya,* but these have not been seen in this century and are presumed extinct in Wisconsin. None of the species on the typical list are confined to the mesic prairies and only a few reach levels of presence there which are greatly in excess of the values for either dry or wet prairies. Among the latter are smooth aster (*Aster laevis*), wild indigo (*Baptisia leucophaea*), *Desmodium illinoense,* rattlesnake master (*Eryngium yuccifolium*), compass plant (*Silphium laciniatum*), (Plate 34), and *Solidago speciosa.*

The grasses are numerically no more important in the mesic prairies than in the dry prairies, but they grow much more vigorously and reach much greater total heights. On relatively limited areas where big bluestem (*Andropogon gerardi*) and Indiangrass (*Sorghastrum nutans*) are more dominant than usual, yields of 7000 to 9000 pounds of dry material per acre (*ca.* 8500 kilos per ha.) have been measured, although the average yield from mesic prairie on good sites is more often in the range of 3000 to 4000 pounds per acre (*ca.* 4000 kilos per ha.). It appears that one of the main factors permitting these great yields is the high content of nitrogen-fixing legumes in the prairie community. A floristic analysis of the total flora of the mesic prairies reveals that the *Leguminosae* comprise 7.4 per cent of all species. This compares to values of 2.4 per cent and 1.3 per cent in comparable sites in the southern and northern forests, respectively. Four other families, in combination with the legumes, include one-half of the total species. These are the *Compositae* with 26.1 per cent, *Gramineae* with 10.2 per cent, *Labiatae* with 3.9 per cent, and *Liliaceae* with 3.5 per cent. Again we see that the daisies far outnumber the grasses in terms of variety of species.

The distribution of blooming seasons in the mesic prairies is essentially like that in the dry prairies, with slightly fewer species blooming before June 15 and after August 15 and slightly more in midsummer. The mesic prairies warm up more slowly in early spring; hence they have little representation of the prevernal group such as pasque flower and buttercup. Characteristic spring flowers of the mesic prairies include shooting star (*Dodecatheon meadia*), bedstraw *(Galium boreale)*, golden puccoon (*Lithospermum canescens*), downy phlox (*Phlox pilosa*), spiderwort (*Tradescantia ohiensis*), prairie oxalis (*Oxalis violacea*), and prairie violet (*Viola pedatifida*). Of these, all but the bedstraw, puccoon, and phlox die down by midsummer in the same manner as the spring ephemerals of the deciduous forests and in response to similar changes in the environment. Butler (1954) found that the average height of the canopy of the mesic prairies increased to 58 cm by August. This canopy is so dense as to reduce light intensities near the soil surface to 30 foot-candles or less.

The midsummer aspect of the mesic prairies is one of a continuously changing carpet of color, largely of yellows and purples in various shades (Plate 35). Among the most showy members of this assemblage are the composites, with tickseed (*Coreopsis palmata*), blazing star (*Liatris aspera*), yellow coneflower (*Ratibida pinnata*), black-eyed Susan (*Rudbeckia hirta*), the rosin weeds and compass plant (*Silphium* sp.), and purple coneflower (*Echinacea pallida*). The last is not common except on the prairies of Walworth, Rock, and Green counties; but in these areas it plays a major role in the summer display, with its yard-high crown of large, golden-brown, conical flower heads each supporting a reflexed ring of long, bright purple rays. Other species blooming in summer are the white-flowered rattlesnake master (*Eryngium yuccifolium*), flowering spurge (*Euphorbia corollata*), the purple-flowered leadplant (*Amorpha canescens*), tick-trefoil (*Desmodium illinoense*), and prairie clover (*Petalostemum purpureum*).

In the autumn, a large number of asters, goldenrods, and sunflowers add to the flowering parade, but while these individually are colorful or even brilliant, somehow their effect is weakened by a diluting or curtaining action of the tall grasses. Such dominants as big bluestem, Indiangrass, and prairie panic grass (*Panicum virgatum*) begin to elongate their flowering stems about mid-August and these attain full growth in late September or early October. Full growth means an average height of 1.8 meters for big bluestem with the best stands attaining 2.5 to 3.0 meters in exceptionally favorable years (Plate 36), while the other two species range from 1.2 to 1.8 meters. The early settlers had great difficulty in finding cattle which had strayed into the prairies at this season, even when they searched on horseback because the animals were completely hidden by the tall swaying grasses.

A certain geographical variation exists between the mesic prairies in different regions of the prairie-forest province. Thus the prairies in Racine and Kenosha counties, located on the poorly drained upland soils of the Elliot-Beecher soil series, have a number of species not found elsewhere in Wisconsin. Included are such colorful species as *Phlox glaberrima*, *Liatris spicata*, and *Allium cernuum*. A number of other species are found only in eastern Wisconsin but they do not have a common western boundary, some extending farther westward than others. Thus the prairie dock (*Silphium terebinthinaceum*) is an abundant and striking member of the prairie south and east of Sauk County; but it is absent to the north or west. The purple coneflower (*Echinacea pallida*) is restricted to the area south and east of Dane County. On the other hand, the northwestern prairies have such species as *Astragalus caryocarpus*, *Psoralea argophylla*, and *Symphoricarpos occidentalis*, which are lacking elsewhere. Most of these regional differences are of a minor nature and the general physiognomic aspect of the mesic prairies is quite uniform throughout.

Orport and Curtis (1957) found that the mesic prairies had a number of characteristic soil microfungi, including *Fusarium oxysporum*, *Penicillium nigricans*, *P. janthinellum*, *P. variable*, *Papulospora* sp., and *Acrostalagmus* sp. No studies of other biota in the mesic prairies are available.

The small size and scattered nature of the present remnants preclude accurate studies of the populations of larger animals. General observations indicate that the major mammals today are moles, mice, skunks, and badgers, but their presence is no doubt influenced by the nature of the surrounding lands. One animal of high fidelity to the mesic and wet prairies is a species of ant which builds large mounds 1 to 1½ feet high by 2 to 3 feet in diameter. These mounds are a very conspicuous sight when the prairies have just been burned over for then they are exposed to full view. The number of mounds may reach 40 or 50 or more per acre. Since they are active only for a limited number of years and are replaced by new mounds, a very considerable percentage of the total area may be worked over in a relatively short time. The moles, skunks, and badgers also move large amounts of soil in their activities and contribute greatly to the internal instability of the prairie.

Structure

In a number of mesic prairies in Rock County, Green (1950) made both frequency and density determinations of the forbs by means of randomly distributed one square meter quadrats. Wagner (1951) examined both grasses and forbs in five mesic prairies in Columbia, Dane, Green, and Rock counties by the subdivided quadrat method later used by Anderson, as mentioned under the xeric prairies. This information, plus that from a number of other P.E.L. studies of individual stands, conveys a useful picture of the structure of mesic stands and at the same time it can be used to study the types of spatial distribution present. The ten species with highest frequency × presence index are *Euphorbia corollata* (6450), *Stipa spartea* (4002), *Helianthus laetiflorus* (3480), *Aster ericoides* (3344), *Rosa* sp. (3276), *Aster laevis* (3113), *Panicum leibergii* (2852), *Andropogon gerardi* (2842), *Ratibida pinnata* (2720), and *Coreopsis palmata* (2584). With the exception of the *Euphorbia* and *Andropogon* none of these are on Anderson's list of the top ten on the xeric prairies. Of especial significance is the high rank of needle-grass (*Stipa spartea*) as the second highest species. On a density basis, the species found by Green in Rock County with the highest numbers of individuals per square meter were *Aster laevis* (7.54), *Aster ericoides* (6.52), *Coreopsis palmata* (5.41), *Helianthus occidentalis* (3.84), *H. laetiflorus* (3.19), *Euphorbia corollata* (2.36), *Heliopsis helianthoides* (2.35), *Tradescantia ohiensis* (2.16), *Phlox pilosa* (1.73), and *Echinacea pallida* (1.47). The total density of all individuals of all species of forbs was 88.1 per square meter, therefore these ten species contributed 41.5 per cent of the total.

Aggregation in the mesic prairie, as measured in several ways, varies from nearly random to extremely aggregated. Groups of species at the two extremes are listed in Table 16, together with their indices of aggregation as determined by the F/f test (frequency % in 1 M^2 quadrats/frequency % in 1/16 M^2 subquadrats, Anderson, 1954), the D/d test (actual density/expected density, Curtis and McIntosh, (1950), and the A/F test (abundance/frequency %, Whitford, 1949). With minor exceptions, these indicated comparable ratings. It is of interest that all but one of the extremely aggregated species are characterized by the possession of active vegetative reproduction by means of short

Table 16
Aggregation of mesic prairie plants

	Indices of aggregation		
Species	D/d (1)	A/F (2)	F/f (16:1) (3)
Aggregated species			
Helianthus laetiflorus	9.38	0.35	14.48
H. occidentalis	24.02	1.74	14.25
Galium boreale	11.6	8.08	14.43
Coreopsis palmata	15.03	0.58	10.00
Antennaria neglecta	20.50	7.23	12.31
Phlox pilosa	11.53	0.85	12.91
Aster ericoides	11.65	0.34	12.04
A. laevis	12.99	0.39	9.42
Random species			
Polytaenia nuttallii	1.35	0.17	.....
Eryngium yuccifolium	1.28	0.07	10.00
Desmodium illinoense	1.23	0.15	9.31
Ambrosia artemisiifolia	2.31	0.08	5.80
Potentilla arguta	2.19	0.21	4.98
Lactuca biennis	.....	....	1.59
Asclepias syriaca	1.75	0.11	.....
Cirsium discolor	1.41	0.09	.....

(1) *D/d*—Ratio of actual density to expected density for random distribution. Random = 1.0

(2) *A/F*—Ratio of abundance to frequency. Random = 0.02

(3) *F/f*—Ratio of frequency in large quadrats to frequency in 16 small, included quadrats. Random = 16.0

rhizomes. The exception is *Phlox pilosa,* which has no known means of vegetative increase. On the other hand, the nearly random species are annuals (*Ambrosia artemisiifolia*), biennials (*Lactuca biennis, Cirsium discolor*), or non-spreading taprooted perennials (*Eryngium yuccifolium*). The presence of such rhizome producers as the *Asclepias* species and *Euphorbia corollata* in the random group is to be explained because these species spread so far and so fast that any clones which they might produce are not distinguishable by quadrats of the size used.

Of interest is the fact that four of the most aggregated species are known or suspected to produce antibiotic chemicals which repress the growth of other species within their colonies. The two species of *Helianthus* have been shown to produce autotoxic substances as well; flowering is restricted to the peripheral stems of the colony which have recently grown into new soil while the central stems remain in a vegetative condition (Curtis and Cottam, 1950). *Coreopsis palmata* shows similar behavior in the field but its activity has not been checked experimentally. *Antennaria neglecta* has the antibiotic but not the autotoxic effects. Extracts of its leaves have been shown to be repressive to the growth of a number of other species. A number of other Composites are known to possess this antibiotic activity (Grümmer, 1955; Knapp, 1955). This form of chemical competition naturally is effective in maintaining an aggregated type of dispersion. The purity of the resulting clones makes them far more conspicuous than others of equal density which are intermingled with a number of other species.

Many species of the mesic prairie tend to grow in loose colonies which are not detected by the methods used in the foregoing studies. Indeed, this tendency for development of large societies or clans of one species separated by considerable distances from the next assemblage of the same species was noted by many early observers of the original prairie. It must have been even more striking on the broad expanses of undisturbed prairie than it is on the present tiny remnants. Short (1845) noted this gregarious habit on the early prairies of Illinois:

> Its leading feature is the unbounded profusion with which a few species occur, in certain localities, [rather] than the mixed variety of different species occuring everywhere. Thus from some elevated position in a large prairie the eyes take in at a glance thousands of acres literally empurpled with the flowering spikes of several species of *Liatris,* in other stations . . . a few species of yellow flowering *Coreopsis* occur in such profuse abundance as to tinge the entire surface with a golden burnish. This peculiarity of an aggregation of individuals of one or more species to something like an exclusive monopoly of certain localities obtains even in regard to those plants which are the rarest and least frequently met with; for wherever one specimen was found, there generally occured many more in the same immediate neighborhood.

Environment

The mesic prairies, as indicated earlier, are on flat or gently rolling land forms. The level sites are frequently on glacial outwash with a stratified and very porous subsoil of sand or gravel while the undulating sites may be on glacial till of recessional moraines or on residual or loessial soils on the rolling surfaces of dolomitic bedrock. The soil profile consists of a surface or A_1 layer which is black in color and very rich in organic matter. At increasing depths, this organic content gradually diminishes until it fades out altogether in the lower B levels 12 to 24 inches or more below the surface. The parent material may be unaltered loess, glacial till, or decomposed rock. In the Racine-Kenosha area, where mesic prairies developed on lacustrine deposits of former beds of Glacial Lake Chicago, the parent material is rich in clay minerals which impede internal drainage to a certain extent. These soils belong to the Elliott-Beecher-Morley soil series. Elsewhere in the glaciated area the mesic prairie soils are in the Parr, Waupun, Warsaw, and Waukegan soil series, while Driftless Area prairies are largely on Tama, Downs, and Muscatine series, with the exact kind at any spot depending on such characters as depth of solum, topography, and nature of parent materials (Hole and Lee, 1955). In terms of drainage, the mesic prairies usually assume a position on mid-catena although the catena relationships of the flat outwash lands are not immediately evident.

The black topsoil layer of the mesic prairies is very rich in nutrients and is slightly acid to neutral in reaction. The calcium content is not as high as in the xeric prairies but is considerably above the levels in most forest soils. The high organic matter content is reflected in the water-retaining capacities which ranged from 51 to 90 per cent in the P.E.L. studies, with an average of 67 per cent for the top 6 inches of soil. These values slowly and evenly declined with depth and reached minimum values of from 20 to 30 per cent at depths of 36 to 48 inches (Partch, 1949; Wagner, 1951). No extensive investigations of the organic matter content of Wisconsin mesic prairie soils have been made, but a few analyses indicate that the total amounts may be relatively great, in the neighborhood of 60 tons per acre (135 metric tons per ha.). One of the major reasons for the continued high agricultural productivity of these soils is this high percentage of organic content, since the favorable structure with highly stable soil aggregates is largely due to activities of the organic matter. Many of the mesic prairies have been in continuous crop production for over a century but they still retain a deep black color and are clearly visible from a distance during spring plowing time.

The macroclimate of the mesic prairies, determined from weather stations located on or near the original prairies, is given in Table XIV-6. Microclimate studies have not been adequate to give a complete picture of the conditions within the prairie. Studies in the tall grass prairie of the University of Wisconsin Arboretum indicate that the growing season in the prairie begins two to three weeks earlier than in the hardwood forest. On the other hand, killing frosts occur sooner in the autumn than they do in the forest, so that the total period for growth for frost-sensitive species may be of about the same length. Most of the prairie species, however, are strongly frost resistant, with some of the late-blooming asters and gentians unharmed by nighttime temperatures as low as 12°F. This frost resistance may be correlated with the high levels of potassium in the soil. Summer temperatures in the prairie may greatly exceed those in the forest. The average July soil temperature at 2 P.M. at a depth of one inch was 82°F. over a three-year period. An adjacent oak forest showed an average of 71°F. under the same conditions. This differential was lowered to 5° at a height of 6 inches above the soil and to 3.5° at the 36-inch level. Maximum and minimum readings of July air temperatures at the 6-inch level showed that the prairie had a much greater daily range in temperature, with an average maximum of 91.5°F. and an average minimum of 56.7°F., for a range of 34.8°F., compared to 75.7°F., 59.1°F., and 16.6°F., respectively, for the oak forest. In the cool summer of 1951, the maximum prairie temperatures in July exceeded 100°F. on over one-half the days and this figure would no doubt be higher in a hot summer. During this same period in 1951, the maximum temperature in the forest exceeded 85°F. only on three occasions and never reached 90°F.

Relative humidities at the same season averaged 66 per cent at the one-inch level in the prairie and 57 per cent in the forest, but at 36 inches the prairie was drier, with a reading of 50 per cent compared to 61 per cent in the forest. The high humidity near the soil surface in the prairie was no doubt influenced by the thick layer of non-compacted mulch consisting of old leaves and stems of grasses. Despite these higher surface humidities, evaporation in the prairie was greater than that in the forest, because the higher temperatures more than compensated for the differential in relative humidity. Scattered readings for 24-hour intervals during rainless periods in the summer months showed a fourfold to fivefold increase in evaporation in the prairie (Selleck and Schuppert, 1957).

LOWLAND PRAIRIES
Composition

The prevalent species in the two lowland prairie segments of the compositional gradient are shown in Tables XIV-7 and XIV-9, together with the other species which reach their highest Wisconsin levels in these communities. There is a great difference in the floristic richness of the two segments, with 186 species and a species density of 44 in the initial, wet prairies, and 252 species and a species density of 62 in the intermediate, wet-mesic prairies. The leading grass dominants in the wet-mesic group include big bluestem (96.8 per cent presence), blue-joint grass (*Calamagrostis canadensis* 74.3 per cent), sloughgrass (*Spartina pectinata* 74.3 per cent), and wild rye (*Elymus canadensis* 64.5

per cent), whereas the wet prairies are dominated by bluejoint (91.0 per cent presence), sloughgrass (19.0 per cent), big bluestem (68.2 per cent), and prairie muhly (*Muhlenbergia racemosa* 36.4 per cent). A number of the prevalent species reach their state-wide maxima in these prairies, as indicated in the tables, but relatively few of the major species are of high fidelity for the type. Most of the major species occur in either the mesic prairies or in such other wet soil communities as the fen or the sedge meadow. Perhaps the most typical species of the lowland prairies are New England aster (*Aster novae-angliae*), bottle gentian (*Gentiana andrewsii*), yellow stargrass (*Hypoxis hirsuta*), Kansas gayfeather (*Liatris pycnostachya*), Turk's-cap lily (*Lilium superbum*), sloughgrass (*Spartina pectinata*), and prairie dock (*Silphium terebinthinaceum*). Of these, only the gayfeather is restricted to the prairie region of the United States; the others extend south and east to the Atlantic Ocean in moist, open sites. The same is true for the majority of the modal species in Tables XIV-7 and XIV-9.

A floristic analysis of the lowlands prairie species shows that 51 families are represented in each segment, with one-half of the total species in five families in the wet-mesic prairies and in the same five plus the *Liliaceae* in the wet prairies. The five are *Compositae, Gramineae, Leguminosae, Umbelliferae,* and *Labiatae*. As in the mesic and xeric prairies, the spring flowering species include only about one-third of the total, while the autumn bloomers are about one-fourth of the total. The wet soils are very slow to warm up in the spring, so the prevernal flora is very poor or totally lacking in some instances. One of the few plants to bloom in May is the valerian (*Valeriana ciliata*). By June, growth is greatly speeded and a number of showy species make their appearance, including downy phlox (*Phlox pilosa*), shooting star (*Dodecatheon meadia*), yellow stargrass (*Hypoxis hirsuta*), Seneca snakeroot (*Polygala senega*), puccoon (*Lithospermum canescens*), and golden alexanders (*Zizia aurea*). The midsummer and autumn displays, like those in the mesic prairies, are dominated by the composites, with the genera *Aster, Solidago, Silphium, Cacalia,* and *Helianthus* prominent.

One interesting feature of the wet-mesic prairies is relatively high content of species which reach their optimum development in the mesic or xeric forests of the state. Altogether, 19 species were noted in the P.E.L. studies with this behavior, including such species as *Allium tricoccum, Anemone quinquefolia, Geranium maculatum, Polemonium reptans, Prenanthes alba, Smilax herbacea,* and *Vitis aestivalis*. It is possible that the high level of available soil moisture compensates in some way for the greater evaporation and insolation of the prairie and allows these forest plants to maintain themselves, but other explanations may be involved.

The moist soil prairies bear close relationships to the fens and the sedge meadows. By some investigators, the fen is thought to be merely a variant of the wet prairie with a specialized environment including an internally flowing supply of bicarbonate-rich water. The index of similarity based on the typical species is indeed high (54 per cent), but the two have been kept separate in this study because of certain unique characters of the fens. The sedge meadows are dominated by members of the *Cyperaceae* rather than the *Gramineae* but they have many species in common with the wet prairies and frequently grade into them at the base of the drainage catena. Many of the accounts of the early travelers and land surveyors failed to distinguish between the wet prairie and the sedge meadow and this fact must be kept in mind when interpretating the early records.

According to Orpurt and Curtis (1957), the microfungi of the lowland prairies included *Clasdosporium herbarum*, *Myrothecium striatisporum*, *Penicillium javanicum*, *Cylindrocarpon* sp., *Mortierella* sp., and *Cephalosporium* sp. The best account of the animals of the moist prairies is given in Hawkins' report (1940) on the history of the University of Wisconsin's Faville Grove Prairie Preserve in Jefferson County. Among the birds, the most typical species are the prairie chicken, the upland plover, the Brewer's blackbird, the bobolink, the dickcissel, the Henslow's sparrow, and the savanna sparrow. The massasauga or prairie rattlesnake is nowhere common in Wisconsin but the most frequent habitat is the wet prairie.

Structure

Relatively few studies of the quantitative structure of moist prairies have been made in Wisconsin. Frequency values are available for a number of stands but no density determinations are on record, nor are there any reports of detailed studies of aggregation. On the basis of the frequency x presence index, the ten leading species of the wet-mesic prairies are: *Ratibida pinnata* (4320), *Andropogon gerardi* (3880), *Helianthus grosseserratus* (3321), *Solidago rigida* (2926), *Aster azureus* (2623), *Rosa* sp. (2430), *Comandra richardsiana* (2310), *Spartina pectinata* (2072), *Pycnanthemum virginianum* (2016), and *Rudbeckia hirta* (1776). Insufficient data are available from the wet prairies to give a comparable index rating for this community.

Aggregation appears to be of the same order as that in the mesic prairies. One factor which contributes to the lack of spatial homogeneity is the varying moisture supply caused by slight differences in microrelief (local elevation). Most of the wet prairies have an uneven surface, with hummocks or small patches of soil which are 6 to 12 inches higher than the intervening land. The improved aeration in these small rises is enough to greatly influence the species growing there, therefore a mosaic of different combinations results. Frequent members of the low spots include blueflag (*Iris shrevei*), water hemlock (*Cicuta maculata*), and sloughgrass (*Spartina pectinata*) while the hummocks will support asters (*Aster azureus* and *A. laevis*), phlox (*Phlox pilosa*), prairie dock (*Silphium terebinthinaceum*), and Sullivant's milkweed (*Asclepias sullivantii*). Bluejoint grass (*Calamagrostis canadensis*) and many other species appear indifferent to these minor altitudinal gradients and are found throughout the prairie.

Environment

The moist prairies are usually located on lowlands subject to inundation by heavy rains or by floodwaters from nearby streams. Beds of extinct glacial lakes are a favored site in eastern Wisconsin, while in the Driftless Area, the moist prairies are largely confined to the regions between floodplain forests and the uplands bordering the larger river valleys. All such habitats are likely to accumulate cold air which drains from the hills on still nights. This effect is particularly noticeable during the summer when dense blanket-fogs are likely to form over the prairies. The fog droplets collect on leaf surfaces and must contribute a considerable amount of moisture to the already wet site. In the morning, after the sun has dispelled the fog, local air humidities within the vegetational canopy remain very high, so that by noon there is an almost unbearable wave of hot, sultry air. The lush and rapid growth of the low prairie plants reflects these tropical conditions. In fact, there are very few places in the tropics which combine the high moistures and high temperatures of the wet prairies of the Mississippi valley in midsummer.

The same air drainage which lowers temperatures below the dew point in summer also is responsible for late spring and early autumn frosts; many people have discovered this who have tried to grow tomatoes or other tender crops on old prairie lands. The native prairie plants, however, are largely frost resistant and do not appear to be adversely affected by relatively short growing seasons.

The soils of the moist prairies occupy the lowest positions on the catenas. They are rich in organic matter and in nutrient salts; they show an impeded internal drainage by the presence of a gley layer 2 to 4 feet below the surface. Prominent groups include the Ashkum and Bryce series in the Racine-Kenosha area, the Elba, Kokomo, and Brookston series in the remainder of the glaciated area, and the Garwin and Judson series in the Driftless Area (Hole and Lee, 1955). The top layers of the soil frequently approach a peat in structure, with a very high content of partially decomposed organic matter derived from the fibrous grass roots. This is shown by their low specific gravity and high water-holding capacity. As in the other prairie soils, the black color of the top layer gradually decreases in intensity with increasing depth, but there may be an abrupt change to bluish-gray or even white when the gley layer is reached.

A number of lowland prairies are known that possess an unusual internal water system. These are most common along the Rock River in Jefferson County but are also found in other counties. When a pipe with a well point at its end is thrust into the ground along the border between the prairie and the river marshes, water gushes forth from the pipe, often to a height of 2 or 3 feet above the surface of the ground. These artesian sources are called flowing wells and many of them have been producing a steady stream of water for one hundred years or more. They appear to result from a perched water table, suspended by the impervious layers of lacustrine clays laid down under glacial lakes. When the dam of the lake was broached by the stream of the valley, that stream became recessed to a level considerably below the original lake bed. The rain and run-off waters accumulating on the level surface of this now terrace-like plain are held under pressure and jet forth through the well points which enter the pool. It is to be expected that this internal pressure system also maintains a constant upward movement of water into the overlying prairie soil; thus it keeps the prairie plants well supplied with moisture even during drought periods.

GEOGRAPHICAL RELATIONS OF WISCONSIN PRAIRIES

In contrast to the situation in forest ecology, the prairies of North America have received considerable phytosociological attention, with quantitative floristic studies available for most portions of the grassland area. Much of this work has been done by J. E. Weaver and his students at the University of Nebraska. Weaver has summarized most of this information in two recent books *North American Prairie* (1954) and *Grasslands of the Great Plains* (1956). Another center of ecological work on the prairie is Illinois, where Gleason, Vestal, and their students added much to our knowledge. A large portion of the research on grasslands has been conducted from the viewpoint of yield, drought relations, or other special relations, therefore the reports are not complete accounts of the total floristic composition. It is common for the prairie investigators to give great or exclusive attention to the grasses, in much the same way that the trees are emphasized in forest studies, with the result that information on the non-grass species is frequently incomplete.

A few regional studies have been made in which a number of stands of a given type have been examined by comparable methods and for which presence and frequency data are available. In most regional studies, only one or a few stands of each type in the area were used, so that no information on presence is given. Because of this, the general comparisons of Wisconsin prairies with those of other regions are made on the basis of the number of species listed in each study which are present on the P.E.L. Wisconsin lists, and these are expressed as a percentage of the total number found. In this way, studies of a few stands may be used as well as broad studies of many stands. Where adequate information is available, comparisons are made by the index of similarity $(2w/a + b)$, as used in earlier chapters.

One of the features immediately noticeable in a comparison of Wisconsin prairies with those of other areas is the familiar pattern of decreasing similarity with increasing distance (Table 17). Thus, the prairies in adjacent portions of Illinois, Iowa, and Minnesota have 75 per cent or more of their species in common with those on the Wisconsin lists, whereas these values drop to 30 or 40 per cent in the true prairie states, and they drop to values below 20 per cent in the high plains and the Palouse region of Washington and Idaho. Several geographical trends are apparent in the way the similarities change. Thus, the Ohio prairies have higher values than their location might suggest and the same is true along a band extending southwest through Missouri to Oklahoma. Another exceptional band follows the edge of the prairie-forest border northwestward into Canada, where the grasslands of the Peace River country of Alberta (at 59°N., 119°W.) have a surprising 30 per cent of their species in common with those of Wisconsin (at 44°N., 90°W), over 1600 miles away.

Examination of the ranges of the typical Wisconsin species reveals several interesting groupings which partially explain some of the above trends. As Butler (1954) pointed out, the plants of the lowland prairies of Wisconsin generally have ranges which extend toward southeastern United States, while the dry prairie plants range to the southwest.

Table 17
Geographical relations of Wisconsin prairies

Location of study	No. of species listed	% in common with Wisconsin stands
NE. Illinois (Vestal, 1914)	95	95.7
Ann Arbor, Michigan (Gleason, 1917)	26	92.3
Peoria, Illinois (Brendel, 1887)	92	90.2
Wilton, Iowa (Shimek, 1925)	156	83.9
Central Iowa (Conard, 1952)	233	78.1
Clay Co., Kansas (Shaffner, 1926)	48	70.8
Hancock, Illinois (Kibbe, 1952)	247	68.4
Oklahoma (Smith, 1940)	47	65.9
S. Manitoba (Shimek, 1925)	131	58.0
Ohio (Jones, 1944)	157	52.3
Central Saskatchewan (Coupland et al., 1953)	85	41.2
W. North Dakota (Hanson et al., 1938)	57	36.8
Sand Hills, Nebraska (Pool, 1914)	101	36.6
NW. Alberta (Moss, 1952)	139	30.2
E. Colorado (Hanson, 1955)	89	22.5
SW. Saskatchewan (Coupland, 1950)	94	20.2
S. Alberta (Moss et al., 1947)	260	20.0
S. Montana (Wright et al., 1948)	50	16.0
W. Idaho (Daubenmire, 1942)	77	10.4

The differences in affinities of the floras of our wet and dry prairies have considerable significance in any interpretation of prairie history because they indicate that all of the grasslands could not have had the same origin. The large Alleghenian meadow element in the wet prairies would indicate that this community migrated into Wisconsin from the southeast along with the hardwood forest of similar relationships. Environmental conditions favorable for such a migration would no doubt be very different from those suitable for the entrance of a xeric flora of southwestern desert or plains origin. Thus, it appears reasonable to believe that the expansion of prairies in Wisconsin was not a single, unified movement of a homogeneous plant formation, but rather a gradual entrance of a variety of elements, with each arriving during separate climatic regimes favorable to that particular group. Certain similarities in structure of the various communities and in their environmental tolerances enabled some of the component species to diffuse from one element to another and thus impart a certain degree of local uniformity to the whole. Thus the bluestem grasses, leadplant, flowering spurge, and other broad tolerance species came to assume prominent roles in all of the local prairies, regardless of their origin. The prairies in other Midwest states have had a similar history, but since they are differently located with respect to source regions and since they have been differently affected by postglacial climatic changes, the relative composition of their flora is correspondingly different. Thus Gleason (1901) reported that 48 per cent of the 415 species he listed for the prairies of Illinois were of southeastern origin. The prairies of Iowa, however, have a higher content of southwestern species (Shimek, 1911), as do those of Minnesota (Moyer, 1900), since they are closer to that source region and have been under the influence of dry climates more frequently and for longer periods than regions to the east.

Detailed comparisons are possible between the xeric prairies of Wisconsin and those of Illinois because of the thorough study of the latter by Evers (1955). A total of 251 species was reported for 36 stands in Illinois, compared to 247 species in the P.E.L. stands. The index of similarity, based on presence percentages, is 55.9 per cent, which indicates a considerable uniformity in behavior of the component species in the two areas. Five of the ten leading species in Illinois were also in the group of ten leading species reported by Anderson. Evers noted that most of the Illinois xeric prairies, which he called "Hill prairies," were located on steep bluffs, facing southwest, and that these sites were characterized by a very deep layer of loess. He attributed the xeric nature of the habitat to this loess blanket and felt that the structure of the prairie was related to the loess. Anderson, however, found that the very similar prairies of Wisconsin were associated with very thin soils over bedrock and that when deep pockets of loess were encountered, the prairies changed greatly in composition. Herein lies a good example of the dangers of single factor causation in explanations of why plants grow where they do.

The mesic prairies of Wisconsin closely resemble those of the areas to the west. In connection with the preparation of a report on the continuum nature of the entire group of prairies (Curtis, 1955), a series of prairies was studied in northeastern Iowa and southeastern Minnesota. The index of similarity of this group with the Wisconsin mesic prairies was 67.5 per cent. Conard (1952), in his book on the vegetation of Iowa, listed the plants seen on prairies in seven countries in central Iowa. Of the 233 species he found, 78.1 per cent are also on the Wisconsin lists; the entire group had an index of similarity of 59.1 per cent, based on presence values.

An extensive study of 100 stands of upland prairie was conducted by Weaver and Fitzpatrick (1934) in Nebraska, Missouri, and Iowa. On the basis of the 142 most important species listed by them, an index of similarity of 49.8 per cent was obtained by comparing their presence values with those of Wisconsin mesic prairies. These same figures yielded a highly significant correlation coefficient of 0.516 when compared to the first 65 stands studied in Wisconsin (Curtis and Greene, 1949).

The wet prairies seem most closely related to those of Illinois and Indiana, although truly adequate bases for comparison are lacking. One of the difficulties is the failure of most investigators in those states to distinguish between wet prairie and sedge meadow. The Illinois prairies, in particular, were extensively developed on a planosol topography which was almost perfectly flat and had very poor internal drainage; thus they were inundated in the spring and frequently remained wet all year. Early accounts of the Illinois prairies such as those of Short, Brendel, and Gleason all describe a flora with a much higher content of hydrophytes than anything ever found in Wisconsin, although they never indicated whether the area was dominated by sedges or by grasses. Prominent among the essentially aquatic plants they list as common on the prairie are bogbean (*Menyanthes trifoliata*), sundew (*Drosera rotundifolia*), arrowleaf (*Sagittaria latifolia*), and numerous reeds, sedges, and rushes. Gleason (1901), for example, lists 32 species of sedge, which would imply that the areas approached sedge meadow rather than real prairie.

SUCCESSION

Our knowledge about prairie successions is mostly derived from studies of secondary succession, such as the succession produced when prairie invades an abandoned field or a road cut. Autogenic succession at a primary level is poorly understood but it almost certainly differs in many respects from the comparable process in forests. In the first place, shade tolerance plays a minor role in the prairie, not because low light is never a factor, but because solid canopies produced by a given set of dominants are rarely if ever produced. The very great importance of burrowing animals with their newly deposited mounds of mixed subsoil has no exact counterpart in the forest. In effect, there is a continual, internal succession going on in the prairie at all times, with new mounds being invaded by pioneer species tolerant of extremes in moisture and nutrient supply and able to withstand full sunlight in their seedling stages. As the mounds become mature and are gradually eroded away to base level, these pioneer plants are replaced by others

Plate 35.—View in mesic prairie, showing forbs and grasses, with compass plant (*Silphium laciniatum*) in foreground.

Plate 36.—Big bluestem (*Andropogon gerardi*) and Indiangrass (*Sorghastrum nutans*) along path in University of Wisconsin Arboretum, Madison.

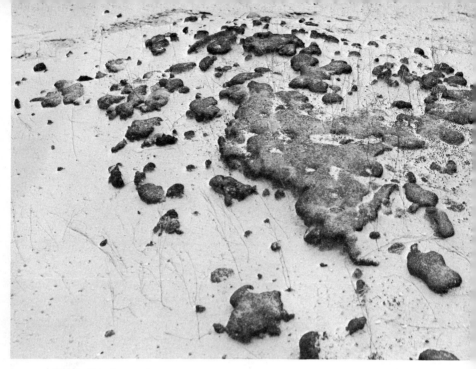

Plate 37.—Moss hummocks on sand barrens.

Plate 38.—*Scleroderma geaster* and Hudsonia *(Hudsonia tomentosa)* growing in desert pavement.

Plate 39.—View of blowout in sand barrens from windward end. Near Arena, Iowa County.

Plate 40.—Excavation of grass tussocks by sandblast action of wind.

Plate 41.—Hudsonia carpet on stabilized floor of blowout. Near Spring Green, Sauk County.

Plate 42.—Bracken-grassland, showing slow invasion by balsam fir (*Abies balsamea*) on periphery. Northern Highland State Forest, Vilas County.

with more exacting needs and finally by the most conservative prairie species which need a more even supply of soil materials and whose seedlings are able to grow in reduced light. Thus, the prairie consists of a large proportion of micro-gap-phases, each of rather short duration. Any given spot of mature prairie is likely to be destroyed and replaced by pioneer plants at any time. The over-all composition of a large area of prairie may remain stable and constant, but any small area within it would show violent fluctuations with time.

Under present conditions, the soil mounds of ants and mammals are frequently invaded by exotic weeds from nearby agricultural fields; these weeds include campion (*Lychnis alba*), sweet clover (*Melilotus albus*), and quackgrass (*Agropyron repens*). Among the native species which pioneer in such spots are the vigorous, weedlike annuals and biennials including ragweed (*Ambrosia artemisiifolia*), fleabane (*Erigeron strigosus*), prairie thistle (*Cirsium discolor*), sticky catchfly (*Silene antirrhina*), and evening primrose (*Oenothera biennis*). These are followed by a number of short-lived perennials, which include wild rye (*Elymus canadensis*), vervain (*Verbena stricta*), black-eyed Susan (*Rudbeckia hirta*), prairie fivefinger (*Potentilla arguta*), bergamot (*Monarda fistulosa*), hairy aster (*Aster pilosus*), and common milkweed (*Asclepias syriaca*). The subsequent invasions are naturally dependent in large part upon the moisture relations of the surrounding prairie and on the variety of seed supplies available, with a spot in the middle of a catena covered by prairie from top to bottom more variable than one in the center of a large, uniform area of a single prairie type.

Some indication as to the relative conservatism or climax status of the important prairie species can be derived from the work of Anthoney (1937) in Rock County, Thomson (1940) in central Wisconsin, and Gould (1937) in Dane County. These investigators plotted the existing distribution of plants along roadsides in relation to the known boundaries of former prairies, as shown by the surveyors' records. They reported a group of species which appeared to be totally confined to the former prairie area, including such plants as the grama grasses (*Bouteloua hirsuta*, *Bouteloua curtipendula*), thimbleweed (*Anemone cylindrica*), downy gentian (*Gentiana puberula*), yellow puccoon (*Lithospermum incisum*), compass plant (*Silphium laciniatum*), prairie dock (*Silphium terebinthinaceum*), and rattlesnake master (*Eryngium yuccifolium*). Another group had spread to varying distances from the prairie borders but remained concentrated around these areas. It included prairie dropseed (*Sporobolus heterolepis*), pasque flower (*Anemone patens*), prairie smoke (*Geum triflorum*), the prairie clovers (*Petalostemum* sp.) which had moved only small distances, silkyaster (*Aster sericeus*), blazing star (*Liatris aspera*), stiff goldenrod (*Solidago rigida*), and needlegrass (*Stipa spartea*) which had moved much farther. Finally a group of species were noted whose present range bore little relation to the original prairies. Included here were such widespread types as big bluestem (*Andropogon gerardi*), spiderwort (*Tradescantia ohiensis*), switch grass (*Panicum virgatum*), orange milkweed (*Asclepias tuberosa*), and the baptisias.

Observations made on this basis have a low degree of reliability, because at least some of the species were originally growing in the areas surrounding the prairies, as members of the savannas or open forests. Furthermore, the fact that some of the species were absent from the non-prairie areas may have been due to the lack of suitable conditions along the roadsides and cannot be taken as an indication of inherent conservatism. Thus, when the reputedly very conservative rattlesnake master (*Eryngium yuccifolium*) was introduced in a few places in the University of Wisconsin Arboretum, it literally exploded, making solid stands several acres in extent, almost to the exclusion of all other plants. The blazing stars, yellow coneflowers, and prairie docks have done the same.

Judging by all the evidence, both from distribution studies and from the prairie re-establishment experiments (Greene and Curtis, 1953) it would appear that the legumes as a whole are the group which spreads the least and is the most susceptible to destruction by disturbance. Among other very conservative plants would be the following: *Asclepias sullivantii, Aster oblongifolius, A. ptarmicoides, Cacalia tuberosa, Callirhoe triangulata, Castilleja sessiliflora, Cirsium hillii, Kuhnia eupatorioides, Lilium philadelphicum, Panicum leibergii, Parthenium integrifolium, Sporobolus heterolepis*, and *Tofieldia glutinosa*.

Secondary successions on abandoned agricultural fields have been studied by Thomson (1943) in Juneau County and by Buss (1957) in Dunn County with very similar results. Both of these areas are on very sandy soils and both are near the contact zone of prairie and northern pine-hardwood forest. During the first five years after abandonment, the fields were dominated by annual weeds, which included ragweed (*Ambrosia artemisiifolia*), sandbur (*Cenchrus pauciflorus*), horseweed (*Conyza canadensis*), and foxtail grass (*Setaria glauca*), with a prominent display of the biennial sand primrose (*Oenothera rhombipetala*) in the third and fourth years. A number of perennial prairie plants invaded the area during this same period but they did not become dominant until the second five years. Included were *Agrostis scabra, Andro-*

pogon scoparius, Euphorbia corollata, Lespedeza capitata, and *Cyperus filiculmis*. The prairie community reached its peak development from 10 to 20 years after abandonment, with such additional species as *Helianthus occidentalis, Liatris aspera, Baptisia leucophaea,* and *Coreopsis palmata*. In the absence of fire the prairie never develops any further, since jack pine and Hill's oak have entered the stands by 15 years and, together with such forest shrubs as sweet fern (*Myrica asplenifolia*) and blackberries (*Rubus* sp.), they have exerted dominance by shading at 35 years.

No comparable studies have been made in either mesic or wet prairies in Wisconsin, largely because such lands are too valuable to abandon. Some indication of the rates, however, may be obtained from the work on old fields in the University of Wisconsin Arboretum, where intensive artificial introductions of species has resulted in a fair to good stand of prairie on sandy soils after 14 years but only a poor to fair stand on the heavy soils after a 24-year period. In the latter area, a control block has remained unseeded and unburned during the entire period. During that time dominance passed from mixed exotic weeds to quackgrass sod and finally to a bluegrass sod which has remained solidly in control. No invasions by prairie plants took place for 20 years but after 23 years a few individuals of big bluestem, Indiangrass, and prairie clover began to appear. It would appear that a very long period will be necessary before a complete prairie community will be established here, probably of the order of 100 years or more, unless succession is aided by fire.

These long intervals for complete establishment of prairie on mesic sites are supported by the work of Shimek (1925) in Iowa, who found that a railroad cut made in 1854 had not fully returned to its original condition by 1924, after 70 years, in spite of the fact that it had been burned every year or so during that period. In North Dakota, Whitman et al. (1943) found great differences between original and secondary prairie after 35 years and they postulated that 60 years or more would be necessary for complete recovery.

ORIGIN OF THE PRAIRIE

The origin of the Midwest prairies has captured the interest of observers from the time of the first explorers to the present. The European background of these men had conditioned them to the viewpoint that trees were the natural covering of any non-arctic lands left free from disturbance, so the treelessness of the prairies was a source of wonder, which called for some clear-cut explanation. Had the United States been settled by peoples from the Old World deserts or steppes who landed in Texas and spread northward, they perhaps would have marveled at the forests and been puzzled by the "grasslessness" of these communities. As it was, a variety of theories were proposed to account for the lack of trees. Some of the theories were rather fantastic, and even the most reasonable were shown to be inapplicable in one place or another. No single explanation has yet been advanced that adequately accounts for the actual distribution of grasslands and forests, nor can it ever be hoped for because the very complex nature of both communities clearly implies a multifactoral relationship with the environment.

An examination of all of the published theories reveals a rather amazing uniformity in the accounts written in the very early period by men who saw the prairies before great changes had taken place in the Indian populations (Sauer, 1950). These early writers were almost unanimous in their conclusion that prairie fires were responsible for the absence of trees. In later years, after the Civil War, a variety of other explanations were offered, which included those dealing with soils, climate, bison, and other factors. These explanations were made by men who had never had the opportunity of witnessing an uncontrolled fire on an extensive prairie. This change in viewpoint, based on the firsthand experience of the observer or his lack of such experience, can be seen in the early accounts of prairies in Wisconsin and neighboring states. Gleason (1913) and Anderson (1954) have summarized many of the early descriptions and some of the following excerpts are from these sources.

The occurrence of fires and the tremendous extent of the affected area were adequately treated by many writers. Latrobe, as quoted by Anderson, described a fire in 1833 as seen from the top of Maiden Rock Bluff on the Wisconsin shore of Lake Pepin:

As we looked upon the summit early in the morning, across the troubled surface of the lake . . . a dense column of smoke from the opposite gave us intimation that the Prairies were on fire. The spread of the conflagration on the low grounds opposite, which drew our attention at intervals during the day, continued unabated; and as evening approached, other columns of smoke springing up in all directions, both on the summit of the opposite range of mountains and in the valleys at their feet, showed us that the Indians had taken advantage of the driving wind to fire the country for a great many miles inland. . . . At sunset, the flame seemed to have gathered full strength, and to have reached a long tract of level grassy prairie nearer the shore, upon which it then swiftly advanced, leaving a black path in its trail. . . . In one place the progress of the fire, effectively checked by a small river, died away or edged over the country with slower progress. In another, after being seemingly choked, it would burst forth with redoubled fury, sending bright jets of flame far on the wind. . . . We calculated at this time that the fire spread over a tract nearly twelve miles in length, while the distant glare upon the clouded horizon showed that it was raging far inland. The whole evening the lake, the Maiden's Rock, the clouds, and the recesses of the glen were illuminated by the flames, while, gaining the rank growth on the borders of the lake and the brow of the distant mountains, the country opposite blazed like tinder in the wind: and from the summit of the Maiden's Rock, which we had again ascended before we retired to rest, the scene was fearfully grand.

Lyman Foot (1836) made the following remarks on Indian summer in the region around Fort Winnebago in Columbia County: "Frosts have already put a stop to vegetation. The leaves have fallen, annual plants have become dry, and the fields, the swamps, the forests and the prairies are set on fire by Indians and hunters. The smoke arising from them is abundantly sufficient to produce all that peculiar redness of the sky so common to Indian summer."

George Catlin, the famous painter of Plains Indians, described the fires in 1842:

The prairies burning form some of the most beautiful scenes that are to be witnessed in this country, and also some of the most sublime. Every acre of these vast prairies (being covered for hundreds and hundreds of miles, with a crop of grass, which dies and dries in the fall) burns over during the fall or early in the spring, leaving the ground of a black and doleful colour.

There are many modes by which the fire is communicated to them, both by white men and by Indians—par accident; and yet many more where it is voluntarily done for the purpose of getting a fresh crop of grass, for the grazing of their horses, and also for easier travelling during the next summer, when there will be no old grass to lie upon the prairies, entangling the feet of man and horse as they are passing over them.

Over the elevated lands and prairie bluffs, where the grass is thin and short, the fire slowly creeps with a feeble flame, which one can easily step over; where the wild animals often rest in their lairs until the flames almost burn their noses, when they will reluctantly rise, and leap over it, and trot off amongst the cinders where the fire has passed and left the ground as black as jet. These scenes at night become indescribably beautiful, when their flames are seen at many miles distance, creeping over the sides and tops of the bluffs, appearing to be sparkling and brilliant chains of liquid fire (the hills being lost to the view), hanging suspended in graceful festoons from the skies.

Proof that this custom of burning the grasslands was no recent innovation is shown by the account of Cabeza de Vaca who wrote about southeastern Texas in 1528: "[The Indians] go about with a firebrand, setting fire to the plains and timber so as to drive off the mosquitoes, and also to get lizards and similar things which they eat, to come out of the soil. In the same manner they kill deer, encircling them with fires, and they do it also to deprive the animals of pasture, compelling them to go for food where the Indians want" (Stewart, 1951).

The relation of these fires to the relative location of prairie and forest is shown in a number of early descriptions, including that of Lothrop (in Anderson, 1954) writing of Kenosha County:

After the first frost, in the autumn of 1835, had killed the millions of tons of grass west of us, we began, at Pike River, to see the rising smoke at a distance. The Indians probably had fired the prairies as early as they could for hunting purposes. It was sometime in the latter part of September. We began to see the advancing fire towards evening on the prairie, three miles west of us; and, before twelve o'clock, it became a serious affair. The wind was from the south-west, and pretty strong, and the fire progressed rapidly. . . . The roaring terror came through the woods with awful grandeur. Large trees, as well as all smaller vegetation, quickly fell before the ruthless invader.

An anonymous author, quoted by Gleason (1913) gave this account of the intensity of a prairie fire as seen in 1828:

The flames advanced very rapidly, continued to spread, and before they had arrived opposite to the place where I stood, formed a blaze of fire nearly a mile in length. How shall I describe the sublime spectacle that then presented itself? I have seen the old Atlantic in his fury, a thunderstorm in the Alps, and the cataracts of Niagara; but nothing could be compared to what I saw at this moment. The line of flame rushed through the long grass with tremendous violence, and a noise like thunder; while over the fire there hovered a dense cloud of smoke. The wind, which even previously had been high, was increased by the blaze which it fanned; and with such vehemence did it drive along the flames, that large masses of them appeared actually to leap forward and dart into the grass, several yards in advance of the line. It passed me like a whirlwind, and with a fury I shall never forget.

Loomis, also quoted by Gleason, concluded in 1825, that "the heat and fury of the flames driven by a westerly wind far into the timbered land . . . has no doubt for many centuries added to the quantity of open land found throughout this part of America." Lieutenant Ruggles (1835) writing from Fort Winnebago in Columbia County, Wisconsin, said:

Among the elevations of this country, we often meet with extensive prairies. The origin of prairies is doubtless attributable to the extensive fires which scour the whole country when vegetable matter has become dry; and I believe it is the prevailing opinion among men of observation that this is the principal, if not the only cause. We have constant evidence of the operation of this cause around us—the country is very thinly wooded, and it is still diminishing; the dry and decayed trees, are often felled to the ground by the flames, and the most flourishing arrested in their progress.

Further evidence of the role of fire in preventing the growth of forest is given by writers of a later date, who had a chance to see the effects of several decades of protection from fire. Thus Owen (1852) described the situation in the Driftless Area:

The timber in the whole mineral district, on both sides of the Mississippi, grows in those situations which are least exposed to fire, and to the blast of winds which sweep over the extensive prairies. . . . The annual fires which have undoubtedly been kept up by the aborigines for ages past have also, no doubt, . . . kept open our vast western prairies; for those parts of the western country which were originally prairie, and where the fires have been kept out for twenty years or more, are now covered with thick groves of small trees. The American aspen, in the whole district of mineral lands, seems to be the pioneer tree which first invaded the prairie. In many places we see copses of this tree in the broad prairie, like little islands in a vast lake. And almost everywhere in the prairie we see little shoots of it of one year's growth, which would soon be trees were it not for the annual fires. When once the prairie sward has been broken by this kind of tree, others come in one after another; the prairie changes to the thicket; and in a few years it becomes the vast wilderness, the boundless contiguity of shade.

The State Geologist Wight (1877) made the same point for the region in northwestern Wisconsin: "Again, fire has killed the trees over wide areas on which grass was growing, exhibiting before our eyes nature's simple method of converting woodland into prairie. The reverse process is just as simple. When prairies are no longer swept over by fire, timber springs up, reconverting prairie into woodland. Grass, with fire as its ally, can beat timber. Timber can beat grass when it has no fire to fight."

Some doubts as to the efficacy of fire as the sole cause of prairies were expressed in 1877 by that remarkable observer, Moses Strong, in his description of the Upper St. Croix district in northwestern Wisconsin:

Analagous to the "barrens," are the areas known as "brush prairie" and simple "prairie." These are covered with a scattered growth of the shrubs that are usually associated with the more open timber of the region, but fully developed trees are absent. On some of these "prairies," however, young trees are springing up, and bid fair, if undisturbed, to attain the usual size. These have been appealed to as examples of prairies returning to forest, since annual fires are no longer permitted to ravage the region. So far as these areas are concerned, the appeal seems to be well taken, save that we might, perhaps, justly dissent from the use of the term "prairies" as applied to them; for there seems to be no evidence that these ever were prairies in the sense of being completely and compactly covered by prairie grasses, to the exclusion of all shrubs and stubs of arboreous plants, as is the case with true prairies. They rather appear to have originally been open forest areas, which, on account of the character of the soil, were especially subject to dryness, and thus to the destructive action of annual fires; while moister adjoining areas escaped. On the cessation of the destructive agent, they appear to be returning to their normal condition. It will be observed that even under this hypothesis, the primary conditions are those of soil and moisture, and that annual fires are impotent without them. Otherwise, these so-called prairies should have covered the whole region, instead of being confined to circumscribed areas.

Strong clearly points out two features or salient points of the fire theory which had received inadequate attention by the early writers. In the first he noted that great areas of the prairie region in the Middle West were actually occupied by trees, although these trees were reduced to "grubs" or brush by the continual fires. The species involved were mostly oaks, especially bur, black, and Hill's oak. From a distance these "brush prairies" were indistinguishable from the true prairies, since the prairie grasses exceeded the oaks in height and protected them from view. The true prairies, on the other hand, were devoid of all true tree species, although some woody plants in the form of shrubs were present on these as on all grasslands. The Rock Prairie in Rock County, the Arlington Prairie in Dane and Columbia counties, and the upper portions of the Military Ridge Prairie in the Driftless Area are examples of true prairie, while much of the prairie land in Waukesha and Dodge counties and in the central Wisconsin sand area has been occupied by brush prairie. In general there is a correlation between topography and prairie type; the true prairies are largely restricted to extensive flat plains or gently undulating uplands, and the brush prairies are most common on hillsides, glacial moraines, and other uneven grounds. An exception must be made in the case of the very dry, thin soil prairies such as those studied by Anderson, where truly treeless grasslands may be found on extremely steep or clifflike hill-slopes with a southwest exposure. Obviously, fire or its lack will have very different consequences on the two types of prairie. Brush prairie remnants are known on railroad rights of way in Green and Dane counties which have been burned annually for one hundred years, yet their content of oak grubs remains essentially constant. Controlled fires on the University of Wisconsin Arboretum prairies have also indicated that the oak brush originally present is not reduced in density by frequent burnings. In fact, no evidence exists to indicate that an area occupied by suitable species of oak can ever be denuded of its oak component by fires, regardless of their frequency or intensity.

The second major point made by Moses Strong concerns the obvious fact that fires may be influential in stimulating prairie in the grassland region, but similar fires in forest regions result not in prairie but in barrens or scrub. He says, "the primary conditions are soil and moisture [i.e. climate] and the annual fires are impotent without them." Gleason made the same point when he said that in order to have a prairie fire, there must first be a prairie. No account of the origin of prairies can be adequate which does not take into consideration the two kinds of prairie and the fact that prairies exist only in certain areas of the country.

Theories concerning the treelessness of the prairie which do not invoke fire include those of Whitney (1876) and others who believed the soil was responsible. Whitney was convinced that the fineness of the soil was detrimental to the growth of trees. He pointed out, "children are born and grow up without ever having seen a fragment of stone, a boulder, or even a pebble large enough to throw at a dog," in many places on the prairie and from this he concluded that prairies and fine soil are causally related. Other theories involved the bison as the responsible agents, who kept down the trees by trampling. More recently, some investigators have proposed the idea that trees were absent from the prairies because of the absence of symbiotic fungi which are thought to be essential for the growth of trees (Wilde, 1954). Obviously, such theories could account at best only for the true prairies. Even there, however, the absence of the fungi is due to the absence of trees as hosts. The real problem is why both trees and their associated fungi are lacking, since there is no conceivable mechanism whereby an area of forest could be deprived of its mycorrhizal fungi without first destroying the trees. Furthermore, since all prairies contain woody shrubs which are infected with root fungi and since all prairie herbs have root fungi of their own, it is necessary to assume a completely unproved degree of specificity (Curtis, 1939) between tree and fungus in order to accept even the basis of their theory. The historical accounts of the invasion of true prairie by aspen clearly indicate that all trees are not affected by the supposed relationship, which was originally proposed on the basis of afforestation tests by conifers not showing any phytogeographic relationship to the prairie.

Climatic explanations for the observed distribution of grasslands have become more numerous in recent years. Shimek (1924) and Transeau (1935) presented the case for atmospheric evaporation as the basic factor, whereas Borchert (1950), in a very thorough account, clearly demonstrated that the prairies were closely correlated with the mean paths of the major air masses and further, that the incidence of drought cycles was much higher in the prairie region than in the adjacent forests.

From a complete review of all of the foregoing evidence, the following statement as to the nature of the prairie-forest border appears to be reasonable. Plant communities dominated by grasses can exist on a large scale only within a particular climatic regime, one that is characteristized largely by an excess of potential evaporation over available rainfall. Within the area with this climate, soils and topography are of minor importance. Toward the humid border of the prairie region, climatic conditions tolerable by both grasslands and forests may exist. In this ecotonal region, certain soils and certain topographies may favor one formation over the other. On the majority of sites, however, the presence or absence of grassland is determined by the presence or absence of burning, with fire favoring grass and repressing trees. In the still more humid regions to the east, grasslands can occur only on special topographically dry sites, with forest in some density covering the bulk of the land, regardless of fire.

This general statement fails to account for the presence of true prairie versus brush prairie within the ecotonal region of the prairie-forest province of Wisconsin. Some of the early accounts as well as certain very recent experimental evidences give a lead as to the cause of this differentiation. The proposal is here made that the nature of the prairie is related to the nature of the preceding plant community, that true prairie originated from old climax forest lands while brush prairies were derived from pioneer or intermediate stands. Such a theory is made necessary by the fact that oak trees and their woody associates cannot be destroyed by burning, as discussed above, and therefore an oak woods can never be converted to true prairie by the agency of fire. The simplest means whereby the oaks may be eliminated from a particular stand is by suppression from the shade cast by a succeeding maple forest. Such a succession is most likely to go to completion on flat or gently rolling lands, because the maples have the greatest chances of producing a solid, light-tight canopy under such circumstances. On hillsides and moraines, however, succession would be slower and less likely to go to completion, because of the poorer water supply and because of the entry of light from the side. Such forests, therefore, would tend to retain at least a portion of their oak complement for long time periods.

Following a period of climatic or other conditions which favored forest growth and during which successions had proceeded to their maximum extent consistent with the topography, a change, to drier conditions, to more frequent periods of drought or lightning, or to the entrance of Indian tribes with new hunting habits, could produce circumstances where a succession of fires or even one single fire would destroy the forest trees and expose the soil to full light. Since the Brown Forest soils of the mesic forest have nearly the same physical characteristics as the mesic Prairie soils, any mesic prairie plants in the vicinity would be able to enter the burned area and successfully compete there with the shade-demanding herbs of the maple forest. The same invasion would take place in the areas of oak forest, except that there the oaks and certain of the oak forest understory species would be able to regenerate from underground parts and to maintain themselves as members of the new community.

That such a process can actually occur is indicated in this quotation from Caton, written in Illinois in 1870:

In the majority of instances, no doubt, a new growth of trees takes the place of the old [following fire], but such is not always the case. Mr. Daniel Ebersal, of this city, who is a good observer, and of undoubted veracity, informs me that many years ago, on the Vermilion River, a fire occurred under his own observations, which utterly destroyed root and branch an entire hard-wood forest, and that the entire burnt district was directly taken possession of by the herbaceous plants peculiar to the prairies, and that in a very few years it could not be distinguished from the adjoining prairies, except by its greater luxuriance. The testimony of Mr. J. E. Shaw, who has resided upon the prairies of Illinois for more than fifty years is equally to the point. He assures me that he has known many forest districts entirely burned over and every living thing upon them destroyed. Generally they were replaced with trees similar to the former growth; but that sometimes the prairie herbage takes and maintains possession. He cites an instance on his own farm, where, forty years ago, when he took possession, there was a forest of large trees, which was destroyed by a fire, when a part of the burned district was again covered with trees, and a part was taken possession of by the prairie grass, and in a comparatively short time could not be distinguished from the adjoining prairie.

I venture to ascertain, that a thousand witnesses may be found still living who can state particular instances of the same kind.

In my early wanderings over the wild prairies it several times occurred, when approaching a body of timber, that I met in the prairie grass charred remains of forest trees, perhaps half a mile or more from the edge of the wood, and I have in no instance inquired of one who had similar facilities for observation who did not remember having observed the same thing.

A more recent instance is that to be seen on the Crex Meadows Conservation Area in Burnett County in northwestern Wisconsin. A series of experimental burns were made there by Norman Stone, District Game Manager. The purpose was to convert a large area of twenty- to thirty-year-old jack pine and Hill's oak to openings suitable for prairie chicken and sharp-tailed grouse. In one instance, the initial burn was made in 1952. Wet weather prevented further fires until the spring of 1956. An examination of the land in the fall of 1956 revealed an amazing development of prairie, after only two fires. The ground was dominated by prairie grasses, especially big and little bluestems and Indiangrass. Altogether, 54 species of prairie plants were found on the area, including such heavy seeded species as lupine (*Lupinus perennis*), and redroot (*Ceanothus ovatus*). Oak grubs and a few forest plants were also present. In a still more spectacular case, the initial fire occurred in the spring of 1957. It was an exceedingly hot crown fire which totally destroyed the tops of the trees. Examination of the area in July of that year revealed over 70 prairie species, most of them in bloom or in bud. The list included the prairie onion (*Allium stellatum*), leadplant (*Amorpha canescens*), prairie birdfoot violet (*Viola pedatifida*), and many other conservative prairie species. A casual examination of unburned control areas of identical forest revealed only a small fraction of the total list of prairie plants. The exact mechanism responsible for this dramatic reappearance of the prairie remains in doubt. The relative importance of seedling establishment versus the resurrection of stunted holdovers which persisted under the dense shade of the pine canopy for more than three decades is unknown at present. A detailed study of this area and its future extensions would certainly be very rewarding.

Thus it appears probable that the mesic prairies moved eastward by a series of jumps, each following a catastrophic destruction of the pre-existing forest. If that forest had developed sufficiently close to a climax condition so that it would have eliminated the oaks, then a true prairie resulted, whereas if the forest contained any proportion of oak (other than red oak) then a brush prairie was likely. Continued fires would occur because of the ease of combustibility of the prairie grasses, and also because the lay of the land with respect to wind exposure and water barriers which would allow the first fire would also favor succeeding fires. Any change in incidence of fire, whether due to climatic or ethnic change, might allow the forest to regain temporary control of the brush prairie, but would not greatly influence the true prairie.

The only other theory which can account for the origin of true prairies in the prairie-forest border region would be one which postulated that these areas had never been occupied by trees in postglacial times but had been invaded initially by prairie and had remained that way ever since. The fossil pollen record (Chapter 22) indicates this was probably not the case. In fact, a good argument can be advanced to the effect that many of the true prairies were originally spruce flats during the period of boreal forest dominance in early postglacial times, and that these spruce flats were directly invaded by climax hardwoods and remained that way until about 3000 B.C., when they were replaced by prairies. The sequence of soil changes involved in this clisere succession, from Podzol to Podzolic, to Brown Forest, to Grood or Prairie soil is interesting to contemplate, involving as it does the origin of one soil from another, rather than from parent material.

The case for the very dry prairies and the very wet prairies may be quite different from the above account for the wet-mesic, mesic, and dry-mesic prairies. Anderson postulated that the dry prairies were present in the Driftless Area through the Pleistocene period and their origin therefore may be so ancient as to be undecipherable. It is clear that these prairies are in no way dependent upon fire for their maintenance, since moisture conditions are so severe that they prevent the growth of trees. The fact that the flora of the dry prairies is the most distinct of any community in Wisconsin, as shown by its ratio of modal prevalents to total prevalents of 87.1 per cent is undoubtedly related to this severe microclimate, as well as to absence of a frequent back-and-forth oscillation of grassland and forest.

The wet prairies, on the other hand, show little evidence of stability. In the absence of fire, they are quickly invaded by trees, including aspens, willows, cottonwoods, ashes, elms, and oaks. They are definitely of a temporary nature and could have originated from lowland forests at any time in the past. It is possible that the major areal expansion of wet prairies was contemporaneous with the expansion of mesic prairies on the uplands, because the fires on the latter would spread into the lowlands as well, but the lowland flora may have been present from an earlier migration and may have existed in small, meadowlike openings in the forest. The high content of forest plants in the lowland prairies may indicate a recent date of origin for some of them.

UTILIZATION AND CURRENT MANAGEMENT

More than any other Wisconsin plant community, the prairie has been subject to direct destruction for agricultural purposes. From an original area of over two million acres, the prairie has been reduced to an almost vanishing remnant probably no greater than several thousand acres in total, with no single tract of original prairie larger than forty acres known today. The most widespread type of site for the remaining prairies is that provided along railroad rights of way, where a narrow strip between roadbed and boundary fence frequently remains in essentially undisturbed condition. This situation is particularly common in the southeastern portion of the state, where the railroads were laid out before the land had been completely devoted to crop production. Subsequent maintenance operations by the railway crews consisted mainly of an annual burning of the strips to keep down woody plants and to reduce the chances of accidental fires. This treatment did not differ greatly from that normally afforded the prairies by Indian burning, so little change has taken place in the community composition. The linear nature of the remnants, however, combined with many breaks in contiguity caused by earth fills and cuts on uneven topographies has resulted in a high degree of isolation of the remnants, so that chances for normal migrations and re-entrances of temporarily depleted species are greatly reduced (Curtis and Greene, 1949). In consequence, each remnant is likely to show less than a normal complement of species, although a series of stands along one railroad usually supports the entire flora typical of the region.

Some of the wet prairies were maintained for many years as mowing meadows; they provided a reserve source of hay for those drought years when the uplands failed to produce. Such use has greatly decreased in recent years, and the former prairies have been turned over to the production of specialty crops like onions, potatoes, or lima beans, after the lands have been drained. The Faville Prairie Preserve of the University of Wisconsin, located near Lake Mills in Jefferson County, is such a prairie. It was mowed at intervals from 1840 to 1945, when the University gained control through a gift by a public-spirited citizen. Since then, the prairie has been controlled by burning, but drainage of surrounding lands has greatly changed the water relations on the prairie and has initiated processes of deterioration because no typical dry prairie species are present in the vicinity to take advantage of the new conditions.

In the early days, prairies were utilized as pasture for cows and horses. Great differences in response to this grazing were exhibited by the various types of prairie. The wet prairies were least resistant, with a virtually complete replacement of the original flora by such exotics as bluegrass, redtop grass, and European canary grass within two or three years. The mesic prairies were almost as sensitive, with complete replacement by bluegrass, orchard grass, quackgrass, dandelions, and clover within eight to ten years. The dry prairies, on the other hand, were much more resistant to grazing, not only because of inherent differences in the behavior of side-oats grama grass (*Bouteloua curtipendula*), and little bluestem (*Andropogon scoparius*) to the grazing pressure (Neiland and Curtis, 1956) but also because none of the exotic grasses could successfully compete under the xeric conditions. As will be discussed later in Chapter 23, some dry prairies on thin soil hillsides are known which have been grazed continuously for over a century but they are still dominated by the two grasses mentioned above and still contain a number of their typical forbs, including the pasque flower (*Anemone patens*). The contribution made by these grazed dry prairies to the total agricultural economy is small because of their small aggregate size, but for the particular sites involved, no other method of use would be of equal value.

CHAPTER 15

Sand barrens and bracken-grassland

NATURE, LOCATION, AND COMPOSITION OF SAND BARRENS

On the higher terraces of the Wisconsin and Mississippi rivers and at a few other places south of the tension zone, the oak barrens give way to an open, prairie-like vegetation. In their undisturbed condition, these open plains would probably have qualified as dry-mesic or dry prairies, but disturbances have occurred which have changed their plant cover to something very different from a prairie. These changes have had various initial causes but the result in each case has been the movement of the sandy soil by wind. The new vegetation develops on the wind-influenced areas, either on the blowouts from which the sand is removed, or the dunes or sand deserts on which it is deposited. As will be seen later, these inland dunes possess a very different flora from the dunes along Lakes Michigan and Superior and must be discussed separately. The sand barrens are most closely related to the dry-mesic prairies and to the oak barrens. No quantitative studies of the type have been made in Wisconsin, although Gleason (1910) and Vestal (1913) have reported detailed investigations in the adjacent areas of northern Illinois. The following discussion is based on a presence survey of 20 stands in Wisconsin, made by J. T. Curtis and H. C. Greene (Figure XV-1).

The prevalent species in the P.E.L. stands are listed in Table XV-1 and the other species reaching a peak in the sand barrens are also given in the same table. Of the prevalent species which attain an optimum here, all but *Gnaphalium obtusifolium* and *Koeleria cristata* are restricted to sandy habitats while several of the other species are essentially confined to the sand barrens, including *Croton glandulosus, Krigia virginica, Talinum rugospermum, Festuca octoflora, Cyperus filiculmis, Linaria canadensis, Calamovilfa longifolia,* and *Rhus aromatica*. A floristic family analysis shows five families with over 50 per cent of the total species (*Compositae,* 23.5 per cent; *Gramineae,* 15.7 per cent; *Rosaceae,* 5.2 per cent; *Leguminosae,* 3.3 per cent; *Ericaceae,* 3.3 per cent). Of interest are the relatively high values of the *Euphorbiaceae* (2.0 per cent), *Cistaceae* (2.0 per cent), and *Cactaceae* (1.3 per cent), which reflect the extreme aridity of the habitat. The phenological pattern of the sand barrens resembles that of the prairies, with a midsummer peak and with about equal numbers of spring and autumn bloomers. Twenty-four species of shrubs were recorded in the P.E.L. stands, but their low importance in the community is indicated by the fact that only one of them, *Hudsonia tomentosa,* is on the prevalent species list.

Among the other biota of the sand barrens grasslands, the cryptogamic plants play an unexpectedly important role. Certain blue green algae form circular, gelatinous masses on the surface of the loose sand and are of great importance in stabilization. The hairy cap moss (*Polytrichum piliferum*) also produces sand binding mats (Plate 37). Several fleshy fungi somehow obtain sufficient water and nutrients from the sand to develop their fruiting bodies. Prominent among them are the leathery earthstars of the genus *Geaster* and the unusual *Scleroderma geaster* (Plate 38). Lichens are numerous on areas that have been stabilized for some time; these include *Cladonia rangiferina, C. cristatella,* and other similar forms.

Among the animals, the reptiles are of unusual prevalence, with a number of snakes, such as the hog-nosed snake and the blue racer, and one of the few Wisconsin lizards, the six-lined lizard. The lark sparrow, a common bird of the western plains, is largely restricted to the sand barrens grassland in the state. Vestal (1913) indicated a number of characteristic insects in his detailed study of Illinois sand barrens, but no comparable work has been done in Wisconsin.

SAND BLOWS

When the prairie sod which originally existed on the sand barrens was broken by the plow, the loose sands of the underlying soil were free to move with the wind. In the areas around Muscoda in Grant County and north of Arena, in Iowa County, for example, early attempts at farming the land were quickly doomed by the excessive wind erosion which followed. Frequently the wind caused simply a shifting surface, with as much sand deposited as was picked up on the average. In some places, however, the sand was removed to a considerable depth, and was deposited in dunes against fence rows, clumps of cacti, or other wind barriers. The troughs thus formed continued to grow, both in area and depth, while the dunes similarly increased in size (Plate 39). Such a hole in the barrens surface is called a blowout. Several are known in the Wisconsin River valley which cover more than 40 acres, with the leeward dunes enveloping another 10 acres or so. The sand in these river terraces was originally deposited by outwash waters from the melting glaciers. Included in the sand is a very small percentage of small stones or gravel, evenly spread through the sand matrix at the rate of 60 to 75 pebbles per cubic meter. The winds have sufficient velocity to pick up the sand grains but they cannot lift the stones. When the blowout has reached a certain depth (about 2 meters in the Arena area) a sufficient quantity of these stones has been concentrated on the surface to protect the underlying sand from further erosion. The surface layer of pebbles (Plate 38) is called a "desert pavement" from its resemblance to similar layers in the deserts of southwestern United States. The blowout can get no deeper after this pavement is formed, but it can continue to spread, even into areas which have had no artificial disturbance, since the sandblast action of the winds quickly kills the plants on the edge of the blowout (Plate 40). On occasion the blowout can penetrate into the scrub oak savannas which frequently border the barrens by removing the sand from the roots of the trees.

After the bottom of the blowout has been stabilized by the desert pavement, a number of plants can become established. One of the most common is the xerophytic, evergreen shrub (*Hudsonia tomentosa*) shown in Plate 41. This plant has an extensive root system and its small, hard, scalelike leaves can resist the abrasive action of the blowing sand. Once they are firmly established, the *Hudsonia* bushes frequently act as a protected center in which other plants gain a foothold, such as the three-awn grass (*Aristida basiramea*) and several sedges in the genus *Cyperus*. Later succession includes many of the typical species of the sand barrens flora.

The dunes are sometimes invaded by sand cherry (*Prunus pumila*) but the usual early colonizers are grasses which include little bluestem (*Andropogon scoparius*), switch grass (*Panicum virgatum*), or sand grass (*Calamovilfa longifolia*). These grasses are quite successful at stabilizing the dune surface and preventing movement of the sand where they grow. If they are not smothered by new additions of sand from an expanding blowout, they may rather quickly provide a continuous canopy over the surface and serve to protect invading seedlings of many other species. Sometimes the new plants are typical members of the sand barrens grassland, but not infrequently the stabilized dunes are taken over by trees from the oak savannas. At both Arena and Muscoda, the extensive jack pine barrens now found on such dunes were formed in postsettlement times.

This process of blowout and dune formation can occasionally lead to some unusual plant combinations. For instance, near the old town of Helena in Iowa County, a mixed stand of river birch and jack pine can be seen. Upon examination it is noted that the pines are on small dunes while the birches are on the bottoms of old blowouts which were close enough to the water table to provide the moisture necessary for the birch.

ENVIRONMENT OF SAND BARRENS

The sand barrens grasslands are Wisconsin's closest approach to true desert. On a hot day in July, the approximation to desert conditions is sufficiently good to satisfy almost anyone but a native of Arizona. Frequent measurements of surface soil temperatures in midsummer have given values of 140°F. to 155°F. with one reading of 162°F.

No records for total evaporation have been made, but they must be very high indeed. The severity of these microclimatic conditions is combined with a very low surface water supply and an almost total lack of available nutrients to produce an extremely severe and exacting environment. Analyses of the sand show only a trace of calcium and nitrate nitrogen, 17 p.p.m. of available potassium, 30 p.p.m. of available phosphorus, and 5 p.p.m. of ammonia nitrogen. The acidity is not great, with an average of pH 6.2. The water-retaining capacity varies between 20 and 30 per cent.

The plants of the sand barrens have a number of different adaptations which enable them to survive under the rigorous conditions. A number of ephemeral annuals are present which develop their flowers and fruits quickly in the spring when the sand is still cool and moist, then die down and pass the summer in the form of seeds. Included in the group are *Silene antirrhina, Draba reptans, Festuca octoflora, Linaria canadensis, Krigia virginica,* and *Specularia perfoliata.* Other species have developed water storage organs which presumably allow the plants to carry over the benefits of water from rains to the drought periods which follow. The prickly pear cactus (*Opuntia compressa*) shown in Plate 56 is one of the most notable of these. This species is actually more widespread in the cedar glades, but it never reaches the density levels there that it does in the barrens, particularly in pastured areas where the cows serve as active agents of dispersal. The water is stored in the enlarged stems of the cactus, as the true leaves fall off soon after they are formed. In the fameflower (*Talinum rugospermum*), an endemic species of the sand barrens of Wisconsin and the immediately adjacent areas of neighboring states, the leaves are thickened and adapted for water storage. This plant has the interesting habit of opening its flowers only in the late afternoon after the heat of the day has passed. An unusual method of water storage is exhibited by the sand milkwort (*Polygala polygama*), which produces normal racemes of bright purple flowers at the tips of the leafy stems and also swollen, white, cleistogamous flowers on underground branches. The prime purpose of these underground flowers is water storage because no seeds are set in them.

A number of species are known as sclerophylls because they have leaves that are small, hard, and tough. The *Hudsonia* shrub already mentioned is one of these, as are *Selaginella rupestris, Polygonella articulata,* and *Polygonum tenue.* It is thought that these plants achieve drought resistance through a reduced transpiration rate but no experimental measurements are available. A number of species have a very hairy covering on their leaves which is believed to act in a similar manner, but again no direct evidence is available. Included in the group are *Gnaphalium obtusifolium, Hieracium longipilum, Chrysopsis villosa, Anaphalis margaritacea,* and *Froelichia floridana.* Most of the remaining species have deep tap roots or deep feeding roots of a fibrous structure which can penetrate to the moist layers 2.5 to 3.0 meters below the surface. Finally there is a group of plants which has no obvious modifications for either increasing intake of water or preventing excessive outgo, yet they grow successfully side-by-side with the other species and must contend with the same conditions. Among these apparent defiers of the habitat are birdfoot violet (*Viola pedata*) and eyebane (*Euphorbia maculata*).

GEOGRAPHICAL RELATIONS OF SAND BARRENS

The sand barrens of Illinois have received considerable attention in the past. Hart and Gleason (1907) and Gleason (1910) studied the sand regions of the Illinois River valley and made frequency and density determinations on the grass-covered barrens, on the blowouts, and on the dunes. Of the 28 species listed in the 1907 paper as most characteristic of the area, 20 species or 71 per cent are found on the Wisconsin lists as well. The total barrens flora of 113 species reported in 1910 had 79 per cent in common. On a frequency basis, Gleason found that the most important species were *Leptoloma cognatum, Ruellia ciliosa* (not found in Wisconsin), *Ambrosia psilostachya, Bouteloua hirsuta, Koeleria cristata,* and *Andropogon scoparius.* A less detailed but more extensive study of other regions in Illinois by Sampson (1921) showed 81 per cent of the species in common. Sand plains in more remote areas retain a relatively high degree of similarity. Thus, Britton (1903) listed 75 species for the North Haven sands of Connecticut, of which 48 per cent were common to the Wisconsin community. To the northwest, Shimek (1925) listed 96 species for the sand barrens of Manitoba of which 48 per cent were also on the Wisconsin lists. The famed Sand Hills of Nebraska also show considerable similarity, with 41 per cent of the species listed by Frolik and Shepherd (1940) in common. It is of interest that Gleason reported 40 per cent of his Illinois species to be on an earlier list of the flora of the Nebraska area made by Rydberg.

UTILIZATION AND MANAGEMENT OF SAND BARRENS

The obvious dangers of erosion and the low fertility and moisture levels of most of the sand barrens reduce them to essentially worthless lands for agricultural purposes. Attempts are being made in the Wisconsin River valley and elsewhere to plant the barrens to pine trees, but the low fertility greatly reduces the chances of securing an economically successful crop. The trees, however, will stabilize the sand and may improve the water-holding capacities of the sand to a point where a following generation of hardwoods may become established. Little effort has been devoted to the stabilization of the blow sands by means of grasses and other sandbinders, largely because of the low valuations placed on the land and the lack of great dangers to surrounding areas. In the long run, it is probable that a prairie flora would do more to enrich the soil than a pine or oak forest; it would certainly require less attention and expense.

NATURE, LOCATION, AND COMPOSITION OF BRACKEN-GRASSLANDS

During the course of the P.E.L. studies in northern Wisconsin, a number of places were seen on the uplands where no trees were present and where the community was apparently dominated by bracken fern (*Pteridium aquilinum*). Many of these sites showed a number of old pine stumps scattered throughout the area, and accordingly were assumed to be secondary communities resulting from logging and subsequent fire. Within the past several years, however, a closer examination of these bracken communities revealed that they possess a high degree of homogeneity with a fairly definite combination of species. Furthermore, a number of examples were discovered which had no pine stumps or other evidences of a former forest covering. The actual dominants of the community include many grasses as well as the ubiquitous bracken. The soils tend to be melanized, with a fairly deep incorporation of organic matter and little evidence of a highly leached A_2 layer. It appears that these bracken-grasslands trace back to the grassy openings and "prairies" reported by a number of the earliest explorers (Norwood, 1852) of the region and that some of them, at least, are natural grasslands of the same nature and origin as the southern prairies. The very recent recognition of this fifth major type of grassland in Wisconsin means that little information is available at present. Much more field work must be done before the exact nature and limits of the type are fully understood. The following discussion is based upon a preliminary study of 27 stands by the author, with the aid of Grant Cottam and J. L. Habeck in some of the locations (Figure XV-2). Both the sand barrens and bracken-grassland stands are shown in the map at the opening of this chapter. The conclusions should be regarded as tentative and subject to future revision, particularly as to the relations of original grassland and secondary extensions into burned-over forest land.

As understood at present, bracken-grasslands are restricted to the region north of the tension zone and seem best developed in the area from Marinette to Bayfield counties on the north and from Brown to Rusk counties on the south. On the basis of the index of similarity, the three most closely related communities in Wisconsin are the Pine Barrens (38 per cent), the Northern Dry Forests (32 per cent), and the Sand Barrens (25 per cent). Since all of these figures are very low, the bracken-grassland appears to be a relatively discrete community.

The prevalent species are listed in Table XV-3 along with the other species which attain an optimum here. The index of homogeneity is fairly high, but the major evidence for the distinctness of the community is seen in the high number of prevalent species (56 per cent) which achieve maximum presence in the bracken-grassland. This figure is exceeded only in the dry prairies.

The bracken fern is the most widespread species of the group. It frequently attains high levels of density, and forms an almost unbroken canopy of fronds in some places (Plate 42). In no sense can it be considered as an indicator of the community, however, since it is also very common in a wide range of communities throughout the state. Among the native species, the grasses, *Agropyron trachycaulum* var. *unilaterale, Bromus kalmii, Danthonia spicata,* and *Oryzopsis asperifolia,* and the forbs, *Anaphalis margaritacea, Arabis glabra, Aster ciliolatus,* and *Viola adunca* are the most characteristic of the common species, while among the rare species, *Festuca saximontana, Oryzopsis pungens, Erigeron glabellus* (in northwestern counties), *Senecio pauperculus,* and *Spiranthes gracilis* are apparently at their optimum.

One of the unusual features of the bracken-grasslands is their high content of exotic species. Eighteen such plants were found; these include some of considerable frequence such as hawkweed (*Hieracium aurantiacum*), wild lettuce (*Lactuca scariola*), timothy (*Phleum pratense*), the two bluegrasses (*Poa compressa*) and (*P. pratensis*), sheep sorrel (*Rumex acetosella*), and mullein (*Verbascum thapsus*). The mullein is conspicuous because of its height but it is present only in small numbers. The hawkweed, timothy, and bluegrasses, however, may be very abundant and may attain complete dominance in local areas. The hawkweed is especially important because of its apparent production of a potent antibiotic which is effective against most other species except the bracken fern. When the hawkweed gains control of a local spot, it is readily replaced only after it has been weakened by the shading action of the fern. This may take many years since the bracken is very frost sensitive; its fronds do not appear before mid-June and they may be killed back as early as mid-August, thus offering a long period of full light to the species growing under it.

Four families contain more than 50 per cent of the 158 total species and these four (*Compositae*–24.7 per cent, *Gramineae*–15.2 per cent, *Rosaceae*–7.0 per cent, and *Leguminosae*–4.4 per cent) are also in leading positions in the true prairie communities. The very high value of the grasses is especially noteworthy because it is higher than in any of the prairies and is exceeded only in the sand barrens. The dominance of the grasses is most apparent in late spring, before the bracken fronds unfold, since most of the grass species bloom before June 15. Shrubs are represented by relatively few species (15.0 per cent of total) but two kinds may be present in very high densities—the sweet fern (*Myrica asplenifolia*) and the blueberries (*Vaccinium angustifolium* and *V. myrtilloides*). In fact, the blueberry crop to be obtained from these bracken-grasslands approaches that from the pine barrens in size and is likely to excel it in regularity and in sweetness of the berries, probably because of the better water-holding capacity and nutrient content of the grassland soils.

Equal numbers of species bloom in the spring before June 15 and in the summer between June 15 and August 15 (37 per cent), and the remaining 26 per cent bloom in the autumn. Prominent in the spring display are ladies' tobacco (*Antennaria* sp.), ragwort (*Senecio pauperculus*), and sand violet (*Viola adunca*). Few very colorful species are present in the summer, but the autumn display of asters and goldenrods rivals that of the true prairies. The most common aster (*Aster ciliolatus*) is among the most beautiful of the upland members of the genus, with large heads of a deep, glowing purple.

No details on the other biota of the bracken-grasslands are available. Little evidence was seen of the activities of burrowing mammals except for the pocket gopher which is abundant in the westernmost stands, as far east as central Bayfield County. The white-tailed deer favor the community in the twilight hours, especially after hot summer days when they may be seen in numbers on the grasslands along fire lanes and back roads. One of the favorite recreations of many of the vacationers who visit the region is driving slowly along these little used pathways in the early evening and trying to see as many deer as possible. This use, combined with daytime blueberry picking, has served to give the northern grassland a wide popular knowledge that is not matched by a current scientific understanding.

ORIGIN OF BRACKEN-GRASSLAND

There can be little doubt that the initial cause of the bracken grasslands was fire, since there is no reason to believe that the area was not occupied completely by forest during the postglacial boreal period. At what time the first grasslands appeared is open to question, but the most probable period would be during the maximum expansion of the southern prairies (see Chapter 22). In subsequent times, the grassland no doubt fluctuated considerably in the relative portion of the region it has occupied. Probably the greatest expansion has occurred in postsettlement times, especially since the heyday of the logging industry. The stations mentioned earlier as having scattered pine stumps are known locally as "stump prairies." They were produced by severe fires in the slash following lumbering, fires which frequently destroyed or reduced the humus in the soil. In some cases, these degraded sites were invaded by members of the bracken-grassland from nearby original stands, rather than by aspen, white birch, or pin cherry as was most usually the case. The possibility also exists that some of the stump prairies were originally pine savanna, as described by early travelers, and that the understory had been continuously present. The wide spacing of the remaining stumps supports this alternative, but firm conclusions are subject to limitations because of the differential rates of decay of red pine and white pine stumps.

Regardless of initial origin, it is clear that the bracken-grasslands do not need continual fires for their maintenance. Many of the areas have not been burned for over 20 years, yet they show no traces of invasion by trees; several are known which have been artifcially planted on more than one occasion, with total failure of the transplants. Foresters of the region are wont to explain this failure of trees in the grasslands as the result of "frost pockets." Many of the existing areas are located in depressions in the pitted outwash topography. The bottoms of the basins usually are rich in grasses and sedges, with the bracken fern dominant on the surrounding slopes and with a solid wall of forest beginning on the crest of the uplands. Such a steep-walled basin acts both as an efficient radiator of heat and as a collector of cold air drainage, with the possibility of lowering temperatures below the freezing point on any clear night during the year. Such frost conditions are highly probable in early June and after the middle of August. Aspens are frequently seen around the peripheries of these depressions; they have either their lower leaves or all of their leaves nipped by these local frosts. However, little evidence exists to show whether or not they are killed by the cold, since simple defoliation by other agents, such as the tent caterpillar, invokes no permanent harm. The tree most commonly seen invading these frost-prone areas is balsam fir (*Abies balsamea*) (Plate 42), which is known to be the least subject to frost damage of any conifer tree of the region.

In spite of the apparent activity of frost, however, this explanation cannot be the sole reason for the maintenance of the bracken-grasslands, since many of the larger areas are on level plains or rolling uplands where microclimatic frost due to air drainage is not a problem. One possible explanation for such sites which may be operative in the frost pockets as well is antibiotic production by the grassland flora. As mentioned earlier, *Hieracium aurantiacum* is particularly potent in this respect, while such species as *Agropyron trachycaulum*, *Antennaria neglecta*, *Poa compressa*, *Helianthus occidentalis*, *H. giganteus*, *Erigeron glabellus*, and *Aster macrophyllus* are known or strongly suspected to be active. An experimental demonstration of the nature and extent of this phenomenon is badly needed since the success of future reforestations may hinge upon this knowledge.

The bracken-grasslands are found on a variety of soil types, varying from loams to loamy sands or fine sands. They are not often seen on the coarse sands favored by the jack pine barrens. Characteristic soil series include the Antigo, Cloquet, Omega, Marenisco, Milaca, and Vilas groups. The topography is level to rolling, with good internal drainage except in the bottoms of frost pocket depressions, where a gley layer may be present within 24 inches. The few soil analyses which are available for this specific community reveal a considerable variation in accord with the age of the grassland, with the stands of recent, secondary origin differing in their organic content from the stands of older grassland. Wilde, et al. (1949) discuss the general soil types which include the bracken-grasslands and some of the pine barrens as well under the heading of "melanized sands." These are characterized by the dark color and depth of the A_1 layer. The leached A_2 layer is thin or lacking and appears to grade slowly into the B and C layers, much as in a true prairie soil. The reaction varies around pH 6.0 in the surface layer, which has a relatively much higher content of bases than the underlying parent material, thus reflecting the pumping action of the grasses. According to these authors, the melanized sands supported a thin forest of red pine, scattered white pine, and scrub oak, although they indicate that large areas have been reduced to grass with scrub oak sprouts. Regarding the occurrence of these grassland soils in a forest region, they say:

> The development of unleached sandy soils, with incorporated "mull" humus right in the heart of the podzol region, is a biological phenomenon of considerable interest. Ordinarily, the prerequisites for the development of mull soils include an abundant supply of bases, a favorable position of the ground water table, or forest cover of soil-conserving hardwoods such as hard maple, beech, basswood, and white ash. None of these factors is instrumental in the development of melanized sands in Wisconsin. . . . the dryness of the soil and general deficiency of mineral nutrients are responsible for the comparatively low density of virgin stands of red pine and jack pine. This condition, in turn, permits the existence of an understory of light-demanding oaks, accompanied by a fairly dense cover of xerophytic grasses and some less exacting legumes. This is the biological makeup which, in all probability, has led to the constant maintenance of a mildly acid soil reaction, the development of an active population of soil organisms preventing the accumulation of raw humus, and the incorporation of organic matter, largely the dead roots of herbaceous vegetation, into the soil.

GEOGRAPHICAL RELATIONS OF BRACKEN-GRASSLANDS

No descriptions of bracken-grasslands in Minnesota are available for comparison with the Wisconsin community but Gleason (1918) has given an account of similar groupings in northern Michigan. He paid particular attention to the secondary "stump prairies" which had developed since the logging operations and the subsequent fires. Of the 36 species for which he lists frequency values, 72 per cent also occurred in the Wisconsin community. These included such prevalent species as *Anaphalis margaritacea, Hieracium aurantiacum, Phleum pratense, Poa compressa, P. pratensis, Pteridium aquilinum, Rumex acetosella,* and *Verbascum thapsus.* The major differences were in the lack of native grasses and in the lower complement of prairie plants in the Michigan region. In fact, Gleason called the community the "blue grass association" because of the great dominance of the *Poa* species. Regarding its permanence, he says:

It is not known how long the blue grass association may persist. It is still well developed in numerous places in the aspen region, which was formerly occupied by pine forest and was lumbered over thirty years ago. . . . If the development of the surrounding forest is restricted by further clearing or by fire, the bluegrass association probably continues indefinitely.

Another similar community is the aspen park land of the Prairie Provinces of Canada or some of the open grasslands of the same region. For example, Moss (1952) listed 154 species for the *Agropyron-Stipa-Carex* community of the Peace River country of NW. Alberta (56°N., 119°W). Of these, 41 or 27 per cent are on the list for Wisconsin bracken-grasslands. This Alberta community is an open grassland and has no bracken fern. The leading grasses, however, were very similar to those in Wisconsin—*Agropyron trachycaulum, Bromus ciliatus, Calamogrostis canadensis, Danthonia* sp., *Koeleria cristata, Poa pratensis,* and *Schizachne purpurascens.* The major difference was in the importance of *Stipa* (four species) and *Carex* (nine species) in the Alberta grassland. Coupland and Brayshaw (1953) reported 85 species for the fescue grasslands in the aspen park land of western Saskatchewan, of which 41 per cent were on the Wisconsin list. The similarities to the Wisconsin area are enhanced by the greater importance of species of *Festuca* and the decreased dominance of *Stipa* and *Carex* in Saskatchewan. However, the absence of bracken fern as an important component in most of the Canadian types means that the summer aspect would be very different in the two regions.

The bracken-grasslands are thus seen to have a relatively close affinity to northwestern areas. The northwestern element of their flora is enriched by a number of species from the true prairies to the south, and by a few species of northern forest relationships. The exotic element is of particular interest, since its magnitude is not duplicated in any other native Wisconsin community. European weeds can become established in any community, either grassland or forest, if suitable soil disturbances occur, but their stay is limited and their place quickly taken by native species. Very few exotics have managed to assume a permanent position in the vegetation of Wisconsin (see Chapter 21). In the bracken-grassland, however, a large number of exotic species appear to have become fully integrated with the native flora and are able to compete with it successfully in the absence of any disturbance except fire.

The bracken-grassland thus gives every promise of developing into a new plant community—one with a species composition of mixed origin that has come into existence only in the last 50 or 75 years following the tremendous regional disturbances caused by lumbering. Whether or not it has reached equilibrium or will lose some members and gain others, only time will tell. A detailed study of the life-history characteristics of the exotic component, particularly of the factors involved in reproduction and population maintenance, would be of great interest for a deeper understanding of the evolution of plant communities.

UTILIZATION AND MANAGEMENT OF BRACKEN-GRASSLANDS

The open grasslands are used chiefly as sources of blueberries and for game management purposes. Extensive areas have been planted to trees in recent years, sometimes with apparent success, but often with total failure. Foresters in general have agreed that the frost pockets are not worth the expense of afforestation and some believe that not all of the other upland grasslands should be devoted to tree production. The latter view is biologically sound, because the grasslands were a normal part of the original landscape and the total biota, including the insect populations, has become adapted to their presence. Until we understand the possible importance of a certain amount of edge effect in governing the balance between various forest pests and their natural predators, it behooves us not to destroy the remaining openings and their edges.

Part 5

Savanna and shrub communities

CHAPTER 16

Savanna

GENERAL NATURE, TYPES, AND LOCATIONS

When the Spaniards began their explorations of the islands in the Caribbean, they found many types of vegetation totally unfamiliar to them. One of the most striking was a peculiar combination of grassland and forest, in which the bulk of the land was occupied by grasses and a few shrubs, but which also had widely spaced tall trees, frequently of a single species at a given place. The native Carib Indians called such landscapes "savanna." This has become the general name in many European languages for any similar plant community where trees are a component but where their density is so low that it allows grasses and other herbaceous vegetation to become the actual dominants of the community. In the tropics, savannas are largely independent of climate and are found in forest regions on soils and topographies of unusual characteristics (Beard, 1953). In most of the temperate zone, however, savannas are best developed in the climatic belt separating the grasslands from the forests. Their intermediate structural nature thus parallels the intermediate climatic conditions under which they are found (Dyksterhuis, 1957).

In Wisconsin, savanna constituted one of the most widespread communities in presettlement times. A number of more or less distinct types were present, which differed in the kind of trees or the kind of understory, or both. The most familiar, as well as that occupying the greatest area, was the oak opening. This savanna type was present throughout the prairie-forest floristic province, south and west of the tension zone (Figure XVI-2). The usual combination on the uplands was that of bur oak (*Quercus macrocarpa*) and mesic prairie, although black oak (*Q. velutina*) and white oak (*Q. alba*) were sometimes present, while swamp white oak (*Q. bicolor*) and wet-mesic prairie formed the community on low grounds. A rough estimate of the area occupied by the oak openings in the early 1800's is 5,500,000 acres.

Another type of savanna occurred on the sand plains of central Wisconsin, on the outwash terraces of the Wisconsin and Sugar rivers, and at other places with a very sandy soil. These were locally known as scrub oak barrens. The dominant tree was Hill's oak (*Q. ellipsoidalis*) although true black oak was sometimes present. The understory was largely dry-mesic prairie or sand barrens grassland. Approximately 1,800,000 acres were so occupied in presettlement times. North of the tension zone, this type gradually gave way to jack pine (*Pinus banksiana*) barrens, although the scrub oaks remained in the mixture throughout the northern counties (Figure XVI-4). The greatest areas of pine barrens were in Marinette, Oneida, Adams, and Juneau counties, and in the northwestern counties in the bed of hypothetical glacial Barrens Lake (Aldrich and Fassett, 1929). The understory was either sand barrens grassland or depauperate bracken-grassland. Pine barrens originally covered about 2,300,000 acres.

In the south a fourth type of savanna was present. This was the very specialized cedar glade, which occupied local areas of steep topography, especially on limestone ledges and clifflike slopes in the Driftless Area and on abrupt knolls of limestone drift in the Kettle Moraine and other glacial moraines. The tree member was red cedar *(Juniperus virginiana)*, while the understory was dry prairie. No firm estimates of the total area of this type are available, but it undoubtedly was very small, probably only a few thousand acres.

The unusual nature of the savannas called forth almost as much discussion from early observers as did the prairie, for there were many descriptions and attempts at explanation. In spite of this, the first general summary was written as recently as 1946, by A. B. Stout, who confined his attention to the bur oak openings of southern Wisconsin. In the P.E.L. program, Cottam (1949) studied the spacing and composition of the oak openings in Dane and Green counties on the basis of surveyors' records. He investigated the changes that have taken place in the openings in the past century. Brown (1950) also compared the situation when Wisconsin was originally surveyed with present conditions in the jack pine barrens of Adams County. The locations of the savanna stands upon which the material in this chapter is based are shown on the map at the opening of this chapter.

The major P.E.L. phytosociological study of Wisconsin savanna was made by Bray (1955), who examined 59 stands of oak opening, oak barrens, cedar glades, and other types in southwestern Wisconsin. One of the difficulties encountered in Bray's study was the very great scarcity of stands which remained in their original condition. As will be seen later, the oak openings quickly change to closed oak forest in the absence of fire, so any area which has been protected has long since lost its savanna qualities. Many areas can be found which retain their original trees at their original spacings but this preservation has been accomplished by continuous grazing, which has destroyed the understory. Beyond question, an oak savanna with an intact groundlayer is the rarest plant community in Wisconsin today. In the entire P.E.L. study, no stands on either wet or wet-mesic sites were found which were not pastured; hence no information is available on these types, except for their tree compositions.

Bray was able to assign adaptation numbers to the southern savanna trees on the basis of their mutual occurrences, in much the same way as had been done in the earlier forest studies. When the stands were arranged according to a summation index based on these adaptation numbers, a compositional gradient was achieved which appeared to be related solely to moisture rather than to combined moisture and light as in the upland or lowland forests. The behavior of the major species when plotted against the compositional gradient was comparable to the situation found in the true forests.

No similar information is available for the pine barrens, so no ordination is possible. The information consists of a presence study of trees and herbs in 32 stands of true jack pine savannas. Some of the bracken-grasslands in the north have islands or scattered trees of aspen and are very similar to the aspen park land savannas of the Prairie

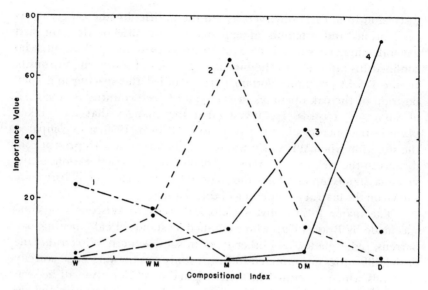

Figure 19.—Behavior of major trees along moisture gradient of the southern savannas (after Bray, 1955). 1, swamp white oak (*Quercus bicolor*); 2, bur oak (*Quercus macrocarpa*); 3, Hill's oak (*Quercus ellipsoidalis*); 4, red cedar (*Juniperus virginiana*).

Provinces of Canada, but no detailed studies have been made. As mentioned above, none of the wet or wet-mesic savannas in the south were intact in their groundlayers, so that a division of the southern gradient into equal segments for study purposes is not possible. Because of these incompletenesses in both north and south, it has been deemed best to pool the information for all stands of each of the four major types and to present their average composition as a basis for further discussion.

OAK OPENINGS
Composition

Perhaps the first mention of Wisconsin oak openings is that by Charlevoix (1761), who described the ". . . immense prairies, interspersed with small copses of wood, which seem to have been planted by hand." A little later Jonathan Carver (1781) reported that "only a few groves of hickory and stunted oaks covered some . . . [of the prairies]." Keating, in 1824, spoke of "thin woods, which gradually disappeared," and were replaced by prairies. Owen (1848), in the Driftless Area wrote, "We have clumps of trees, disposed with an effect that might baffle the landscape gardener, now crowning the grassy height, now dotting the green slope with partial and isolated shade." Bayley (1954) writing of eastern Wisconsin in 1845, described the appearance of the trees in the following passage:

The Oak openings are covered with trees about as far apart as in a common orchard. The Burr oaks have tops somewhat resembling the apple tree in form but preserving all the stiff appearance of the "gnarled oak." The White Oaks are taller and more valuable for timber. The roots of both these strike down so that they are not in the way of the plough, and it is no uncommon thing to see large fields of heavy wheat, with the trees all standing, but girdled to prevent the foliage shading the ground.

A more eloquent description of the savanna was given by Hoyt (1860) in the account accompanying his vegetation map:

The "Openings" are a peculiar feature of Wisconsin and Minnesota, not being characteristic of any other State. They are of two kinds—the Burr Oak and the Black Oak. The Black Oak Openings belong to the sandy regions, and are not marked by any considerable agricultural capacity. The Burr Oak Openings, however, are among the most productive portions of the State, being especially adapted to the continued production of wheat. They are, moreover, the most beautiful portions of the varied and picturesque surface of the country. Grouped here and there, like so many old orchards, on the summit of a gentle swell of land, or on the border of marsh, prairie or lake, there is nothing in the whole catalogue of American sylva that equals these Burr Oaks for the charming, homestead-like expressions they give to the landscape. The timber they furnish is brittle and of but little worth, except for fencing and fuel; still, abounding as they do in what would otherwise be a prairie country, and constituting so charming a feature of Wisconsin scenery, they possess a value which is beyond computation.

That Hoyt was wrong in his restriction of the oak openings to Wisconsin and Minnesota is indicated in this passage from Fennimore Cooper's novel, *The Oak Openings,* written in 1848 from information supplied by notes made in 1812 at Kalamazoo, Michigan:

The country was what is termed "rolling," from some fancied resemblance to the surface of the ocean, when it is just undulating with a long "ground-swell." Although wooded, it was not, as the American forest is wont to grow, with tall straight trees towering toward the light, but with intervals between the low oaks that were scattered profusely over the view, and with much of that air of negligence that one is apt to see in grounds, where art is made to assume the character of nature. The trees, with very few exceptions, were what is called the "burr-oak," a small variety of a very extensive genus; and the spaces between them, always irregular, and often of singular beauty, have obtained the name of "openings," the two terms combined giving their appellation to this particular species of native forest, under the name of "Oak Openings."

These woods, so peculiar to certain districts of country, are not altogether without some variety, though possessing a general character of sameness. The trees were of very uniform size, being little taller than pear-trees, which they resemble a good deal in form; and having trunks that rarely attain two feet in diameter. The variety is produced by their distribution. In places they stand with a regularity resembling that of an orchard; then, again, they are more scattered and less formal, while wide breadths of the land are occasionally seen in which they stand in copses, with vacant spaces, that bear no small affinity to artificial lawns, being covered with verdure. The grasses are supposed to be owing to the fires lighted periodically by the Indians in order to clear their hunting-grounds.

The spacing of the openings was indicated by Bayley who said, "The Oak openings, prairies, and marshes are so situated and of such extent that most farms of 160 acres embrace a part of each." Similarly, Guernsey and Willard (1856) described the situation in Rock County:

This county is about equally divided between prairie and oak openings. . . . Groves are interspersed through the prairies at intervals besides which, points of openings jut into the prairie at different places, thus rendering the outlines of the prairie unequal, and at the same time bringing wood and timber within a short distance to all its settlers; in fact, few men on the prairies live more than three miles from timber.

These accounts are largely in agreement regarding the size, spacing, and shape of the trees in the oak openings. Generally speaking, all of the trees in a particular stand are of the same size and age, although the age may vary from forty or fifty years to over two hundred years. The trees are open-grown, with large lower branches which frequently sweep close to the ground. Bray found that the shape was related to the density of the trees, with the ratio of crown height to crown diameter ranging from 1.25 in stands where the trees were so far apart as to exert no mutual interactions, to 2.40 in stands where at least some of the trees were within crown diameter distances from each other. He studied the same species in close-grown forest stands for comparison and found the ratio to be as high as 3.03 in such circumstances.

The density per acre is the most variable of all characteristics of the oak openings. Purely for convenience an arbitrary limit is set between savanna and forest at 50 per cent canopy coverage. In other words, if more than one-half of the ground area is shaded by the trees at noon in midsummer, the stand is classed as a forest rather than a savanna. A similar criterion was used by Clements (1928) in his discussion of savannas: "The transition [from savanna] to forest or woodland is usually gradual, and it is impossible to draw a sharp line between the two. However, it is a simple matter to distinguish the general areas from each other. As long as the trees or shrubs are far enough apart so that their shadows do not touch, the grassland or scrub remains in control." At the other extreme, the situation is more difficult. In the P.E.L. studies, a minimum of one tree per acre was the criterion used to separate savanna from prairie, but this is highly artificial and may be biologically misleading, since savanna lands were recorded by the land surveyors where the trees averaged over 200 feet apart and where many trees were as far as a quarter of a mile from a witness corner. Such very open savannas would be classed as prairies in this study, but their physiognomy as seen by a traveler and their behavior with respect to fire would be very different from those of a true prairie. As a matter of fact, all of these savannas are related to brush prairie rather than true prairie.

The oak openings not only varied in average density but also in the spatial distribution of the trees, with extreme aggregation the rule. As indicated in the early accounts, the trees were arranged in isolated patches, or groves, or as tongues, or irregular peninsulas of widely spaced trees projecting from denser forest into the open prairie. The beauty of such a disposition of tree and grass must indeed have been unique (Plate 43).

Far and away the most common arboreal species in the oak openings was the bur oak. In southwestern Dane County and adjacent portions of neighboring counties, white oak was frequently the main or the sole tree (Cottam, 1948). It is suspected that introgression from the bur oaks had conveyed an extra degree of fire resistance to these white oaks, since their bark appears thicker than normal. Black oak was frequently found, but most other species were rare, including shagbark hickory (*Carya ovata*), large-toothed aspen (*Populus grandidentata*) and black cherry (*Prunus serotina*) (Table XVI-1). Altogether, 18 tree species were recorded in the oak openings.

The true dominants of the oak opening are grasses rather than trees, as the ground vegetation forms the community matrix in which the trees are conspicuous but incidental members. The prevalent grasses and forbs are listed in Table XVI-2, together with the other species which reach their Wisconsin optimum in this type. As can be seen, the flora is a mixture of prairie and forest species, with the prairie types predominating. A number of species which are widespread on the prairies actually reach a higher level of importance in the openings, but no species are known which are confined to them. The closest approach would be *Besseya bullii,* whose total range is confined to the area of the Middlewest savannas and which is much more common in the openings than in the prairies.

The intermediate nature of the oak opening is shown by a floristic analysis of the groundlayer, which shows that seven families comprise over one-half of the species. The three leading families are the *Compositae* (21.5 per cent), *Gramineae* (10.7 per cent), and the *Leguminosae* (7.9 per cent), which had the same position in the prairie, but the *Ranunculaceae* (5.6 per cent) are much more important in the savanna. The shrubs are also much more prominent than in the prairie, with a sum of presence of the prevalent species of 616 or 22.6 per cent of the total. This compares to 9.5 per cent in the mesic prairie and an average of 18.9 per cent in the southern upland forests. Of the prevalent species in Table XVI-2, 16 species reach equal or higher levels in the prairies, and the same number have an optimum in the southern forests. Only one, *Heliopsis helianthoides,* reaches its optimum in the oak openings, while the remainder attain their highest levels in other savannas or in the bracken-grasslands. A similar mixture of grassland and forest plants was reported for the savannas, *Waldsteppen,* of Russia by Krylov (1932).

Structure

The average frequencies of the prevalent species are shown in Table XVI-2. On the basis of the frequency-presence index, the ten most important species in the oak openings are *Amphicarpa bracteata* (3250), *Euphorbia corollata* (3052), *Amorpha canescens* (2670), *Galium boreale* (2622), *Monarda fistulosa* (1940), *Rosa* sp. (1890), *Cornus racemosa* (1725), *Corylus americana* (1428), *Apocynum androsaemifolium* (1320), and *Andropogon gerardi* (1320). The fact that four of these are shrubs emphasizes the intermediate nature of the savanna flora. No studies of groundlayer density nor actual state of aggregation have been made in Wisconsin oak openings (nor elsewhere so far as known). Bray (1955), however, conducted a thorough investigation showing how trees contribute to the obvious lack of randomness in the openings. For this purpose, he laid out a number of transects, extending in the four cardinal directions from the base of isolated trees, and studied the changes in the understory as they were related to the shade cast by the canopy of the trees. In a series of openings on mesic or dry-mesic sites, chosen to reduce differences in factors other than light, it was possible to calculate an average quadrat frequency, according to three classes of light intensity—0-1000 foot-candles, 1000-10,000 foot-candles, and over 10,000 foot-candles. The areas of the first class were immediately beneath the canopy, while the last were never shaded during midday

Table 18
Relation of certain common oak opening species to light intensity (after Bray)

	Average frequency in quadrats with indicated light		
	0–1000 F.C.	1000–10,000 F.C.	10,000+ F.C.
Light species			
Amorpha canescens	28	61	77
Andropogon gerardi	21	35	61
A. scoparius	8	13	56
Aster ericoides	19	22	29
A. sericeus	2	20	47
Bouteloua curtipendula	9	18	47
Helianthus laetiflorus	24	38	77
Panicum leibergii	16	35	55
Petalostemum purpureum	3	5	38
Stipa spartea	21	37	63
Intermediate species			
Anemone cylindrica	28	31	7
Aster azureus	37	41	4
Commandra richardsiana	17	41	34
Coreopsis palmata	37	47	45
Euphorbia corollata	37	60	49
Galium boreale	43	55	23
Monarda fistulosa	37	49	22
Rosa sp.	27	30	23
Smilacina stellata	21	25	17
Solidago ulmifolia	40	43	17
Shade species			
Amphicarpa bracteata	61	50	15
Apocynum androsaemifolium	19	14	7
Aralia nudicaulis	36	17	0
Aster laevis	40	29	0
Carex pensylvanica	33	30	5
Cornus racemosa	59	38	7
Corylus americana	41	26	38
Geranium maculatum	23	6	2
Helianthus strumosus	53	47	22
Ratibida pinnata	29	24	4

hours. The results for some of the prevalent species are given in Table 18. It is of interest that all but two of the ten most important species reach maximum values in the intermediate or low light classes. This is related to the tree densities and the consequent proportions of the stand area in each of the three light classes. Bray found that 37.5 per cent of the readings were in the lowest light class, 34.9 per cent in the intermediate, and 27.6 per cent in full sun. Naturally, these figures vary from stand to stand, depending on the number of trees present.

Origin and maintenance of the oak openings

Several writers have reviewed the information bearing on the questions, how the oak openings were formed, and how they are maintained (Gleason, 1912; Stout, 1944; Cottam, 1949; and Bray, 1955). They are in agreement that oak openings were derived from pre-existing forests and do not represent a migration of bur oak trees across an open prairie. They further agree that the agency responsible for this degradation was fire. These conclusions are based largely upon historical evidence of the frequency and effects of fire and the results of artificial fire protection, but they also take into account evidence derived from soil studies. Generally similar conclusions have been reached by many writers concerning similar savannas on the forest-steppe boundary in Russia (Denisov, 1948; Popov, 1953; Borisova *et al.*, 1957).

The early writers all commented upon the role of fire. Thus Bayley, on the basis of his 1845 observations said: "The fires sweep over these [oak openings] as over the prairies every year, but the thickness of the bark is such that the trees are rarely killed by it. Where the fire is kept off, these openings soon become covered with a vigorous growth of young timbers." Knapp, too, (1871a), reported:

A large crop of grasses and other plants, annuals and perennials, grow here, many of which are gay flowered; these formed a thick coat for the devouring fires that ran over the country in former times, and caused much of its openings character. . . . The only trees that could withstand the fires were those so covered with an incombustible cork that the heat of the prairie fires could not penetrate to and destroy the cambium. The oaks, hickories, and some poplars could alone do this.

Theron Haight (1907), in his *Memoirs of Waukesha County*, also commented on the importance of fire in the maintenance of the openings:

. . . including magnificent oak-openings over the whole region except at the margins of the many lakelets and streams where maples, birches, lindens and willows flowered, . . . on the crests of the loftiest hills where the vegetation was sparse, and on a few patches of prairie in Eagle, Waukesha and Summit with their rank grasses swaying in the summer winds like the waters of the sea, and annually involved in the great fires which swept over the whole of the country, keeping it clear of shrubbery and underbrush, though not injuring the noble oaks which shaded the greater portion of the country.

Not all of the early observers were agreed that fire was solely responsible. Owen (1852) thought wind action was also important, as shown in this passage:

> The country . . . is prairie, with oak openings. For about half the distance the growth of timber is very stunted; indeed, along the whole distance we saw no large trees. The dwarfish character of the timber does not appear to depend here upon the sterility of the land: it is true, the soil is siliceous, having been derived, in a measure, from the destruction of the sandstones . . . but it is not deficient in organic matter, and produces better crops than its appearance at first view would indicate. . . . Here I noticed that the most stunted trees grow in the more exposed situations. This seems to indicate that the strong winds which often sweep over the extensive prairies, exercise a considerable influence in retarding, and even suppressing, the growth of timber.

When the fires were stopped by the new settlers, a very rapid change took place in the oak openings. Within a decade, the openings became filled by saplings and brush and within twenty-five to thirty years, dense oak forests were present. This was noted by Irving (1880) in the central Wisconsin region: "In many places, regions once prairie have been invaded by a timber growth, which has come in since the settlement of the country, having been in former times checked by the annual prairie fires. Very large areas in Adams County, for instance, which are now covered with a thick growth of small oaks, are said to have been open prairies at the time of the first settlements." Sargent, in his *Tenth Census Forest Report* (1884), also described the changes: "The forest area has somewhat increased in the prairie regions of the state [Wisconsin] since its first settlement and the consequent decrease of destructive prairie fires . . . and the oak forests on the uplands have been . . . losing their open, park-like character by the appearance of a young growth which has sprung up among the old trees." John Muir, in his autobiographical account of his boyhood in Columbia County (1913) also described this phenomenon in the following passage: "As soon as the oak openings in our neighborhood were settled, and the farmers had prevented running grass-fires, the grubs grew up into trees, and formed tall thickets so dense that it was difficult to walk through them and every trace of the sunny 'openings' vanished."

One of the most significant features of this closing of the oak openings was the seemingly spontaneous generation of black oaks in the new growth. Frequently a pure stand of black oak would spring up among the widely spaced bur oak veterans although there might not be any mature black oaks for miles around. Some of the early observers attributed this type-conversion to a mass seeding of the openings by flocks of passenger pigeons, each of which was presumed to have brought a single black oak acorn across the prairies to the bur oak grove and then dropped it, in concert with his flock fellows. Actually, no such far-fetched explanations are necessary, as the black oaks were there all of the time, but were growing as oak brush or grubs among the prairie grasses, in much the same fashion as has been described earlier for the brush prairies. When the fires stopped, the oak brush developed into mature trees that characteristically had two or three trunks per tree.

This oak brush was first mentioned by Carver in 1781 who considered it a separate species: "Like that it bears an acorn, but it never rises from the ground above four or five feet, growing crooked and knotty" (Bray, 1955). John Muir gives a graphic account of these repeatedly burned oak plants:

When an acorn or hickory nut had sent up its first season's sprout a few inches long, it was burned off in the autumn grass fires; but the root continued to hold on to life, formed a callus over the wound and sent up one or more shoots the next spring. Next autumn these new shoots were burned off, but the root and calloused head, about level with the surface of the ground, continued to grow and send up more shoots, and so on, almost every year until very old, probably for more than a century, while the tops, which would naturally have become tall broad-headed trees, were only mere sprouts seldom more than two years old.

The name grub is a local expression, etymologically related to the Old German, *grubben,* to dig. Grubs were best known to the settlers who were forced to laboriously grub them out by hand, using a grub hoe or similar heavy instrument, since the underground parts were too tough for the plows then in use. The grubs were not wholly in the oak genus, since one of the worst trouble-makers for the settlers was the shrub, redroot (*Ceanothus americanus*), which frequently produced a huge underground burl-like root stock. The briar from which briar pipes are made is a Mediterranean plant of similar lifeform, induced by fire like the Wisconsin grubs.

The tops of the grubs were killed back by the frequent fires, but the roots continued to grow, sometimes forming masses or surface plates 3 feet or more in diameter (Plate 44). Both black and Hill's oak as well as bur oak produce grubs readily, but the bur oak may get away to safe tree size with a chance period of protection from fire of only 12 to 15 years, while trees in the black oak group are never safe, regardless of the period of protection. Thus, on the University of Wisconsin Arboretum, a destructive ground fire in 1947 killed the trunks on a large number of ninety-year-old Hill's oak trees, but all of these put out numerous shoots from the root crown. Two further accidental fires in the succeeding five year period consumed all traces of the original trunks, but the fires merely stimulated further shoot growth. These plants are now typical grubs again, with no surface indications that they are at least one hundred years old. Indeed, there is no good reason why they may not be a thousand years old, since the same process may have been repeated a number of times since the germination of the original acorn.

Cottam (1949) studied in detail a forest in Dane County which had originated by this process of grub maturation. The original, open-grown oak trees were still present, at the rate of seven per acre. They still had traces of their former low branches, now present as broken and hollow stubs. The trunks of the younger oaks derived from the grubs were all the same age; the trees started in the period, 1845 to 1855. This new growth was present at the rate of 136 trees per acre and consisted of single trunks or groups of two or three trunks on a single root, but all trees had a typical, forest-grown form with a straight trunk devoid of lower branches (Plate 13). The new trees had overtopped the original oak-opening trees and these were dying as a result of suppression. On the basis of Cottam's study and other observations in similar forests, it appears that the veteran savanna bur oaks survive this overtopping in healthy conditions for 80 years or so. After that, wood-rot fungi entering the shade-killed lower branches bring about a rather rapid deterioration of the main trunk, so that by 100 or 110 years, most of the veterans are gone, snapped off by high winds and winter storms. These huge, open-grown trees tell a graphic story of an interesting phase of Wisconsin's history. It is unfortunate that many consulting foresters are so imbued with textbook silviculture theory as to invariably recommend that forest owners remove these derisively named "wolf trees" so that more valuable (*sic*) trees may take their place. It is highly probable that an active appreciation of the history of their forests as dynamically changing biotic complexes would in the long run be of more value to the owners.

OAK BARRENS

Composition

Several of the early writers mentioned that the bur oaks and white oaks of the heavy soil openings were replaced by black oaks on the sandy areas, but few detailed descriptions of the type exist. Most of the comments refer to the jack pine barrens found on similar sites in the north. Actually the two types are closely related and intermediate mixtures of both oak and pine are widespread in central and northwestern counties. For purposes of this discussion, oak barrens are considered to be those savannas which have black oak or Hill's oak as their most prominent tree and in which jack pine is absent. As such, they are located entirely in the prairie-forest province south of the tension zone. They are prominent on the outwash-filled valleys of the Wisconsin River from Portage to Arena and the Sugar River in Green County, and on the sandy uplands of Marquette and Waushara counties. Elsewhere, they are limited to local areas of sand derived from underlying sandstone bedrock, such as many places in the Driftless Area.

The composition of the tree community in the 14 P.E.L. stands (Figure XVI-1) is given in Table XVI-4. Considerable difficulty was experienced in separating black oak from Hill's oak in these barrens since extreme introgression has occurred, but most cases were resolved in favor of black oak. This should not be taken too seriously; it would probably be better to combine the values and let them refer to the entire hybrid complex. Actually, both species react similarly to the main environmental factor, fire, in that their trunks are equally fire-sensitive and they are equally capable of sending out basal sprouts. Because most of the trees are of grub origin, they usually have multiple trunks and rarely achieve the great size and gnarly grandeur of the oak opening burs. They also tend to reach higher densities, with 20 or 30 trees to the acre or more.

The prevalent groundlayer species and the modal species of the oak barrens are given in Table XVI-5. The ten leading species by the frequency x presence index are *Euphorbia corollata* (3600), *Amorpha canescens* (2300), *Corylus americana* (1420), *Comandra richardsiana* (1320), *Rosa* sp. (920), *Smilacina stellata* (916), *Pteridium aquilinum* (878), *Smilacina racemosa* (860), *Helianthemum canadense* (850), and *Fragaria virginiana* (750). Only four of these were in the leading group of ten in the oak openings, but the general floristic resemblance of the oak barrens understory to the oak openings is relatively high, with an index of similarity of 66.9 per cent. The main differences lie in the reduced importance of the grasses and the increase in representation of the *Ericaceae* and *Asclepiadaceae* in the barrens flora. Shrubs have a less prominent place under the scrub oaks, with a sum of presence of only 443 among the prevalent species, or 15.5 per cent of the total.

The scrub oak barrens appear to be more closely related to the prairie than to the forest, perhaps as a result of the periodic removal of all canopy coverage by severe fires. Of the prevalent species, 28.6 per cent reach higher presence values in the prairie, as against only 16.3 per cent in the forest, while of the twelve species with optimum values in the barrens itself, all are more frequent in the prairie than the forest. Some of the species which reach numerical superiority in the barrens are rather surprising because they are commonly considered to be very characteristic prairie plants; these include especially *Aster linariifolius*, *Lespedeza capitata*, and *Lithospermum canescens*.

The oak barrens are found on very sandy soils, but only a few of the species on the prevalent species list can be considered to be sand plants in the sense of being restricted to such habitats. The major sand species are *Artemisia caudata*, *Aster linariifolius*, *Helianthemum canadense*, *Lupinus perennis*, *Tephrosia virginiana*, and *Viola pedata*.

The origin of the scrub oak savannas is the same as that of the oak openings—degradation of prior forests by fire. Maintenance is also by fire, but with the major difference that the tree component is likely to be completely destroyed at rather frequent intervals.

PINE BARRENS

Composition

In the section on northern pine forests (Chapter 11), it was pointed out that considerable areas north of the tension zone were covered with pine barrens, rather than closed forest. Sweet (1880) for example, described the community as he saw it in the course of his geological surveys:

The "barrens" have a timber growth exclusively their own. The trees are either scrub pines, or black-jack oaks [*Q. ellipsoidalis*], averaging in diameter about three or four inches and in height not over fifteen feet. In some places, as on the sand hills of the barrens, the trees are at considerable distances from each other. . . . On the sides of the barrens, and in low places, quite large groves of Norway pines [*Pinus resinosa*] are frequently found.

These barrens are true savannas, in that the dominant plants are grasses, forbs, and shrubs, with a scattered stand of trees. The most usual tree is jack pine, although red pine may be the main species in unusual cases. Hill's oak is usually present as a grub or as a scattering of larger trees. This relation is shown in Table XVI-7, which summarizes the tree information from the P.E.L. studies (Figure XVI-3). The oak seen there has a high presence but a relatively low importance value. Altogether, seven tree species were encountered. Others which are sometimes seen include white pine, pin cherry, and white oak.

The pine barrens are concentrated in the areas of very sandy soil, with four separate regions of major concentration (Figure XVI-4). Those found in the old Barrens Lake area of the northwest counties have been described by Murphy (1931):

> The Northwestern Pine Barrens of Wisconsin is a long, narrow strip of sand where coniferous forests and open expanses of sweet fern and grassy barrens dwarf into insignificance the few evidences of man's present occupancy and use of the land. . . . The grassy and sweet fern barrens bear no mark of present utilization, but are desolate open tracts where only an occasional charred stump, a cluster of jack pines, or a scrub oak bush, breaks the monotonous sweep of the rolling, thinly clad ground surface.
>
> Almost every year forest fires sweep sections of the Barrens, leaving their record in ghost-like dead trees, or, if several fires have occurred, barren tracts dotted with charred stumps. . . . In most areas the fire hazard is greatest in May and October. In the Barrens, May is by far the worst of these two, since in the fall the sweet fern is still green and the scrub oak still has its leaves. In the spring, on the other hand, the oak leaves are on the ground and the sweet fern is dry and brown—an ideal bed in which a fire may travel. . . .

The pine barrens of other areas presented a similar aspect, according to many early accounts and more recent descriptions written before 1935.

The prevalent groundlayer species are given in Table XVI-8. The outstanding feature of the groundlayer in the pine barrens is the extraordinary development of shrubs. This is shown by the fact that the sum of presence of the shrub species on the typical list is 41.2 per cent of the total, far higher than for any other community in Wisconsin. Two of the shrub species, redroot (*Ceanothus ovatus*) and huckleberry (*Gaylussacia baccata*), reach their maximum Wisconsin levels in this community; but the blueberry (*Vaccinium angustifolium*) is of even greater importance. In fact, the blueberry is the plant most firmly associated with the pine barrens in the minds of Wisconsin citizens, because of the ample and free crops of this delicious fruit that are to be had for the picking. Murphy (1931) reports that the Indians deliberately fired the barrens as a measure to improve the berry crop. Another shrub which is highly characteristic of the barrens is the sweet fern (*Myrica asplenifolia*). The pungent odor arising from this plant on a hot summer day is a never-to-be-forgotten symbol of the barrens.

The 134 species found in the barrens are distributed in 48 families, of which these 5 contain over one-half of the total: *Compositae*–23.9 per cent, *Gramineae*–10.4 per cent, *Rosaceae*–8.2 per cent, *Liliaceae*–6.7 per cent, and *Ericaeae*–6.0 per cent. Approximately equal numbers of the prevalent species bloom in the spring and the summer (44.0 per cent and 49.6 per cent of the total sum of presence, respectively) while only a few bloom in the autumn (6.3 per cent). The very early spring species are largely of prairie affinity and include such plants as pasque flower (*Anemone patens*), golden puccoon (*Lithospermum canescens*), and wild lupine (*Lupinus perennis*). As in the scrub oak barrens, few of the species are obligate sand plants, even though the pine barrens are on very sandy soils. Included in this limited group are *Leptoloma cognatum*, *Myrica asplenifolia*, *Ambrosia psilostachya*, *Aristida basiramea*, *Froelichia floridana*, *Polygonella articulata*, and *Prunus pumila*.

Structure

Little accurate information is available concerning the structure of the pine savannas. Brown (1950) studied an area in Adams County by using the information in the surveyors' records. He found that the densities usually varied from 2 to 8 trees per acre, with an average intertree distance of about 100 feet. Frequently the jack pines are present in island-like groups, with a circle of young trees surrounding an older central tree (Plate 45). This extreme aggregation in the tree layer is not particularly noticeable in the understory. A number of the major species have effective means of vegetative reproduction, but the resulting colonies are not so dense nor so conspicuous as are the clone formers in the open prairies. Perhaps the major reason for this is the very high density and even distribution usually displayed by the blueberry (*Vaccinium angustifolium*), which sometimes exerts almost total dominance, making a blueberry heath rather than a grassland. The huckleberry (*Gaylussacia baccata*) occasionally replaces the blueberry, particularly in the southernmost barrens.

Origin

As in the first two savanna types, there is no doubt that the immediate cause of a pine barren is fire. In this case, soil and topography are major contributing factors, since it is essential that the fires be repeated at such short intervals as to prevent the active reseeding of jack pine from its serotinal cones. Suitable conditions for such frequent fires are provided by the very dry sands and by the flat or gently rolling surface of the barrens. Many of the early observers were aware of these interrelationships, which are shown in the following quotations. Chamberlin (1877), for example, had this to say about the barrens of eastern Wisconsin:

The plains that lie along the margin of the Kettle Moraine, and were formed by the escaping glacial waters, are of less uniform character. In some instances they were mainly formed of gravel derived from the crystalline rocks. . . . In other instances, they are composed quite largely of sand, and have a low degree of fertility, while in other cases, almost every ingredient was washed away except purely silicious sand, and the resulting soil is extremely poor. This is the character of some of the "barrens." Some of the so-called "barrens," however, of this and other regions of Northern Wisconsin, are not due so much to extreme sterility of soil, as to successive fires which have destroyed not only the growing vegetation, but also the vegetable mold, which would otherwise preserve the moisture, and enhance the fertility.

Sweet (1880) thought both fire and soils were important in northwestern counties:

The area occupied by this [sandy] soil is known as the "Barrens" of Bayfield and Douglas counties. It is nearly coextensive, as far as this district is concerned, with the great moraine and Kettle range, extending to the southwest from the Bayfield peninsula. This soil is made up, to a great depth, of light colored, rounded grains of quartz, frequently mixed with fine gravel. There are exceedingly few boulders scattered over the country occupied by this soil. It owes its origin, of course, to the drift. This soil is of the poorest quality. It sustains but a scant growth of stunted scrub pines. It contains no vegetable mould, and only here and there bears a tuft of grass. Fires have passed over the "Barrens" recently, during very dry seasons, and destroyed in many places the little timber that formerly existed on the soil. Now there are areas of eight or ten miles in length, notably along the east bank of the upper Brule, upon which there are no trees or shrubs, and very little vegetation of any kind. These treeless areas contain no streams, and very few lakes, while the remainder of the sandy district, although showing no streams, has an abundance of small lakes. I think, however, that the scant vegetation of the "Barrens" is not mainly due to a lack of water—for the trees are no larger about the lakes than elsewhere—but to an insufficient supply of proper plant food in the soil.

Moses Strong (1877) also postulated a dual control for the mixed barrens of the St. Croix region:

The timber occupying these tracts is peculiar and does not justify the application of the term "barrens." Some portions are covered with scrub pine to the exclusion of all else save underbrush. Most nearly similar to these are the patches of Norway pine. Other areas are covered with burr, black, and even white oak bushes, with occasional trees of these species. With these are associated the common white poplar, or trembling aspen, which is the most widely prevalent and abundant arboreous species. Curiously enough, the great toothed poplar is not uncommonly associated with it. There are also areas where white pine occurs associated with both poplars named; with the three species of oaks—though the burr oak is less common in this group—with the soft maple, and with scrub and red pines, forming a very strange association of plants that usually seek quite diverse conditions of soil and moisture. Probably the explanation is to be found in the contribution which the diabases and melaphyrs—the prevalent crystalline rocks of the region—have made to the soil, by virtue of which it is less purely siliceous than most soils of similar physical appearance. . . .

It seems also an open question whether the meagerness of vegetation on the so-called "barrens" is not largely due to the repeated fires, which not only destroy the life of the vegetation best adapted to such a soil, but also consume the dead vegetable matter whose accumulation would otherwise cover and enrich the soil.

Brown (1950) compared the conditions in Adams County in 1851 with those a century later. He found that densities had increased tremendously, from 2 to 8 trees per acre originally to 160 to 250 trees per acre today. Surprisingly, he found almost no change in percentage composition, with jack pine, Hill's oak, and bur oak having almost exactly the same relative representation in the dense forest as they had in the savanna. The trees he found were mostly twenty to thirty years old, thus reflecting the better fire protection which was initiated in the 1920's. No jack pines and only a few bur oaks were old enough to be relics of the original savanna, which probably had been regenerated several times between the surveyors' visit and the onset of fire protection. Most of the pine barrens of Wisconsin have undergone similar density changes and their place is now taken by dense jack pine forests. One of the important consequences of this replacement of savanna by forest is the great reduction in open lands needed by such birds as the prairie chicken and the sharp-tailed grouse. The dismal future of these favorite game birds in the face of an all-encompassing system of fire protection has been eloquently portrayed by Hamerstrom, Hamerstrom, and Matson in their classic publication, *Sharptails into the Shadows* (1952). Not only the grouse harvest but the blueberry crop as well have become victims of the bureaucratic dictum, that since most forest fires are the source of economic loss, therefore all fires are bad and must be prevented at any cost. This dogma has been supported by such an intensive propaganda campaign that there is danger of its being accepted as truth. On the contrary, the facts plainly indicate that fire is a normal environmental influence in the life of the forest; the evolution of such fire adaptations is clearly shown by the serotinal cones of jack pine. Fires have been burning in northern Wisconsin for at least 10,000 years and will continue to burn for another 10,000 unless artificially stopped. The forests of the region are adapted to this situation and the normal complements of species as we know them can exist only if it is continued. The unnatural elimination of any major environmental factor will cause severe readjustments in the vegetation. In this case, the adjustment will likely be in the direction of climax hardwoods at the expense of the economically more valuable conifers, but upsets in insect populations and their normal predators, and in other complicated biotic interrelations introduce a large element of uncertainty. The logical solution to the problem lies in the acceptance of the quantitative evaluation of fire rather than its complete condemnation. By all means, uncontrolled fires sweeping over large acreages are a tremendous danger to life and property and must be prevented by the best possible methods. On the other hand, controlled fire, burning when and where desired, can be used as a valuable tool in both silvicultural and game management operations. Such use should be actively encouraged and promoted, rather than hindered by outmoded taboos, to the end that the health and well-being of our forests and other nonagricultural lands can be raised to their optimum levels.

CEDAR GLADES
Composition

The cedar glades of Wisconsin are of little significance from the standpoint of area covered, but they nevertheless are of great botanical interest. They are restricted to the prairie-forest province and are found within that area only on very special sites. Two types of habitat are favored, a steep hillside of thin loess over limestone or quartzite bedrock, or a gravelly glacial moraine. In both types, some form of fire protection from the southwest must be present, either a bare rock cliff, a stream, a lake, or some other effective barrier. Such areas are common in the Driftless Area near the contact zone of the hard Platteville-Galena dolomites and the easily eroded St. Peter sandstone beneath. The sandstone forms the cliff, and the capping dolomites provide a substrate for the glade. In the glaciated area, suitable conditions are much more infrequent; they are largely restricted to the lake district in the Kettle Moraine in Waukesha County and to several recessional moraine systems, such as those at Rock Lake near Lake Mills in Jefferson County. This peculiar restriction of habitats is occasioned by the extreme fire-sensitivity of the dominant tree in the glades, the red cedar (*Juniperus virginiana*). The cedar glade is thus seen to be very different from the other Wisconsin savannas, but it is a savanna nevertheless, because it is a grassland community with a scattered stand of trees.

Bray (1955) studied the trees on seven glades (Figure XVI-1) and his results are summarized in Table XVI-10, which clearly shows the predominance of red cedar. Of the ten other tree species, the basswood is worthy of special comment. This tree appears to belong to the species *Tilia americana* but it differs in many respects from the common representative of that taxon as usually seen in the mesic forests of the region for it has smaller leaves, a slower growth rate, and a different branch habit. Further study may show it to deserve varietal rank. Another frequent associate of the red cedar is the white birch, particularly in the lower Wisconsin River valley, where its brilliant white trunks stand in striking contrast to the somber green of the cedar.

Sometimes the cedars develop into dense, pure stands which lose all resemblance to a savanna and take on the internal aspect of a northern hemlock forest (Plate 47). One of the best examples of such a cedar forest is to be seen on Observatory Hill, a quartzite monadnock south of Montello in Marquette County near the boyhood home of John Muir.

The groundlayer of the cedar glades is very similar to the dry prairie in composition, as seen in Table XVI-11, which shows the prevalent and the modal species. The glades are floristically rich (species density of 57) and more uniform than the other savannas, with an index of homogeneity of 62.7 per cent. Like the dry prairies, they have a large number of modal species which achieve a state-wide optimum in the type (31.5 per cent of the prevalent species). The ten leading species on the basis of the frequency-presence index are: *Andropogon scoparius* (3800), *A. gerardi* (2390), *Euphorbia corollata* (1920), *Amorpha canescens* (1790), *Anemone cylindrica* (1720), *Aquilegia canadensis* (1660), *Bouteloua hirsuta* (1660), *Arenaria stricta* (1640), *Solidago nemoralis* (1480), and *Petalostemum purpureum* (1370). Of these, 6 were also on the list of the ten highest species of the dry prairies.

Shrubs are of less importance in the cedar glades than in the other savannas, with only 12.8 per cent of the total sum of presence of the prevalent species. Among the total shrubs present are two close relatives of red cedar—the low juniper (*Juniperus communis* var. *depressa*) and the horizontal or creeping juniper (*Juniperus horizontalis*, Plate 48). The latter is especially interesting because of its great similarity to red cedar in everything but growth-form and because of its very peculiar range. Individuals, intermediate between the upright red cedar and the creeping juniper, are common in the cedar glades of Wisconsin (Fassett, 1945) and may be expected wherever the two species grow together. The creeping juniper is found abundantly on the dunes along Lakes Michigan and Superior. Elsewhere in Wisconsin, it is very rare and is restricted to a very narrow belt running from New Glarus, in Green County, northward 75 miles to Grand Marsh, in Adams County. This belt is from one to five miles wide. It has an eastern border at the margin of the glacial drift so that the entire belt is in the Driftless Area in the region where periglacial effects would have been most extreme during glaciations. Any explanation of this highly unusual distribution pattern must consider that the species is known from Iowa and Minnesota on similar sites at the western border of the Driftless Area.

One hypothesis which would fit this distribution pattern regards the creeping juniper not as a true species but as a cold-induced mutation of the red cedar which affected only the growth substance distribution in some individuals and caused them to grow flat as lateral branches rather than to grow erect with a central trunk. The frequent intermediate individuals could either be hybrids or partial retrogressive mutations. The extreme sensitivity of the prostrate shrubs to both fire and competition would prevent their invasion of the deep glacial soils; the poor seed production of the decumbent plants would tend to reduce their ability to spread westward onto apparently suitable sites in the interior of the Driftless Area. The poor seed production would increase the chances that any individual which did succeed would tend to be swamped-out by backcrossing with the more numerous red cedars of normal growth form already present in the invaded area. This rather far-fetched hypothesis can be subjected to test by suitable transplantation and genetic experiments. If validated, it could provide valuable information on conditions within the Driftless Area during glacial times.

Structure

No detailed information on the structure of cedar glades is available from the P.E.L. studies. The glades are usually so small and on such precipitous sites that they discourage the taking of necessary measurements. A common pattern is shown in Plate 46, where a forest of oak and birch, present on the surrounding areas of deeper soils, is gradually invading the cedars. These, in turn, are most numerous around the periphery of the glades and become more scattered in the central area which is usually the steepest and most directly aligned toward the southwest. The red cedars in most of the glades are small trees, although in a few places they reach a height of 40 feet and a diameter of 12 inches or more. Growth is usually very slow because of the thin soils; trees that are less than 2 inches d.b.h. at fifty years are the rule rather than the exception.

The over-all aggregation of the understory is extreme, not because of any special tendencies toward clone production, but because of the great differences in light intensity beneath the evergreen cedars and the open spaces between them. Shrubs play a prominent role under the oaks in an oak opening, but the shade of the cedars is so great that it excludes most shrubs and allows only a few forest plants to thrive. Among the species frequently seen under the cedars are hog peanut (*Amphicarpa bracteata*), columbine (*Aquilegia canadensis*), wild geranium (*Geranium maculatum*), and pale corydalis (*Corydalis sempervirens*).

Origin

The other Wisconsin savanna types originated by fire from preexisting forests, in which certain species especially adapted to resist fire were selected from the original mixed populations and were more or less purified by subsequent repetition of the fire. In the case of the cedar glade, this explanation cannot be true, because the red cedar is extremely fire-sensitive and no evidence exists to indicate that the sites were ever forested. It appears that a glade could be formed by the invasion of red cedar into a pre-existing dry prairie, but such a hypothesis would need to explain where the cedars came from and why they alone entered the prairie to form the pure stands we find today. These difficulties are so great that we must admit that we have no adequate explanation for the origin of the glades, nor for the factors involved in their maintenance. Some observers have suggested that the Indians may have been responsible for the glades. This is based on the fact that the glades along the Wisconsin River and its major tributaries are located on just those promontories needed to gain a long view up and down the rivers and that many are known to have been the sites of Indian signal fires in historical times. It is conceivable that the Indians introduced the cedars to those vantage points, but their purpose in doing so is not clear, unless the cedar in some way was of an advantage in starting the fire, either by providing shelter from drafts or by providing a source of easily combustible tinder. Probably so complicated an explanation is unnecessary, since red cedar seeds are known to be transported readily by birds, who consume the fleshy, berry-like fruit coat without injuring the seeds. A random scattering of seeds by this agency could result in the inoculation of the present glade sites, where successful cedar seedlings would have been protected by the fire barriers mentioned earlier. Seedlings beginning on the ordinary uplands amid oak savanna would have been destroyed by the frequent fires on those sites. This hypothesis receives some support from the modern observations of cedar seedlings on open sites far removed from the nearest glade. The wandering movements of flocks of cedar waxwings would be especially likely to bring about such a result.

GEOGRAPHICAL RELATIONS

The oak opening as a name for a savanna type is largely of middle western origin but the community so designated is of much wider geographic distribution in temperate America. Almost identical savannas are found in Michigan and Minnesota and from Ohio to Missouri and Nebraska, at least so far as their trees are concerned. The understory probably differs as much as the prairies in the same states, but the necessary information for detailed comparison is not available. Towards the south, in the barrens of Kentucky (McInteer, 1946), Oklahoma, and Texas, the species of oaks gradually change, as does the physiognomic aspect, until the shinnery is reached, with its knee-high, rather dense cover of brushy oaks mixed with prairie flora (Osborn, 1942). To the east, the oak openings originally extended to the Atlantic seaboard, usually with white oak replacing bur oak as the most important tree. The nature of the understory of these eastern savannas probably cannot be determined at this late date, but their widespread occurrence has been described in detail by Day (1953). Norman Taylor (1923) reported oak savanna with a high proportion of prairie plants in the groundlayer for the Montauk region of Long Island. Near Poughkeepsie, New York, McIntosh has found a number of open-grown relics of the original oak opening, while on the estate of Dr. J. W. Tukey near Princeton, New Jersey, there is a beautiful example of an oak opening now grown up to mesic forest. The open-grown white oaks are mammoth trees 4 feet in diameter, surrounded by sugar maples, beech, and other mesic trees dating back to pre-Revolutionary times. Similar instances can be seen from Virginia to New England.

One of the differences between the oak openings of Wisconsin and those in states to the west and south concerns the presence or absence of a shrub border between the trees and the surrounding prairie. In Wisconsin, the only references to such a border strip are by Knapp (1871a) and by Chavannes (1940). Most of the other early observers commented upon the openness of the savanna and the ease with which a horse and wagon could be driven between the trees. In Illinois, Iowa, and Nebraska, a strip of brush composed of hazelnut (*Corylus americana*), dogwood (*Cornus racemosa*), or other species is commonly present on the periphery of the groves; the strip varies in width from 15 feet to as much as 100 feet. No good explanations have been advanced for the origin of such a border, nor for its mode of maintenance.

The jack pine barrens of Wisconsin have their close counterparts in central Michigan and on the Anoka Sand Plains and elsewhere in Minnesota. Closely related jack pine savannas are found in the Prairie Provinces of Canada, especially in Manitoba and Alberta (Moss, 1955). Similar areas were also present in eastern Ontario and in New York State. The famous pine barrens of New Jersey greatly resemble the Wisconsin barrens in appearance and behavior, but the major tree is pitch pine rather than jack pine and many of the most prominent shrub members of the community are different, although the same families are represented. On the basis of Stone's study (1911), only 8.3 per cent of the 386 Jersey barrens plants are in Wisconsin, but such important genera in both areas as *Agrostis, Asclepias, Aster, Euphorbia, Lechea, Lespedeza, Panicum, Solidago,* and *Vaccinium* are represented by closely related species. The open short-leaf pine areas of Missouri and Arkansas bear some resemblance both to the jack-pine barrens and the cedar glades of Wisconsin but the relationship is quite remote.

The cedar glades have been studied ecologically only in Tennessee (Freeman, 1932; Quarterman, 1950) and Missouri (Erickson *et al.,* 1942; Kucera and Martin, 1957), but there is a series of such glades located along the Mississippi from Minnesota to Missouri. Studies in them would no doubt reveal a gradually lessening similarity, since the Missouri glades have only 36 per cent of their species in common with Wisconsin while the Tennessee stands show a similar resemblance. It is generally considered that the cedar glade is predominantly an Ozark community in origin and ancestry. This is supported by the great development of the type in the Ozark region and particularly by the rather large number of endemic species located on the glades of that area. None of these rare glade species have been found in Wisconsin, although some may be present since the glades have received little botanical study.

The savannas of Wisconsin bear little relation to most of the savannas of tropical America, although a number of species in common occur between our jack-pine barrens and the fire-controlled pine savannas of the high mountains of Haiti (Marie-Victorin, 1943; Curtis, 1947) and other West Indian islands; these include *Pteridium aquilinum, Lycopodium clavatum, Osmunda cinnamomea, Malaxis unifolia, Fragaria vesca, Poa compressa,* and a rather impressive number of genera are represented by closely related species: *Andropogon, Rubus, Eupatorium, Spiranthes, Myrica, Hieracium, Lobelia, Asclepias, Oxalis, Hypoxis, Sisyrinchium, Panicum, Verbena,* and *Habenaria.* The low altitude tropical savannas are quite different and are related to alternating seasons of excessive moisture and extreme dryness (Beard, 1953). A gley layer prevents internal drainage during the rainy season and also prevents upward movement of water during the dry season. Only forbs with underground storage organs, grasses, and a few trees can withstand these extremes. No such edaphic savannas are present in Wisconsin nor elsewhere in the Middlewest. The planosol prairies of Illinois and Iowa have the same general moisture conditions, but no trees with the necessary adaptations are present.

UTILIZATION AND CURRENT MANAGEMENT

As indicated earlier, the savannas of fire origin show great successional changes in the absence of fire, so most of the original area of these types has long since been converted to closed forest. Many oak openings and scrub-oak barrens retain their original tree cover because of intensive pasturage of the understory, but intact examples of either are extremely rare. The jack-pine barrens remained until about 30 years ago, largely because of their low value for agricultural purposes, but almost perfect fire protection and wide-scale reforestation since then have nearly exterminated the community. The only savanna which has remained relatively undisturbed is the cedar glade, largely because it occupies lands too steep to be utilized for other purposes. In fact, the physiognomic glade type has been greatly extended in recent years, because of the ability of red cedar to invade overpastured, open hillsides. Such invasion has greatly extended the area of the glades along the Wisconsin River bluffs near Lodi in Columbia County, in the Kettle Moraine in Waukesha County, and at many sites in the Driftless Area. The new areas resemble the original glades in the density and spacing of the cedars, but, of course, differ greatly in the composition of their understory.

CHAPTER 17

Tall shrub communities

NATURE AND LOCATION

Many of the plant communities of Wisconsin have a high content of shrubs and these may reach high densities in such types as the open bog and the pine barren. In all cases discussed so far, however, the shrubs have been low species which were competing with the herbs for light, or else they have been subordinate members of forests. Two communities are known in which the dominant layer is composed of tall shrubs, 1.5 to 3.0 meters in height, which attain a canopy coverage of from 50 to 95 per cent. These tall shrub communities are intermediate stages in the succession from wet prairie, fen, or sedge meadow to lowland forest, or conifer swamp. They are widely distributed throughout the state wherever these moist grasslands are present and may occasionally cover extensive areas of land. Although successional in nature, these tall shrub communities may persist in unchanged condition for long time periods; hence they are deemed worthy of study as separate types.

Wet-ground shrub communities have received little study in America and have not received any generally accepted name, either in the vernacular or in ecological terminology. In England, they are usually termed "carr" (Tansley, 1939), while transitional stages in their development are called "fen carr" to show their origin. In this country, Cain and Slater (1948) have applied the name "shrub fen" to the tall shrub communities in Michigan. Mature carr in England may contain trees or shrubs which reach tree size by commonly accepted standards. None of the Wisconsin shrub species have this potentiality, so one of the communities is here designated as shrub-carr to clearly differentiate it from the European type. The shrub-carr is considered to be a wet-ground plant community dominated by tall shrubs other than alder with an understory intermediate between meadow and forest in composition. The other community is similar in aspect but is dominated by an almost pure stand of tag alder (*Alnus rugosa*). The two types have different ranges; the alder type is north of the tension zone and the non-alder is south of it. The northern community is widely recognized and is known locally by the name "alder thicket." This name is here retained, so the two tall shrub assemblages will be discussed as southern shrub-carr and northern alder thicket. The P.E.L. information available on these communities is very limited, and consists of presence data for only 10 carrs (Figure XVII-1) and 12 thickets (Figure XVII-2). These data were collected by J. T. Curtis with the aid of Grant Cottam and H. C. Greene, in the stands shown on the map at the opening of this chapter.

SOUTHERN SHRUB-CARR

Composition

The prevalent species are listed in Table XVII-1, which also shows the other modal species. The low index of homogeneity (51.7 per cent) and the low number of modal prevalent species (17.8 per cent) testify to the intermediate nature of the community. The true importance of the shrubs is not well shown on this floristic basis, since only 14.3 per cent of the total sum of presence is contributed by these plants. No adequate sampling methods are known for tall shrub communities. Basal area is essentially meaningless, as is density. The only valuable measure would be cover, but the determination of cover in plants which are head-tall is a difficult and frustrating procedure. In the absence of quantitative data, it must suffice to point out that the shrubs are the true dominants, and they are in control of the environment. The most important single species is the red osier (*Cornus stolonifera*) but the willows collectively are of greater significance. Included in the group are *Salix bebbiana, S. discolor, S. interior,* and *S. petiolaris.* Other shrubs of local importance include the silky dogwood (*Cornus purpusi*), the currant (*Ribes americanum*), the red raspberry (*Rubus strigosus*), the elderberry (*Sambucus canadensis*), and the nannyberry (*Viburnum lentago*). Both herbaceous and woody vines are numerous and greatly increase the difficulties of travel by foot through the community. Prominent among these lianas are virgin's bower (*Clematis virginiana*), wild cucumber (*Echinocystis lobata*), woodbine (*Parthenocissus vitacea*), black bindweed (*Polygonum convolvulus*), and poison ivy (*Rhus radicans*).

The prevalent species are largely those with sedge meadow affinities (Table XVIII-3), but the less frequent species include many plants found in the moist hardwoods, especially such shade species as Jack-in-the-pulpit (*Arisaema triphyllum*), bedstraw (*Galium aparine*), touch-me-not (*Impatiens biflora*), bloodroot (*Sanguinaria canadensis*), and Gleason's trillium (*Trillium gleasoni*).

Stability

The shrub-carr is a normal stage in the primary hydrosere successions around lakes and ponds which follows the sedge meadow stage and in-turn is followed by the initial wet hardwood forest. Under such circumstances, it usually is present only as a narrow band or zone up to 20 or 30 meters wide. When fires or other agencies extend the sedge zone to larger areas, the extension often takes place at the expense of both the shrub and forest communities. In Jefferson County and elsewhere in eastern Wisconsin, many ancient glacial lake beds had developed to the stage of nearly pure tamarack swamps, as described in Chapter 12. A number of these were converted to mowing meadows in early postsettlement times by cutting and burning, which was followed by annual mowing. This treatment produced secondary sedge meadows which often were of considerable size, up to 2000 or 3000 acres as in the London and Jefferson marshes, and in parts of the Scuppernong Marsh. These meadows resembled the brush prairies because the woody shrubs, largely holdovers from the previous forest, were present throughout but were kept down by the mowing. When this mowing practice ceased, as it largely did in the 1930's, the shrub stubs grew up into a mature shrub-carr, with an almost solid canopy of tall shrubs over the entire area. Tree invasion of these shrub communities is very slow and after twenty years is evident only around the periphery, where cottonwood, willow, elm, and green ash are to be seen overtopping the shrubs. On the basis of a limited number of observations, a rough assumption would place the life of a large shrub-carr at a minimum of fifty years. During this time there is apparently a slow elimination of the sedge meadows species and a gradual invasion of forest understory. A number of species in addition to the shrubs were able to maintain themselves from the tamarack swamp stage through the meadow stage into the carr stage. These include the small yellow ladyslipper (*Cypripedium parviflorum* Salisb.), the showy ladyslipper *(C. reginae)*, the crested fern (*Dryopteris cristata*), and one of the asters (*Aster puniceus*).

ALDER THICKET

Composition and environment

The alder thicket is a common community along streams and around lakes in the regions north of the tension zone and has a few isolated stands in the south. Like the shrub-carr, it may occasionally occupy large areas on lands formerly covered with conifer swamp. The most apparent difference between the two tall shrub communities is in the dominant species; they range from almost pure stands of alder (*Alnus rugosa*) in the north with all other species of minor importance to a mixed dominance of willows and dogwoods with no alder in the south. The understory also shows considerable differentiation, the alder thicket having a high homogeneity (60.5 per cent) and a large number of prevalent species which reach optimum levels there (41.9 per cent). Both the prevalent species and the other modal species are shown in Table XVII-3.

As in the southern carr, the floristic list fails to do justice to the importance of shrubs in the alder thicket, since the sum of presence of shrubs on the typical list is only 14.3 per cent of the total. No measurements of actual dominance were made for the same reasons given for the south, but the alders clearly bear the same relation to the community as the trees do in a forest. True vines are not especially prominent, but clambering plants with weak stems which more or less support themselves by reclining on their stronger neighbors are numerous. Especially noticeable are the bedstraws, of which four species are present; the very harsh *Galium asprellum* is the most widespread and characteristic of these. Other plants with this habit include *Campanula aparinoides*, *Potentilla palustris*, and *Poa palustris*.

The genus *Alnus*, to which the tag alder belongs, has received considerable autecological study, especially by McVean (1956) in England, because of its nitrogen fixing properties and its important role in colonization of recently deglaciated areas. No quantitative measures of nitrogen fixation for *Alnus* have been made in Wisconsin, but the growth of certain associated species believed to have high nitrogen requirements indicates a considerable activity. In Alaska, Crocker and Major (1955) reported that the alder thicket stage in the successions which followed the recession of the ice at Glacier Bay showed an accumulation of nitrogen in the forest floor and top layers of the mineral soil at the rate of 4.9 grams per square meter per year.

As in the southern carr, the alder thicket flora is composed largely of species from the sedge meadow and from the swamp forest (Table XVII-3), but its higher content of modal prevalents may be related to the special soil conditions created or accentuated by the alder. The usual site for alder is along streams or in other places where the soil water is in movement. Wilde and Randall (1951) examined the ground water in alder thickets in Price and Forest counties. They found reactions from pH 7.1 to 7.7, specific conductivities from 14.2 to 21.0 Mhos, total alkalinities from 75 to 98 p.p.m., free oxygen contents up to 5.4 p.p.m., and oxidation-reduction potentials from 49 to 89 mv, thus showing a nutrient rich, non-stagnant environment. The soils are usually mucks and may be relatively shallow over glacial tills or shallow to deep over lacustrine peat beds. The northern alder thicket is more likely to be alkaline than any other community of the region, but not all stands are in this condition, since some of them show acidities as great as pH 4.8.

Stability

The alder thickets along streams and in other positions of primary succession appear to have a high degree of stability. Those derived from prior forest by burning, however, are much more liable to be replaced by trees. As mentioned in the chapter on swamp conifers, several instances were seen where young stands of white cedar were present under alders which had developed on old tamarack-black spruce sites. The optimum soil and soil water requirements of alder and white cedar are very similar and the secondary alder thicket may be the normal

Plate 43.—Oak opening of bur oaks (*Quercus macrocarpa*) near Albion, Dane County.

Plate 44.—Bur oak grub, showing large basal plate.

Plate 45.—Pine Barrens near Nekoosa. Wood County.

Plate 46.—Cedar glade, with surrounding xeric forest of oak (*Quercus* sp.) and white birch (*Betula papyrifera*). Dane County.

Plate 47.—Cedar glade which approaches a cedar (*Juniperus virginiana*) forest in density. Dane County.

Plate 48.—Creeping juniper (*Juniperus horizontalis*) on sandstone cliff. Dane County.

Plate 49.—Southern sedge meadow on lake shore. Waukesha County.

Plate 50.—Hummocks of *Carex stricta* in a grazed stand of southern sedge meadow.

mode of origin of cedar swamps. Not all alder thickets are invaded by cedar, however, as many instances were noted where black ash, red maple, mountain ash, and large-toothed aspen were present. A more thorough study of the alder thicket and its successors is needed before any firm understanding is reached. The above remarks are to be construed as tentative only.

GEOGRAPHICAL RELATIONS

Cain and Slater (1948) studied a tall shrub community in southern Michigan which they called a *Betuletum pumilae potentillosum* because of the prevalence of bog birch (*Betula pumila*) and shrubby cinquefoil (*Potentilla fruticosa*) in its early stages. Other shrubs appeared in slightly drier locations; these shrubs included two dogwoods and five willows. Of the total of 56 species listed, 63 per cent are on the Wisconsin list of shrub-carr flora and include 12 species from the prevalent group. No other clearly comparable studies were discovered for neighboring regions. Many studies of carr vegetation in England reveal surprising similarities, not only in actual species but especially in related species of similar behavior in the same genera. Thus Tansley (1939) reports 91 species for the carrs of Esthwaite and East Anglia, of which 19 are also in the Wisconsin community. Of the 64 genera in England, 43 are in common with Wisconsin and include such important groups as *Carex, Cirsium, Galium, Lycopus, Mentha, Rumex, Stachys, Thelypteris,* and *Viola*. The alder thicket has been studied extensively in Germany. Oberdorfer (1957) reported 93 species, of which 12 occurred in Wisconsin, including *Dryopteris cristata, Caltha palustris,* and *Osmunda regalis*. There were 32 genera in common, among them *Angelica, Cirsium, Cornus, Epilobium, Galium,* and *Potentilla*. Rubner (1954) reported similar findings in a detailed study of 280 stands in Bavaria.

UTILIZATION AND CURRENT MANAGEMENT

The tall shrub communities receive perhaps the least economic attention of any Wisconsin community of comparable size. Most stands are totally unused for any direct product except wild game. The shrub-carr, in particular, is in an active state of expansion on the wet lands bought or leased by the Conservation Department for Public

Hunting and Fishing Grounds. Many of these areas were used as mowing meadows prior to state control but have since developed into dense thickets as a result of protection. The continuing expansion of the hunting grounds program probably means that shrub-carr as a community type will also continue to increase in importance. It might be well if the actual game bird productivity of the type could be accurately assessed before many additional areas are allowed to progress to it, since some doubt appears to exist on the question. In general, land managers find it is easier to prevent a succession than to reverse it, and the old adage of "better safe than sorry" may well apply in this case.

Part 6

Lesser communities

CHAPTER 18

Fen, meadow, and bog

FENS

Nature and location

The word fen means marsh or bog, but it has been given a more restricted connotation by many English ecologists. Tansley (1939) used it to refer to a "soil-vegetation" type in which grasses and forbs are the dominant plants and in which the soil is on organic peat that is either neutral or alkaline in reaction. He differentiated it from "marsh" which has a mineral soil and "bog" which is on acid peat. In this country, the term has been employed mostly in Iowa (Conard, 1952; Hayden, 1943) and other middlewestern states. As understood in this region, a fen is a grassland on a wet and springy site, with an internal flow of water rich in calcium and magnesium bicarbonates and sometimes calcium and magnesium sulfates as well. Frequently the fen is on a hillside overlooking an existing or extinct glacial lake, whence the local names, "perched bog" and "hanging bog." It is most closely related to the wet prairies and to the sedge marshes (usually called sedge meadows). This is shown in Wisconsin by indices of similarity between fen and wet prairie of 54 per cent and between fen and southern sedge meadow of 59 per cent.

The fens in Wisconsin derive their alkaline waters from subterranean sources originating in dolomitic limestones; hence their distribution is limited by the distribution of such rocks. Nowhere are they common, but most frequently they are seen in the southeastern counties in the area of Niagara limestone and along the Prairie du Chien (Magnesian) escarpment to the west. A particularly favorable location was formerly to be found around the lakes of the Madison area and several good fens still exist there. The Wingra fen in the University Arboretum is one of the finest examples known in the state.

No detailed quantitative investigations have yet been undertaken in the Wisconsin fens. The following discussion is based on the very insecure foundation of a presence study of only six stands, made by the author with the aid of Dr. H. C. Greene. Since great botanical interest attaches to the fens, it is thought advisable to present the findings here, although they necessarily must be regarded as tentative and subject to considerable revision.

Composition

The fens studied had an average species density of 41. Of the 115 total species found, the prevalent forms are listed in Table XVIII-1 together with the other modal species which show a peak presence in this community. The homogeneity of 61.3 per cent is rather high, as is the number of prevalent species which reaches an optimum in the fen (51 per cent). Both percentages indicate a relatively uniform and distinct set of species; but when the flora list is examined, it is seen that that the community is made up of an unusual combination of groups of different origins (Table XVIII-1). Thus there are plants of northern affinities, like *Scirpus cespitosus, Aster umbellatus, Betula sandbergii, Menyanthes trifoliata,* and *Solidago uliginosa,* plants of the prairies like *Andropogon gerardi, Asclepias sullivantii, Equisetum laevigatum, Liatris pycnostachya,* and several others, as well as many plants of general distribution in southeastern United States, that include a number of so-called Atlantic Coastal Plain plants (McLaughlin, 1932) like *Gerardia paupercula* and *Bidens connata.* The fen is to be considered a hybrid community where the unusual combination of environmental factors has sorted out and retained suitably adapted species from each of the major formations as they passed by in postglacial times.

A floristic analysis of the fens emphasizes this unique composition. The eight families which contain one-half of the species are *Compositae* (17.4 per cent), *Gramineae* (8.7 per cent), *Labiatae* (5.2 per cent), *Cyperaceae* (4.3 per cent), *Umbelliferae* (4.3 per cent), *Scrophulariaceae* (4.3 per cent), *Salicaceae* (3.5 per cent), and *Rosaceae* (3.5 per cent). The *Leguminosae* which are so prominent in the prairies are of greatly reduced importance here. Shrubs make up only 8.7 per cent of the total flora, with half of the species in the willow genus. There is a relatively equal distribution of the flowering times through the three seasons. Among the spring bloomers highly characteristic of the fens are marsh marigold (*Caltha palustris*), white ladyslipper (*Cypripedium candidum*), blueflag (*Iris shrevei*), and valerian (*Valeriana ciliata*). The summer display includes swamp milkweed (*Asclepias incarnata*), swamp thistle (*Cirsium muticum*), and Kalm's lobelia (*Lobelia kalmii*). A number of asters and goldenrods appear in the autumn and include such characteristic species as *Aster lucidulus* and *Solidago riddellii*, but there is a group of four fall-blooming species which seems to represent the fen at its best. These are the ladies'-tresses orchid (*Spiranthes cernua*), fringed gentian (*Gentiana procera*), grass-of-parnassus (*Parnassia glauca*), and meadow milkwort (*Polygala sanguinea*). Many of the fens are mowed for hay in August, but this treatment simply reduces the average height of the canopy and allows these four low-growing species to be displayed to optimum advantage.

Stability

When the fens are protected from external disturbance, they tend to develop into a shrub-carr by the increase in size of some of their component dogwood and willow shrubs and by the invasion of other woody plants. It is presumed that the natural agent maintaining the fens was fire, which could burn the mulch and top growth of the community with little danger to the peat beneath because of the steady water supply. In postsettlement times, many of the fens were used as mowing meadows since they provided a reliable source of hay. The mowing accomplished the same end as the fire with respect to repression of the shrubs but it no doubt had different effects on the species composition, since those species which came into bloom at the time of mowing would gradually lose out in favor of earlier or much later species. Most of the existing fens have had a past history of mowing so the composition as presented in Table XVIII-1 is probably biased.

When both the mowing and the fires are stopped, the shift in dominance to woody plants is usually rapid, but the exact rate is considerably influenced by the size of the local rabbit populations. The willows, dogwoods, and bog birches (*Betula glandulosa* and *B. sandbergii*) are all highly palatable to these rodents and their annual growth may be completely pruned back during the winter by the animals. In the Wingra fen of the University of Wisconsin Arboretum, the plants in a fenced exclosure showed very great differences within three years from those in the rabbit-infested areas outside the fence. Eventually, however, enough stems escaped the browsing and grew to a size no longer subject to attack so that the area outside the fence resembled that inside. This enlargement in size and increase in numbers of the shrub members exerted a profound effect on the light demanding forbs of the community. A study of population changes in the white ladyslipper (Curtis, 1946) showed a great decrease in density and in flower production as the shrubs began to take over. Experimental control of the shrubs allowed the orchids to regain most of their initial numbers within five years.

Geographical relations

The Wisconsin fens are closely related to those of neighboring states. Hayden (1943) listed 59 species as characteristic of the fens of Palo Alto County, Iowa, of which 54 per cent were on the Wisconsin list. A more complete account of the Iowa fens was provided by Conard (1952), who summarized his own findings and those of other local investigators. He reported the fen soils to be very alkaline (pH 8.5 to pH 9.5), with a total solute content of the soil water in excess of 1100 p.p.m. Sulfur bacteria were active in the springs of some fens and these had prominent deposits of tufa. Of the 45 species he considered most typical of the Iowa fens, 41 species or 91 per cent are also found in the Wisconsin community. Conard also gave brief summaries of fens in New York, Pennsylvania, Ohio, and Missouri. Cain and Slater (1948) reported coverage and frequency values for a fen bordering Sodon Lake in southern Michigan. This community, which they mellifluously named the *Dryopteridetum Thelypteridis eleocharosum*, contained 33 species, of which 77 per cent are on the Wisconsin list. It is interesting that there is a rather high degree of similarity between the Wisconsin fens and those of England as reported by Tansley (1939). Of the 75 species listed by him, no less than 16 or 21 per cent are on the Wisconsin list, while 24 genera are represented by closely related species. Included in the species-in-common are *Menyanthes trifoliata, Phalaris arundinacea, Caltha palustris, Juncus effusus, Mentha arvensis,* and *Stachys palustris.*

Other more recent English floristic studies showing similar resemblances include those of Holdgate (1955) and Poore (1956). Gorham (1954) and Gorham and Pearsall (1954) analyzed the ground waters in a number of English fens. They reported alkaline conditions and a high calcium content in some areas but an acid reaction and a low calcium content in others. In Germany, the closest assemblage seems to be that described by Oberdorfer (1957) on the trollius-reed community. Only 25 of 70 genera were in common with Wisconsin, but these included such typical groups as *Angelica, Caltha, Cardamine, Galium, Liparis, Parnassia,* and *Valeriana.*

SEDGE MEADOW

Nature and location

The sedge meadow is here understood to be an open community of wet soils, where more than half the dominance is contributed by sedges rather than grasses (Plate 49). As such, it is closely related on soils of similar moisture to fens, bogs, and wet prairies among the other open groups, and to the shrub thickets and wet forests of the closed communities. Towards wetter conditions, it grades into cattail and reed marshes or other emergent aquatic groups. It usually occupies a very low position on the regional soil catenas. The ground may be flooded in the spring or after heavy summer rains but it typically lies just above the permanent water table. The soil is either a raw sedge peat or a muck produced by decomposition of such peat, and is frequently incorporated with mineral matter deposited by overwash from the surrounding uplands. Water is always plentifully present and never a limiting factor by its lack. Excess water, however, may induce difficult conditions for many plants because of the disturbed oxygen relations. The sedge meadow soils are frequently in a reducing condition and may reach extreme conditions in this respect, with the production of methane or other highly reduced "marsh gases."

The community is represented in all regions of the state, although the Driftless Area naturally has fewer places of suitable moisture except along the larger rivers. In the glaciated country, sedge meadows are found in extinct lake beds, around the shores and banks of existing lakes and streams, and in depressions in pitted outwash or moraine topography. In the north, most of the meadows are small and are restricted to the shores or banks of lakes or streams (Plate 51); the larger extinct lake beds are usually occupied by conifer swamps. In the south, however, many of these lake beds have been invaded by the sedges, so that much larger bodies of the community are present. This is easily seen in Figure XVIII-2, which depicts only the larger meadows. Their total area in presettlement times was 1,134,000 acres, or 3.24 per cent of the land surface. The meadows north of the tension zone are very similar to those in the south, with an index of similarity of 62 per cent. Nevertheless, there are a number of important species which are restricted to one or the other province. Because of this, the two groups of meadows will be discussed separately.

Although all of the meadows are wet, some are wetter than others, and it was at first thought that they could be arranged in an ordination based upon this moisture gradient. Several trials, using the combined data from all stands throughout the state, showed that this could indeed be done, but the results gave a separation only among the southern stands, with all the northern group appearing at the wet end of the ordination. In the south, sedge meadows grade slowly into wet prairies as moisture decreases, but in the north such a decrease favors the invasion of alder thicket, so that relatively dry open meadows are rare or non-existent.

Another trial based upon a pH gradient again resulted in a geographic correlation, with all of the alkaline meadows in the south and most of the acid meadows in the north. It appeared obvious from these two ordinations that the northern and southern meadows would need to be treated separately if compositional gradients of maximum meaning were to be obtained. Unfortunately, the amount of information available from the two areas separately is insufficient to justify such a calculation. Therefore, the data from northern and southern meadows have been combined into two average groups and the discussion is based on these averages. Altogether, 79 stands were investigated in the P.E.L. study, largely by J. T. Curtis and H. C. Greene. These stands are shown in the figure reproduced at the opening of this chapter. Of these, 35 were north of the tension zone and 44 south. In addition to this presence study, more detailed frequency analyses are available for a number of stands. Other workers who have reported on sedge meadows in Wisconsin include Stout (1914), Pammel (1907), Costello (1936), Frolik (1941), and Partch (1949). The study by Stout is of particular interest because it was the first detailed quantitative phytosociological investigation of any Wisconsin plant community.

One of the great difficulties involved in the study of any sedge meadow is the near impossibility of identifying most of the dominant sedges when they are not in flower or fruit. Since such reproductive parts are present only for a short period in early summer, this means that meadows visited at other seasons will be incompletely sampled with respect to these important constituents. An adequate analysis of the meadows on a state-wide basis would require many years of work that would be concentrated in the flowering season. In the P.E.L. studies, an attempt was made to collect all sedges with reproductive structures. The resulting list is probably a fair indication of the species which may participate in the sedge community, but the quantitative analyses of presence or frequency are of limited value. The following discussion of sedge meadows may be likened to a description of a forest type which concentrated on the groundlayer but neglected the trees. The information on the non-sedges is reasonably accurate and probably adequate, but the information on the sedges is fragmentary and largely of floristic value only. Since there is little likelihood that this situation will improve in the foreseeable future, we must proceed with what information we have.

Southern sedge meadows—composition, structure, and environment

The 44 southern stands had an index of homogeneity of 51.1 per cent. The prevalent species are listed in Table XVIII–3 with the other modal species. Only five of the prevalent species reach their statewide optimum in the community. The seven leading families as recorded in the data are: *Compositae* (20.3 per cent), *Cyperaceae* (8.1 per cent), *Gramineae* (7.6 per cent), *Labiatae* (5.1 per cent), *Rosaceae* (4.6 per cent), *Polygonaceae* (3.9 per cent), and *Umbelliferae* (3.9 per cent); but the *Cyperaceae* are no doubt under-represented. At best, however, they would probably not exceed the *Compositae*, which lead here as in most other open communities. Shrubs are in low numbers, with only 6.6 per cent of the total species. The high moisture conditions of the spring season prevent a rapid warming of the soil, so the spring flowering species are few in number. Of the prevalent non-sedge species, less than one-fourth bloom before June 15. The earliest is cuckoo flower (*Cardamine bulbosa*), which usually opens the last week in May. Nearly one-half of the group blooms in summer and the remaining quarter in autumn. Prominent forbs of widespread distribution which lend character to the sedge meadows include swamp milkweed (*Asclepias incarnata*), Joe-pye weed (*Eupatorium maculatum*), boneset (*Eupatorium perfoliatum*), meadow rue (*Thalictrum dasycarpum*), and angelica (*Angelica atropurpurea*). The meadow aster (*Aster lucidulus*) is particularly conspicuous in the autumn. Plants of lesser stature which seem characteristic of the community include water horehound (*Lycopus americanus*), meadow bedstraw (*Galium obtusum*), and marsh fern (*Thelypteris palustris*). Differential species which separate the southern meadows from those in the north are given in Table 19. Many of the southern indicator species are also important in the wet prairies (Table XIV-9) while the northern group has close affinities with the

Table 19

Differential species of southern and northern sedge meadows

Southern species	P% in South	P% in North	Northern species	P% in South	P% in North
Anemone canadensis	55	23	Aster puniceus	9	51
Angelica atropurpurea	41	9	A. umbellatus	2	29
Aster lucidulus	32	3	Bromus ciliatus	2	37
A. novae-angliae	23	3	Campanula aparinoides	27	63
Caltha palustris	25	9	Cicuta bulbifera	5	31
Cardamine bulbosa	32	0	Dryopteris cristata	2	31
Galium boreale	27	0	Galium trifidum	7	23
Helianthus grosseserratus	39	9	Glyceria canadensis	0	40
Lathyrus palustris	55	17	Polygonum coccineum	7	23
Naumbergia thyrsiflora	30	11	P. sagittatum	0	54
Pilea pumila	32	6	Salix petiolaris	2	29
Spartina pectinata	52	3	Scirpus atrovirens	11	63
Viola cucullata	43	6	Scutellaria galericulata	2	29
			Solidago graminifolia	7	49
			S. uliginosa	2	34
			Spiraea tomentosa	2	26
			Triadenum virginicum	0	20

bog flora (Table XVIII-8). One feature of the two types not clearly indicated in the table is the great increase in importance of species of *Scirpus* in the north. Many of these are present in the south but do not attain high levels of density there.

The structure of the southern sedge meadows can be described on the basis of the information given by Stout (1914), Costello (1936), the P.E.L. studies of Partch (1949), and others. In his remarkably clear and detailed paper, Stout divided a sedge meadow near Madison into three associational groupings, a *Caricetum,* a *Lycopus-Caricetum,* and a *Calamagrostis-Caricetum.* The division was based upon floristic grounds and upon correlated soil differences. Information on frequency and density was obtained from a transect strip of 3450 quadrats; dominance by weight was determined from 46 quadrats of 5 square feet, spaced along the transect at 50-foot intervals, and harvested in the last half of August, 1907.

A clear indication of the importance of sedges in these meadows is given by Stout's data. Of a total of 52,377 plants counted in the transect, 33,989 or 63 per cent were species of *Carex*. An additional 24 per cent were grasses. Two species in combination, *Carex stricta* and *Calamagrostis canadensis*, accounted for 58 per cent of the total number. In making his counts, Stout enumerated the individual culms of the sedges and grasses, since there was no way to determine the limits of individual clumps. That this did not distort the results greatly is shown by the frequency and dominance figures. *Carex stricta*, for example, had frequencies of 80 per cent or more in most of the 50-foot segments of the transect while *Calamagrostis* varied from 60 to 100 per cent. On a dominance basis, the importance of the pair was accentuated, with 43 per cent of the total weight contributed by *Carex stricta* and 24 per cent by *Calamagrostis canadensis*. The total yield per acre varied among the different sub-communities, with a value of 3650 pounds dry weight per acre for the *Lycopus-Caricetum*, 5500 for the true *Caricetum*, and 7100 for the *Calamagrostis-Caricetum*. *Carex* species made up 87 per cent of the total weight for the *Caricetum* but only 51 per cent for the *Calamagrostis-Caricetum*. The total number of species was greatest in the *Lycopus-Caricetum* (60) and least in the true sedge area (18), with mints, composites, and ferns adding greatly to the former group. *Lycopus* species contributed 2237 individuals to the sub-group named for the genus and only 264 individuals to the other two subgroups.

Costello (1936) studied a number of sedge meadows near Milwaukee and reported data on quadrat frequencies as determined by 1 meter quadrats. Combining his data with those of Partch and the author, a more general picture of Wisconsin meadows may be obtained. The average values in Table XVIII-4 are based on 21 stands, all in counties south and east from Madison. The results are very similar to those of Stout, and the same two species are in positions of overwhelming dominance. On a sum of frequency basis, the same four families are in the lead as were found from the total number of species in the larger presence study, but the order is different—*Cyperaceae* (108), *Labiatae* (54), *Gramineae* (46), and *Compositae* (32).

The very great importance of *Carex stricta* in many if not most of the sedge meadows has led to its detailed study by a number of workers. Stout gives the following description of its manner of growth:

In the ground there is a well developed root stock system. The wire like rootstocks grow about six inches below the moss layer and below the roots of many of such secondary species as may be present. The terminal buds turn up and develop aerial branches in the form of clumps of culms. By continued growth and branching from the bases of their culms a thick dense cluster of culms is produced. When a culm produces fruit it dies. Two types of growth and branching of rhizomes are thus to be distinguished. One for spreading and reaching new territory and the other for the immediate production of aerial culms. A root stock grows for a considerable distance, often to a length of more than a foot, then its terminal bud sends up aerial leaves and the formation of a tussock is begun. Further creeping rhizomes arise from the tussock and thus the two types of branching provide for vegetative spreading and for leaf and flower production.

Two types of roots are developed. Long, cylindrical, mostly unbranched roots develop from the area of short compact internodes at the base of the culms and push downward into the cold saturated peaty soil to a depth of six or eight inches. These are essentially soil roots. Fine fibrous roots also develop at the base of the culms but grow upward to the surface where they form a mass of finely divided rootlets.

Stout also discusses the interrelations of the sedge and of bluejoint grass (*Calamagrostis canadensis*), which is its almost constant companion:

The plants that give character to this marsh vegetation possess, we may say, the same general appearances and habit of growth and hence should be competitors. Yet if we analyze their adaptations further we are forced to give up any such general conclusion.

To illustrate: There are, in spite of the same general grasslike habit of growth, marked anatomical differences already mentioned between *Carex stricta* and *Calamagrostis canadensis*. Yet . . . there is a tendency for both to occupy the same territory. One would say at first thought that these two species are close competitors yet the mode of root and rhizome growth, the vertical stratification, and the seasonable development really provide two different environments for the two plants altho they may stand side by side.

A plant society may be not so much a collection of plants of various species which are adapted to the same conditions as an association of species which are adapted in a different fashion to the same locality.

The conditions of competition in such a dense population already so fully occupying the space affords a most interesting field for the study of development. Practically every inch of the ground is occupied and hence the question of seed dispersal is of little importance in the spread of species. The struggle is chiefly between species very similar in structure and adaptation. This struggle is severe beneath the ground for the rhizome development admits of steady persistent spread and gives rather permanent possession even after the most favorable conditions have ceased to exist. Yet it is noticeable that there is a definite grouping of species with definite areas of best development and marked zones of contact. Slight differences in moisture content, in soil composition, and slight elevations or depressions are associated with change of species. In a habitat with such uniform conditions as is here found and with so many species of similar structure, slight differences in environment are correlated with the individual peculiarities of the various species most strikingly.

Costello, who applied the name "tussock meadow" to those sedge communities dominated by *Carex stricta*, also discussed the growth habits of the species. He considered the cespitose form of the tussock to be of great advantage in the competition with other species:

The perennial nature of the tussocks and the considerable age to which they attain are factors which help to bring about their permanent occupancy of an area. The mass of dead leaves and the matted roots and rhizomes which are added to the soil each year tend to create a fibrous peat which is favorable to the continued growth of the sedge.

The cepitose habit contributes to the dominance of *Carex stricta* in still another way. It constitutes an admirable adaptation to a fluctuating water level. During that portion of the growing season when the water is well above the surface of the ground, the leaves and stems are seldom if ever submerged. Aeration of roots and rhizomes presents no problem, owing to their well developed aerenchyma. With the lowering of the water table and the consequent decrease in moisture supply in the region of the fibrous roots, the soil roots in the zone of saturation supply the necessary moisture.

Costello studied the internal anatomy of the sedge and found that the leaves had well-developed motor cells which aided them in folding during temporary dry periods. Both the roots and rhizomes had abundant sclerenchyma or masses of hard-walled cells on their outer surfaces and at the same time had a considerable amount of aerenchyma or air passages in the internal portions. He explained this anomaly as follows:

The prominence of mechanical tissue in both the root and the rhizome would seem to indicate xerophytism, while the abundance of aerenchyma indicates hydric tendencies. This condition, however, does not appear to be anomalous when the water relations of *Carex stricta* are considered. As a rule both roots and rhizomes are exposed to the atmosphere for a portion of each growing season. Adequate protection from desiccation is afforded by a sclerenchyma layer in the outer part of the cortex. At other times the root and rhizome systems are submerged. The utility of an aerating system then becomes apparent. The presence of additional mechanical tissue in the stele explains to some extent the difficulty with which tussocks are uprooted or toppled over, either by artificial means or as the result of natural forces.

The sedge clumps or tussocks are known colloquially as "hummocks" in much of Wisconsin where they are a familiar feature of the landscape in the dairying regions. They are particularly noticeable in heavily grazed meadows (Plate 50), where the cows have removed the covering plant material and have accentuated the clumps by compaction of the soil between them. Dawkins (1939) studied the formation of tussocks by the similar *Schoenus nigricans* in England and reported that they showed a slightly more regular spacing than would be expected in a random distribution. No detailed studies of aggregation have been made in Wisconsin sedge meadows but the somewhat regular distribution of the *Carex stricta* tussocks is apparent in the plate. The other major dominant, bluejoint grass, also seems to be nearly random in most meadows. The lesser species display two different types of aggregation. Some of the taller forms tend to grow in large colonies with few or no individuals scattered between them. This habit is often shown by angelica (*Angelica atropurpurea*), water hemlock (*Cicuta maculata*), and especially by the beggar-ticks (*Bidens coronata*), which frequently forms a solid sheet of gold in early September over areas 50 to 100 feet in diameter. Micro-aggregation is imposed upon the shorter species by the tussocks, with many species confined to the spaces between them and others growing epiphytically on the clumps themselves.

The soils of the southern sedge meadows are largely in the Badoura, Carlisle, and Kokomo (formerly Clyde) series in the glaciated area and the Arenzville series in the Driftless Area. They may be on deep deposits of peat, 6 to 20 feet, or they may be on shallow layers of peat or muck. The lower strata may be reduced, blue, lacustrine clays, whitish marls, or, more rarely, sand. The water-holding capacities of the peats are high, averaging 270 per cent, while those of the mucks are lower, averaging 165 per cent. Most of the soils are rich in calcium and range in reaction from pH 7.2 to pH 8.5. Potassium is generally low, while available phosphorus may be very high. A severe deficiency in boron, copper, and other trace elements becomes quickly apparent when the soils are placed in agricultural production. This lack is probably reflected in the native vegetation as well, but we have no information whatever on the requirements of native plants for these rare elements.

Wilde and Randall (1951) studied certain chemical characteristics of the ground water in a sedge meadow in Richland County. They found a high specific conductivity of 65.0 Mhos $\times 10^{-5}$ and a corresponding high total alkalinity of 332 p.p.m. Both values were the highest for any waters sampled in 27 sites in Wisconsin. The free dissolved oxygen was present only to the extent of 0.10 p.p.m., which was reflected in a low oxidation-reduction potential or Eh of 31 mv.

A few records of the microclimate in sedge meadows suggest that temperatures, evaporation, and growing season length are lower than in the surrounding uplands, whereas moisture addition by condensation may be higher. Air drainage is responsible both for the shortened growing season and for the increased condensation by night fogs in the summer. A sedge meadow in the University of Wisconsin Arboretum in Madison had a killing frost during the first week in September in every year from 1950 to 1957, although the nearby uplands were spared until the end of September or even the end of October in some years. Costello studied the evaporation rates in a meadow near Cedarburg in Ozaukee County in the summer of 1929. He found a water loss of 9.1 ml. per day from his atmometers placed at a height of 20 cm in the tussock meadow. In comparison, an adjacent upland pasture lost 36.1 ml., while a nearby beech-maple forest showed a loss of about 10 ml. for the same period.

Northern sedge meadows—composition, structure, and environment

The prevalent species found in 35 stands of sedge meadow located north of the tension zone are shown in Table XVIII-6. A total of 177 species was found, with an average species density of 29 per stand. Eight of the prevalent species or 27.6 per cent reach a statewide optimum in this community. Four of these, *Aster puniceus, Glyceria canadensis, Solidago graminifolia,* and *S. uliginosa,* are on the list of differential species which separate the northern from the southern meadows. The great prevalence of *Scirpus* species in the north has already been mentioned. Other species which seem highly characteristic of the northern sedge meadows include *Polygonum sagittatum, Potentilla palustris, Cicuta bulbifera, Epilobium strictum, Galium asprellum, G. labradoricum, Gentiana rubricaulis, Triadenum virginicum,* and *Prunus nigra.* The six families that include 50 per cent of the species are *Compositae* (15.8 per cent), *Cyperaceae* (9.0 per cent), *Gramineae* (9.0 per cent), *Rosaceae* (6.8 per cent), *Labiatae* (6.2 per cent), and *Ranunculaceae* (4.0 per cent). Shrubs make up 10.7 per cent of the total species, with only one, *Spiraea alba,* on the list of prevalent species. The lack of spring flowering species is more pronounced in the north than in the south, with only one-eighth of the total sum of presence of the prevalent species contributed by such plants. About one-half of the group bloom in summer with the remainder in the autumn, especially such composites as *Aster, Eupatorium,* and *Solidago.*

No quadrat analyses of any sort have been reported for the community, so nothing can be said of structure, spacing, or yield. Some studies of the environment have been made, but they are incomplete in their coverage. The soils of the northern meadows are largely on the Greenwood, Houghton, Maumee, Spalding, Newton, Keowns, Cable, Spalding, or Adolph series. (Hole and Lee, 1955). This large range is apparently to be differentiated on the detailed nature of the underlying strata and may not be indicative of ecologically meaningful properties of the surface layers. This is shown by the relatively high uniformity of the plant communities on the various soils. The peat of the northern meadows is frequently in a raw fibrous state and is less often in the decomposed muck condition so common in the south. Water-holding capacities of the peats analyzed in the P.E.L. studies averaged 320 per cent. The acidity was highly variable, ranging from pH 4.9 to pH 6.6. Ground water analyses in two meadows in Marinette County by Wilde and Randall showed an average specific conductivity of 43.2 Mhos $\times$ 10^{-5}, a total alkalinity of 162 p.p.m., a dissolved oxygen content of 0.97 p.p.m., and an oxidation-reduction potential (Eh) of 100 mv. No measurements of microclimatic factors have been made in the northern meadows, so far as is known.

Stability

The sedge meadow is a normal stage in primary hydrosere successions, and appears when the soil or peat surface is at or just above the water level. It follows the emergent marsh community of reeds or cattails where the water is above the soil and is followed by a hydric shrub community of willows, dogwoods, or alders as drier conditions are produced. In the north, sedge meadows are essentially restricted to such sites, and often consist of narrow strips between streams or lake shores and a fringing alder thicket (Plate 51). In the south, meadows are found in similar places but they also occur in much more extensive tracts of many hundreds of acres. Examination of the peat in such large meadows usually reveals tamarack logs, or cones, or other evidences of former occupation by conifer swamp, or other forest. Fragments of charcoal in the peat provide the clue as to the origin of these stands—they were caused by fire at some time in the past, most probably at the time of expansion of the prairies. This relation of southern sedge meadows to fire is dramatically shown in Jefferson County. This is a region of glacial drumlins, oriented in a north-south direction, with low basins between them. Zicker (1955) mapped the original vegetation of the county from the surveyors' records. She found that in the western portion of the county the drumlins were covered with oak savanna or prairie while the low grounds were composed of sedge meadows. To the east of the Crawfish and Rock rivers on identical topography, however, the uplands were covered with mesic forests of maple and basswood and the lowlands supported tamarack swamps. Fires sweeping eastward from the prairies of Dane County on the prevailing westerly winds were stopped by the rivers, which accounted for the differences in the vegetation on the two sides.

Further evidence as to the influence of fire on the sedge meadows can be seen when they are artificially protected. Those areas which are on relatively dry sites are quickly converted to shrub-carr or alder thicket, usually within 10 to 20 years. Only the wettest meadows are able to resist this invasion and only these can be considered as non-fire communities. Costello (1936) reported an instance near Thiensville where rapid degeneration of the sedge community occurred following an artificial lowering of the water table through drainage. Frolik (1941) discussed in some detail the various reactions of the sedge meadows of Dane County to fire, drainage, and grazing. He reported that fire was a natural feature of the environment which served to prevent the encroachment of shrubs and trees. When the meadows had been partially or completely drained, however, fire had serious or disastrous effects, because the underlying peat beds became ignited and continued to burn or smolder for long periods of time, sometimes in excess of two years. Such subterranean fires completely destroy the sedge community and the productivity of the underlying soils. The areas of deep peat burns are almost universally taken over by solid stands of nettle (*Urtica dioica*), which remain in control for a long time. An area of this type in the University Arboretum, burned in 1933, is still completely dominated by nettles after 24 years, while another area, reported to have been burned in 1915, is just beginning to be invaded by *Helianthus, Solidago,* and *Thalictrum,* after 42 years. On rare occasions, peat fires may occur in natural meadows during years of severe drought and reduce the level of the ground to a point below the level of the water line that prevails in normal seasons. When normal water levels return, such areas are reinvaded by emergent aquatics and then by new sedge meadows, without the intervention of a nettle stage. It is highly probable that many or most of our extensive areas of sedge have been affected this way at more than one interval in their history. The successive layers of charcoal fragments now stratified in the peat can be explained in no other way.

Geographical relations

The sedge meadows have received considerable attention from plant ecologists. Both Stout and Costello reviewed the early literature and pointed out the many studies made in Europe on closely similar types. *Carex stricta* and several other sedges, as well as many other meadow plants, are circumboreal in distribution and produce comparable communities of similar behavior in Europe and Asia. Oberdorfer (1957) in South Germany reported a number of slightly different sedge communities including one dominated by *Carex elata* which resembles the southern sedge meadows of Wisconsin. Fifteen of 60 species and 33 of 44 genera were in common between the two regions and included such typical plants as *Caltha palustris, Eleocharis palustris, Mentha arvensis, Poa palustris,* and *Scutellara galericulata.* Other European meadow studies included those of Sjörs (1950) in Sweden, Raabe (1949) in Schleswig-Holstein, Tallos (1954) in Hungary, Motyka and Zawadzki (1953) in Poland, and Larin et al. (1956) in the Kalingrad region of Russia. Closer to home, sedge meadows have been investigated in several neighboring states. Sherff (1912) reported on the community in the Skokie Marsh near Chicago and gave special attention to the distribution of underground organs of the meadow flora. Of the 113 species he recorded, 84 species or 74 per cent are on the Wisconsin list. Sampson (1921) in a general study of the grasslands of Illinois, reported 57 species in his sedge community, of which 73 per cent are also in Wisconsin meadows. Pool (1914) described the "meadow formation" in the Sand Hills region of Nebraska, and listed 87 species in three subsidiary communities of the assemblage. Of these, 59 per cent are on the Wisconsin lists. The sedge community reported by Moss (1952) for the Peace River region of Alberta had 26 per cent of its species in common with Wisconsin, and similar degrees of relationship are shown by several sedge meadows that have been studied in New England and the Middle Atlantic states.

Utilization and current management

In past years, the sedge meadows were economically important as mowing grounds for the production of marsh hay. This was used to some extent for feed and bedding for farm animals, but the major market was as insulating and packing material for natural ice harvested from lakes and rivers in the wintertime. In the days before electrical refrigeration, every village and town had its barnlike icehouse where the year's supply of ice would be stored. In the vicinity of Milwaukee, large quantities of ice were needed for the chilling and shipment of ice-cold beer to Chicago and other centers of consumption. The sedge meadows were mowed in August after the small grain harvest and before corn-picking time. In wet years, the mowing was frequently delayed until the ground was frozen, in late November or early December. The midsummer plants thus had an occasional chance to set seeds and retain their position in the community. This use of marsh hay has largely disappeared. The only demand for the product at present is for horticultural mulching purposes because its weedless nature makes it a highly desirable blanketing material for the protection of tender garden bulbs and perennials. A rather specialized use for some stands of rather pure sedges is in the making of grass rugs. The famous Crex Meadows in Burnett County were once extensively used for this purpose by a large rug company in Minnesota (Murphy, 1931) and other local areas have been similarly utilized.

At present, the major demand for areas of sedge meadow comes from two diametrically opposed interests, the commercial truck gardeners and the game managers. The former group would like to drain all of the meadows and put them into the production of such specialty crops as mint, lettuce, onions, and carrots. They realize that such use is a form of mining, since the muck soils of the drained marshes have only a short and predictable life before they are completely oxidized or blown away and the sterile subsoils are exposed, but they consider this to be little different than the exhaustion of a lead mine, quarry, or oil well, and something to which their rights as land owners entitle them. The game managers and conservationists, on the other hand, would like to see the meadows protected and undrained, so that a supply of game animals could be maintained for the recreation-demanding urban citizens. An important benefit to be derived from this last use is the continued activity of the meadow as a water-conserving reservoir which ultimately would benefit farmer and city man alike and might far outweigh the value of the game by-product. On many of the lands acquired for game production purposes, the managers have failed to realize that the sedge meadow is not a permanent community but needs to be burned or mowed to be perpetuated. The resulting invasion of shrubs and trees is converting these areas to forest, which may well be neither productive of game nor efficient as a water reservoir.

OPEN BOG
Nature and location

The word bog has a number of meanings in the English language; most of them eventually trace to a particular physical condition of the land. To offer insecure footing and the chance of sinking in or becoming mired or "bogged" down is one such meaning. In ecology, the word has been given a more restricted connotation and refers to a soil-vegetation complex in which a rather specialized group of herbs and low shrubs grow on a wet, acid soil composed of peat. The bog as thus defined is a common feature of the glaciated landscapes of the entire Northern Hemisphere and has a remarkably uniform structure and composition throughout the circumboreal regions. In Wisconsin, the open bog is very closely related to the wet conifer swamps (index of similarity–60 per cent) and is usually a long-lived stage in the primary hydrosere succession leading to that community (Table XVIII-9). It is also related to the northern sedge meadows with which it has certain features in common. For purposes of recognition in the field, a bog is differentiated from a sedge meadow by the possession of a nearly continuous carpet of sphagnum moss on the groundlayer.

In Wisconsin, open bogs are most numerous in the region north of the tension zone, although a few isolated relic stands are known in the south, particularly in the Kettle Moraine district of Waukesha and Walworth counties and in certain recessional moraines in Jefferson and Columbia counties. The bogs are almost always located in pitted outwash or in kettle depressions usually associated with an active or extinct glacial lake. No good examples of bogs on the uplands are known in Wisconsin, although they may occur on such locations in northern Minnesota, Ontario, and other regions to the north.

No detailed phytosociological investigation has been made of the Wisconsin bogs, although several of them have received considerable specialized attention, particularly Hope Lake Bog in Jefferson County, Cedarburg Bog in Ozaukee County, and Forestry Bog in Vilas County. The following discussion is based largely on a presence study of 17 stands made by the author (Figure XVIII-3).

Composition, structure, and environment

A total of 119 herbs and shrubs was recorded in the 17 stands. The prevalent species are shown in Table XVIII-8 with the modal species. The index of homogeneity of 53.7 per cent is rather low, but the presence of a core of characteristic species is shown by the fact that 55 per cent of the prevalent forms are also modal, and most of them have high presence values. The two most important families are the *Ericaceae* and the *Cyperaceae,* each with 9.2 per cent of the total species. The ericads are of greatest importance in lending character to the community since they form an almost continuous canopy over the surface of the bog whereas the other species grow beneath them or in the rare openings between them. The most important of these bog shrubs are bog Rosemary (*Andromeda glaucophylla*), leatherleaf (*Chamaedaphne calyculata*), bog laurel (*Kalmia polifolia*), and Labrador tea (*Ledum groenlandicum*), all of which are evergreen. The deciduous blueberries (*Vaccinium angustifolium* and *V. myrtilloides*) are also prominent in many bogs. Non-ericaceous shrubs such as the bog birch (*Betula glandulosa*) and bog holly (*Nemopanthus mucronata*) help to swell the total shrub component of the bog to the very high value of 54.6 per cent of the total sum of presence of the prevalent species.

The bogs, like the sedge meadows, remain cold in the spring but in spite of this the bog flora is predominately spring flowering, with 64 per cent of the sum of presence for the prevalents contributed by species which bloom before June 15. Many of these bloom in April or early May from buds set the preceding summer. Most of the bog shrubs belong in this category. Leatherleaf and bog laurel have been seen in full bloom while their roots were still embedded in frozen soil. All of the remaining species bloom in the summer, with none of the prevalent species flowering after August 15, and only a few of the rare species blooming at that time.

The sedges are represented by a number of genera, including *Carex, Cladium Dulichium, Eriophorum, Rhynchospora,* and *Scirpus* (Greene, 1953). Of these, the cotton grasses of the genus *Eriophorum* are the most conspicuous with their waving tufts of cottony, white, fruit heads. Orchids of many species are present, but they never reach important densities. The rose pogonia (*Pogonia ophioglossoides*) and the grass pink (*Calopogon pulchellus*) are perhaps the most characteristic orchids, although the red ladyslipper (*Cypripedium acaule*) is more widespread.

The plants most closely associated with open bogs in the minds of most field botanists are the insectivorous species. The great inherent interest in these forms and their high degree of fidelity to the community are responsible for the opinion. The most spectacular member of the group is the pitcher-plant (*Sarracenia purpurea*), which catches insects in a complicated tubular or trumpet-shaped leaf partially filled with water. Stiff hairs pointing downward prevent any insect which falls in the water from escaping. The sundews (*Drosera* species) are much less conspicuous but equally interesting. They have a rosette of small leaves, each expanded into an oval or elongate blade at the end of a long petiole. The blades are beset by thick glandular hairs, each with a glistening drop of highly viscous fluid at its tip. Insects landing on the hairs are trapped by the sticky liquid and are then slowly enclosed by a folding or wrapping movement of the blade, after which they are digested by proteolytic enzymes secreted by the hairs. The bladderworts (*Utricularia* species) are commonly present in open water pools in the bogs. They capture mosquito larvae and other aquatic animal life by a type of box-trap, a specialized leaf, that is a hollow ovoid structure equipped with a trap door at one end. The door is triggered by a set of long hairs near the opening.

The presence of these insectivorous species in bogs is usually explained by the low available nitrogen content of the peat soils and the consequent need of additional sources of nitrogen which is met by these plants in rather unusual fashion. No experimental proof of the explanation has been forthcoming to date.

The vertebrate animals of an open bog near Rhinelander in Oneida County were studied by Jackson (1914). He reported the swamp sparrow as the dominant bird with the song sparrow and redwing blackbird as common associates. Among the mammals, the mink, muskrat, meadow mouse, masked shrew, and short-tailed shrew were typical, while the leopard frog, common toad, and gartersnake were representative of the cold-blooded vertebrates.

No studies of structure or of degree of aggregation have been made in Wisconsin bogs. Environmental analyses have also been relatively few, as compared with the activity in other states and countries. One of the first intensive studies of a single plant community, including its quantitative floristic composition and detailed environmental relations, was conducted on bogs in Germany by Lorenz in 1858. Since then there have been many other studies in both Europe and America, most of which have been devoted to an elucidation of the environmental causes of the rather unique bog flora. Much of this work has been reviewed in several extensive papers by Transeau (1905), Dansereau and Segadas-Vianna (1952), and in the detailed studies of Baas-Becking and Nicolai (1934) in the Netherlands, and Conway (1949) in Minnesota. In Wisconsin, environmental details of soil or climate have been reported by Cox (1910), Rhodes (1933), and Wilson (1939).

A typical open bog has a surface layer of living sphagnum over a layer of dead and loosely compacted sphagnum peat. At greater depths the peat becomes more compacted and may or may not be of sphagnum origin. Many bogs show alternating layers of sphagnum and sedge peat which indicate changed conditions in the past. Not infrequently, layers of marl or sand will be found interspersed with peat and muck layers. Rhodes (1933) thought such layers indicated alternating periods of total submergence in the past, since marl is deposited only in open water. Of 14 bogs whose profiles were analyzed in the P.E.L. study, not a single one was composed of a continuous deposit of sphagnum peat.

The rate of production of peat is almost impossible to determine with accuracy. Frequently one bog will be found on a peat bed 40 to 50 feet thick and another similar bog across the road under slightly different circumstances will show a depth of only 5 or 6 feet, although both presumably originated at the same time. Leisman (1953) investigated the phenomenon of peat accumulation in the Itasca region of Minnesota and concluded that the sedge mat ring of *Carex lasiocarpa* between the sphagnum mat and the open water was accumulating at the rate of 0.6 inches per year for a total of 5,477 pounds of organic matter per acre per year in the uppermost 4 inches. Decomposition and compression of a given year's deposit as it becomes buried naturally reduce the thickness, so that calculations based upon deep sections show much slower rates than this. Various estimates in the Great Lakes Region range from 100 to 800 years per foot. A peat bed, 50 feet deep, in a bog in the Valders moraine in Vilas County (age, *ca.* 9000 years) must have accumulated at a minimum rate of 180 years per foot, barring possible periods of interruption.

The surface peat (below the green sphagnum) is very acid, with pH readings from 3.5 to 4.5. Water-retaining capacities of this peat are tremendously high, ranging from 800 to 1300 per cent on a dry weight basis. Nutrient analyses of raw sphagnum peat show a high total nitrogen content but a very low supply of available nitrogen. The other nutrients are very poor as well, although the low specific gravity of the peats makes comparison difficult. Wilde and Randall (1951) found the ground waters in three bogs in northern Wisconsin to range in specific conductivity from 5.0 to 6.1 Mhos $\times$ 10^{-5}, in free oxygen from none to 0.2 p.p.m., and in oxidation-reduction potential from -153 to -364 mv. None of the waters had any detectable carbonate alkalinity.

Microclimatic studies have been largely limited to temperature investigations. The use of many open bogs as sites of commercial cranberry fields has focused attention on their susceptibility to summer frosts because the ripening cranberries are sensitive to frost and must be protected by flooding. A special service of the United States Weather Bureau is devoted to the prediction of nighttime temperatures in the cranberry bogs and these predictions are broadcast by radio daily for the benefit of cranberry growers. Thus the temperature microclimate of a rather specialized and areally insignificant Wisconsin plant community is better known to millions of Wisconsin residents than any other environmental feature of any other community. Without going into detail about these bog temperatures, it is sufficient to state that they are low on the average; furthermore, they are subject to frequent periods of very low readings, with frost possible on any night in the year. Differentials of 20°F. to 30°F. between temperatures at the bog surface and on nearby uplands are common. From the standpoint of the native plants, however, the low night temperatures are probably not as significant as the high daytime temperatures, since all are adapted to the frosts and uninjured by them. During the midsummer days, surface temperatures on the sphagnum, particularly on those species which develop a deep red color, frequently reach 95°F. or 100°F. or more, yet temperatures at the root level within the mat rarely exceed 60°F., and are usually between 45°F. and 55°F. The differential between the transpiring surfaces in the leaves and the absorbing surfaces in the roots is thus 40°F. to 50°F. or more. In early spring, conditions are still more rigorous, with temperatures below freezing at the roots and temperatures of 75°F. to 85°F. at the surface. Since water absorption is much reduced at low temperatures, the bog plants are actually growing under conditions comparable to those of xeric habitats with an actual deficiency of water. The markedly xeromorphic adaptations of the bog shrubs have been assumed to have survival value under these circumstances.

The steep temperature gradients from surface to subsoil are due to the great insulating qualities of the sphagnum moss, with its large airfilled chambers in the leaves. Sphagnum is also responsible in large part for the great acidity and low nutrient content of the bog soils, due to its ability to preferentially absorb and hold cations from its surroundings. Some investigators believe that toxic organic compounds are produced by the sphagnum and by the other members of the community as a result of fermentive respiration under conditions of low oxygen supply. The bog environment is thus seen to be an excellent example of multiple factor causation. Low temperature, initially induced by topographic air drainage, is favorable to the development of a sphagnum mat which accentuates the temperature effect and changes the nutrient and oxygen status of the ground water. This favors a select group of companion species, all of which work together to maintain the conditions and prevent encroachment of other plants through antibiotic activity.

Stability

As indicated earlier, sphagnum bogs are one stage in the primary hydrosere succession from open-water lake, to conifer swamp, and eventually to climax mesic hardwood forest. The stage is very long-lived, and, in terms of actual stability, far less liable to fluctuation than most of the stages which precede or follow it. This long duration of stability is no doubt one of the major reasons for the great stand-to-stand homogeneity exhibited by the bogs. As explained in Chapter 12, the bog frequently originates on a floating mat of *Carex lasiocarpa* or other sedges by the invasion of sphagnum and the various bog shrubs. In the absence of disturbance, the bog mat gradually thickens, fills the entire lake bed, and becomes solidified in the process (Plate 52). Sooner or later, it is likely to be invaded by tamarack and black spruce, which rather quickly transform the area to a swamp forest. Under natural conditions, the lack-of-disturbance requirement is rarely met. Periodic fluctuations in the water table, correlated with weather cycles, make the surface layers dry enough to support surface fires and these fires destroy the advance forest reproduction as well as the bog flora. Upon return of high water, the bog community re-enters, frequently on a greatly increased area, gained at the expense of the surrounding conifer swamp. If natural or artificial drainage has lowered the water table, a considerable volume of the surface peat layers may be removed by the fires. Under such circumstances, a hybrid community of open bog and sedge meadow may develop. The famed Powell Marsh in Vilas and Iron counties is of this nature. It was formed in an extensive but shallow lake bed and had gone through the succession to conifer swamp on at least one occasion in the past, as shown by peat analyses. Fires reduced the peat to an average depth of 2 to 5 feet and allowed the development of the muskeg-like community now present. The ground is covered with a sphagnum mat 6 to 12 inches in thickness; the mat lays over alternating thin bands of sedge peat and sphagnum peat containing much charcoal. Growing in the sphagnum mat and sharing dominance with it is an almost pure stand of *Carex oligosperma*. Scattered throughout is a layer of bog shrubs, largely leatherleaf (*Chamaedaphne calyculata*). The remaining flora is very limited and is composed of about equal parts of typical bog and sedge meadow species. Several other similar areas are known, especially in the central Wisconsin sand plains. By the definition used in the P.E.L. study, they would be classed as open bogs, due to the presence of a sphagnum layer, but they are clearly intermediate in nature. A similar community in Germany has been called the *Caricetum lasiocarpae* "association" (Oberdorfer, 1957).

Geographical relations

The widespread homogeneity of the sphagnum bogs in circumboreal regions has already been mentioned. Analysis of many species lists made in bogs from North Carolina to Washington and from Quebec to Alberta and Alaska reveals about the same pattern of decreasing similarity as that shown by the conifer swamp, with nearby areas in Minnesota (Conway, 1949) and Michigan (Gates, 1942) having over 80 per cent of their species in common with Wisconsin, whereas remote areas have fewer of the same species. Jerome bog (Buell, 1946) in North Carolina, for example, had only 18 per cent similar species, including such abundant plants as *Chamaedaphne calyculata, Sarracenia purpurea,* and *Pteridium aquilinum.* Six bogs in northwestern Alberta studied by Moss (1953a) had 54 per cent of their species in common, and included the leading dominants *Chamaedaphne calyculata* and *Ledum groenlandicum.* The open bog type has been studied intensively in Japan (Jimbo, 1941; Mizushima and Yokouchi, 1953), in Russia (Katz, 1926; Regel, 1947), and in Europe (Osvald, 1925; Ludi, 1935). The general aspect of the community over this wide area is quite uniform, with a ground covering of sphagnum moss with associated sedges, sundews, and shrubby ericads. The latter, however, are present in much less variety and abundance than in the American bogs.

Utilization

Many former areas of open bog have been converted to cranberry marshes but the original community is totally destroyed in the process. The only other economic use of sphagnum bogs is made by the mossing industry, largely in the central counties. The top layers of sphagnum are harvested, dried in the sun, and shipped to cities for use in the florist trade and for general horticultural purposes. Little is known about the rate of recovery of the harvested bogs nor of the nature of the higher plant succession that occurs on them, but such a study would hold great interest for an understanding of the dynamics of the bog community in general.

CHAPTER 19

Aquatic communities

NATURE, TYPES, AND LOCATION

All of the communities discussed so far may be classed as terrestrial, although several of them occurred on very wet sites. The differentiation between terrestrial and aquatic is based largely on the average relation between the soil surface and the surface of the water table. If the soil is above the water for most of the year, the site is called terrestrial, and if the water is above the soil for most of the year, it is aquatic. On this basis, there is clearly a gradient of conditions, from those land surfaces so high as to be beyond any influence of the water table (dry prairies, cedar glades, many dry forests) to those soils so deep below the water surface as to support only submerged aquatic plants. The sedge meadows, bogs, and some lowland forests are on the terrestrial side of the dividing line; they are submerged for considerable periods in the spring but emersed for most of the growing season. Intermediate conditions sometimes can be seen in the field where terrestrial and aquatic criteria are about equally fulfilled on an annual basis, but such places are surprisingly rare. Rather, there is usually an evident dividing line which separates the truly aquatic from the terrestrial. The aquatic community present in Wisconsin at this line is the marsh, an open, herbaceous grouping dominated by cattails (*Typha* sp.), reeds (*Scirpus* sp.), or other essentially grasslike plants. In the southern states, certain forest types may occupy this first aquatic stage, including the cypress and tupelo swamps and others; but no woody plants are present in Wisconsin which can grow on permanently submerged sites. The relatively sharp dividing line between marsh and meadow is largely the result of autogenic processes. When siltation or water level changes occur which lower the water level in the marsh to a critical value, the marsh is invaded by sedges. The rapid growth of their roots and rootstocks soon results in a buildup of the soil surface to a level above the water, thus accentuating the transition. In the following discussions, attention will be concentrated on the true marshes and the communities of deeper water.

With increasing depth of water, the composition of the marsh communities gradually changes, since some species are unable to grow in deep water. The change is usually accompanied by an increase in number of species which float on the surface or have floating leaves. Eventually a depth is reached which prohibits all emergent plants. The floating species may or may not extend beyond this limit into deeper water. Beyond the edge of all emergent and floating species, the community consists entirely of submerged plants. These may extend to very deep water, the limit being set by light intensities and hence by the clarity of the water. The record in Wisconsin is held by a rather simple community of two species of moss which is found at a depth of 20 meters in Crystal Lake in Vilas County (Fassett, 1930).

The submerged aquatic communities are found in lakes, where a depth zonation is particularly evident; they are also found in rivers and streams, where the intensity of the water currents often restricts the emergent or floating forms and places certain limitations on the kind of submerged plant which may grow there.

Aquatic plants of all types are greatly affected by the chemical nature of the water in which they grow (Pearsall, 1918) and this in turn is influenced to a considerable extent by the geological nature of the water source. Some lakes and streams are dependent largely upon rain water for their supply; hence they contain very little in the way of dissolved nutrients. Other water bodies are fed by springs which originate in deep rock strata. The waters here are likely to be rich in various compounds leached from the rock during slow seepage of water. These differences in water chemistry are unusually noticeable in Wisconsin because of the unusual distribution of rock types. The limestones are essentially confined to the south, and the largely insoluble siliceous rocks are in a position to affect water supplies only in the north. As a result, there is a compound gradient, ranging from cold, soft waters in the north, to warm, very hard waters in the south. The aquatic communities naturally reflect this gradient. Fortunately, there are a few lakes in the north with hard waters resulting from calcareous glacial tills, so the differential action of temperature and chemistry can be at least partially understood.

Streams and streamside marshes are present throughout Wisconsin, but lakes and lacustrine marshes are limited mostly to the glaciated portions of the state. This is shown on the map at the opening of this chapter, which depicts the aquatic stands in the P.E.L. study and also indicates the approximate boundary of the Cary ice-sheet. Many famous marshes, like the Horicon Marsh and the Grand Marsh, are located in extinct glacial lake beds. Others are found on the shores of existing lakes. The lakes themselves are widely but not randomly distributed, with local concentrations in the area of the Waukesha County lakes, the Waupaca area, the Vilas County area, and the northwestern sand barrens.

A special branch of ecology called limnology is devoted to the study of aquatic biota of inland waters and their environmental relations (Hutchinson, 1957). Wisconsin has played a leading role in the development of limnological science, largely as a result of early activities of the Wisconsin Geological and Natural History Survey. In recent years the limnological work has been decentralized. A. D. Hasler, of the hydrobiology laboratory of the Department of Zoology, has emphasized the ecology of fishes and experimental limnology, and the Lakes and Streams Committee has sponsored programs chiefly concerned with pollution and the causes of nuisance blooms of algae. Much of the work has been concentrated on Lake Mendota in the south and on some of the Vilas County lakes in the north, with detailed investigations of water chemistry, bacteriology, physics, and meteorology, as well as accompanying studies of phytoplankton and aquatic animals. The submerged aquatic communities of seven or eight lakes were investigated in detail from the standpoint of standing crop, but extensive phytosociological studies of aquatic plant communities have been undertaken only very recently. As a result, our knowledge of the nature of submerged communities, their distribution, and variation is rather limited and is largely confined to the shallow water communities of hard water lakes. Prominent investigations include those of Denniston (1921), Rickett (1921), and Andrews (1946) on Lake Mendota; Rickett (1924), Wilson (1941), and Potzger and Van Engel (1943) on other specific lakes; and especially the state-wide P.E.L. study of Natelson (1954).

The emergent communities of the marshes are of considerable interest from a game management viewpoint, since they are of great value as duck breeding habitats. The Wisconsin Conservation Department has studied the marshes on an intensive scale. Most of our knowledge of these communities has been summarized by Catenhusen (1944) and Zimmerman (1953). The artificial lakes created by damming streams and old drainage ditches on public hunting grounds have also been studied by the game management research division of the Conservation Department. River and stream communities are poorly understood in Wisconsin, and only a single stream, the Brule, has received attention (Evans, 1945; Thomson, 1944). It is probable that the files of the Stream Pollution Division of the State Board of Health contain valuable ecological data, but they have not been published.

The great amount of background information available from all of this varied limnological research should be of great value in phytosociological work in the future. It is unfortunate that such work was not instituted at the start of the program because it would have provided a basis for interpretation of many of the results which is now lacking. The present discussion is an attempt to synthesize those portions of the available information which are reasonably comparable, but it should be looked upon only as a starting point for a proper phytosociological investigation.

EMERGENT AQUATIC COMMUNITIES

Composition

The marshes of Wisconsin are rather poor in species compared to most terrestrial communities. In the very limited P.E.L. studies, only 68 species were found, while in a random selection of 40 marshes from Zimmerman's study, only 12 additional were recorded, which made a total of 80 species. Zimmerman was interested mainly in the dominant plants which contributed cover of value for wildlife and his published records do not include the data for all minor species. Never-

Plate 51.—Northern sedge meadow along stream, with alder (*Alnus rugosa*) thicket at margins. Vilas County.

Plate 52.—Open bog surrounding a bog pool. Northern wet forest invading from the margin. Near Powell, in Iron County.

Plate 53.—Shaded sandstone cliff in Perrot State Park, Trempealeau County.

Plate 54.—Exposed sandstone cliff with relic red pine forest, Governor Dodge State Park, Iowa County.

Plate 55.—Relic stand of white pines (*Pinus strobus*) on cliff. Jonesdale, Iowa County.

Plate 56.—Prickly pear cactus (*Opuntia compressa*) on exposed cliff. Green County.

Plate 57.—Ferns, mosses, and lichens on sandstone cliff.

Plate 58.—Beach at Trout Lake Biological Station, Vilas County, showing linear arrangement of plants on strand. Mature pine forest on dune in background.

theless, it is probable that the actual total is less than 100 species. This paucity of the flora is noticeable in individual stands, which have an average species density of only 11 and in which large areas may exist with only three or four species present.

The prevalent species of the entire group of marshes (Figure XIX-1) are listed in Table XIX-1 with the remaining modal forms. It is evident that the sedge family (*Cyperaceae*) is of great importance, but the genus *Carex*, so prominent in the sedge meadows, is here replaced by the bulrushes (*Cyperus*) and spike rushes (*Eleocharis*). The cattails (*Typha*), arrowheads (*Sagittaria*), and bur-reeds (*Sparganium*) are also of high importance. A floristic analysis shows that five families contain over 50 per cent of the total species—*Cyperaceae* (22.5 per cent), *Gramineae* (12.5 per cent), and *Polygonaceae, Labiatae,* and *Compositae* (5.0 per cent each). The grasses are represented by numerous species, mostly of low presence, which include such characteristic forms as wild rice (*Zizania aquatica*), giant reed grass (*Phragmites communis*), rice cut-grass (*Leersia oryzoides*), and sloughgrass (*Spartina pectinata*). The genus *Scirpus* includes the great bulrushes with long, slender, cylindrical, stemlike shoots (hard-stem bulrush—*Scirpus acutus,* soft-stem bulrush—*S. validus,* and slender bulrush—*S. heterochaetus*) and the triangular-stemmed leafy bulrushes (three-square bulrush—*S. americanus,* river bulrush—*S. fluviatilis,* and wool grass—*S. cyperinus*).

The figures in Table XIX-1 show the over-all state-wide occurrence of the major species, but they do not indicate the extent of variation in the community. Two important environmental variables have a great influence on the makeup of any particular marsh—the solute content and the depth of the water. Zimmerman reported the hardness of the water in each of his stands on the basis of parts per million of calcium carbonate. The values ranged from 16 p.p.m. to 215 p.p.m. When his stands are placed into three groups on the basis of this hardness gradient (0-50 p.p.m., 50-150 p.p.m., 150 p.p.m. or more) and the average presence calculated for the major species in each group, it is seen that certain species have a pronounced peak of occurrence in each group while others appear to be uninfluenced by water hardness. This is shown in Table 20 which lists a number of the more important species according to their behavior on the gradient.

It would be possible to make a similar classification on the basis of water depth, since it is apparent that the species differ greatly in their tolerance limits for this factor. Unfortunately, there is a lack of information on depth variation in the stands available, so no quantitative alignment is possible. From limited observations, it appears that the plants in the deepest water include *Scirpus acutus, Typha angustifolia,* and *Zizania aquatica;* with *Equisetum fluviatile, Carex trichocarpa, Pontederia cordata, Leersia oryzoides, Sagittaria latifolia, Scirpus americanus,* and *Typha latifolia* in medium water depths; and *Carex aqua-*

Table 20

Emergent aquatics by water hardness classes

	% presence in		
	50 p.p.m. or less ($CaCO_3$)	50–150 p.p.m. ($CaCO_3$)	150 p.p.m. or more ($CaCO_3$)
Species with optimum in soft water			
Glyceria canadensis	8	6	0
Sagittaria latifolia	83	62	53
Scirpus cyperinus	17	6	0
Spartina pectinata	25	12	0
Species with optimum in medium hard water			
Scirpus validus	33	65	54
Sagittaria rigida	8	53	23
Eleocharis acicularis	42	47	31
Species with optimum in very hard water			
Iris virginica	17	24	39
Leersia oryzoides	8	18	39
Scirpus acutus	42	82	100
S. americanus	25	47	54
S. fluviatilis	8	35	54
Species with no pronounced optimum			
Eleocharis palustris	17	18	15
Pontederia cordata	58	59	46
Typha latifolia	67	71	77
Zizania aquatica	58	65	62
Phragmites communis	42	35	54

tilis, Calamagrostis canadensis, Eleocharis elliptica, Scirpus validus, S. cyperinus, Phragmites communis, and *Sparganium eurycarpum* in the shallow waters. No exact limits can be set because the waters are usually much deeper following the spring runoffs than they are in late summer. Furthermore, most of the species grow quite well in waters less deep than the deepest they can tolerate; hence they may be found in various mixtures. Frequently, however, a given stand will show relatively narrow ranges of depth for each species, although these may differ to some extent from stand to stand. More detailed observations are clearly necessary before a valid classification can be constructed.

Structure

Some impression of the structure of a cattail marsh can be obtained from Table XIX-2, which shows the average density and frequency for the common species in three marshes in the vicinity of Madison. All were in areas of very hard water, and all had approximately the same depth of water. The *Lemna* and *Utricularia* are insignificant members of the community, in spite of their high densities and frequencies. *Eleocharis elliptica* and *Cicuta bulbifera* also contribute very little to the dominance because of their small size. Estimated major dominance, in weight or cover, is shared by the cattail (*Typha angustifolia*), the three-square bulrush (*Scirpus americanus*), and the sedge (*Carex aquatilis*), with important contributions by arrowhead (*Sagittaria latifolia*) and giant reed grass (*Phragmites communis*).

The important members of Wisconsin's marsh communities have a number of autecological characteristics in common. They usually possess bulky underground parts, spread vegetatively by means of rhizomes, and have a marked development of internal passageways for gas transfer. Laing (1940), in Michigan, has studied a number of the species experimentally and finds that they possess the ability to carry on anaerobic respiration, with the production of alcohols and other intermediate metabolites. When the leaves are fully developed above the water, there may be a considerable internal, diurnal interchange of gases, with the oxygen produced in the leaves by photosynthesis during the day diffusing into the roots and rhizomes at night, and the carbon dioxide produced by respiration in these organs moving to the leaves. These special adaptations tend to be restrictive because normal growth cannot take place in atmospheres of the usual oxygen content. Therefore, the plants able to grow in deep water are at a disadvantage in shallow water, particularly in those seasons when the water table drops below the soil surface. Thus *Typha* is less able to compete in areas which occasionally dry out than is *Sparganium* and both are eliminated by *Carex* when these dry periods become frequent.

The vigorous rhizome development of many species tends to produce a strongly aggregated or colonial type of dispersion. In a mature cattail or bulrush marsh this may not be very apparent, because the colonies have merged with each other. When a new species invades a marsh, however, the colonies may be very striking. In the University of Wisconsin Arboretum, for example, the narrow-leaved cattail (*Typha angustifolia*) was introduced around 1910. It has hybridized with the original broad-leaved cattail (*T. latifolia*) to produce a number of intermediate offspring (Fassett and Calhoun, 1952). Each of these has produced a nearly circular colony which is clearly evident from a distance, due to differences in color and shape of the fruits. The individual cattail plants produce rhizomes which are between 15, and 20 cm in length, and the colonies have been spreading radially at about this rate. The bulrush (*Scirpus americanus*) produces shorter rhizomes, which average about 5 cm per year, while the giant reed grass (*Phragmites communis*) has much longer annual growths, which average 40 cm.

Very similar communities of emergent aquatics are found throughout the north temperate zone. In Europe they are usually placed in the order *Phragmitetalia*. Oberdorfer (1957), for example, lists 37 species in his *Scirpus-Phragmites* community in South Germany of which 9 are found in Wisconsin, including *Phragmites communis, Typha latifolia, Alisma plantago-aquatica, Acorus calamus, Eleocharis palustris*. Twenty genera of the total of 31 are in common with Wisconsin, and include such major groups as *Scirpus, Carex, Sparganium, Sagittaria, Iris,* and *Cicuta*.

The marshes of Wisconsin provide a favored habitat for a large variety of animals. Among the birds, the ducks are particularly prominent and include such common species as the mallard, black duck, blue-winged teal, and shoveller. Other common gallinaceous birds include the coot, Florida gallinule, and several rails. Among the songbirds, the red-winged blackbird, yellow-headed blackbird (in the south), the short-billed marsh wren, and the swamp sparrow are typical. A number of mammals frequent the marshes, but the most important of these is the muskrat. This species builds houses of considerable size, constructed of mud, laced together with various plant stems, especially bulrushes and cattails, that have been harvested in the vicinity. A number of marsh species are used as food, including arrowhead and cattail, but the influence of the muskrat on these two species is very different. A number of marshes with high animal populations are known where the cattail is either absent or present in very low densities on sites that are apparently very suitable for their growth. These particular marshes contain an almost pure stand of arrowhead. It has been found that the cattails stage a recovery during years of low muskrat levels but when the muskrat numbers build up the cattails are again reduced to low densities. No quantitative figures are available on concurrent changes in arrowhead populations, but in aspect-dominance at least there appears to be little change. It is clear that the muskrat can exert a great influence on the nature of the plant community; this has a tendency to retard succession or perhaps cause an actual retrogression in some cases.

SUBMERGED AQUATIC COMMUNITIES OF LAKES

Nature of submerged community

The usual pattern of distribution of aquatic plants in Wisconsin lakes is the familiar one of zonation correlated with water depth; the emergent species are on and near the shore, the floating species in deeper water, and the submerged species in the zone nearest the center of the lake. This concentric pattern is rarely perfectly expressed and may be totally lacking on large lakes with wave-swept shores or on smaller lakes of fluctuating water level with broad sandy beaches. In the latter cases, the submerged plants comprise the sole aquatic vegetation. In spite of the apparent homogeneity of this submerged environment, the communities are very diverse in their composition and show surprisingly great variation within short distances. The major environmental controls are water depth as related to light intensity, water chemistry, water movement, and nature of the substrate. Various intensities or qualities of these factors can interact in a variety of ways to influence the local composition of the community. As a result, a single lake may contain a number of relatively homogeneous stands, each with a different species makeup, depending on depth, nature of adjoining shore line, and degree of protection from waves. These differences have not always been appreciated in past research, so that data gathered from several distinct groupings often have been lumped into an over-all average for an entire lake. By analogy, the results are as meaningless as would be a comparable average of the yield from a mixed area of prairie, cedar glade, mesic hardwood forest, and weed patch which happened to be on a single farmer's land.

Alone among the environmental variables, the chemical nature of the water tends to be uniform throughout a given lake. This factor is a complex one and involves the amount and nature of dissolved materials, including oxygen, the essential nutrient elements, bicarbonate ions, and many organic compounds (particularly tannins and other complex, deeply colored substances). In the absence of clear indications as to the relative importance of these components, it has been customary to measure the chemical factor by indicative tests of either the hardness or the total conductivity. The former concentrates attention on the bicarbonate content and thus it involves the calcium concentration and the pH or acidity, while the conductivity is correlated with the total inorganic solutes. It is obvious that neither of these measures provides an adequate indication of the organic compounds, which may be of great ecological significance. This is particularly true of such trace materials as vitamins or antibiotics that are active in very low concentrations (Hasler and Jones, 1949).

As indicated earlier, there is a geographical aspect to the chemical factor, with the very hard, alkaline waters of high conductivity largely confined to the limestone regions of the south and the very soft, acid, low conductivity waters most numerous in the north. This regional separation and the great biological difference of the two extremes is well shown by some of the studies of standing crops by the Wisconsin Geological and Natural History Survey. Thus Rickett estimated the total crop of submerged plants in Lake Mendota (1921) and in Green Lake (1924), and Wilson obtained similar figures for several lakes in Vilas County (1935). On the basis of 221 samples of one square meter each in Lake Mendota and 309 samples in Green Lake, Rickett reported an average crop of 178 grams dry weight per square meter in Green Lake and 202 grams in Lake Mendota. These correspond to 1590 pounds per acre in Green Lake and 1804 pounds in Lake Mendota. In contrast, Wilson found crops as low as 0.08 grams per square meter in Silver Lake and a maximum value of only 0.52 grams in Little John Lake. These values are equal to only 0.71 and 4.64 pounds per acre, respectively. The southern lakes thus contain from 300 to 2500 times as much plant material per unit area at any given time as do the northern lakes.

Data are not available for plant densities, but if they were, they would show no such great differentials between the North and the South. In fact, some of the northern lakes of lowest yields might well outnumber the rich southern lakes in terms of number of individuals per unit area. The great difference in weight is largely a result of differences in life-form. This aspect of aquatic communities was studied in detail by Fassett (1930). He recognized four groups of species of greatly differing appearance.

The first was composed of plants with stiff leaves in a basal rosette or on short unbranched stems. It included such species as *Isoetes macrospora, Sagittaria cuneata, Eriocaulon septangulare, Elatine minima, Gratiola aurea,* and *Lobelia dortmanna.*

A second group had long, flexuous stems, and compound or flexuous leaves, suspended in the water. It contained *Potamogeton natans, P. amplifolius,* and many other *Potamogeton* species, *Najas flexilis, Anacharis canadensis, Ceratophyllum demersum,* and *Myriophyllum* species.

The third division included those species with a horizontal rhizome on the lake bottom connected by flexuous petioles to leaves which floated on the surface, such as *Sparganium fluctuans, Vallisneria americana, Polygonum natans, Brasenia schreberi,* and *Castalia odorata.*

The last group had the emergent aquatics like *Scirpus, Eleocharis, Pontederia,* and others already discussed in the marsh communities. The first group was essentially confined to the very soft and very clear water lakes with sandy bottoms, as Crystal and Weber lakes in Vilas County, whereas the other three were found in the slightly eutrophic lakes with harder water, as Trout, Muskellunge, and Wildcat lakes, also in Vilas County. The southern lakes also have the last three groups, and especially the second.

One group of species in the second category is of especial importance in the dynamics of lake ecology. These are the stoneworts, macrophytic algae belonging to the genus *Chara.* These multicellular algae have a very high content of calcium carbonate which makes them brittle and rough to the touch. Schuette and Alder (1929) analyzed some of the samples collected in Green Lake by Rickett and found a calcium carbonate content of 65 per cent of the dry weight. On the basis of Rickett's figures for total crop of this alga in Green Lake, Schuette and Alder calculated that the total *Chara* population in the lake returned 993 metric tons of calcium carbonate per year. When it is realized that the amount of this compound within the plant body at any one time is only about one-tenth of that actually produced during the season, it appears probable that the total contribution of *Chara* to the marl deposits may reach 10,000 metric tons per year in the entire lake or about 1000 pounds per acre. Other submerged aquatics are also active in this process. Ruttner (1953), for example, reported that 100 kilograms of *Anacharis canadensis* could precipitate ·2 kilograms of calcium carbonate per day. Such tremendous production goes far to explain the extensive marl beds beneath many active and extinct glacial lakes in southern Wisconsin and how the latter attained their extinction.

The depth factor is important in submerged aquatic communities almost entirely because of its relation to light intensity, since the pressure component is probably never of significance. The absorption of light by lake water has been studied intensively by Birge and Juday (1932) in many Wisconsin lakes. No meaningful correlations can be drawn between depth and actual light intensity because the color of the water and the amount of suspended and colloidal organic material greatly influence the relationship. Some of the stained water lakes with bog margins have less light at a depth of ½ meter than some clear water lakes have at a depth of 20 meters (Curtis and Juday, 1937). Largely because of this variation, a community with a particular light requirement may be found at a depth of 0.6 to 0.9 meter in one lake and at 2.4 to 3.0 meters in another. It is essential, therefore, that phytosociological sampling be confined to particular contours and not to transects which extend out from the shore through several depths (Swindale and Curtis, 1957).

Many of the submerged plants remain active throughout the entire year, existing in a quiescent condition beneath the ice in winter. Others die down in summer or early autumn and pass the winter as buds or "turions," which are often specialized branches with greatly compressed internodes. These begin to elongate in early spring, often before the ice is melted, and produce new mature plants with roots and normal branches. Few direct observations on the role of seed reproduction have been made, but true seedlings are usually very rare and are far outnumbered by the turions. In a thorough study of the relations of populations of invertebrate animals to the submerged aquatic plants of Lake Mendota, Andrews (1946) showed that there were great changes in the weight of the standing crop of such major species as *Vallisneria americana*, *Potamogeton crispus* and other *Potamogeton* species during the course of a growing season. These differences in seasonal development mean that the community will have a succession of dominants, as determined by weight, throughout the year, much as the aspect-dominance in forest or prairie changes with season. A phytosociological study, therefore, must use frequency or presence measurements as an important tool, since these are not subject to such seasonal variation.

Composition

The P.E.L. studies of submerged aquatic communities were conducted by Natelson (1954—see Swindale and Curtis, 1957). She investigated 54 stands, using the quadrat method, with each sample 300 square inches (1930 sq. cm) in size (Figure XIX-2). Each stand was tested for homogeneity by appropriate statistical methods and each was at a uniform depth, over a uniform substrate, and opposite a uniform shore. The data from 555 separate quadrats were recorded on marginal punch cards. From the cards, the number of times each species occurred jointly with every other species was determined (Cole, 1949). An index of joint occurrence was calculated by expressing the number of quadrats in which two species occurred together as a percentage of the number of quadrats of occurrence of the less common species of the pair. Such an index was obtained for all species and the indices used to arrange the species in an order based on similarity of behavior (Guinochet, 1955). The species which had high indices of joint occurrence were placed close together while those with low indices were placed far apart in the order. Table 21 shows the list

Table 21

Joint occurrence groups of submerged aquatics
(Arranged in order from plants of very soft waters, Group 1, to plants of very hard waters, Group 4)

1	3
Myriophyllum tenellum	Potamogeton pusillus
Elatine minima	Vallisneria americana
Juncus pelocarpus	Anacharis canadensis
Potamogeton epihydrus	Zosterella dubia
Eriocaulon septangulare	Bidens beckii
Gratiola aurea	Najas flexilis
	Potamogeton natans

2	4
Potamogeton robbinsii	Potamogeton pectinatus
Eleocharis acicularis	Myriophyllum exalbescens
Sagittaria graminea	Potamogeton friesii
Potamogeton amplifolius	Potamogeton richardsonii
Isoetes macrospora	Ceratophyllum demersum
Myriophyllum alterniflorum	Potamogeton illinoensis
Potamogeton praelongus	Potamogeton zosteriformis
Potamogeton gramineus	Ranunculus aquatilis

of 29 species as arranged by the joint occurrence index, with an arbitrary division into four approximately equal segments. These segments or groups are comparable to the indicator groups as used in the ordination of Wisconsin prairies (Chapter 14) and are similar to the adaptation numbers of trees as used in constructing the compositional gradients for forests. In the case of the submerged aquatic communities, the relative frequency of each species in a given stand was weighted by its joint occurrence index number and the weighted values for all species in the stand were summed to give a stand index whose magnitude was a measure of all the species composing the stand. When all of the stands in the P.E.L. studies were treated in this manner, it was found that their indices ranged from 100 to 400, the possible limits. As in the forests and prairies, it was then possible to visualize the behavior of each species separately by plotting its importance (as

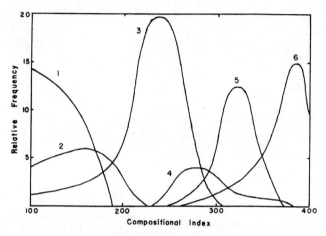

Figure 20.—Behavior of typical submerged aquatic species along compositional gradients based on joint occurrence groups (after Natelson, 1954). 1, *Elatine minima;* 2, *Potamogeton epihydrus;* 3, *Eleocharis acicularis;* 4, *Potamogeton praelongus;* 5, *Zosterella dubia;* 6, *Myriophyllum exalbescens.*

measured by quadrat frequency only) in each stand against the index number of the stands. The curves in Figure 20 were obtained in this way. As can be seen, the various species produce a series of overlapping curves, each with its own modal point and width-of-tolerance range. No clear-cut groups of species of similar behavior are apparent. The submerged aquatic communities follow a continuously changing compositional gradient, very similar to those demonstrated by other methods for the forests and prairies.

When various features of the soil and water environment are plotted against the stand-gradient, a number of distinct correlations are evident. Stands with high indices have high water conductivities, and a high content of organic matter, calcium, magnesium, and nitrate ions, a high pH, and less sand in their substrates, whereas low index stands are the opposite. The magnitude of these differences is shown by the following figures. The conductivity ranged from 30 Mho $\times$ 10^{-6} in stands with an index of 100 to 621 Mho $\times$ 10^{-6} in a stand with an index of 394. The calcium carbonate content of the soil increased from a trace to a maximum of 95 per cent, and the pH increased from 5.4 to 8.0. No definite correlations were found for phosphorus, potassium, or ammonia in the soil.

A compilation of the floristic data from the P.E.L. studies and those data reported by Rickett, Fassett, Wilson, and others for definite stands in Wisconsin lakes shows a total of 51 species. The average species density per stand is only seven, so the list of prevalent species is very short. In Table XIX-3, these prevalent species are given with a list of all other species attaining a presence of 10 per cent or more. The most impressive feature of these lists is the overwhelming predominance of members of the family *Najadaceae*, which includes the genera *Najas* and *Potamogeton*. These two genera are represented by 21 species whose sum of presence is a remarkable 46.7 per cent of the total sum of presence of all species. No other community in Wisconsin even approaches this record for single family domination. Other families of importance in the submerged aquatic communities are the *Halorigaceae*, the *Hydrocharitaceae*, and the *Lentibulariaceae*, each with 5.9 per cent of the total species, and the *Alismaceae, Ceratophyllaceae, Isoetaceae,* and the *Ranunculaceae,* each with 3.9 per cent. Of these seven, all but the last is confined to aquatic habitats.

Planktonic communities

The files of the Wisconsin Geological and Natural History Survey are literally bulging with data on the free-floating phytoplankton populations of many Wisconsin lakes; they include counts made at different seasons and at various depths, and have been gathered largely by Birge and Juday and their co-workers. Most of the information was obtained in the course of their studies on the "total productivity" (actual standing crop) of lakes and was not recorded in a manner suitable for phytosociological analysis. Many problems exist in the study of phytoplankton communities; these largely evolve from the tremendous population changes exhibited by the member species during the growing season and the resultant difficulties in properly sampling the community. The algae, which are almost the sole members of the group, may build up to very great densities during short time intervals. A sample taken on any given date may have as high as

97 or 98 per cent of its individuals represented by a single species. A week later, that species may be reduced to 0.1 per cent or less of the total and a different species will be in dominance. These periodic explosions are further complicated by the fact that the high populations at a given time may be confined to a fairly small depth stratum within the lake and may also show a highly aggregated distribution in lateral directions. This situation depends upon the nature of the water currents and the average wind velocity in recent days. Similar seasonal and spatial variation occurs in the zooplankton as well (Johnson and Hasler, 1954). Both kinds of plankton may be adversely affected in population size by competitive influences of higher submerged aquatic plants; hence they may vary in response to seasonal changes in the latter (Hasler and Jones, 1949). All of this means that sampling must be widespread and adequately stratified in three dimensions and must be repeated at close intervals throughout the year. Only under such circumstances will an adequate picture of the density and dominance relations of the community be obtained. Needless to say, the expenses of such a program have prevented its utilization.

Loeffler (1954) studied these sampling problems in a P.E.L. investigation on Lake Mendota and arrived at the very reasonable conclusion that frequency determinations would avoid most of the above problems in much the same way that frequency must replace density and dominance measurements in grasslands and other communities where data for the latter are difficult or impossible to obtain and are difficult to interpret. An extension of Loeffler's conclusion based upon the experiences of the present book, would indicate that simple presence, if obtained in a large enough number of lakes, would provide an adequate basis for a satisfactory description of plankton communities and their variations.

The available information in records of the Natural History Survey is all based upon extensive counts of dominant species in the samples; the rare or apparently rare species are lumped together or omitted entirely from the records. Accordingly, it is impossible at this time to present a co-ordinated account of the phytoplankton communities of Wisconsin, in spite of the great amount of time and effort expended on them in the past. This unfortunate situation forms one of the strongest arguments for the procedural concept in ecology which states that extensive sociological studies should precede intensive productivity or behavior studies in any given region. The community framework so obtained provides a means of interpretation of what may otherwise become meaningless and wasted efforts.

RIVER AND FLOWAGE COMMUNITIES

The aquatic communities of rivers and streams are commonly discussed separately from those of lakes (Tansley, 1939). This convention results from the fact that rapidly flowing waters exert definite limitations on the species that occupy them. It is obvious that these limitations become less effective in a slowly moving waters and may cease to exist in streams that have been damned by beavers or other natural agencies. In recent years in Wisconsin, many man-made dams have impounded large bodies of water on rivers and streams. These impoundments are locally called "flowages" for some obscure reason. They usually differ in no important way from the drainage lakes of the same regions and should be combined with them for studies of plant communities. Some of the impoundments, particularly in central Wisconsin, were made by damming old drainage canals that had been injudiciously used to lower the water table on extensive peat lands of old bogs or sedge meadows for agricultural purposes. Oxidation or actual burning of the peat during the drained period produced a substrate which was quite different when reflooded from that present under natural lakes. The plant communities of these artificial, peat-bed lakes has received little study, but the available information clearly indicates a degree of sterility and a lack of plant colonization which is quite different from the situation in normal lakes. In the absence of published reports on the flowage communities little more can be said of them.

The status of our knowledge of the communities of Wisconsin's streams is only slightly better than that of the flowages. A single stream, the Brule in Douglas County, has received detailed attention by Thomson (1944), Evans (1945), and Fassett (1944). Fassett studied the aquatic flora of the Mississippi River before the navigational dams were built in the 1930's, but his reports are not available. It is known that the sloughs and backwaters of the Mississippi and the lower Wisconsin rivers contain a number of species not found elsewhere in the state; these species include *Nelumbo lutea, Hibiscus militaris, Azolla caroliniana,* and several others. A complete diagnosis of these large river communities must await future studies.

CHAPTER 20

Beach, dune, and cliff communities

NATURE AND LOCATION

The rather diverse habitats to be treated in this chapter have several features in common. Their substrates have no true soil but are composed of loose sand or bare rock. In consequence their plant covering is rather sparse and shows little plant-to-plant interaction. The plants to be found on lake shores and on cliffs have very little in common and are not to be regarded as related communities. In fact, there is only small reason for regarding them as communities at all, in the sense of an integrated assemblage of organisms with mutual interactions. Rather, they are specialized groups of species, selected from the flora by the very specialized environmental conditions of the site. As such, they are worthy of separate discussion, even though they are not wholly equivalent to the other communities treated in earlier chapters. The beaches and lake dunes are confined to the larger lakes, with best development on Lakes Michigan and Superior. Many of the interior lakes in sandy districts also have beaches and dunes but these are usually of small extent and rarely are continuous around the entire lake. In the lake district of northwestern Wisconsin, particularly in the area of Glacial Lake Barrens in Douglas and Bayfield counties, a number of lakes normally had very wide sandy beaches. During the low water stages of the 1930's, these beaches were 50 to 100 feet wide or more and were occupied by a well-developed plant community, which included many rare species of unusual range affinities. Since 1950, however, a period of excessively high water has set in which has completely submerged all of the beaches. In consequence, the beach communities have been totally destroyed and their composition can be reconstructed only from the partial descriptions provided by McLaughlin (1932).

The lake dunes are found immediately inland from the beach, at a level beyond the reach of the highest storm waves. They are fed by sand from the beach which is borne by winds off the lake. An extensive dune system is found along Lake Michigan; but the size of the individual dunes in Wisconsin rarely approaches that to be seen along the Indiana and Michigan shores of the lake.

The cliffs are a heterogeneous group of sites with only geomorphological unity. They are best developed on the Niagara and Magnesian escarpments and in the recessed valleys of the Driftless Area. Certain gorges on the fall line south of Lake Superior, the quartzite monadnocks throughout the state, and a few river channels in northwestern Wisconsin provide other cliffs. All stands of beach, dune, and cliff communities in the P.E.L. study are shown on the map at the opening of this chapter.

BEACH COMMUNITIES

The beach or strand communities of the Great Lakes have received considerable attention at the hands of plant ecologists. The most complete early account was made by Cowles (1899) for the beaches of Lake Michigan. Bruncken (1910) followed Cowles' method of treatment in his discussion of the beaches of Milwaukee County as did Gates (1912) in his investigation of the Kenosha area. Various floristic aspects of the southern Wisconsin beach communities were reported by Oehmcke (1937) and Shinners (1940). No studies are on record for the Wisconsin beaches on Lake Superior nor for the beaches of the inland lakes. The following discussion is based on the work of the above investigators plus a P.E.L. presence study of a number of stands made by J. T. Curtis and H. C. Greene (Figure XX-1).

Cowles divided the Lake Michigan beaches into three separate areas—the lower, the middle, and the upper. The lower beach is under the constant influence of waves and tides and consists of a hard-packed, wet sand, a gravel or shingle, or a bare rock substrate essentially devoid of higher plants. The middle beach is similar but is affected by waves only during storms and may be quite dry for long periods. A small but highly characteristic group of seed plants grows in this area, usually at very low densities. The upper edge of the middle beach is usually marked by a row of driftwood and other flotsam cast up by the waves. The upper beach is above the direct action of waves except in the most severe storms, but is influenced by blowing sand and spray from the lower areas (Plate 59). It tends to be very dry in its surface layers although water is present within several feet of the surface in sand or gravel areas. The upper beach has a relatively rich assortment of higher plants, although the total density is much lower than in normal terrestrial communities.

The beaches of Lake Superior resemble those of Lake Michigan in microtopography but differ considerably in vegetation, with a greatly reduced number of species except in completely protected bays. In fact, long stretches of Lake Superior beach, often several miles in length, can be found which have not a single seed plant except for a few stragglers in areas of buried organic debris. The inland lakes may have the same three types of beach present, but the usual lake has only two—the lower and upper, with nothing that is directly comparable to a middle beach (Plate 58). The plants on the upper beach have a greater tendency to grow in linear micro-aggregations than do those on the Great Lakes. Frequently, a series of lines will be present, each containing a different assortment of species and each resulting from a different storm earlier in the growing season.

A considerable number of the beach plants are annuals or else perennials which develop from such detached and floatable parts as rhizomes or tubers. Because of the capriciousness of storms and the vagaries of high and low water stages, the beach community is in a constant state of flux, with relatively rich assemblages present in some years and almost nothing in the same location in other years. In view of this great instability, it is not surprising that the homogeneity of the community is very low, with only a single species having a presence over 50 per cent. On the other hand, the number of species attaining their optimum presence on the beaches is relatively high; this indicates the very specialized nature of the environment. There are four of these modal species on the prevalent species list (Table XX-1) and 32 on the total flora list.

The strand communities of Lakes Michigan and Superior have a number of species which are not found on the inland lakes or else are very rare inland. Included in the group are many of the most characteristic strand plants, such as sea rocket (*Cakile edentula*), beach pea (*Lathyrus maritimus*), *Euphorbia polygonifolia, Juncus gerardi* (plus six other species of *Juncus*), *Potentilla anserina,* and *Satureja glabella,* with several species originally adventive from the west (*Corispermum hyssopifolium, Cycloloma atriplicifolia,* and *Salsola kali*). The inland beaches, on the other hand, have a group of species which are of minor importance on the Great Lakes. Especially prominent are members of the mint family (*Lycopus* sp., *Galeopsis tetrahit, Scutellaria* sp., *Mentha arvensis,* and *Mimulus ringens*), the genus *Bidens* (five species), and the genus *Polygonum* (six species).

Both types of beach have a large number of exotic weed species and native species which act as weeds in disturbed places in general. Included in the first group are *Avena fatua, Asparagus officinalis, Brassica arvensis, Galeopsis tetrahit, Panicum capillare, Salsola kali, Sonchus arvensis, Stellaria graminea,* and *Thaspium barbinode;* the native weeds include *Acnida altissima, Asclepias syriaca, Bidens frondosa, Cerastium vulgatum, Erigeron annuus,* and *Xanthium strumarium.*

Bruncken (1910) determined the density of several species on the beaches of Milwaukee County. He found the most common species to be sea rocket (*Cakile edentula*), but it was present only at the rate of 6.7 plants per 100 square meters. All other species were represented by less than 2.5 plants per 100 square meters. Similar determinations in the P.E.L. study showed an average of only 1.3 plants of *Cakile edentula* per 100 square meters, on beaches from Door to Kenosha County; but this figure is misleading since the plant is highly aggregated. Some stretches of beach show a random distribution and comparatively high density, and others are nearly devoid of the plant. It was noticed that the high densities were on those areas where herring gull flocks frequently rested, but the nature of the correlation is not apparent.

LAKE DUNES

Sand dunes and their history have been objects of interest and study for many years. Cowles (1899) paid considerable attention to the origin, movement, and eventual stabilization of the Lake Michigan dunes in his famous pioneering study of the processes of plant succession. Very large dunes and dune systems are present at the south end of the lake in Indiana and on the eastern shore in Michigan. Since the prevailing winds of the region are westerlies, the Wisconsin coast has a less extensive development of dunes. Nevertheless, variations in orientation of the coast line and frequent storms from the northeast have permitted the development of a number of dune areas, notably in Kenosha County, near Terry Andrae State Park in Sheboygan County, and at several places in Manitowoc and Door counties. In addition, a minor dune system is present immediately adjacent to the upper beach along most of the shore line. Similar minor dunes are present on Lake Superior and on some of the larger inland lakes.

The processes of origin and movement of the lake dunes are similar to those of the sand blows on the inland sand barrens (Chapter 15), except that the source of the sand is the beach, rather than the substrate of the blowout. Sooner or later the moving dune becomes colonized by plants which tend to reduce the movement and to stabilize the dune. On such stabilized dunes, an orderly process of plant succession may set in, which leads through shrubs and pioneer xerophytic trees to a closed forest. Eventually, a mesic forest may develop on the dune. One of the finest examples of beech-maple-hemlock forest in Wisconsin formerly existed on a 12 meter high dune in Door County. Obviously, therefore, a complete discussion of the plant communities of lake dunes in Wisconsin would need to include a great variety of species groupings. However, since the forest communities differ in no significant way from their counterparts on dry sites elsewhere in the state, the present discussion will be limited to the initial communities—those which stabilize the dunes and make the later successions possible.

Descriptions of these early dune communities have been presented by Bruncken (1910) and Gates (1912), both of whom gave lists of species found. A P.E.L. study conducted in 1955 and 1956 by J. T. Curtis with the aid of H. C. Greene has provided presence information for 20 dune systems, on Lakes Michigan and Superior and on several inland lakes. The species density per stand (22) is much higher than that of the beach community. The index of homogeneity of 45.6 per cent (Table XX-2) is also much higher and approaches that of the normal upland groupings. Over half of the prevalent species are also modal because they reach their highest state-wide levels of presence on the dunes.

Some idea of the structure of typical dune communities can be obtained from Table XX-3, which shows the average frequency and density of the plants at three stations in Ozaukee, Sheboygan, and

Plate 59.—Upper beach, Lake Michigan. Bailey's Harbor, Door County.

Plate 60.—Beachgrass (*Ammophila breviligulata*) on Lake Michigan dunes, near Wilderness Ridge Scientific Area, Point Beach State Forest, Manitowoc County.

Plate 61.—Lake dune stabilized with creeping juniper (*Juniperus horizontalis*). Lake Michigan shore, Manitowoc County.

Plate 62.—Evergreen, heathlike mat of *Hudsonia, Juniperus,* and *Arctostaphylos* on Lake Superior dunes on Wisconsin Point. Douglas County.

Manitowoc counties. By all odds the most important of these dune species is the beachgrass (*Ammophila breviligulata*). In Wisconsin, this plant is confined to the Great Lakes dunes where it is the initial invader and prime stabilizer of the moving piles of sand (Plate 60). It is particularly well adapted to this role by reason of its great drought resistance and its extensive development of long rhizomes which penetrate the sand with ease and anchor it by a well-developed fibrous root system (Laing, 1957). Other species with similar abilities are the grasses *Calamovilfa longifolia* var. *magna*. *Agropyron dasystachyum*, *A. trachycaulum* var. *glaucum*, and *Elymus canadensis*. The beach cinquefoil (*Potentilla anserina*) also anchors the sand, but it spreads by means of aerial runners rather than subterranean rhizomes. Other forbs which are active in holding the sand include the beach pea (*Lathyrus maritimus*), the prairie pea (*L. venosus*), and the lesser false-Solomon's seal (*Smilacina stellata*). When the sand has become stabilized, a number of mat-forming shrubs enter the community (Plate 61). Prominent among them are the two junipers (*Juniperus communis* and *J. horizontalis*), the bearberry (*Arctostaphylos uva-ursi*), and the hudsonia (*Hudsonia tomentosa*). All of these are prominent on the flat-topped, wind-swept dunes on Lake Superior (Plate 62) where they convey a distinct heath-like aspect to the community. Other shrubs common on the dunes are willows (*Salix cordata* var. *adenophylla*, *S. interior*, *S. lucida*, and *S. glaucophylloides*), roses (*Rosa blanda* and other species), buffalo berry (*Shepherdia canadensis*), shrubby St. John's wort (*Hypericum kalmianum*), and poison ivy (*Rhus radicans*). Altogether, 36 species of shrubs were found on the dunes, or 23.2 per cent of the total flora.

A number of tree species are more or less frequently present on the dunes in the form of seedlings or saplings but only rarely as mature individuals. These include white birch (30 per cent presence), pin cherry (25 per cent presence), white ash (25 per cent presence), and several poplars, especially balsam poplar (45 per cent presence), and cottonwood (10 per cent presence). Such introduced species as white poplar (*Populus alba*) and black poplar (*P. nigra*) are also present. These initial trees often show signs of severe damage from the sandblast action of the winds. The first trees to become successfully established on old dunes are mostly oaks in the south, especially Hill's oak, and mixed oaks and pines in the north.

The lake dunes flora, in addition to its endemic or highly restricted species like *Cirsium pitcheri*, *Lathyrus maritimus*, *Potentilla anserina*, *Iris lacustris*, and *Salix cordata* var. *adenophylla*, possesses a peculiar mixture of other species showing different affinities. Thus it has a group of plants which are characteristic of the northern pine forests or the boreal spruce-fir forests. It also has a group of prairie plants, including both grasses and forbs, a few plants typical of southern oak forests, and a considerable element of exotic weed species. Because of this mixture, the total community is quite different from the interior sand barrens grassland which occurs under what appears to be very similar environmental conditions. Considering only the prevalent species (Tables XV-1 and XX-2), the index of similarity is a low 2.3 per cent. Gleason (1907), in his treatment of the inland sand communities of Illinois, also pointed out their low similarity with the dune groupings on Lake Michigan in that state.

A number of the species of both beach and dune are present on marine coast lines as well (Cooper, 1936). Prominent in this group are *Ammophila breviligulata, Lathyrus maritimus, Potentilla anserina, Cakile edentula, Calamovilfa longifolia,* and *Juncus balticus*. The lack of salinity in the inland sites is apparently no barrier to the successful establishment of these species.

No discussion of the lake shore communities of Wisconsin would be complete without at least a mention of the beaches of Door County. At the Ridges Wild Flower Sanctuary near Bailey's Harbor, at Peninsula State Park, and elsewhere in the county, there exists a number of very wide beaches with a gradual transition from the upper beach to the adjacent low dunes. These dunes are frequently linear, paralleling the shore line, and are separated by strips of moist or submerged meadows and sloughs. The variations in moisture and organic matter content are great enough to allow the development of a rich mixture of species which is almost startling in its beauty at many seasons. Perhaps the acme is attained in late May or early June when the ground is literally covered with mats of the bright blue lake iris (*Iris lacustris*) studded with the pink umbels of *Primula mistassinica* and the blossoms of the brilliant purple gaywings (*Polygala paucifolia*). In the opinion of many people of wide experience, no other wild flower display in the entire Middle West is the equal of this.

CLIFFS

A cliff is a geological feature, not a biotic community type. Furthermore, cliffs can be formed in a number of rocks by a variety of erosional processes, so that even from a geological standpoint, the word cliff scarcely reflects a homogeneous unit. Nevertheless, cliffs present environmental conditions which are very different from normal sites and which allow a rather specialized group of plants to survive. For this reason, they are discussed under one heading.

The cliffs of Wisconsin are found on dolomites, sandstones, quartzites, and granitic rocks and are present throughout the state; the greatest numbers are in the Driftless Area. They vary greatly in height, steepness, and direction of exposure, from the 200-foot vertical walls of the Niagara escarpment in the northeast to sloping ledges a few feet high in the nonglaciated regions. They may present a continuous surface exposed to the sun at the tops of ridges, as in the Penokee Range in Iron County, or they may be narrow gorges cut deep in the rock as at Copper Falls and Manitou Falls. One of the most important variables affecting their plant occupants is the presence or absence of a forest cover on or over the cliff. The great range in types of habitat naturally means that the plant communities are also highly varied. It would no doubt be possible to study enough examples of each type to bring some order to the whole, but practical difficulties have prevented this approach. More effort was spent in the P.E.L. studies on the attempt to find suitable cliff sites for investigation than was spent on any other community, and the effort was far less successful. Quarrying operations had ruined many of the easily accessible stations and a high percentage of the others was disturbed by grazing, fire, logging, or other practices. The final count was 38 usable stands (Figure XX-2), which was an insufficient number to give an adequate subsample of each major cliff type.

In an effort to group the data into meaningful segments, each site was classed as to rock type (dolomite, sandstone, igneous) and exposure (exposed or shaded), and presence percentages calculated for the plants in each class. It was found that the type of rock was of far less significance than the type of exposure. For example, the shaded sandstone communities (Plate 53) had an index of similarity of 64.3 per cent with the shaded limestone communities, but an index of only 28.6 per cent with the exposed sandstone communities (Plate 54). Accordingly, all cliffs in exposed situations were lumped together into one group and all shaded cliffs into another. The resulting lists showed only a 36.8 per cent index of similarity. The two groups were still not homogeneous communities, since areas north and south of the tension zone were combined in each. A geographical separation was not possible because of the limited amount of data, but it is believed that the important features of the cliff communities can be demonstrated from the two large groups.

One type of cliff that is well developed in other states but occurs very rarely in Wisconsin is the dripping limestone cliff, where trickling water keeps the rocks permanently wet. A few such cliffs are known on the Prairie du Chien (Magnesian) dolomites in Wyalusing State Park and several other areas near the Mississippi in the Driftless Area. In general, the limestone cliffs in Wisconsin are dry, frequently to the point where no plants whatsoever can grow on them. The sandstones, however, even in what appear to be topographically xeric sites, contain a supply of moisture that moves by capillarity through the rock and allows at least certain plants to become established (Plate 57). The granitic rocks may also be very dry, but water supplies from cracks and crevices are usually adequate for a few plants and such cliffs rarely present the barren aspect of the worst limestone sites.

A study of the P.E.L. results shows that the bulk of the plants present on any given cliff are closely related to the vegetation type present in the vicinity of the cliff. Ledges with a small soil deposit, or with deep fissures containing soil pockets allow almost any species to become established; however the species that are most numerous in the vicinity have the greatest chances. In addition to these matrix species, there are a number of specialized forms which are peculiarly adapted to life on the cliffs and which are rare or absent away from them. These are mostly species which can thrive with very little soil, by sending their roots back into the rock in the tiniest cracks in search of water and nutrients. The ferns are the most prominent group in this specialized cliff flora, but a number of other kinds of plants are represented as well.

The prevalent species of exposed cliffs are listed in Table XX-4 and of shaded cliffs in Table XX-5; both tables also give the other modal species of lower presence. Both communities have about the same species density (23 and 22) and homogeneity (35.2 per cent and 39.7 per cent). Considering only the prevalent species, the ferns comprise 20 per cent of the sum of presence on the exposed cliffs and 27 per cent on the shaded rocks. No other family approaches this degree of predominance. Statements as to the preferred habitat of other ferns as given in the *Ferns and fern allies of Wisconsin* by Tryon *et al.* (1940) would indicate that many more species should be considered a part of the cliff community, although they were not seen in the P.E.L. study. The book should be consulted by those interested in a full list. Several other discussions of the cliffs of Wisconsin which emphasize the ferns are available; these include studies by Hill (1891), True (1897), Pammell (1907), Marshall (1910), and Ellen (1924).

One of the interesting features of the shaded cliffs in the Driftless Area is their high content of endemic species or rare species with disjunct ranges, the nearest neighboring stations being along the south edge of the major continental glaciation in Missouri, Kentucky, or Pennsylvania. Included among these are *Aconitum noveboracense, Dodecatheon amethystinum, Gnaphalium saxicola, Solidago sciaphila,* and *Sullivantia renifolia.* In spite of their restricted range, some of these plants are extremely common on the cliffs and exist in populations numbered in the hundreds of thousands. One of the outstanding sights in Wyalusing State Park is the pink sheet of color spread over the cliffs in late May by the amethyst shooting star (*Dodecatheon amethystinum*).

The exposed cliffs have a much smaller complement of such rare species, although a few species of unusual range have been reported for these habitats. One of the most unusual is the arctic rhododendron (*Rhododendron lapponicum*). This plant is of circumboreal distribution in arctic regions but is found in Wisconsin only at a single station on the river cliffs at the Wisconsin Dells (Fuller, 1946). The only endemic plant of the exposed cliffs is *Commelina erecta* var. *greenei* Fassett.

The exposed cliff communities of Wisconsin show considerable similarities with those of southern Illinois. Winterringer and Vestal (1956) reported a detailed study of the rock-ledge vegetation of that region. Of the 48 seed plants they list as of major importance, 16 or one-third are on the Wisconsin list. The most numerous species in Illinois included such succulent plants as *Sedum pulchellum, Agave virginica,* and others whose range does not include Wisconsin. The major succulent species on the exposed Wisconsin cliffs are *Draba reptans, Opuntia compressa* (Plate 56), and *Talinum rugospermum,* although a number of other species of obvious xeromorphic habit are also present, including *Selaginella rupestris, Polygonum tenue, Cheilanthes feei,* and *Viola adunca.*

CHAPTER 21

Weed communities

NATURE OF DISTURBED HABITATS

All of the communities treated in previous chapters occur in undisturbed habitats or else are subject to what might be termed natural disturbance as induced by fire, wind, wave action, or flooding. They represent a relatively rare condition in the present pattern of land use. The majority of Wisconsin's land surfaces have been greatly modified by man through a variety of artificial disturbances. The plant groupings which occupy these modified areas differ in many ways from the natural communities; these groupings are the subject of the present chapter.

The degree and the type of artificial disturbance of the habitat are obviously subject to great variation and are difficult to classify readily. Adopting the terminology of Hamel and Dansereau (1949), they may be grouped into three divisions for the sake of convenience. If the disturbance of the original community is incomplete and sporadic, the habitat is said to be "degraded." If the original community is destroyed and the destructive agent is repeatedly applied, the habitat is called "ruderal" if the area is not used for the production of economic crops and "cultivated" if the area is used for crop production. The degraded habitats include prairies and forests used for pastures, meadows and fens used for hayfields, and forests used for intensive lumber production. The ruderal sites are extremely varied and include roadsides, spoil banks, railroad yards, trackways, city dumps, stockyards, and other waste places. The cultivated areas are farm lands devoted to field crops, vegetable gardens, orchards, and other intensively used areas that are ordinarily plowed or cultivated at least once each year.

In spite of the great variety of such disturbed habitats, they almost always have one feature in common, namely the presence of a soil layer of some sort, either left over from the original vegetation or produced artificially by mechanical mixing as in fills and banks. This is the major difference between disturbed sites and native primary sites on rock, water, or sand. The open disturbed areas also resemble each other; usually they are exposed to full sun, have violent fluctuations of daily temperature, and show rapid changes in water supply. The degraded forest sites naturally have fewer pioneer characteristics and may differ only slightly from natural conditions in certain situations. Another feature that is frequently present in disturbed habitats is a lack of soil stability, which is expressed as settling movements or as erosional shiftings of the surface. Again, this character is less evident in degraded habitats, especially in pastures or mowing meadows.

NATURE OF SECONDARY SUCCESSION

When a ruderal or cultivated site or severely degraded community is relieved from the causal disturbing agent and allowed to develop by itself, a secondary succession is initiated. The pioneer species are adapted to the disturbed conditions and are largely of the plant type called weeds. In each kind of area the very first plants are mostly species which were present during the period of active disturbance. As the succession proceeds, other weeds invade the area, together with pioneer native species from nearby native communities. With time, the proportion of native species increases until eventually a mature community, characteristic of the region, is developed. The time interval between initial abandonment and final recovery may be very great. Drew (1942) made a thorough study of the revegetation of abandoned cropland in Missouri. He reviewed most of the literature dealing with the process and found that 25 to 40 years were required for the establishment of the dominants of the native community under a variety of conditions in central and eastern United States. None of the investigators ever had observed areas where the complete community was restored, although several thought this might occur within 80 to 100 years.

In Wisconsin, such studies were made by Thomson (1943) and Buss (1956) and were discussed in Chapter 14. They both found an interpolation of what might be called a prairie stage in the secondary succession from cropland to forest. This phenomenon is relatively common in Wisconsin, since widely distributed prairie remnants provide sources of propagules for the invasion of the abandoned areas and since many of the prairie plants are well adapted to grow in such sites. In the absence of an inoculum of prairie species, the forest plants may invade the area directly; the type will depend on the moisture relations of the site and the nature of the surrounding forest remnants.

From all this, it can be seen that an infinite variety of species mixtures may occur in the intermediate stages of secondary successions. The initial stages, although varied, are far less numerous, and the terminal stages are relatively few. For this reason, the following discussion of weed communities of open places will be concentrated on the initial stages, especially on the species groupings to be found on sites with continuous disturbance and on such sites during the first three years after abandonment. The intermediate stages may be of considerable economic or conservational interest, but they necessarily will need to be studied on a strictly local basis, since their variety precludes meaningful generalization.

NATURE OF WEEDS

There is probably no word used to describe plants which is subject to more misunderstanding and varied usage than the term "weed." No definition has been advanced which adequately covers the manifold meanings which are current. To some persons, a weed is any insignificant plant with which they are unfamiliar, such as the wild flowers in a wood lot, the forbs in a prairie, or any other small herbs. To others, it is any undesirable, useless, or harmful plant. Still others consider a weed to be a plant out of place. On this basis, a rare orchid growing in a cranberry field would be a weed, as would any other plant growing where some one person did not want it. All of these definitions really hinge on man's behavior, not on any basic characteristic of the plants themselves. Under them, all species of plants in the world might be called weeds under some conditions. Obviously, no statements of biological significance can be made about weeds under such definitions.

Other descriptions of weeds have been given which combine the human relation with a stated or implied behavior of the plants. These usually relate chiefly to plants which are troublesome in cultivated areas. Blatchley (1920), for example, defined a weed, "A plant which contests with man for the possession of the soil." Another similar definition describes a weed as a green plant (thereby excluding parasitic fungi) which incessantly invades man's cultivated fields and there competes with his crop plants, to the detriment of the latter. The plants of major concern to agriculture, which are usually specified in state and local weed control laws, fit this definition. Such plants are best termed "agricultural weeds."

Not all plants usually considered to be weeds are covered by the agricultural category; it is necessary to delimit and describe certain other kinds of unwanted plants. Those species which are poisonous, either to the touch (poison ivy) or when eaten (water hemlock), cause hayfever (ragweed), or are otherwise actively dangerous to man and his animals, may be differentiated as "noxious plants." The heterogeneous group of unsightly or useless plants which happen to grow where they are not wanted could be called simply "undesirable plants." The plants of the last two categories have almost no botanical characteristics in common, but are delimited solely on the basis of man's physiological and psychological responses and may be subject to reclassification as man's interests change. The tomato was formerly regarded as a poisonous "love apple" and would then have been regarded as a noxious plant.

Still other definitions for weeds have been proposed which are based almost entirely on plant behavior. Harper (1944), for example, defined a weed, "A plant that grows spontaneously in a habitat that has been greatly modified by human action, and especially a species chiefly or wholly confined to such habitats." This definition has much merit because practically all of the species classed as weeds in the agricultural sense plus many of the most troublesome noxious and undesirable plants would be included under it. The ruderal weeds, those plants found in waste places associated with industrial or urban activities, also fit Harper's definition.

One important aspect of all definitions concerns the local nature of their applicability. A plant which is a weed of great economic importance in one region may be an inconspicuous and minor member of a natural community in another region. This feature points up one characteristic of many weeds, namely that they are adventive in the region where they act as weeds. Very rarely does a normal member of a native community become a weed within the geographical range of that community. In fact, a large majority of the worst weeds are exotic adventives, originating from a different continent. In Harper's definition, mention is made of species wholly confined to artificially modified habitats. These are the "obligate weeds" of Hamel and Dansereau (1949), plants whose evolutionary history has been associated with man and his agricultural practices and which do not exist independently in any known natural community anywhere in the world.

In the P.E.L. studies of the weed communities of Wisconsin, it was found necessary to utilize in part a non-floristic criterion for the choice of stations for investigation. The exotic weed species are well known and their presence can be used as an indicator of a weed community, but little prior knowledge existed either as to which native species could act as weeds or to what extent they participated in weed communities. Accordingly, a study was made of any plant assemblage which occurred on a site obviously modified by human action and which contained a representation of those species commonly recognized as exotic weeds, regardless of the nature of the native plant component. A further restriction limited the major study to areas on which the disturbing factor was still active or had been active within the past three years. Thus, the intermediate stages of secondary succession were eliminated from consideration. The stands studied included degraded, ruderal, and cultivated habitats. For convenience in discussion, the degraded sites are separated from the ruderal and cultivated. The latter resemble each other because the original plant cover has been totally destroyed at least once.

Plants which can invade ruderal or cultivated areas have certain characteristics in common. They are very vigorous, in the sense that they are of rapid growth; they have the ability to withstand and surmount high intraspecific competition; they show great tolerance of soil disturbance, partial defoliation, or other regressive influences; and generally they possess a high reproductive potential. Accompanying this vigor is a great genetic variability of both morphological and ecological characters. Many of the species have efficient means of spread and may invade in large numbers. The seeds germinate under a variety of conditions and the seedlings are tolerant of the extreme fluctuations of soil temperature and soil moisture typical of open, disturbed areas. A number of weeds show great reproductive plasticity, in that they may set seeds when still very small, as when growing under extremely rigorous conditions, or they may develop into very large, multibranched individuals before flowering, as on favorable sites or in climatically optimum seasons (Sorensen, 1954).

In spite of this vigor, however, very few weeds are able to persist for long in competition with the native plants of a given region. Those which do are said to be naturalized, in that they may become regular members of native communities. Such truly naturalized species are quite rare in Wisconsin. Notable examples would be Kentucky and Canada bluegrass (*Poa pratensis* and *P. compressa*) in prairies and other non-forest communities, devil's hawkweed (*Hieracium aurantiacum*) in the bracken grasslands, and redtop (*Agrostis stolonifera*) in wet prairies and fens. Naturalized species in forest communities are practically non-existent in Wisconsin. The closest approach is probably the tartarian honeysuckle (*Lonicera tatarica*) which appears to be successfully established in undisturbed wood lots around the larger cities, but the evidence is incomplete and is based on observations of insufficient duration. The related Japanese honeysuckle (*L. japonica*) is a well-known invader of closed forests in the southern states.

WEED COMMUNITIES OF OPEN PLACES
Nature and location of communities

The communities to be discussed here are those present in cultivated and ruderal areas. The cultivated areas are extremely variable with respect to soil and cultural practices; they range from rich truck-crop gardens on drained peat and muck lands, through mesic upland sites with the typical corn-oats-hay rotation of Wisconsin's dairy farms, to marginal fields on dry sand where an occasional crop of corn is attempted. The particular crop that is present on a field in any given year is associated with a more or less distinct group of weed species which are adapted to the seasonal sequence of operations employed with that crop. Thus the oat fields regularly show a high density of foxtail grass (*Setaria* sp.) and ragweed (*Ambrosia artemisiifolia*) in late summer after the oat crop has been harvested. The corn fields in the prairie-forest region commonly have high populations of butterprint (*Abutilon theophrasti*), flower-of-an-hour (*Hibiscus trionum*), horseweed (*Conyza canadensis*), and crabgrass (*Digitaria sanguinalis*), all of which can develop best in the hot weather of midsummer after the final cultivation of the corn. The hay stage of the rotation, usually a mixture of brome grass and alfalfa or timothy and redtop, may be relatively free of prominent weeds in the first year, but frequently is invaded in later seasons by mustard (*Brassica nigra*), Canada thistle (*Cirsium arvense*), quackgrass (*Agropyron repens*), or fleabane (*Erigeron strigosus*). The seeds, roots, or rhizomes of all these species and many others are present in each field at all times, but they develop into important populations only in the years when their associated crops are present. The weeds present in the first year after abandonment of a field which has been in the corn-oats-hay rotation naturally are influenced by the crop last present, but after two or three years, these differences become less important. A similar association between weeds and crops has been reported for Nebraska by Norris (1935) and has been studied frequently in Europe (Demianowiczowa, 1952; Oberdorfer, 1954).

The ruderal areas include roadsides, especially those which have been recently reconstructed with numerous earth cuts and fills; railroad properties that include switchyards, depot grounds, and the ballast of the tracks; stockyards with their rich organic debris; and a variety of "waste places" where the soil has been scraped, piled, or otherwise mechanically moved.

Both cultivated and ruderal areas are found in all parts of Wisconsin as is shown in the figure reproduced at the opening of this chapter. The P.E.L. study included 129 stations, distributed in 65 counties, with data available only on simple presence in most cases.

Composition

In the previous communities, some type of ordination of the data was usually possible, based either on behavior of the dominants, as in forests, on the use of topographic indicator species, as in the prairies, or on an index of joint occurrence, as in the submerged aquatic communities. Ellenberg (1952) was very successful in his use of a behavioral classification of German weed communities. It was based largely on their response to soil factors, while Raabe (1949) used joint occurrences as a basis for constructing an ordination of the weed communities of east Holstein. In the local weed communities, none of these methods was readily applicable, although the last is theoretically possible. The high species density and the great total number of species in the Wisconsin data, however, would call for an excessively heavy amount of calculation. Accordingly, a less rigorous method of analysis was employed in this preliminary treatment; the stations were grouped into various categories based on environment or past treatment, and presence percentages were calculated for each species in each category.

By this means, it was possible to obtain at least a general understanding of the behavior of the major species.

In the initial P.E.L. weed study, Lindsay (1953) showed that a number of prominent Wisconsin weeds were essentially restricted to the areas either north or south of the tension zone, in much the same fashion as the native members of undisturbed communities (Chapter 1). By means of a frequency survey of selected species in each county, Lindsay demonstrated that ox-eye daisy (*Chrysanthemum leucanthemum*), buttercup (*Ranunculus acris*), hawkweed (*Hieracium aurantiacum*), and sow thistle (*Sonchus arvensis*) attained high levels of abundance only in the north, while the reverse was true for crabgrass (*Digitaria sanguinalis*), barnyard grass (*Echinochloa crusgalli*), and bristly foxtail (*Setaria verticillata*). A number of species, including Canada thistle (*Cirsium arvense*), Kentucky bluegrass (*Poa pratensis*), and curly dock (*Rumex crispus*) appeared to be uniformly distributed throughout the state.

In view of these findings, the data from the later P.E.L. presence study were first divided into two groups, one north and the other south of the tension zone. When the resulting lists were compared, an amazingly high index of similarity of 76 per cent was obtained. Analysis of the figures showed that many common species were equally represented in both areas. However, a number of species, including those mentioned by Lindsay, showed a differential behavior. These are shown in Table 22, which is divided into two groups on the basis of soil texture. The exclusive northern species are *Chrysanthemum leucanthemum, Anaphalis margaritacea, Sonchus arvensis, Hieracium aurantiacum,* and *Ranunculus acris,* and the southern exclusives are *Oenothera rhombipetala* on sands and *Abutilon theophrasti, Setaria viridis, Parthenocissus vitacea,* and *Matricaria matricarioides* on heavy soils. A number of other species of lower average presence could be added to these lists.

Some species react strongly to the soil texture but appear to be indifferent to the climatic factors associated with the tension zone. Some of these are shown in Table 23, which is based on the same two general soil classes as before. A more complete separation of the southern weeds based on a combined soil and moisture gradient is shown in the community compositions listed in Tables XXI-1, 2, and 3. The moist, nitrogen-rich soils are those of muck gardens, heavily fertilized areas near barnyards, pea vineries, stockyards, and other places receiving an overflow of nutrient rich waters. The mesic heavy soils include silt

Table 22

Weed species with markedly higher presence in north or south on soils of same texture

SANDY SOILS			HEAVY SOILS		
Species	N.	S.	Species	N.	S.
Achillea millefolium	72	26	Achillea millefolium	60	32
Agrostis stolonifera	72	13	Agrostis stolonifera	70	36
Conyza canadensis	91	66	Chrysanthemum leucanthemum	40	5
Erigeron annuus	62	13	Erigeron strigosus	60	32
Lychnis alba	76	18	Phleum pratense	100	50
Oenothera biennis	86	35	Sonchus arvensis	50	0
Potentilla norvegica	62	22	Verbascum thapsus	60	27
Trifolium hybridum	72	18	Hieracium aurantiacum	50	0
Chrysanthemum leucanthemum	43	0	Ranunculus acris	30	0
Anaphalis margaritacea	33	0			
Hieracium aurantiacum	57	5			
Cenchrus pauciflorus	10	40	Amaranthus retroflexus	30	55
Euphorbia corollata	10	63	Asclepias syriaca	40	73
Lepidium campestre	5	53	Capsella bursa pastoris	10	50
Physalis heterophylla	5	40	Chenopodium album	50	100
Silene antirrhina	14	44	Abutilon theophrasti	0	36
Oenothera rhombipetala	0	35	Lactuca scariola	20	68
Froelichia floridana	5	35	Setaria viridis	0	27
			Parthenocissus vitacea	0	32
			Matricaria matricarioides	0	23

loams and clays, with the bulk of the stations on loessial soils of old prairies or oak savannas. The sands include glacial outwash, river terrace, and aeolian sands. A comparable set of tables cannot be given for the north because of the absence there of a sufficient number of stations on nitrogen rich sites. The northern species are given separately for mesic heavy soils and dry sands in Tables XXI-4 and XXI-5.

One of the interesting aspects of the distributions is the decrease in relative content of exotic species with decreasing moisture and nutrients (Table 24). Thus, considering the presence values for each soils group of the southern communities separately, exotic (extra-continental North America) species contribute 64.3 per cent of the total sum of presence in the nitrophilous group, 58.1 per cent in the mesic heavy soil group, and only 32.9 per cent in the sands. Another ecoclinal shift is seen in the proportions of annuals, biennials, and perennials, with the first two increasing and the last decreasing as moisture and nutrient supplies decrease. The very high value of 52.4 per cent annuals in the sandy soil weed communities of southern Wisconsin greatly exceeds

Table 23

Weeds which are indifferent to north-south gradient but react to soil texture

Species	Heavy Soil		Sandy Soil	
	S.	N.	S.	N.
Poa pratensis	55	60	32	29
Chenopodium hybridum	23	40	5	14
Cirsium arvense	50	70	5	14
Bidens frondosa	27	30	4	5
Bromus inermis	14	20	0	0
Medicago lupulina	55	50	13	19
Melilotus officinalis	36	40	9	10
Ambrosia psilostachya	5	0	13	24
Monarda punctata	0	0	8	5
Rumex acetosella	5	20	48	67
Aristida basiramea	0	0	4	14
Eragrostis spectabilis	0	0	5	14
Mollugo verticillata	0	0	49	57
Oxybaphus nyctagineus	0	0	9	14
Potentilla argentea	0	0	5	10
Salsola kali	0	0	13	10
Leptoloma cognatum	0	0	5	10
Cycloloma atriplicifolia	0	0	18	5

that of any other community and approaches the values recorded for certain Mediterranean communities.

The weeds of railroad yards are treated separately, because the disturbing agencies are continuously active and because the cinders so widely used for ballast and roadways present a soil environment quite different from that in other ruderal areas. Fresh cinders have a high content of sulfur-containing compounds and other soluble, more or less toxic, materials. They also have extremely rapid internal drainage but

Table 24

Summary of weed communities—prevalent species only

Soil	Sum of presence			
	Annuals	Biennials	Perennials	Exotics
Southern—Moist, N_2-rich	35.5%	8.2%	56.3%	64.3%
Southern Heavy	38.0	13.4	51.3	58.1
Southern Sandy	52.4	22.2	25.4	32.9
Northern Sandy	33.5	16.7	49.8	53.4
Northern Heavy	25.1	12.0	62.9	64.0
Railroad Yards	38.3	13.7	48.0	62.7

a fair water-holding capacity. All of these factors in combination appear to simulate the growing conditions found on the sub-saline soils of the western Plains. One result is the occurrence of many species adventive from the West. These may be either firmly established local populations or ephemeral occurrences maintained by repeated inoculation from cattle cars or other eastbound railroad traffic. Considering all of the railroad yards as a group, a prevalent species list constructed in the usual manner is given in Table XXI-6 along with less common species which reach an optimum in the yards. Of this group of 74 species, 18 per cent originate in western United States, 61 per cent are true exotics, and the remainder are native Wisconsin species. Among the western species are *Agropyron smithii, Iva xanthifolia, Helianthus petiolaris, Triplasis purpurea,* and *Ratibida columnaris.*

Floristic analyses of each of the types of weed communities show a rather unusual uniformity, with one-half of the species contributed by the same families in each community. Typical values are shown by the railroad weeds, where the *Compositae* had 18.6 per cent, the *Gramineae* 14.8 per cent, the *Cruciferae* 7.7 per cent, the *Polygonaceae* 6.6 per cent, and the *Leguminosae* 4.9 per cent of the total species. In addition to the crucifers and polygonums, the chenopods, amaranths, mints, and caryophs all attained unusual levels of prominence. With respect to phenology, the weed communities show an approximately even division between spring, summer, and autumn. The spring levels are possible in spite of the high percentage of annuals since many of these are actually winter annuals which germinate in the previous autumn and are large enough to flower early in the following growing season. *Draba reptans, Thlaspi arvense,* and some other crucifers are among the earliest plants to flower, commonly appearing in early April.

An examination of the combined data for the 129 weed communities of the P.E.L. study, grouped into three southern and two northern types which were based on soil plus the railroad communities, reveals a number of truly ubiquitous species which reach high levels of presence in all types. One of these of considerable economic interest is the lesser ragweed *(Ambrosia artemisiifolia)*. This is found in suitable sites in all counties of the state, and has presence percentages of 80 or more in all but the northern sandy soils. The lower counts of atmospheric ragweed pollen noted in the north are due to the low percentage of the land in a disturbed state, not to any inability of the plant to grow there. In fact, in suitable places in the far northern counties, plant densities per unit area exceed anything to be found in the south. A number of other, less common species usually associated with southern communities also show a relatively high presence in disturbed sites in the north; these include fleabane (*Erigeron strigosus*), gumweed (*Grindelia squarrosa*), carpetweed (*Mollugo verticillata*), evening primrose (*Oenothera biennis*), and black-eyed Susan (*Rudbeckia hirta*). The major ubiquitous weed species are quackgrass (*Agropyron repens*), ragweed (*Ambrosia artemisiifolia*), milkweed (*Asclepias syriaca*), lamb's quarters (*Chenopodium album*), horseweed (*Conyza canadensis*), wild lettuce (*Lactuca scariola*), sweet clover (*Melilotus albus*), timothy (*Phleum pratense*), plantain (*Plantago major*), bluegrass (*Poa pratensis*), black bindweed (*Polygonum convolvulus*), foxtail (*Setaria glauca*), dandelion (*Taraxacum officinale*), red clover (*Trifolium pratense*), and white clover (*Trifolium repens*), all of which occur in every type with a presence of 20 per cent or more, but usually over 50 per cent.

Among the species of more restricted occurrence, it is possible to select groups of relatively common species which are particularly characteristic of certain sites. Thus, a group of nitrophilous species, frequently found in moist places high in nutrients, would include burdock (*Arctium minus*), marihuana (*Cannabis sativa*), crabgrass (*Digitaria sanguinalis*), the introduced strain of goosegrass, not the native ecotype found in mesic southern forests (*Galium aparine*), motherwort (*Leonurus cardiaca*), cheeses (*Malva neglecta*), pineapple weed (*Matricaria matricarioides*), catnip (*Nepeta cataria*), purslane (*Portulaca oleracea*), black nightshade (*Solanum nigrum*), and nettle (*Urtica dioica*). The same species occur in similar habitats in Europe (Marthaler, 1937; Tüxen, 1950).

An indicator group for the upland loessial soils of the prairie-forest province would include butterprint (*Abutilon theophrasti*), whorled milkweed (*Asclepias verticillata*), shepherd's purse (*Capsella bursa-pastoris*), chickory (*Cichorium intybus*), yellow oxalis (*Oxalis stricta*), wild parsnip (*Pastinaca sativa*), and green foxtail (*Setaria viridis*). A typical group of indicators of weed communities on sandy soils throughout the state might include perennial ragweed (*Ambrosia psilostachya*), sandbur (*Cenchrus pauciflorus*), winged pigweed (*Cycloloma atriplicifolia*), cottonweed (*Froelichia floridana*), carpetweed (*Mollugo verticillata*), jointweed (*Polygonella articulata*), silvery cinquefoil (*Potentilla argentea*), sheep sorrel (*Rumex acetosella*), and sticky catchfly (*Silene antirrhina*).

Indicator species for northern weed communities are pearly everlasting (*Anaphalis margaritacea*), ox-eye daisy (*Chrysanthemum leu-*

canthemum), hemp-nettle (*Galeopsis tetrahit*), hawkweed (*Hieracium aurantiacum*), button buttercup (*Ranunculus acris*), bladder campion (*Silene cucubalus*), and sow thistle (*Sonchus arvensis*).

WEED COMMUNITIES OF FORESTED AREAS

The total species complement on degraded sites is naturally very large, since it includes a varying percentage of the original community members plus the weedy species which invaded following the disturbance. Detailed studies of a particular type of disturbance in a particular community are needed for an adequate understanding of the role of the two groups of species. Such studies have been made in Wisconsin only for a few communities under the influence of grazing or mowing. Limited observations of other degrading agencies allow only a tentative discussion of the nature of the invading species.

Forested areas, particularly in southern Wisconsin, are subject to a number of disturbing influences. The most widespread is grazing by dairy cattle, which will be discussed later. Selective logging is another common agent of degradation. The manipulations attendant upon cutting the trees, lopping the branches, skidding the logs to the nearest road, and other operations cause soil disturbance and destruction of understory species, as well as changes in light intensity caused by opening the canopy. In many wood lots, a portable sawmill is set up and the logs sawed to rough lumber right on the spot. Sawdust piles and bark accumulations add to the disturbance of these local areas. Examination of wood lots which had recently undergone a heavy harvest revealed that a variety of weed species were capable of invading the disturbed forests, but that relatively few were able to reach high population levels or to maintain themselves for many years. The soils exposed by the logging operations tend to be rich in organic matter and nutrients while the light intensities are practically never as high as in a truly open site. The successful weeds, therefore, must be tolerant of rich substrates and shade. In general, the nitrophilous weeds of open communities furnish the greatest number of forest weeds, including *Arctium minus, Galium aparine, Leonurus cardiaca, Glechoma hederacea,* and *Urtica dioca.* In larger openings or around sawmill sites, the mullein (*Verbascum thapsus*) may reach very high density levels, especially in northern forests. Other species common in disturbed forests are *Bidens frondosa, Chenopodium album, Erigeron annuus, Oxalis europaea, Plantago major, Prunella vulgaris, Isanthus brachiatus, Polygonum aviculare,* and *Solanum nigrum.*

The degree of persistence of the forest weeds appears to be influenced greatly by the nature of the forest, although detailed successional studies have not been made in Wisconsin. Pioneer stands on either dry uplands or wet floodplains tend to hold the weeds longer than near-climax stands on mesic sites. In the latter, several native members of the community quickly take over the openings, especially touch-me-not (*Impatiens pallida*) and wood nettle (*Laportea canadensis*). In less mesic areas, the weeds frequently give way to a greatly increased shrub layer in which blackberries and raspberries are prominent members. A number of the weeds, however, are able to persist indefinitely if some soil disturbance is continued. This is particularly noticeable along footpaths, which frequently are lined with *Arctium*, *Leonurus*, and *Nepeta*, even though a continuous canopy is present overhead. In mesic forests, the light intensities are too low for these species, or indeed for any exotic weeds. Severe disturbance of the soil in maple-basswood forests is not followed by weed invasion, unless there is a correlated increase in light intensity. Apparently no weed, either native or exotic, has a sufficiently great shade tolerance to permit existence under a mesic forest canopy in Wisconsin.

Several introduced woody plants, both trees and shrubs, are able to establish themselves in disturbed hardwood forests and persist for a number of years. Among the trees, the most frequent are box elder (*Acer negundo*), black locust (*Robinia pseudoacacia*), Norway maple, (*A. platanoides*), mulberry (*Morus alba*), and common apple (*Pyrus malus*), while the shrubs include the Asiatic honeysuckles (*Lonicera tatarica* and *L. morrowi*), the buckthorns (*Rhamnus cathartica* and *R. frangula*), and the barberry (*Berberis vulgaris*). The last has been subject to an intensive eradication campaign for years, because it is the alternate host for the fungus causing wheat rust, but it has managed to survive in fair numbers, partly because the disturbances attendant upon eradication were ideally suited to encourage re-establishment. The buckthorn plays a similar role in oat rust disease and has been the object of eradication in recent years, without much success.

PASTURE COMMUNITIES

Prairie pastures

Degradation of a native plant community by grazing is a very common phenomenon in Wisconsin, "America's Dairyland." All communities that are neither too steep nor too wet for the cows may be subject to this type of disturbance. Mesic prairies are so valuable for cropland that they are plowed rather than grazed, so no information is available on the nature of a grazed mesic stand. Wet prairies, however, are commonly used for pasture, as are dry prairies on hillsides and other places not suitable for crop production. Sedge meadows also may be employed for grazing, particularly during dry seasons. On this entire moisture gradient of open grasslands, it appears that the native species are most persistent at the two extremes and most susceptible to replacement toward the mesic center. This is due to the nature of the replacing weeds, few of which are adapted to extreme conditions.

In the case of the sedge meadows and very wet prairies, tussock sedge (*Carex stricta*), bluejoint grass (*Calamagrostis canadensis*), sloughgrass (*Spartina pectinata*), and reed canary grass (*Phalaris arundinacea*) are able to resist grazing to a high degree, although N. C. Fassett was of the opinion that most of the reed canary grass in pastures was a strain introduced from Europe and not the native Wisconsin ecotype. A number of asters and goldenrods are able to persist or even expand on such sites under heavy grazing, largely because of their low palatability. Other native species which respond to grazing by an increased relative density are *Cassia marilandica* (in the Wisconsin river valley), blue lobelia (*Lobelia siphilitica*), sneezeweed (*Helenium autumnale*), vervain (*Verbena hastata*), ironweed (*Vernonia fasciculata*), and buttercup (*Ranunculus fascicularis*).

Prairies on wet-mesic, mesic, and dry-mesic sites are literally wiped out by grazing, sometimes with almost unbelieveable rapidity. About 1940, a large tract of virgin wet-mesic prairie adjacent to the Faville Prairie Preserve of the University of Wisconsin Arboretum, in Jefferson County, changed ownership and was subjected to heavy grazing under cattle and horses by the new management. During the first summer, the animals proceeded to attack the species of highest palatability, with the very abundant plants of *Silphium terebinthinaceum* and *S. laciniatum* sought out like hidden candy at a child's birthday party. By the end of the second year, no prairie species whatever were visible in the closely cropped sward, although many were no doubt present as underground roots or rhizomes. The obvious dominants were redtop grass (*Agrostis stolonifera*) and Kentucky bluegrass (*Poa pratensis*). Other portions of the same prairie which had been pastured for a number of years were dominated by the bluegrass, with lesser amounts of redtop and quackgrass, plus an occasional native plant of ironweed, sneezeweed, common milkweed, and high populations of the exotic dandelion, Canada thistle, and white clover. No traces of the prairie grasses originally present nor most of the prairie forbs were to be seen.

This total destruction of the original prairie dominants by grazing is to be explained by comparing their growth habits with those of the invading exotics. The native grasses produce erect growing points which are clipped off by the cattle (Neiland and Curtis, 1956). The exotic species like bluegrass, however, have creeping rhizomes which send up individual leaves. Removal of the tips of the leaves causes no permanent harm, since the tissues are replaced by the meristematic region at the base of the leaf. The exotics thus have a tremendous advantage in the competition for space.

This advantage decreases in the drier prairies because the exotics have a relatively high moisture demand and a number of the native grasses of such sites have a growth form similar to the exotics. These relations were studied in detail by Dix (1955), who used the paired stand technique in the examination of 24 sites; each site consisted of an ungrazed dry prairie and an adjacent grazed prairie under similar conditions of soil, slope, and exposure (Plate 63). Frequency determinations were made for all species in each pair of stands; 40 quadrats were used in each half of each pair. By computing the average differences in quantity of each plant in the grazed versus ungrazed areas, Dix was able to arrange the species along a grazing gradient from those which were present only in grazed areas to those present only on ungrazed areas. The latter include those species most sensitive to grazing which are eliminated readily from the community by the grazing. Such species are said to be "decreasers" because their densities decline in direct proportion to the intensity or length of grazing time. Species toward the center of the gradient include a number of native forms which at least temporarily benefit by grazing, their populations expanding as the decreasers decrease. Such plants are called "increasers." In the exact center of the grazing order are those native species which apparently are uninfluenced by grazing, their populations remaining essentially equal in both grazed and ungrazed areas. Some of the species which are present only in the grazed stands have entered those stands after grazing began. They are called "invaders," and consist of weed species, both native and exotic.

The decreasers Dix found were mostly forbs, including such species as *Aster laevis, Hypoxis hirsuta, Phlox pilosa, Dodecatheon meadia,* and *Lobelia spicata,* many of which are marginally present on the dry prairies and achieve optimum levels in mesic or wet-mesic areas. A few grasses are included in the decreaser group, such as *Sporobolus heterolepis, Sorghastrum nutans,* and *Panicum leibergii,* again mostly species typical of more mesic areas. Among the indifferent species were *Anemone patens, Scutellaria parvula, Arenaria stricta, Solidago rigida, Artemisia caudata,* and *Solidago nemoralis;* all are typical members of the dry prairie community and all are either non-palatable or else protected by a rosette or cushion growth form. The most significant increaser was side-oats grama (*Bouteloua curtipendula*), which achieved high levels of density on the grazed areas, largely because of its growth habit which resembles that of the bluegrasses. Other increasers included *Erigeron strigosus, Asclepias verticillata, Ambrosia artemisiifolia, Verbena stricta,* and *Hedeoma hispida;* all have pronounced weedy tendencies and all are present in ordinary open weed communities of cultivated and ruderal sites.

The invaders were largely exotics, although such native species as *Conyza canadensis, Oxalis stricta,* and *Artemisia frigida* were present. The two most prominent invading grasses were quackgrass (*Agropyron repens*) and redtop (*Agrostis stolonifera*), but neither reached high densities. The two bluegrasses (*Poa compressa* and *P. pratensis*) were important only under light grazing or on sites with especially favorable moisture relations. Most of the invaders were prostrate forbs, either rosettes as the dandelion and plantain, or creeping species with rhizomes or stolons, as the silvery cinquefoil and white clover.

The average composition of lightly and heavily grazed dry prairie pastures, as found by Dix, is given in Table 25. Only the species with a value above 0.15 on the arbitrary density index are included. The most abundant species in the heavily grazed stands are the two bluegrasses; redtop, and side-oats grama and the forbs, dandelion, whorled milkweed, and white clover. The lightly grazed areas contained much higher amounts of side-oats grama, about the same of bluegrass, and a ten-fold increase in little bluestem (*Andropogon scoparius*). The forbs included significant amounts of the original prairie species.

This study by Dix shows in a graphic way the possible complexities of weed communities in degraded habitats. The composition of the disturbed prairie community is related to the intensity of the disturbing factor and the environmental nature of the site. The exotic species are able to enter readily and in abundance under mesic conditions most similar to those in their Eurasian countries of origin but invade with difficulty and only after severe disturbance under very dry or very wet conditions at the margins of their hereditary tolerance.

Table 25

Average composition of lightly and heavily grazed dry prairies

Data from Dix (1955). Values are in special density index units. Only species with index of 0.15 or more in grazed stands are included. Table arranged in order from decreasers to invaders. Exotic species are marked with an asterisk.

Species	Ungrazed	Lightly grazed	Heavily grazed
Aster sericeus	.34	.15	.08
Liatris aspera	.28	.28	.00
Ratibida pinnata	.22	.15	.06
Panicum perlongum	.15	.16	.02
Andropogon scoparius	1.02	.70	.07
Aster azureus	.20	.18	.06
Kuhnia eupatorioides	.20	.11	.24
Solidago rigida	.11	.16	.08
Artemisia caudata	.09	.20	.04
Solidago nemoralis	.27	.43	.29
Bouteloua curtipendula	.80	1.27	.54
Artemisia frigida	.00	.21	.00
Asclepias verticillata	.52	1.42	.90
Poa pratensis*	.55	.87	1.02
Antennaria neglecta	.04	.03	.16
Poa compressa*	.36	1.04	.66
Ambrosia artemisiifolia	.07	.16	.31
Verbena stricta	.05	.16	.30
Euphorbia maculata*	.02	.04	.36
Agrostis stolonifera*	.01	.01	.62
Oxalis stricta	.00	.01	.29
Taraxacum officinale*	.00	.01	.89
Trifolium repens*	.00	.00	.52

Pastures on cleared forest land

Many pastures in Wisconsin occur on old forest lands, on steep or rocky sites, or other locations of low value for crop production. These are particularly numerous in central and northern counties but can be found throughout the state. The only preparation of the land is a removal of most or all of the trees, with no plowing or other manipulation of the soil and no artificial seeding or fertilization. These unimproved pastures are not very productive from an agronomic viewpoint but they sometimes represent the best use of land not suited for anything else. As in other degraded sites, the composition of the plant community on these cleared woodland pastures varies with location and with intensity of grazing. Only a thorough study like that of Dix can unravel all of these variables. An initial step in this direction was made by the P.E.L. presence study of a number of stands scattered throughout the state. All were heavily grazed and very open, with less than 10 per cent canopy coverage by the relic trees. The lists were made in open places, not under the direct influence of these trees. For ease in comparison, a list of common species, prepared in the usual way, is given in Table XXI-7, although it should be realized that this is based on only 12 stands; hence it is an extremely tentative approximation of a complex community. In spite of the diversity of the original forests involved, a surprising number of species reach high levels of presence, thereby giving the community an index of homogeneity of 49.0 per cent. Every single species on the list can be considered an invader, since none would have been present in the original forests. Some are native species, but the great majority are exotics, with 75.4 per cent of the total sum of presence contributed by this group. The 100 per cent presence value for Kentucky bluegrass is a good indication of the important role played by this species, which is almost the sole dominant in most pastures, with relative density figures in excess of 80 per cent in many areas. Other species with high density as well as high presence are *Taraxacum officinale, Trifolium repens,* and *Agropyron repens.*

A number of the original forest species persist in the grazed pastures, but not in sufficient frequency to be classed with the prevalent group. Prominent among these in some areas are mayapple (*Podophyllum peltatum*), windflower (*Anemone quinquefolia*), hazelnut (*Corylus americana*), strawberry (*Fragaria virginiana*), bracken fern (*Pteridium aquilinum*), and crowfoot (*Ranunculus abortivus*); the last three often act as increasers.

When grazing is exceptionally heavy, or when the bluegrass sod is attacked by June beetle larvae, the pastures may be composed largely of tall forbs of which the mullein (*Verbascum thapsus*) is most common, but with such other coarse species as *Verbena urticifolia, Cirsium vulgare, Conyza canadensis, Cannabis sativa,* and *Ambrosia trifida* forming an almost impenetrable jungle-like thicket 4 to 7 feet tall. The forage production by such an assemblage is almost nil and the chances of natural recovery without suitable renovation practices are equally small. Such a field is a sure indicator of a shiftless and irresponsible farmer.

A number of species reach best development in grazed forests on certain kinds of soils and may be considered as indicators. Thus the viper's bugloss or blue devil (*Echium vulgare*) is most common on the lacustrine clays near Lake Michigan and in the bed of glacial Lake Oshkosh, where it is frequently accompanied by such other members of the borage family as species of *Cynoglossum, Lappula, Symphytum, Myosotis,* and *Lithospermum,* all exotics from Eurasia. Other species on heavy soil pastures include *Chrysanthemum leucanthemum, Artemisia absinthium, Cichorium intybus, Hemerocallis fulva,* and *Tragopogon pratensis.* It is of some interest that Baker (1937) found *Chrysanthemum leucanthemum* to be restricted to mowing meadows and absent from adjacent grazed pastures in his famous study of certain commons near Oxford, England, which had been continuously grazed or mown since 1085 A.D. It is possible that the Wisconsin strains represent a different ecotype. Sandy soil pastures are characterized by *Monarda punctata, Verbena stricta, Plantago purshii, Polygonum tenue, Potentilla argentea, Silene antirrhina, Danthonia spicata, Rumex acetosella,* and *Oenothera rhombipetala.* Pastures made from lowland hardwood stands have many of the same species found in sedge meadow or wet prairie pastures; these include especially *Ambrosia trifida, Helenium autumnale, Lobelia siphilitica, Verbena hastata, Vernonia fasciculata, Agrostis stolonifera,* and *Rumex crispus. Cassia marilandica, Amorpha fruticosa,* and *Acer negundo* are frequent woody invader members of such places, particularly in the Wisconsin River valley, where the river birch (*Betula nigra*) may also be present.

Pastures in closed forests

In many cases, forest lands are used as pasture with no preparation whatsoever. The philosophy behind the practice seems to be that cows enjoy a shady woods on a hot summer's day and that this contentment will outweigh the lack of food in the woods. Usually the pastured forest is adjacent to an open pasture, either improved or not, so the cows can forage and rest in the shade. In any case, the practice is very widespread, with 80 per cent or more of the wood lots in all southern counties so employed. The weed communities of these pastured forests differ from those of cleared woodlands largely because of the decreased light intensity and its resultant selection of shade tolerant species (Plate 64). The major difference is the great decrease in importance of bluegrass, which is highly intolerant of shade. In fact, no grasses achieve dominance in most of the stands. Rather, there is a mixture of invading forbs and remnants of the original understory. The invaders are similar to the invaders of wood lots degraded by lumbering activities or other disturbances, but more species of low or decumbent growth form are important in the grazed areas, including such species as chickweed (*Cerastium vulgatum*), plantain (*Plantago major*), dandelion (*Taraxacum officinale*), and white clover (*Trifolium repens*).

A number of the rather delicate spring ephemerals of the mesic hardwood forests are able to withstand an amazingly heavy amount of grazing. These include such species as spring-beauty (*Claytonia virginica*), Dutchman's-breeches (*Dicentra cucullaria*), and troutlily (*Erythronium albidum*). The explanation is that they have completed their life cycle before the cows are turned into the woods for the summer season and can persist through the grazing period as rootless corms or bulbs deeply buried in the ground, and hence they are protected from actual damage by the hooves of the cattle.

MOWING MEADOWS

Still another type of degradation of native communities is brought about by the practice of mowing for hay. The native grasslands usually so employed are the fens, wet prairies, and sedge meadows. As indicated in the discussion of those communities in earlier chapters, some of the stands of wet prairie and fen studied had been mowed at intervals in the past; hence a comparison of the mowed versus unmowed condition is difficult or impossible to make at this time. Obviously mowing would reduce the competitive abilities of some species which are cut down at the height of their flowering period, but the extent of this damage is unknown. Sedge meadows are also used as mowing fields, particularly the less hydric stands. Examination of a number of meadows which had been regularly employed in this way revealed only a few invaders. These included *Agrostis stolonifera, Sonchus arvensis, Phleum pratense, Cerastium vulgatum,* and *Poa pratensis*. Other invaders of sedge meadows, listed by Costello (1936) for the area around Milwaukee, include *Echinochloa crusgalli, Rumex crispus, Pastinaca sativa, Solanum dulcamara, Plantago major,* and *Linaria vulgaris,* although some of these may have been favored by grazing rather than mowing.

SUMMARY OF WEED COMMUNITIES

In reviewing the rather heterogeneous communities that develop on disturbed sites, it is possible to discern certain general patterns.

Plate 63.—Paired stands of grazed (at right) and ungrazed dry prairie (at left). Conspicuous forbs in grazed area are gumweed (*Grindelia squarrosa*) and prairie goldenrod (*Solidago rigida*). Green County.

Plate 64.—Grazed and ungrazed stands of southern dry forest, showing nearly complete suppression of shrubs and saplings under grazing. Dane County.

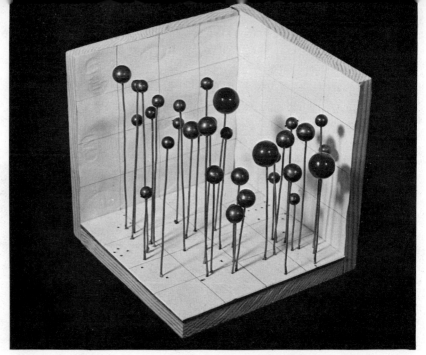

Plate 65.—Model of three-dimensional ordination of stands in southern upland forest. The balls represent the amounts of red oak (*Quercus borealis*) in each stand, with the size of the ball proportional to the basal area per acre.

Plate 66.—Dry-mesic prairie at Midway Prairie Scientific Area, La Crosse County, showing typical location of current prairie remnants on the strip between a railroad and a paralleling highway.

In degraded areas, the degree of replacement of the original components is largely proportional to the amount of disturbance. When the degradation proceeds to the point of total destruction, the area is invaded by plants with pioneer tendencies, that are able to germinate and grow under conditions of variable moisture and nutrient supply. The nature of the new community is influenced by the environmental features of the area and by the type and frequency of repetition of the disturbing action.

In Wisconsin, very wet sites are invaded or held largely by the native species normally present in such areas. Very dry places are similarly occupied by native species, although there may be a large adventive element from the drier areas of the states to the west and south. The mesic sites, either forests or grasslands, are likely to be invaded by a mixture of native weeds and exotic weeds, with the exotics largely coming from areas of similar soil and climate in Europe and Asia where high human populations and consequent soil disturbances have existed for thousands of years and where actual selection and evolution of weedy ecotypes of various Eurasian species have occurred (Anderson, 1952). This exotic element is highest in ruderal areas and in cultivated areas closely associated with cities and towns, such as vegetable gardens and other areas rich in available nitrogen. Not all of the species in these intensively disturbed areas are of Eurasian origin, as a number are American species of such naturally disturbed sites as river banks (*Phytolacca decandra*—see Sauer—1952) or those possibly related to ancient Indian agricultural practices as the amaranths (Sauer, 1950) or giant ragweed. Other native weeds include species normally present on sand dunes, lake beaches, or other areas constantly subject to natural soil disturbance. The net result of this mixture of native and exotic species is a series of weed communities differing more or less from those to be found in other regions.

Few quantitative, phytosociological investigations of American weed communities have been reported, but from the abundant state lists of weeds and other general treatments it can easily be seen that the compositions vary greatly with location. Thus in weed lists from the Dakotas and the Prairie Provinces of Canada, the western American element is more strongly represented than it is in Wisconsin. There is also an increased segment from the steppes and deserts of Russia and Asia Minor and a decreased amount of north European species. Towards the east from Wisconsin, the dry element, both native and exotic, decreases, with a corresponding increase in European species, particularly maritime forms and members of the crucifer family. Toward the south and southwest, an increase is noted in annuals of Mediterranean origin plus an increased representation of species from the dry regions of Mexico and other tropical American countries. Thus the composition of weed communities is seen to be a local phenomenon, which changes with increasing distance in the same manner as the composition of conifer swamps, oak forests, or other native communities.

Part 7

The vegetation
as a whole

CHAPTER 22

Postglacial history

TYPES OF EVIDENCE AVAILABLE
Macrofossils
The best evidence concerning the past history of vegetation is obtained by the study of fossil remains of that vegetation, preserved in such a way as to show sequential changes with time. These remnants may be macrofossils, such as logs, cones, seeds, or leaves; or microfossils, largely pollen grains, trichomes, or stone cells. Macrofossil deposits have been found in Wisconsin in many places, frequently in the operation of well-drilling, but the geological nature of the deposit in which they occur is usually in doubt. In a few cases, however, extensive deposits have been found in unequivocal positions with respect to their glacial history. The most famous of these is the Two Creeks Forest Bed, located near the junction of Manitowoc and Kewaunee counties in the bluffs at the shore of Lake Michigan. The deposit was first de-

scribed by Goldthwait in 1907 and has been studied at frequent intervals thereafter by Thwaites (1953), Wilson (1932) Culberson (1955c), and others. The so-called forest bed consists of well-preserved portions of tree logs, as well as many mosses, with inclusions of pollen grains, protozoans, and molluscs. The assemblage was formed sometime between the withdrawal of the Cary and Mankato substages of the Wisconsin glaciation and the subsequent advance of the Valders substage and was contemporaneous with the Alleröd period in European glacial chronologies (Flint and Deevey, 1951).

A large forest of presumably the same general age covered an area of at least 500 square miles in the bed of Glacial Lake Oshkosh in Winnebago County and adjoining areas (Lawson, 1902). Numerous logs from the forest have been exposed in wells, road cuts, and clay pits, but no other types of plant remains have been identified. No other geologically-defined macrofossil deposits have been studied in Wisconsin, but investigations in Minnesota (Cooper and Foot, 1932, Rosendahl, 1948), Iowa, and Illinois can be used to give inferential information about Wisconsin.

Microfossils

The most widespread type of postglacial fossil and the one most studied is that of plant pollen as preserved in peat beds and other sedimentary deposits. The pollen grains of many plants are transported by wind action and many have adaptations such as balloon-like wings to aid in this movement. In addition, most pollen has a membranous coat of unknown composition on the outer surface which makes it exceedingly resistant to decomposition. As a result of these two factors, we find that pollen from some members of nearby plant communities is moved to lake beds where it settles to the bottom and remains. As the lake bed fills in with organic peat deposits (Chapter 12) and changes to an open bog, conifer forest, or sedge meadow, pollen continues to be deposited on the surface and is gradually buried by the ever-deepening surface layer of mosses or sedges. The result is a layered or stratigraphic deposit, with the oldest layers on the bottom and the youngest on top. Examination of the pollen content in successive layers thus can give information about the vegetational changes, if any, that have occurred in the region adjacent to the deposit. This is usually done by removing a small core from the deepest portion of the basin with a special instrument in such a manner that samples of known depth throughout the full thickness of the deposit can be segregated in an uncontaminated condition for detailed laboratory study. Suitable methods are used for recovering the pollen grains, which are then identified by microscopic examination. The results are usually expressed as relative density.

Dating of vegetational changes

Like the macroscopic fossil evidence, the microfossil peat deposits must be of known geological age before they can be properly interpreted. In most cases, the only such knowledge available is the age of the glacial till in which the lake basin was formed. No direct geological correlations with later events are possible in ordinary situations and even the starting date is uncertain since there is no way of knowing when the lake began to fill in. For example, in Vilas County, Crystal Lake has almost no organic bottom deposits, although other basins within one-quarter of a mile have peat beds over 30 feet thick which support dense conifer swamps. To state, therefore, that a particular peat bed is on Cary till merely means that it could not be older than the date of recession of that glacial substage.

The exact age in years of any vegetational change shown by microfossil examination has been a very difficult problem to solve until very recently. The best method available gave the age of the various glacial substages by means of an examination of the depth of leaching in the surface soils on the drift. By this means, Robinson (1950) attempted to date the Cary and Valders tills in Wisconsin at 20,500 and 8,700 years, respectively. The possible variables affecting the rate of leaching, however, make this method very uncertain.

In 1949, Libby and his associates at the University of Chicago developed a method for determining the age of organic materials by measuring their content of radioactive carbon. This element is produced in the atmosphere by cosmic ray bombardment and is present in a small but constant ratio to ordinary carbon. When plants take in carbon dioxide for photosynthesis they acquire radioactive carbon in this same small proportion and lock it up in the form of cellulose or other complex organic compounds. If the plants happen to be preserved for a long time, as in peat deposits or forest beds, then a gradual reduction in the radioactive carbon content takes place, since the half-life of this element is slightly over 5000 years. By comparison of the ratio of radioactive to non-radioactive carbon in the fossil with that in contemporary materials, following isolation and purification of the carbon, it is easily possible to determine the age of the fossil. When the content drops to a certain level, the determination of the ratio becomes difficult and inaccurate. At present, this level is equivalent to 40,000 years, but it may be pushed further back as improvements in instrumentation are made. In any case, it is possible with Libby's method to determine the age of any fossil material at least from the beginning of the Wisconsin stage of glaciation down to the present.

With the radiocarbon method, it should be possible to date each peat layer in which a pollen analysis is made and thus give a complete historical account of vegetational changes. No such complete determinations have been made on Wisconsin material. The Two Creeks forest bed has been dated at 11,400 ± 350 years; thus it sets the interval between the Mankato and Valders substages. A few peat samples from pollen profiles have been dated in adjoining states, mostly in post-Valders time. Extrapolation is rendered very difficult, however, because of the possible non-simultaneity of vegetational changes in remote areas.

At present, therefore, we know the general order of magnitude of the age of each of the major glacial substages, we know the sequence of vegetational changes following each glacial advance, but we do not know the dates at which the significant vegetational changes occurred, except by inference from scanty records in other areas.

PROBLEMS OF POLLEN ANALYSIS

Before presenting a summary of the information which bears on the postglacial vegetational history of Wisconsin, it may be well to point out some of the difficulties inherent in the technique of fossil pollen analysis, since these difficulties may influence the interpretations to be made from the data (Cain, 1944).

One of the most important of these difficulties is that the composition of the pollen rain on the surface of a bog is not proportional to the composition of the surrounding forest (Fagerlind, 1952). This is due to great differences in the quantity of pollen produced by different species of trees, especially the tremendous differences between wind-pollinated and insect-pollinated species, and to differences in effective distance of dispersal as affected by specific gravity and other properties of the pollen. Added to this distortion in the original deposit is another differential factor, that of preservation. There is evidence that pollen from different kinds of trees is subject to differential rates of decay. Differences in palatability of the entrapped pollen grains to mites and other small animals also lead to changes in the proportional representation. The relative density of fossil pollen at any depth, therefore, is not directly related to the forest composition on the adjacent uplands at the time of its deposition but must be adjusted to take the above variables into account.

Such adjustment of data has been the custom in Europe (Faegri and Iverson, 1950), but in the case of Wisconsin investigations, at least, all workers have presented the original figures only and have based their conclusions on them. Fortunately, several studies of the contemporary deposit on bog surfaces have been made in areas where the existing forest cover is known (Kluender, 1934; Fogelberg, 1935;

Table 26

Actual composition of Vilas County forest versus surface pollen composition

Species	% composition of living forest	% composition of surface pollen record	Species	% composition of living forest	% composition of surface pollen record
Abies balsamea	1.2	4	Pinus strobus	3.0	30
Acer saccharum	11.6	1	Quercus borealis	0.5	0
A. spicatum	1.0	0	Thuja occidentalis	9.5	0
Betula lutea	19.8	29	Tilia americana	5.5	1
Fraxinus americana	1.2	0	Tsuga canadensis	44.0	21
Picea glauca	2.0	10	Ulmus americana	0.5	2

and others). In one notable example, Potzger (1942) examined the surface deposit on a small bog located in the heart of an extensive area of virgin forest in Vilas County. The percentage composition of the forest is well known from the studies of Stearns (1951) and the P.E.L. investigations. In Table 26, the actual values are compared with those of the pollen examination. As can be seen, the insect-pollinated species like sugar maple (*Acer saccharum*), and basswood (*Tilia americana*) are greatly underrepresented in the pollen, whereas the spruce (*Picea*) and pine (*Pinus*) are greatly overrepresented. In the case of pine, Fogelberg found a representation of 5 per cent on the surface of Hope Lake bog in Jefferson County, although the nearest pine forests are over sixty miles away.

By studying all examples of this surface pollen versus actual forest composition, it is possible to arrive at certain rough correction factors to be used in adjusting the raw data. These factors for Wisconsin are shown in Table 27 with comparable values for Denmark as given by Faegri and Iverson. The use of the correction factors does not necessarily give the true forest composition, since differential transport and differential preservation are not taken into account. It is believed, however, that the results are much closer to the truth than the original values. One of the great difficulties encountered in the re-calculation of the various analyses was a result of the unfortunate habit of many pollen workers to present their raw data in the form of small-scale bar diagrams rather than as tabulated numbers. Pictorial presentation is an admirable method of reporting trends, general conclusions, and other results of the analysis of data, but is to be deprecated as a means of transmitting the actual data.

Table 27

Correction factors for pollen amounts

Genus	Wisconsin	Europe*	Genus	Wisconsin	Europe*
Betula	×0.67	×0.125	Tsuga	×3.0	....
Pinus	×0.143	×0.143	Mixed Hardwoods		
Quercus	×2.0	×0.5	(Acer, Tilia, Os-		
Picea	×0.33	×1.0	trya, Ulmus, Jug-		
Abies	×1.0	×1.0	lans)	×5.0	....
Fagus	×1.0	×2.5			

* After Faegri and Iversen.

In Europe, pollen analysis has reached levels of precision that appear to minimize the difficulties discussed above. In part, this is due to major botanical differences between the flora of Europe and North America. Over much of the glaciated region of Europe there is only one species in each of the main genera of trees, such as *Pinus, Picea,* and *Tilia.* As a result, pollen grains identified as to genus are at the same time also identified as to species. In North America, the situation is very different since each of the genera listed has many species, each with different ecological requirements. Since it is usually not possible to separate the species of *Pinus, Quercus, Acer, Tilia, Betula, Populus, Salix,* or *Corylus* from their pollen alone, the results on the generic level may be of somewhat limited value in interpretation of ecological conditions that existed when the pollen was deposited. The genus *Quercus* in Wisconsin alone, for example, contains swamp trees demanding high soil moisture like *Q. bicolor,* xeric trees of open sand plains and dry cliffs like *Q. ellipsoidalis,* and mesic trees of relatively high shade tolerance like *Q. borealis*. The genus *Betula* varies from *B. pumila* of cold peat bogs, through *B. nigra* of southern river swamps, *B. papyrifera* of disturbed open places either wet or dry, to *B. lutea* of dense, mesic forests.

A thorough consideration of all types of tree pollen reported from Wisconsin profiles in the light of an extensive knowledge of their behavior in existing forests leads to the following conclusions regarding their use in interpreting past events.

The genus *Quercus* may be used to indicate a continental climate when it reaches a high relative density. All such forests now present in Wisconsin are in Borchert's continental climate belt or the adjoining transition belt. It is true that *Q. borealis* ranges throughout the entire state, but it does not now reach high densities north of the tension zone; hence it probably never did in the past when competing with the same species it now meets in that region.

The genus *Pinus* may be used to indicate the Type I climate of Borchert when it reaches high densities. When both oak and pine are present at intermediate levels, the transitional climatic belt is probable although not certain.

Since all species of *Quercus* and all species of *Pinus* are today successionally subordinate to the shade tolerant trees like *Acer saccharum*, *Tilia americana*, *Tsuga canadensis*, and *Betula lutea* throughout Wisconsin, it must be assumed that long continued dominance by either or both in the pollen record therefore indicates the continued presence of disturbance by fire or some other agent which would prevent the succession from proceeding to completion.

Both species of spruce (*Picea glauca* and *P. mariana*) are nearly confined to the Type I climatic zone today and may be used to indicate that zone even when present in relatively small proportions. The genus *Abies* includes only a single species, balsam fir (*A. balsamea*). This species grows over a wide variety of soil and light conditions but reaches high densities of mature trees only in the far north, almost always in company with *P. glauca*. Therefore, it seems reasonable to combine the data for spruce and fir and to interpret the moderate presence of the combination as an indication of Borchert's Type I climate.

The genus *Betula* as represented in the Wisconsin pollen record is of limited value as an indicator, since no specific identifications were reported. For the same reason, the existing record of *Salix*, *Populus*, and *Corylus* is of little value. Pollen of *Larix* is mostly derived from a local conifer swamp on the surface of the bog. Such swamps are now found throughout Wisconsin and tell nothing of the surrounding conditions.

The trees composing the terminal upland forests throughout the state belong to the genera *Acer, Fagus, Tilia, Ostrya, Carpinus, Tsuga, Juglans,* and *Ulmus.* Although *Acer saccharinum* and *Ulmus americana* in the south are intermediate trees along river bottoms, they rarely would contribute much to the pollen record. On the other hand, *U. americana* is a widespread though minor constituent of the terminal forests on the northern uplands. The whole group, then, may be combined into one unit of mixed hardwoods for most purposes. Since the group now occurs throughout Wisconsin, no great climatic significance can be attached to its presence in the pollen record. Like *Quercus* and *Pinus,* however, it can be interpreted in terms of disturbance. In this case, the group indicates lack of disturbance, and long dominance by this group means long continued absence of disturbance.

Although *Tsuga* is included in the mixed hardwoods group, its records can sometimes be considered separately with profit. It is perhaps the most characteristic member of the current northern hardwood forest and its presence can be used to indicate the occurrence of that forest type. Since data of its occurrence were not recorded from all bogs, however, a more useful indicator of the northern forest is a combination of pine and mixed hardwoods.

In the case of grass pollen, the situation is even worse than for the trees, since not even genera can be distinguished from the pollen alone. It is therefore necessary to combine everything under a single grass family record. At first glance, this would appear to remove grass records from consideration, since grasses of some sort occur in every plant community in the state. However, the fact that the records from many bogs show a marked peak in grass percentages (with values well over 20 per cent of the total) at a certain level in the profile indicates that a community much richer in grasses than the average must have occurred at that time. The logical candidate, of course, is the prairie which is dominated by grasses. Since all of the records also show tree pollen during the grass peak, it seems best to interpret this grass period as one of savannas, rather than one of pure prairie. In all probability, the vegetation during a grass peak consisted of a mixture of prairie, savanna, and closed forest, each on an appropriate site. Since these grass peaks exceeded in magnitude the percentages found near the tops of the profiles, it may be concluded that the proportions of prairie and savanna formerly were greater than they were at the time of European settlement, and therefore, that they indicate Borchert's Type IV or continental climate.

Many other kinds of pollen, such as bog ericads, cattails, certain composites, and sedges may be disregarded because the most reasonable source for their origin is the surface of the bog itself, rather than the surrounding uplands. Detailed analyses of specific herbaceous species which are characteristic of upland communities would be of great value, but they have not been presented for Wisconsin material.

Taking into account the above considerations, all available pollen analyses for Wisconsin have been treated as follows. The number of pollen grains per hundred seen were determined for spruce-fir, pine, oak, and mixed hardwoods. The values were then put on a relative basis, corrected for proper representation by the factors in Table 27 and recalculated to percentages. The values for grass pollen were used as originally given, since no basis for correction is known. All other types of pollen were disregarded. From the adjusted values, new profile diagrams were drawn, and all interpretations were based upon the apparent vegetational changes shown by these profiles.

In spite of the many problems discussed above, considerable confidence in pollen analysis results is gained from the fact that local groups of bogs show rather high uniformity and that bogs which are more remote show progressive changes along understandable gradients. In fact, this reinforcing factor which comes from comparing results of many analyses affords the best reason for devoting serious consideration to pollen analysis as a means of studying vegetational history.

POSTGLACIAL VEGETATIONAL CHANGES IN WISCONSIN

The locations of the bogs which provided evidence for much of the following discussion are indicated on the figure reproduced at the opening of this chapter and also in more detail as Figure XXII-1 in the Appendix. In the Vilas County region a number of other profiles are available, but they do not differ significantly from those shown on the map. Elsewhere in the state, several other bogs have been studied but they have not been used because either the information reported is obviously incomplete or the geological age of the deposit is doubtful. Typical pollen profiles of several of the bogs are shown in Figure 21. The values have been corrected by the factors given in Table 27.

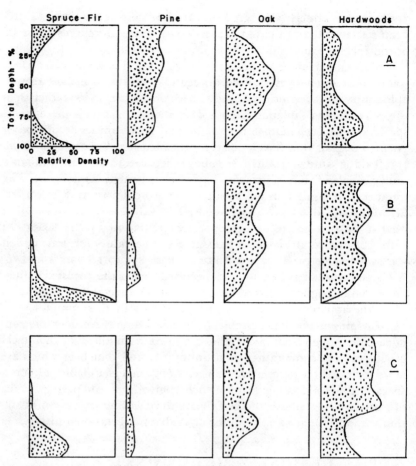

Figure 21.—Typical pollen profiles of Wisconsin bogs. The values for each pollen group have been adjusted by the factors given in Table 27 and are shown on a relative basis. The depth scale (on the ordinates) is arranged with the uppermost layers on the top and deepest layers at the bottom. A, Hayward Bog, Sawyer Co. (Voss, 1934). B, Baraboo Bog, Sauk Co. (Hansen, 1937). C, Hub City Bog, Richland Co. (Hansen, 1939).

Vegetation prior to the Wisconsin stage of glaciation

No direct evidence exists concerning the vegetation prior to the first major glaciation of the Pleistocene, the Nebraskan stage, nor do we have knowledge concerning how much of the state was affected by the Nebraskan ice. The interval between the first and second glaciations, the Aftonian interglacial period, has given a record of its vegetation in Iowa and Minnesota. By inference, the same general sequence probably occurred in Wisconsin. Wilson and Kosanke (1940) showed that the vegetation changed from a boreal forest of spruce and fir at the beginning, to a mixed pine-oak forest with much grass, to mixed hardwoods, and back to spruce-fir in a peat deposit in Iowa. The peat layer was only 6 inches thick; hence it cannot represent the full interglacial period of over 200,000 years. Many of the same tree species were present in another exposure in Minnesota (Nielson, 1935). The Yarmouth interglacial period between the second and third major ice sheets (Kansan and Illinoian) has been studied in Illinois by Voss (1939). He found spruce, fir, and pine, as well as hardwoods. The Sangamon interglacial, between the third and fourth sheets (Illinoian and Wisconsin), was also studied by Voss (1939) in Illinois, and by Lane (1941) in Iowa. The sequence in Illinois was from spruce and fir, to pine, to oak, and to beech, whereas in Iowa the change was from spruce and fir, to pine, to oak, and to grass. Alden (1918) reported numerous encounters with peat, logs, muck, and other organic materials in well-drillings throughout the area of Farmdale glaciation in southern Wisconsin which presumably date to this Sangamon period. Unfortunately, no botanical examination to determine the species that composed the deposits was made.

A consideration of this scant evidence leads to the conclusion that Wisconsin and neighboring states were fully clothed with vegetation during each Pleistocene interglacial interval and that the plant communities which occurred were similar to those now present, at least in so far as their major trees are concerned. The time span involved is variously estimated at 500,000 to 700,000 years, although it may actually be considerably shorter. The Illinoian, or third stage, is now assumed to have been in existence between 130,000 and 85,000 years ago while the Wisconsin, or fourth stage, began about 25,000 years ago and is still in its waning phases (Karlstrom, 1956).

The first substages of the Wisconsin glaciation, the Farmdale (until 24,000 years ago), the Iowan (until 17,000 years ago) and the Tazewell (until 15,000 years ago) have left no known records in Wisconsin on the nature of the vegetation. The intervals between them were very short, but studies in Iowa and Illinois (Voss, 1937) show the presence of spruce, fir, pine, hemlock, and yew (*Taxus*). All but the spruce are currently in existence in the same regions as relic colonies on north-facing cliffs. The Tazewell-Cary interval was likewise very short, since the Cary ice was in existence from 14,000 to 13,000 years ago.

Most of the direct evidence for postglacial vegetational changes in Wisconsin comes from bogs formed in depressions in the Cary and Valders tills. An additional important source of information is derived from bogs formed in stream valleys of the Driftless Area. These bogs originated by the ponding of the streams due to dams of glacial outwash deposited in the Wisconsin and Mississippi basins (numbers 4, 5, and 6 on Figure XXII-1). As studied by Hansen (1939), these bogs give us the best picture of vegetation immediately preceding the maximum advance of the Cary ice. They show a forest of mixed hardwoods, with some oak and pine and a large amount of spruce and fir; the latter two increased in the layers immediately above the bottom and suggested the approach of the Cary ice 30 to 60 miles away. If this spruce and fir pollen was of local origin, then the vegetation in the Driftless Area may have been similar to that found in Douglas and Bayfield counties today. An alternative explanation is that a boreal forest of spruce and fir was growing on top of the nearby ice sheet and the pollen in the peat was derived from this source through the downdraft winds from the main glacial surface. In any case, the large representation of mixed hardwoods with oak clearly rules out a true boreal climate for the Driftless Area.

The Cary ice made many temporary halts followed by readvances. On the southwest side of the city of Madison, for example, a recessional moraine (the Milton Moraine) covered a peat bed and an extensively developed soil profile that had been produced in the interval between maximum advance of the glacier to Verona some 10 miles away (the Johnstown Moraine) and its readvance to this moraine. Similar buried soils are to be found at many places in the state. These resurgences make the exact dating of any peat deposit a very difficult matter, except by direct radiocarbon analysis. The assumption is made in the following discussion that the initial deposits in Cary age bogs are about 12,000 years old.

The vegetation of Wisconsin approximately 12,000 years ago is indicated by the bar diagrams on Figure 22, which show the adjusted percentages of spruce-fir, pine, oak, and mixed hardwood for all bogs in existence at that time and for which complete profiles are available. In view of the prevalence of a spruce-fir maximum throughout the state, it is assumed that any bogs which do not show spruce and fir (numbers 1 and 2, Figure XXII-1) were incomplete. The record is quite clear in its indication of a forest cover of mixed conifers and hardwoods, with spruce and fir partially or totally dominant throughout Wisconsin. There appears to be no way of ascertaining whether this boreal forest was established on the uplands surrounding the bogs or whether it was present on the surface of the nearby glacier.

The Cary and Mankato ice (see Chapter 2) had receded completely out of Wisconsin and at least as far north as the Straits of Mackinac by 11,500 years ago or earlier. The high waters of Early Lake Chicago were lowered as the Mackinac outlet became uncovered. Exposed areas of Cary till with a surface of lacustrine varved clays

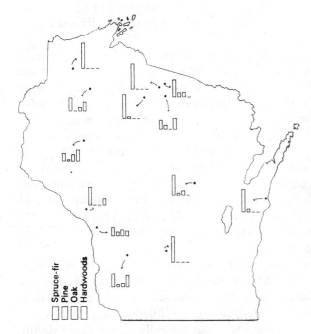

Figure 22.—Forest composition in early post-Cary time, *ca.* 12,000 years ago, as shown by relative fossil pollen content of four major tree groups. Note uniformly high values for spruce-fir pollen. The bars in this and the next two figures represent the percentage content of the tree groups indicated in the legend at a constant time interval in the past.

acquired a layer of sand near the shore line at Two Creeks in Manitowoc County during this process and the moist sand surfaces became invaded by a forest of spruce (Wilson, 1932), which lasted at least 120 years and probably longer. In any event, a spruce forest was growing on the site when the lake water began to rise following the advance of the Valders substage. These high waters flooded the forest and deposited a layer of sand and silt. The spruce trees were killed by the inundation and were knocked over by storms and winter ice formation. Their remains, together with mosses and molluscs on the forest floor, were preserved on the lake bottom and were eventually covered over by till of the Valders ice, some time prior to 11,000 years ago.

A similar situation occurred on the bed of Early Glacial Lake Oshkosh in Winnebago County. There an extensive area of forest developed with spruce, pine, black ash, oak, and basswood as members of the communities. This forest bed (Lawson, 1902), like that at Two Creeks, was inundated by ponded water caused by the advance of the Valders ice and later buried by till or lacustrine deposits from that ice sheet. In both of these cases, the indicated vegetation could be duplicated today within a few miles to the northward of the deposits, so the climate just prior to Valders time was but little different from that existing now.

None of the bogs on the Cary till remote from the outer limits of the Valders ice show any indication of a temporary increase in cold by means of a spruce increase such as might be expected if the readvancing ice were exerting a refrigerating effect on the surrounding areas. This is true in the Waupaca County bog, which is only a few miles from the Valders front and in the Vilas County bogs which are equally close to the Valders ice, as well as in the Sauk and Chippewa County bogs which were 50 to 70 miles away. One possible explanation for this is that the entire area was dominated by a boreal-hardwood forest throughout the period of Cary and Mankato recession, of Two Creeks interval, and of Valders advance and retreat. This seems highly doubtful, but the question will remain unsettled until radiocarbon analyses give us the true picture.

The Valders ice was entirely gone from Wisconsin by 9000 years ago although it was still influencing the high level stage of the Great Lakes known as Glacial Lake Algonquin. The vegetational changes which followed this final glacial retreat differed in southern and northern Wisconsin. The southern bogs typically show a rapid decline in spruce-fir dominance which is accompanied by an increase in importance of mixed hardwoods and sometimes of pine. More or less uniformly, this is followed by a decrease in mixed hardwoods and an increase in oak. Data for grass pollen percentages are available for six of these southern bogs and they uniformly show an increase concomitant with the rise in oak. These changes seem to be a clear indication of a shift in climate in the direction of increased continentality. This time has been called the postglacial "Xerothermic" period by Sears (1942b). Evidences for it are very widespread throughout North America and Europe. Independent dating places the peak of the Xerothermic at about 3500 years ago or around 1600 B.C. On the assumption that the peak in oak and grass and the trough in mixed hardwoods shown in the southern profiles actually were simultaneous in occurrence, Figure 23 has been drawn to show the relative composition of the vegetation of Wisconsin as it appeared about 3500 years ago. The southern bogs are seen to be remarkably uniform, with dominance by oaks or mixed hardwoods and with negligible amounts of conifers. The high grass percentages justify the conclusion that this portion of the state had a vegetation of mixed savanna and forest, probably with extensive areas of interspersed prairie. The fact that

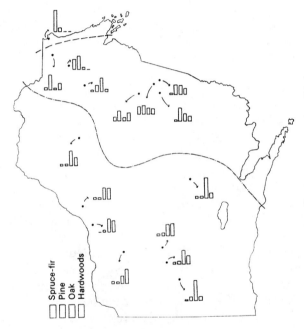

Figure 23.—Forest composition during Xerothermic period, *ca.* 3500 years ago. Note high values for oak and hardwoods south of the prairie-forest border (dot-dash line), of pine north of that border, and of spruce-fir near Lake Superior, north of the boreal forest border (dash line).

the Waupaca and Chippewa (Nos. 7 and 8, Figure XXII-1) county bogs were similar means that the tension zone between the prairie-forest and the northern forest province had moved slightly northward from its present position.

North of the tension zone, the bogs show a rather uniform pattern, which is quite different from that in the south. Here pine achieved a position of dominance or of co-dominance with oak. Both spruce-fir and mixed hardwoods are of reduced importance. A typical profile shows the same rapid decrease of spruce-fir seen in the southern bogs. This is followed by an increase in pine and mixed hardwoods, then a decrease in both with an increase in oak. The oak maximum is much lower in intensity than that in the south, but the assumption is made in Figure 23 that it was simultaneous in the two areas, with a peak at 3500 years ago. The oak peak is most definite in northwestern Wisconsin, in Sawyer and Douglas counties. In fact, oak occurs in the Douglas County profiles only at this time.

In extreme northwest Douglas County (Bog 17, Figure XXII-1), the spruce-fir maximum was maintained throughout the profile, which here dates only from the recession of Lake Algonquin about 7900 years ago. On Figure 23, therefore, the Lake Superior shore is shown in the boreal-hardwood belt, which covered the entire state in Figure 22. The remainder of the north is assumed to be northern hardwoods of the type now to be found along the southern edge of Borchert's climatic Type I or just north of the tension zone. The presence of significant quantities of oak and the frequent occurrence of hickory and walnut throughout the area support this assumption. The shifts in vegetation types in the Xerothermic period in Wisconsin, therefore, can be summarized as a northward lateral shift of continental climates to regions 40 to 60 miles north of their present limits.

It is of considerable interest that this persistence of the tension zone throughout the past 10,000 years has also been reported in Michigan by Potzger (1948) and in New England by Krause and Kent (1944). In fact, if the border of the prairie-forest province in Wisconsin, as shown in Figure 23, is extended across Lake Michigan, it becomes continuous with the tension zone in Michigan which Potzger reported as beginning at the latitude of Muskegon. In all these areas, the Xerothermic period brought about a regression from terminal mesic forests to pioneer forests requiring disturbance for their maintenance, from hardwoods to pine and oak north of the zone, and from hardwoods to oak and savanna south of the zone. In addition, prairies were able to spread and increase their area in southern Wisconsin and southern Michigan. The hypothetical location of the prairie-forest province boundary shown in Figure 23 was arrived at from a consideration of small concentrations of prairie plants which exist today at a few locations in Oconto, Brown, and Manitowoc countries, all of which are remote from the nearest true prairie relic.

In the period since the Xerothermic, a southward shift of the climatic belts has occurred. In the southernmost bogs, no change is apparent in the available data, nor do the profiles differ significantly from the currently existing vegetation. This is shown in Figure 24, which consists of compositional diagrams of subsurface conditions, as determined for the level down from the surface one-tenth of the entire depth of the profile in each bog. The Chippewa and Waupaca county

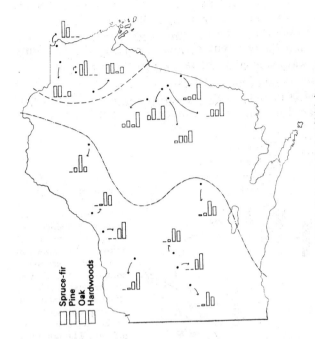

Figure 24.—Forest composition in late postglacial times, *ca.* 500–1000 years ago. Note resurgence of spruce-fir in the northwest and increase in hardwoods in the north central area. Prairie-forest border shown by dot-dash line, boreal forest border by dash line.

bogs show an increase in pine but maintain a dominance of oak and hardwoods, which is consistent with their present day occurrence within the tension zone. The northwestern bogs record a substantial rise in spruce and fir, while the north central bogs show a decrease in oak and a general increase in mixed hardwoods.

The entire picture of post-Cary vegetational change is summarized in Figure 25, which diagrammatically shows the entire profile for each bog according to four vegetational categories. For this purpose, subjective criteria were used to assess the nature of the vegetation; they were based on arbitrary density limits for various pollen groups. The picture would not have been seriously changed had other reasonable limits been used instead. Rather, the relative amount of the profiles occupied by a given community would have been increased or decreased. The vegetation was called boreal-hardwoods if spruce and fir made up 40 per cent or more of the total adjusted pollen density. Such forests today are confined to the regions near Lake Superior and northern Lake Michigan. The conifer-hardwoods type was delimited by a total pollen percentage of pine, spruce, or fir in excess of 20 per cent. Similarly the prairie-forest type of southern hardwoods was judged by a total of oak and mixed hardwoods in excess of 80 per cent. It is possible that certain extensive areas of northern hardwoods in the climax condition would be misjudged by this criterion, since they contain less than 20 per cent conifers (see bog No. 9, Figure XXII-1). The area was assumed to be savanna if the non-adjusted figure for grass pollen exceeded 20 per cent. On this basis, Figure 25 shows at a glance that the boreal-hardwood type was state-wide 12,000 years ago. This was replaced by conifer hardwoods in most stations, but the

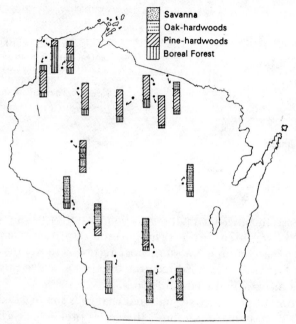

Figure 25.—Summary of postglacial vegetational changes, as shown by typical profiles classified according to the community types shown in the legend as determined by the criteria in the text. Each bar diagram represents an entire profile, with the oldest deposits on the bottom. The dominant pollen type at each level is shown by the coded symbols. Note many stations indicate only two communities, but some have three or four.

conifer-hardwood stage persisted north of the tension zone, while south of the zone it was replaced by prairie-forest hardwoods. Evidence of savanna is present in four of the southern bogs at some place in the profile, assumed to have been about 3500 years ago. In extreme northwestern Wisconsin, spruce-fir remained prominent throughout the profile, either as a continuous dominant or with two or three periods of such dominance.

Briefly summarized, therefore, it can be said that the vegetation of Wisconsin has undergone little or no change in postglacial times in the region near Lake Superior (and probably in Door County on Lake Michigan as well) and that the changes have become greater toward the south, with the major change (following the early boreal forest) one of retrogression from terminal hardwood forests to some type of disturbance forest. The tension zone separating major forest regions today has apparently been operative throughout postglacial times, with a maximum northward displacement of 40 to 60 miles during the Xerothermic period some 3500 years ago. North of the zone, the disturbance forest was pine and oak; south of it, the forest was oak or savanna. The retrogression was due to an increased continentality of the climate which increased the possibility of forest destruction by fire. These fires were probably the result of man (Chapter 23). If not, it is necessary to postulate an increased incidence of dry lightning in the period of increased continentality. In the last several thousand years, there has been a reversal of trend; the movement has been either in the direction of a more moist climate or a decreased incidence of fire or both and has resulted in more stable forests which have an increasing content of terminal hardwoods except in extreme northwestern Wisconsin.

CHAPTER 23

The effect of man on the vegetation

INDIAN EFFECTS

Nature of archaeological evidence

Our knowledge of the possible influence of early Indian populations on the vegetation of Wisconsin is based entirely upon archaeological evidence concerning their cultural habits that has been deduced from tools and other artifacts that have been found and have been associated with their habitations. The archaeologist gathers this evidence by painstaking excavations of old village sites, burial grounds, refuse piles, cave shelters, and other places used by the Indians, and to a lesser extent from artifacts discovered in plowed fields and other places not necessarily connected with a population center. Tools, such as spear points, arrowheads, and other weapons used in the hunt, and fishhooks or harpoon points used for aquatic animals, tell of food-gathering habits. Pottery pieces, both cooking and storage vessels, imply a fairly sedentary population and are usually associated with an agricultural economy. Corncobs, smoking pipes, and other remains give some indication of the kinds of crops raised. Conch shells from Florida, obsidian from the Rocky Mountains, and copper from Isle Royale, found together in a Wisconsin village site, indicate a well-developed trade system and probably a more-or-less stable political system. Traces of palisades around villages, and house plans reconstructed from the patterns of post impressions, can be used to interpret the probable population size of the villages. Charcoal from the fireplaces of the village can be used to date the time of occupancy, using the technique of analysis for radioactive carbon.

Another means employed to learn the time sequence for various cultures is based upon the occurrence of artifacts in stratified layers, as in the debris on the floor of cave habitations, where the oldest materials are on the bottom and the youngest toward the top. A source of strength of the stratigraphic method is the inherent conservatism of all peoples, so much so that a particular culture tends to make one or a few special kinds of arrowheads or pots, or tends to mark their artifacts with certain distinctive symbols, and continues to make the same objects, rather than to experiment with new patterns as the whims of the artisan might dictate. Because of this, it is possible to cross-date the evidence from several archaeological sites, on the valid assumption that two or more areas belong to the same culture if they have all or a majority of their artifacts of the same design or pattern.

Using all of these approaches as well as other more complicated methods, Wisconsin archaeologists have been able to piece together a coherent outline of the Indian cultures of the state, with their distribution in time and space. Only very tentative estimates of population density are possible at this time, except for a few special instances. The available evidence indicates that all parts of the state have been occupied continuously at least since the retreat of the last glaciers, and that populations may have reached high levels at several places.

Prehistoric Indian cultures

The prehistoric Indian culture periods of Wisconsin can be divided for discussion purposes into five segments. The earliest is the Paleo-Indian period. This was a time of very primitive cultures of hunting and food-gathering peoples which is best known from sites in Texas and southwestern United States. Recent radiocarbon dates place the beginnings of the period in Texas at least as early as 37,000 B.C., or well before the Wisconsin stage of the Pleistocene. No datable remains have been found in Wisconsin, but the highly characteristic projectile points of the various facies of the culture have been found here, so it is probably safe to assume that Paleo-Indians were present in Wisconsin during the Cary and Valders glacial substages and perhaps before. In the West, these Indians were associated with mammoths, sloths, camels, horses, and other now-extinct animals and they may well have been hunters of mammoths in Wisconsin.

The second prehistoric period is called the Archaic. In this state the most important culture of the period was the Old Copper culture, so named because of the extensive use of intricate copper tools and ornaments of many forms, which included axes, knives, fishhooks, harpoons, spear points, bracelets, and beads. Such objects are widely distributed, especially in eastern counties. Recent excavations at Osceola Landing, near Potosi in Grant County, and especially at Oconto in Oconto County (Wittry and Ritzenthaler, 1956) have given valuable information as to the habits of this group and, most importantly, have provided a radioactive carbon dating of 5500 B.C. for the Oconto site. Wisconsin Indians were thus using complicated metal tools several thousand years before their counterparts in Egypt and the Tigris Valley began their initial metallurgical trials. The Old Copper peoples apparently were hunters and food-gatherers, since no indications of agriculture have been found. According to the pollen record, they were living in a boreal forest environment in Oconto, but not at Osceola. They may have hunted caribou, elk, whales, and walrus, as well as mammoths, but the total supply of storable foods must have been low and their populations therefore were probably also at a low level. The Archaic period may have lasted from 6000 or 7000 B.C. to around 1500 B.C., with a termination date at the time of the maximum prairie expansion. The new conditions would be expected to bring about changes in culture patterns, either by evolution or by immigration and displacement.

The third prehistoric period is termed the Early Woodland, which may be a misnomer as far as southern Wisconsin is concerned, since the grasslands were a prominent part of the landscape. There is some indication of a shift from nomadic hunting tribes to more sedentary groups in this period, as clay pottery of special design is one of the best indicators of the Early Woodland cultures. The period is thought to have lasted until about 500 B.C.

The fourth period is particularly well marked in the southern and western counties, where new cultures were developed. The approximate dates assigned are from 500 B.C. to 1000 A.D. The Early Woodland types may have persisted in the north and east for longer periods. The most significant feature of this fourth or Middle Woodland period is the extensive introduction of agriculture. The main cultures of the period are the Hopewell and the Effigy Mound Cultures. The first was concentrated along the Mississippi, along Lake Michigan, and in the southern counties. The Hopewellians were a people of high skill in arts and crafts. They made beautiful objects in chalcedony, obsidian, and jasper; these included ceremonial knives and intricate tobacco pipes of polished stone with shell inlays. They raised corn, beans, tobacco, and squash; they understood the art of textile weaving; and they made beads of hammered silver and copper, pearls, and shells. Such a cultural development undoubtedly was based on a highly organized division of labor and could have occurred only under rather high population levels with strong political and religious controls.

The Effigy Mound Builders apparently existed contemporaneously with the Hopewellians. The most famous artifact of these Woodland Indians was the ceremonial or totem mound, built as very lifelike effigies of birds, turtles, snakes, bison, and other animals. This culture attained its acme in Wisconsin, which has literally thousands of these effigies. They are particularly numerous on hill crests overlooking lakes, streams, or springs throughout the southern counties. The mounds were built from hand-carried basket-loads of earth. Their size and their internal structure indicate that they were constructed on open land, free of trees. Whether the grounds were specially cleared for them or whether they were built in prairies is not known with certainty, but the latter is more probable. The Effigy Mound Builders combined hunting and agriculture; they probably migrated back and forth with the seasons according to the availability of game and plant foods (Baerreis, 1953).

The last prehistoric period is termed the Mississippi period. It lasted from about 1000 A.D. to about 1600 A.D. The three cultures known for the period are, the Middle Mississippi, the Upper Mississippi, and the Late Woodland. The latter may well have been a continuous development from the Early and Middle Woodland groups, since agriculture was little practiced, the main economy being one based on hunting and gathering. The Upper Mississippi people were most abundant along the Mississippi river and also in the region around Lake Winnebago. They are thought to have lived in large, permanent villages; they supported themselves by agriculture, hunting, and fishing. They made garden beds of raised mounds of earth, often in intricate curving patterns of parallel rows. Their use of metal was less than that of the Hopewelliāns and their artistic abilities were also less well developed.

The Middle Mississippi culture was represented by a single village, that of the famous Aztalan in Jefferson County (Barrett, 1933). This was apparently an outpost of a high density population of highly civilized Indians centering around the junction of the Ohio and Mississippi rivers. Aztalan was a stockaded village more than 20 acres in size, surrounded by a plastered wall of logs. It included many houses, several temple mounds, and other structures. The Middle Mississippi people were primarily agricultural, although their food supplies were supplemented by hunting and fishing. Metal was used sparingly, but well-designed articles of stone and clay were abundant.

Villages of very similar type were seen in the South by DeSoto in 1541 but they had completely disappeared when Joliet explored the Mississippi, a century later. Some archaeologists are of the opinon that the Spaniards introduced measles and smallpox which completely wiped out the susceptible Indians in their compact village units. The Woodland Indians, in smaller, isolated bands, largely escaped.

Historical Indian cultures

At the time of the arrival of the first white man in Wisconsin, in 1634, there were only three tribes of Indians known to have been living in the state (Ritzenthaler, 1953). These were the Winnebago, who were descendants of the eastern branch of the earlier Upper Mississippi culture; the Menomini, who may have been related to the Late Woodland groups; and the Santee Dakota, who were represented by the Clam River peoples of the Late Woodland cultures.

Tribal wars, aided by guns from the French and the Dutch, caused profound territorial shifts in the Algonkian and Iroquoian groups in eastern states in the seventeenth century. One result was the migration of many eastern tribes into Wisconsin. Included in these newcomers were the Mascouten, Potawotomi, Sauk, Fox, and Chippewa tribes. All but the Mascouten were essentially Woodland types, although agriculture was important in the economy of some of them, particularly those in the southern counties. During the eighteenth century, under French, British, and finally American rule, the various tribes suffered great fluctuations in population levels and often considerable changes in the location of their hunting grounds. Their traditional hunting methods, techniques for food preservation, and use of native herbs for medicinal purposes were all inherited from a former woodland existence, and necessarily were forced to change radically in the new prairie and savanna environment. Some tribes were better able than others to make the necessary adjustments.

The period of important influence of the Indian on the landscape came to an end with the defeat of Chief Black Hawk in 1832. Thereafter, the Indian was confined to small reservations, mostly without adequate supervision or proper regard for resource conservation. The main exception was the Menominee Reservation in Shawano County, where the tribe persevered through great handicaps and finally settled on a sustained harvest method of forest cropping that is one of the finest examples of proper land management in the United States.

Influence of Indians on vegetation

There were five main ways by which the Indians, both prehistoric and modern, influenced the nature of the vegetation in which they lived. These are mostly related to their methods for obtaining food. The hunting tribes could affect the plant communities directly through the use of fire and indirectly through their influence on populations of large mammals. The food-gathering activities usually associated with hunting cultures could have been of direct importance in influencing populations of some of the species harvested. The agricultural tribes exerted a direct destructive effect on the vegetation of the areas used as fields or gardens. The fifth influence was that of plant introduction, both intentional and accidental.

Of these five methods, obviously the first, or fire, was by far the most important, and greatly overshadowed the combined influences of the other four. With this one tool, the Indians changed a very large portion of the entire vegetational complex of Wisconsin. As we have seen in earlier chapters, the oak openings, sand, oak, and pine barrens, bracken-grasslands, true prairies, brush prairies, fens, sedge meadows, shrub communities, and pine forests all owe their origin or maintenance to the repeated presence of fire. To a limited extent, the results could have been obtained by lightning fires, but the known incidence of dry lightning in Wisconsin is totally incapable of explaining the huge areas influenced by fire. It is possible that such storms were more frequent in the past, particularly during the extension of influence of the continental westerlies at the time of prairie expansion, but it is equally probable that during the preceding period of conifer dominance such storms were even less frequent than now. In any case, the presence of nomadic hunting tribes throughout the state in the entire postglacial period means that man-made fires were an important if not the sole cause of the fires.

In nomadic cultures, the hunting grounds were thought to be communal property, so no efforts were made to put out fires except around village sites. Even though they had been possessed of a strong desire for fire control, however, the primitive peoples had no methods for effective fire fighting. Even in modern mechanized societies, no really efficient ways exist for stopping crown fires in tinder-dry forests or the roaring line of a prairie fire under a strong wind. Given the right combination of a severe drought, low humidity, high temperature, and strong wind, there is no reason to doubt that a fire starting in Grant County might sweep all the way to Lake Michigan. It would be stopped only in local areas by lakes, steep leeward cliffs, and the broader stretches of the Rock River. Fires of several hundred thousand acres were commonplace on the western prairies. Forest fires were generally more limited, but such famous conflagrations as the Peshtigo fire of 1871 in Wisconsin and the Tillamook burn of 1933 in the State of Oregon covered hundreds of square miles. The Peshtigo fire swept through Oconto, Marinette, Brown, Door, Shawano, Kewaunee, and parts of Manitowoc and Outagamie counties. Because the influence of a fire was expressed far beyond the locus of original ignition, it is not necessary to postulate high densities in the prehistoric Indian populations in order to account for widespread vegetative effects. A few widely scattered tribes could start enough intentional and accidental fires to keep all but the most protected sites in a retrogressive condition.

Hunting was employed in some degree by all Indian cultures, but agricultural influences were later and more definitely localized in southern counties. A favorite site for cornfields was the floodplain forest, where the trees were killed by girdling and where flood deposits kept the soils constantly supplied with adequate nutrients. Some of the Indian cornfields were surprisingly large, up to several hundred acres in size, but the total land surface so employed was probably negligible except in relation to the river bottom community. The crops grown by the Indians were mainly tropical varieties of Central or South American origin, so there was little chance for escape of the crops to surrounding communities. No clear indications of the nature of the weeds in the Indian fields is available, but they well may have included giant ragweed, various chenopods, and amaranths. There is some evidence that the ragweed was actually encouraged as both a fiber and a food crop.

The native plants used as food varied with the culture, but berries, including the hackberry (*Celtis occidentalis*), and nuts such as walnut (*Juglans*), hickory (*Carya*), acorns (*Quercus*), and hazelnut (*Corylus*) were favorite sources in many cases. The act of gathering scarcely would influence the plant population in these species, since no permanent harm to the plants resulted. The only instance where Indian gathering may have had an adverse effect was the so-called Pomme de Prairie (*Psoralea esculenta*), a prairie legume much prized for its fleshy taproot. A determined search for this species may well have reduced its populations to very low levels, in much the same way as later herb-gathering efforts of the white man nearly exterminated the root-producing ginseng (*Panax quinquefolium*) and goldenseal (*Hydrastis canadensis*).

The final influence of the Indian on the vegetation concerns his actions in plant introduction. No conclusive proof exists that any Indian ever introduced any species into Wisconsin, but circumstantial evidence both here and elsewhere strongly supports the contention that such actions did take place. Day (1953) records a number of instances which seem to be explicable in this way. Among the Wisconsin species which have been described as cultivated by Indians in one locality or another are Canada plum (*Prunus nigra*), white gentian (*Gentiana flavida*), wild leek (*Allium tricoccum*), sweetflag (*Acorus calamus*), and groundnut (*Apios americana*). One interesting example is the Kentucky coffeetree (*Gymnocladus dioica*). The large, hard seeds of this species were used in a sort of dice game by various tribes. As a result, they were carried about when the tribe moved its headquarters, many becoming lost in the vicinity of the villages. At present, the species has a very local distribution in Wisconsin, with each locality at or near the site of an Indian village. Hedrick (1933) reported the same type of distribution for New York State. Other species thought to be introduced into Wisconsin by the Indians include the red mulberry (*Morus rubra*), the giant mallow (*Hibiscus militaris*), and the lotus (*Nelumbo lutea*).

Thus the vegetation of Wisconsin before 1600 was largely a resultant of the physical factors of soil, topography, and climate and the anthropogenic factors produced by a stable or slowly changing series of Indian populations. Much of the land area was directly affected by the fires set by the Indians, but the effect of these fires varied with different soils and macro-climates and may have varied as the climates changed in postglacial times. The only important communities to escape these Indian influences were the mesic forests of both north and south. The frequency of fire and the subsequent degree of response was less on the wet and wet-mesic forest sites than on the dry locations, but it was possible for both to retrogress to open communities of grass or sedge, after which each was about equally susceptible to further fire. A rough estimate based on the vegetation map of Wisconsin would indicate that 47 to 50 per cent of the land surface was directly influenced by this anthropogenic activity in the years preceding settlement by Europeans.

There is no way to ascertain what the vegetation of Wisconsin would have been had there been no Indians. It is probable that the bulk of the land would have been occupied by hardwood forests, but such speculations are as idle as guessing what would have happened had there been no sugar maples, white oaks, or big bluestem grass. All that is important is that a particular geographical region, with a particular geological and climatic history, supported a series of biotic communities, composed of plants and animals, one of the latter having evolved to the point where it could influence the others out of all proportion to its numbers. A quasi-equilibrium was set up amongst these factors, such that the plants and lesser animals best adapted to the imposed conditions were in positions of local dominance, arranged in communities of differing composition largely as a result of local soil and topographic factors.

POSTSETTLEMENT CHANGES

Cessation of fire

The biotic balance was subject to slow change in postglacial times as the climatic patterns shifted and as one Indian culture was gradually replaced by another. With the coming of the white man, however, the tempo of change was accelerated, slowly during the first two centuries and with almost incredible speed during the next century. The shifting tribes of the 1600's and 1700's probably caused a considerable change in impact on various community types. The best evidence for this is seen in the savanna and prairie region, where lands spoken of as barren and treeless by the earliest explorers became covered with brush or young forests when the major settlements were made after 1830. The vegetational records provided by the governmental land survey in the years from 1830 to 1860 reflect the changes that had occurred in the preceding 200 years under the influence of unstable and varied Indian populations, but they do not properly indicate the prehistoric conditions. On the other hand, settlement of the land by Europeans for agricultural or other purposes was largely a postsurvey operation, since title to the lands was granted only after they had been suitably surveyed and recorded. The vegetation as deduced from survey information, therefore, can be thought of as an intermediate stage in the transition from a prehistoric equilibrium between Indians and the land to the modern balance between white men and the land. The following discussion is largely concerned with the changes during the past century, roughly after 1850.

In the early years of settlement, the most important vegetational effects were caused by the elimination of fire, the major Indian agent of control. This was accomplished not only by the removal of the Indians, but by positive efforts at fire prevention by the new occupants. Their small and widely distributed private holdings were guarded against accidental conflagrations and were further protected by the interpolation of roads and plowed fields. As described in earlier chapters, rapid changes took place in the newly protected communities, often with a nearly complete change in species composition. This was particularly true in the brush prairies, pine barrens, and oak savannas, where a closed forest canopy quickly eliminated most of the light-demanding prairie species of the original groundlayer. Most of the current stands of oak forest in the southern counties and the few remaining white and red pine forests in the north date back to this change in land treatment, with trees remarkably close to one hundred years old on the average.

Agricultural utilization

The initial changes due to fire protection were largely of a temporary nature in the southern two-thirds of the state, since the landscape was being rapidly converted to farm lands by the increasing tide of immigrant settlers. The population of Wisconsin increased from 3,245 in 1830, to 305,391 in 1850, and to 1,315,497 by 1880. In the same intervals, the acreage devoted to crop production increased from about 400,000, to 2,900,000, to 15,300,000. The crop acreage leveled off in 1910 at about 21,000,000 acres and has remained relatively constant since then (Ebling *et al.*, 1948). The distribution of this acreage is indicated on the map at the opening of this chapter, which shows the percentage of forest land not in crops in each county. The majority of the cropland is in the prairie-forest province.

The first settlers preferred the forest to the prairie because they were suspicious of the treeless lands, were unable to plow them properly with the tools then available, and were generally short of the capital necessary to build the essential buildings with purchased lumber. This situation changed fairly rapidly when the true richness of the prairie began to be realized and the trend then shifted to settlement of the prairies and savannas in preference to the heavy forests. There was a nearly complete occupation of the prairies by 1880 and the southern savannas were taken up almost as thoroughly. In the savanna country, however, there was a tendency to allow the rougher sections of the farm to develop in forest, as a source of fuel and building materials. One result was the widespread occurrence of farm wood lots, 10 to 60 acres in size on the average, composed of a dense growth of oaks and associated species on lands that were formerly open brushland or widely spaced savanna. Thus it happened that by 1950, after a century of intense utilization, there was more closed forest in many southern Wisconsin counties than there had been at the time settlement began 120 years earlier.

Some impression of the rate of conversion of native communities to farm land can be obtained from the interesting studies of Shriner and Copeland (1904). These investigators presented a map showing the land cover of four townships in Green County as it appeared in 1831, 1882, and 1902. By adding information from the *Wisconsin Land Economic Inventory of 1935* and from aerial photographs of these townships in 1956, it is possible to gain a clear picture of the pace and extent of deforestation. The four townships originally contained 16.8 per cent prairie and 83.2 per cent forest and oak opening. Using the arbitrary criterion of a 50-foot or greater spacing between trees as an indication of openings, the surveyors' records can be interpreted to show that 38.2 per cent of the wooded land was in savanna and 61.7 per cent in closed forest. By 1850 all of the prairie land had been entered in farms, with an almost complete destruction of the original community. By 1882, the savanna area had been reduced to 34 per cent of its original extent and the dense forest to 29 per cent. In Jordan Township, which was 86 per cent oak opening, 12 per cent oak forest, and 2 per cent prairie in 1831, the wooded areas of both types were reduced to an average of 33.6 per cent by 1882, but the remaining acreage contained more wood than the original, due to the conversion of savanna to forest. Tentative estimates, based on known tree densities and calculated basal areas as determined from known ages, indicate that this increase amounted to 20 per cent in terms of total cordwood.

By 1902, the savanna lands had been reduced to 10.2 per cent and the forested lands to 8.8 per cent, with a further lowering by 1935, to 4.7 per cent and 4.3 per cent and by 1956, to 3.2 per cent and 3.6 per cent respectively. Field inspection revealed that 77 per cent of the modern wooded area was heavily pastured in 1956, so that only about 0.2 per cent of the original landscape remained in a semi-natural state.

On the basis of abundant botanical collecting in the period from 1870 to 1900, Shriner and Copeland pointed out that only five species of plants had disappeared from the region in spite of the great reduction in acreage of the original communities. Of these five, two did not really disappear because they have been seen since 1950. In the absence of exact knowledge as to the makeup of the presettlement flora, we cannot determine how many species actually have been exterminated. It appears highly probable that the missing species, if any, would be prairie plants, since the forest remnants of the area appear to contain a full complement of the species to be expected in such conditions. The prairie flora is retaining a precarious foothold along railroad rights of way and may not be greatly reduced in total species, although the densities and hence population sizes are but a tiny fraction of the original.

Other changes brought about by this rapid conversion of the land to farms are numerous. One of the obvious effects is the decreasing size of the remnant wood lots; hence their edge effect is greatly increased. This was shown for Cadiz Township in Green County (Curtis, 1956) where the average wood lot decreased from 91.3 acres in 1882, to 14.3 acres in 1950, but where the average periphery per wooded acre increased from 82 feet to 280 feet in the same period. This increase in relative amount of edge means an increasing degree of xerophytism in the forests, since light penetration and wind movement are both increased. Plants and animals benefiting from these conditions can expand their populations at the expense of the more mesophytic organisms normally present in large tracts of forest. Other changes in the small wood lots may result from their relative isolation which prevents the easy exchange of members from one stand to another.

Various accidental happenings in any given stand over a period of years may eliminate one or more species from the community. Such a local catastrophe under natural conditions would be quickly healed by migration of new individuals from adjacent unaffected areas. In the isolated stands, however, opportunities for inward migration are small or non-existent. As a result, the stands gradually become depauperate in total species and those remaining achieve unusual positions of relative abundance (Curtis, 1956).

The white man removed the major agent of community control employed by the Indians, but he substituted the axe, the plow, and the cow as equally effective controls. He was also much more efficient than the Indian in the matter of plant introduction. As seen in Chapter 21, a whole new set of weed species was brought in and allowed to attain high population levels on the croplands and waste places of farms and cities. Some of the plants used as crops, unlike those of the Indian, were able to escape and become established in a variety of situations. The major species of this type are forage plants like Kentucky bluegrass, redtop, timothy, and other grasses, and white sweet clover, white clover, and other legumes, all adapted to northern climates and continued disturbance. Other crop plants which have become widely distributed are asparagus, the common apple, marihuana, horseradish, buckwheat, parsnip, and carrot.

The agricultural regions of Wisconsin may be regarded as a modified savanna, with the trees growing singly, in small open clumps, or dense groves, usually with an understory of exotic pasture grasses, or rarely with remnants of native forest flora. The spaces between the trees are occupied by a highly artificial and unstable grassland of rather constant composition in which the grasses, corn, oats, and brome are mixed with exotic legumes and a variety of weeds. All are subject to frequent replacement, one by the other, in a crop rotation system. Other open areas are used for pasture and are dominated by exotic grasses and other low herbs tolerant of grazing. The major portion of the area is at once more simple and more uniform than the original land cover, with the varied dominances of the latter now shared by only a small number of species. The remnants of the presettlement communities are individually less complex than they were, but their total contribution to the current regional vegetation confers an over-all floristic richness and complexity exceeding that originally present. New communities have been formed which contain a portion of the native components plus a portion of adventive species. These new communities have rarely reached equilibrium conditions, since they are mostly characteristic of pioneer sites. The indications are clear that only the climax mesic forests have the ability of ever regaining their initial composition under full protection, with all other communities certain to be permanently different, under any foreseeable future treatment.

Timber harvest regions

In the agricultural areas of Wisconsin, the trees of the original forests were looked upon as obstacles to the successful conversion of the land for farms and were destroyed by cutting and burning as rapidly as the energy and resources of the settlers would permit. In the northern portion of the state, however, and in other areas not suited for crop production, the trees were regarded as a resource to be harvested. The harvest began in the 1840's along the Wisconsin, Chippewa, Wolf, and other northern rivers and had reached an output of one billion board-feet or over 8 million solid cubic meters per year by 1869 (Ebling *et al.*, 1948). This increased to a peak of 3.4 billion board-feet by 1899, when Wisconsin was the leading lumber producing state in the nation. The output decreased rapidly thereafter as the supplies became exhausted, falling to about 300 million board-feet by 1934, and fluctuating between 300 and 500 million since then.

One or more fires swept through the slashings on most of the lumbered areas. These fires had very different effects from the earlier fires of the aborigines, since their intensity was heightened by the accumulation of tops and branches that were left on the ground by the loggers, and especially because the prior cutting had removed most or all of the seed trees which ordinarily would have persisted in protected places. The combination of a fire-degraded soil and lack of pioneer conifers like red and white pine meant that the successions following lumbering were very different from those which occurred originally. Huge areas became covered by pin cherry, aspen, and white birch, rather than young pines.

Filibert Roth discussed this situation in northern Wisconsin as seen in 1898:

During forty years of lumbering nearly the entire territory has been logged over. The pine has disappeared from most of the mixed forests and the greater portion of pineries proper has been cut. There is today hardly a township in this large area where no logging has been done. In addition to this, the fires, following all logging operations or starting on new clearings of the settler, have done much to change these woods. Nearly half of this territory has been burned over at least once, about three million acres are without any forest cover whatever, and several million more are but partly covered by the dead and dying remnants of the former forest. . . . In most of the pinery areas proper, the repeated fires have largely cleared the lands of all the heavier debris in slashings. Here are large tracts of bare wastes, "stump prairies," where the ground is sparsely covered with weeds and grass, sweet fern, and a few scattering, runty bushes of scrub oak, aspen, and white birch. . . . Throughout the hardwood districts there is no young growth of pine of any consequence. Some groves of young pine occur on many old and burned-over slashings on the sandy loam districts where settlement has put a stop to the fires. In all pineries proper many thickets of young pine occur which have sprung up during the last 25 years. . . . If protected, these groves soon furnish a considerable quantity of merchantable timber, but under present conditions most of them will be crippled or entirely killed by fires.

This last pessimistic judgment was borne out in the succeeding years, since effective fire protection did not begin until after 1930, so that most of the initial conifer regeneration was destroyed and with it the last traces of seed stock. As indicated in Chapter 11, much of the land

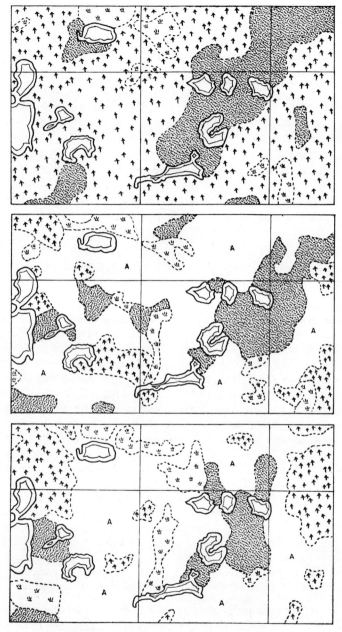

Figures 26, 27, and 28. (Legend at bottom of facing page.)

formerly in northern pines or pine-hardwoods is now in aspen. A typical example of the magnitude of this shift is shown in Figures 26, 27, and 28, which indicate the vegetational cover of the civil township of Plum Lake in Vilas County as reconstructed from the land survey records of 1865, *The Land Economic Inventory of 1928,* just before fire protection was instituted, and the Wisconsin Conservation Department *Forest Inventory of 1956,* after a quarter-century of protection. The great increase in acreage of aspen forest is readily apparent from these maps.

Comparison of effects of Indian and White Man

In their broader aspects, the influences of the aboriginal Indian and the postsettlement European on the vegetation of Wisconsin have been very similar. Each has used tools in conjunction with his efforts to gain a living from the land; these have greatly changed the plant communities over most of the state. Both have produced assemblages of species which were not in tune with the prevailing climate and were unable to persist by themselves in the absence of continuing disturbance. These communities were subject to change in proportion to the changes in the cultural habits and procedures of the dominant human population but in every case were different from the combinations which would have occurred in the absence of man. From a physiognomic viewpoint, the changes wrought by the Indians were at least as great as those produced by white man, with a very large portion of the area converted from forest to savanna, grassland, or other open vegetation. The actions of the Indian, however, tended to leave the soil covered with a dense blanket of living plants of one sort or another, so that soil erosion and water depletion were no problems. The white man, on the other hand, tended to destroy all soil covering, as on his agricul-

---

Figures 26 (top, facing page), *27* (middle), and *28* (bottom).—Forest cover of the civil township of Plum Lake, Vilas County in 1858, 1929, and 1951. Symbols are as follows: Irregular shading, mesic hardwood forest of maple, birch, and hemlock; arrow-like tree symbol, pine forest of white, red, or jack pine; marsh symbol, conifer swamp of tamarack, black spruce, or white cedar; blank with letter A, cutover and burned-over land supporting trembling aspen, white birch, or pin cherry. Notice the great increase in aspen by 1929 and the recovery of pine by 1951. Slight differences in the location of conifer swamps on the three maps are apparently due to differences in interpretation by the three sets of surveyors and are not believed to represent actual vegetational changes.

tural fields. Thereby he opened the land to the evils of erosion, and made other changes in the natural cycle of water runoff, percolation, and storage, so that water supplies gradually decreased. The Indian exerted retrogressive actions on the plant communities but not on the carrying capacity of the land. Insofar as he was able to replace forest with grassland, he actually may have increased the potential productivity, due to the anti-podzolization actions of the grass.

The retrogressive actions of white man have not only affected the biotic communities, but the basic soil and water resources as well, such that the net productivity is less than it was when he first appeared. If the ultimate criterion by which to judge the biological success of a civilization is based on its ability to maintain rather than degrade its environment, then the postsettlement European culture in Wisconsin must be considered inferior to its predecessors. The hope for a more favorable future judgment must rest on superior scientific and technical programs, which can reverse the degradation if they are given proper support.

CONSERVATION AND LAND MANAGEMENT
Nature of conservation

The idea that human populations are dependent on natural resources and that society must act to maintain those resources for the sake of its future is embraced in the word "conservation." The conservation concept has been slow to develop, and has reached a meaningful, operational level largely in the twentieth century. The realization that man, no matter how scientifically advanced, is dependent on natural resources for his continued well-being, both material and spiritual, came to fruition only after the closing of the frontier, when there were no more unexploited lands to be occupied. Before that time, the idea was widely held that food, water, lumber, metals, and fuel were inexhaustible, and when local supplies became scarce, other regions could be tapped for these products. In fact, this erroneous concept is still a dominant feature in the thinking of many countries today.

Permanent occupancy of the land means that the soil and water resources must be conserved and that the land must be managed in the most efficient way possible. No materially well-developed civilization has achieved this goal and it well may be unattainable in the face of constantly increasing populations. Nevertheless, this permanent occupancy is the goal of the conservation movement and the future demands that it be approached as closely as possible.

Complexities of land management

The problems facing the conservationist are very different in Wisconsin from those found in Mediterranean or Oriental regions with a long history of agricultural economy and high population levels. Here the agricultural lands are only a portion of the total, with the remainder in an altered but essentially natural condition. Forestry problems are added to those of soil and water, and recreational land use by residents of neighboring cities is an important part of the total picture. The Wisconsin conservationist has a definite advantage over his colleagues in more ancient centers—he still has examples of the original communities available for study. This is of highest importance because biotic communities are not subject to the experimental methods normally used in science, since they are too complex and are influenced by too many variables. The ordinary techniques used by the agronomist or physiologist in his efforts to understand the behavior of species and the methods which may be used to control them are necessarily based on one or a very few kinds at a time, as these are influenced by one or a very few environmental variables at a time. Through the use of adequate controls, such research can say that a particular plant will respond in a particular way to an increase in nitrogen supply, water, or other important external factors. What it cannot possibly do is tell how a complicated group of micro-organisms, small animals, large animals, small plants, and large plants will do if treated in the same way. The number of interactions possible in such a community is infinitely large and the effects of the treatment may reverberate through the community for decades before an ultimate response is evident. In the meantime, the assemblage may have been subjected to all sorts of uncontrolled influences due to weather, fire, insect outbreak, or new migrations, such that the initial cause cannot be connected with the final effect. In the ecological approach to such problems, advantage is taken of the millions of years of evolutionary experiments by trial and error that have resulted in integrated communities of mutually adjusted species. Changes in the composition of these communities in response to topographically and climatically influenced variables can be studied to gain information on the probable response of the community to artificial manipulation. More important to the well-being of an agriculturally-dependent human society is the information that can be learned concerning methods whereby natural communities maintain themselves without resource degradation or how they actually improve the site during their occupation. Such studies would include the basic nature of the energy transformations in natural systems. Armed with this knowledge, the land manager may be able to duplicate the natural situation, at the least, and to improve upon it in specialized instances, at the best.

Scientific areas

To make use of the ecological approach to conservation, the land manager needs preserved areas of natural communities. Graham (1944) has pointed out that the land cannot sensibly be managed in the absence of reference areas. They are needed for the basic information only they can provide and also as control areas for judging management practices, for determining whether the management results are better or worse than those which would have occurred in the absence of management. The reference areas should include examples of all of the plant communities of the area, preferably with replicate stands in the different geographical sub-regions. It is already too late to preserve some types in Wisconsin, since suitable areas in an undisturbed condition no longer exist. This is particularly true of mesic prairies and oak openings and may soon be true of the pine barrens. All of the other communities exist in sufficient areas and have a sufficiently wide distribution so that samples may be set aside if action is taken in the near future.

Fortunately, Wisconsin has been active in the movement to preserve natural communities for study purposes. Under the guidance of the late Aldo Leopold, the Conservation Commission appointed a Natural Areas Committee in 1942. The Committee was charged with locating suitable areas and arranging for their purchase with the use of Commission funds. This group was replaced by an official body designated by the 1951 State Legislature as the State Board for Preservation of Scientific Areas. The Board was empowered to stimulate the designation of suitable areas for scientific study purposes and to maintain a list of existing areas (Harrington, 1952; Curtis, 1954b). Twenty-eight sites had been entered in the Scientific Area System by 1957. These included 5 prairies, 8 hardwood forests, 12 conifer-hardwood forests and swamps, and several areas containing a variety of communities (Welty, 1957).

Many other examples of native plant communities in a protected condition are to be seen in the State Parks and in some sections of the State Forests designated as wilderness areas. Especially good stands of the various types which are known to occur in these areas are mentioned in the tabulated community summaries in the Appendix.

The Scientific Areas and other reference areas are of value not only in land management research but also in the teaching of biology, nature study, and conservation in the schools and colleges of the state. To be fully effective for this purpose, many more areas are needed to give a better geographic distribution.

University of Wisconsin Arboretum

No discussion of ecological research on reference areas in Wisconsin would be complete without a description of the University of Wisconsin Arboretum in Madison. This 1200 acre area on the shores of Lake Wingra was founded in 1932. Its development pattern was charted largely under the influence of Aldo Leopold, G. W. Longenecker, N. C. Fassett, and other field biologists, who saw the need for a research, a teaching, and a demonstration area for the new science of land ecology. They determined that the Arboretum was not to be merely a collection of trees, as is customary in other institutions of the same title, but rather that major emphasis was to be placed upon a collection of biotic communities, each as complete as possible, and each located in a natural manner with respect to interrelations with each other and the actual distributions of soils and topography. At present, the Arboretum contains original, native communities of prairie, fen, sedge meadow, shrub-carr, lowland hardwood, black oak barrens, and dry and dry-mesic southern hardwoods. In addition, the Arboretum controls Scientific Areas in other locations with native prairies, oak openings, oak forests, and pine forests. Artificially established communities on the Madison area include various prairie types, various pine forests, a spruce-fir stand, conifer swamps, cedar glades, and mesic forests of both southern and northern types. In addition, a beginning has been made on examples of other communities found in adjacent states, including Ohio Valley hardwoods and short leaf pine-oak forests as found in southern Illinois. Additional development of these and other communities is planned as time and very limited funds permit.

In the short period of its existence, the Arboretum has proved to be exceedingly valuable in the training of field biologists. The problems encountered in the attempts at artificial establishment of natural plant communities have demonstrated the complexities of community integration in a way that no other experience could provide.

CHAPTER 24

Interrelations of communities

THE STRUCTURE OF A STAND

The accounts of Wisconsin plant communities given in the previous chapters were designed largely to convey information about the distribution, structure, and floristic composition of the several types. The interrelations of the communities were discussed only to a limited extent, except in those vegetation types which had been shown to form a vegetational continuum. It is the intention here to explore the interrelationships of all types more fully, but before this is done, it may be advantageous to review and summarize some of the findings that have been described earlier with respect to the structure of stands, communities, vegetation types, and especially to their variability.

In considering first a typical stand of any community, it has been shown that the individuals of the component species differ in their spatial arrangement on the ground. These individuals may be distributed at random or they may approach a regular or uniform dispersal, but more frequently they are aggregated into clusters or clumps with relatively large intervening spaces which are devoid of the species. These clusters may be large or small, dense or diffuse, numerous or rare. Furthermore, the kind of distribution may be uniform throughout the stand or it may vary from strongly aggregated in one microsite to nearly random at another place in the same stand. Similar variation in spatial patterning is shown to be typical of many vegetations by Greig-Smith (1957).

Some species may be closely associated with others, since individuals of the two sorts are found together within small lateral distances to a greater extent than would be expected by chance. On the other hand, other species pairs or groups may be significantly disassociated; an individual of one species is rarely found in close proximity to an individual of another species. These differences in local distribution and local association may be attributed to local micro-environments in some cases. Such micro-sites as tip-up mounds, animal burrows, ridge tops, drainage ways, and spot openings in the canopy of forests may favor certain species or groups of species and be detrimental to others. Frequently, the micro-sites show no present-day differences from the surrounding area but are relics of historical gaps created by past disturbances such as lightning-fire or wind throw. Direct interspecific competition in the form of antibiotic or probiotic compounds may be responsible for some association, while indirect competition due to the dense shade of scattered dominants may bring about a non-random dispersion of the groundlayer species. An additional factor influencing the spatial patterns is the difference in immigration rates of species just entering the community from adjacent assemblages. The first individuals to enter usually are dispersed at random, while later stages tend toward aggregation (Whitford, 1949). Some of the immigrants, which are chance adventives from non-related communities, may lack the necessary adaptations to establish themselves permanently; their period of tenure is thus fleeting and highly unpredictable.

From this, we must conclude that stands of any major community are highly variable, differing widely from place to place within the stand and, to a considerable extent, from year to year at the same place.

THE STRUCTURE OF A COMMUNITY

In earlier chapters, communities were delimited by various methods as groups of stands with sufficient characters in common to produce studiable assemblages. The major vegetation types of a floristic province were identified by gross physiognomy, as forests, savannas, or grasslands. Each of these types was split arbitrarily into communities by an objective division of a floristic compositional gradient. The lesser communities were characterized largely by physiography, as beaches, dunes, cliffs, and aquatic areas, although some were separated by the nature of their dominants, as shrub communities or sedge meadows. In spite of these diverse ways of delimiting the communities, it is believed that most of them are comparable to each other in terms of degree of homogeneity. This contention is supported by the data in Table IV-2, which show the index of homogeneity for all communities. Most of these indices range between 50 and 65 (24 out of 34) although the extremes vary from 34.1 to 70.3. Similar indices were reported by Raabe (1952) for a series of communities in Germany. Many of the communities reported above for Wisconsin are comparable to the "alliances" of the common hierarchic classification scheme employed in Europe, rather than to the subsidiary "associations" of that system, although direct comparisons are difficult because of the much greater floristic richness of the American communities. Thus Dahl (1957) reported species densities (average number of species per stand) for a series of communities in south Norway ranging from 5.7 to 40.9, with an average of 18.9. The Wisconsin communities in comparison had species densities ranging from 7 to 75, but averaging 40.4, or more than twice as great. The indices of diversity similarly were much greater in Wisconsin, with a maximum of 70 compared to a top value of 41 in Norway.

In any one of these communities, the stands which compose it show a marked degree of non-uniformity. In the mesic southern hardwoods of Green County, for example, the average index of similarity among the stands is only 64.4 per cent, although all occur under the same general conditions in a small geographical area. In a similar group of mesic stands in Washington and Ozaukee counties in eastern Wisconsin, the stand to stand index of similarity average 60.3 per cent, while a third group in Richland and Vernon counties in western Wisconsin has an index of 61.2 per cent. These relatively low values are due to the chance lack of some species from one or more stands, as seen in the Green County example (Table 7). There is no apparent pattern to this record of presence or absence because groups of species do not appear or disappear together. The situation is in every way comparable to the results obtained from individual quadrats within individual stands and may be thought of as aggregation on a large scale, with colonies or local concentrations, several to many acres in size, that are separated from each other by distances up to several miles.

When the Green County mesic stands, as a group, are compared with the average for stands in eastern and western Wisconsin, the indices of similarity are 55.5 per cent and 53.5 per cent, respectively, while the latter two have a mutual index of 61.1 per cent. In these cases, the low indices result mostly from the lack of several important species in one or two of the regions. Thus, *Trillium recurvatum* is confined to the Green County area, while *Aralia racemosa* is absent but it is widespread in the mesic forest areas of Richland-Vernon counties.

When the stands of these three areas are combined with the other mesic forests of southern Wisconsin, comparisons are possible with stands in other states. As shown in Chapter 6, the resemblances are fairly high for neighboring areas but become progressively poorer with increasing distance. In many other kinds of communities, this decrease in similarity with increasing distance is more pronounced than in the mesic hardwood forest.

From this, it may be concluded that any given community shows a great stand to stand variability, with greatest resemblances to be expected only in local areas and with progressively greater floristic changes with increasing distance, such that remote areas on opposite peripheries of a community range show very low floristic similarities (as the conifer swamps in Chapter 12).

THE STRUCTURE OF A VEGETATION TYPE

Frequently, it is possible to arrange the stands of a given vegetation type along a compositional gradient called a vegetational continuum. This has been done for the southern hardwood forest (Chapter 5), the northern conifer-hardwood forest (Chapter 9), the boreal forest (Chapter 13), the prairie (Chapter 14), the hardwood savanna (Chapter 16), and the submerged aquatic communities (Chapter 19). The ordinations were based on the joint occurrences of dominants, on the relative presence of groups of indicator species, or on other information. They commonly show correlations with various factors of the environment, notably soil moisture and light. Many studies in other parts of the world have shown a similar continuous gradation in the composition of communities and many authors have concluded that vegetation forms a continuous rather than a limited set of discontinuous and discrete entities, each with its own describable characteristics by which it may be separated from its relatives. A number of early studies illustrated the gradual rise and fall in the frequency of various species in related communities arranged along an environmental gradient. The "turnip" diagrams of Yapp, Johns, and Jones (1917 have been much used to show these gradual changes. More recent studies based on better quantitative measures have given similar results. Thus Matuszkiewicz (1947), who used the multiple measure correlation methods of Czekanowski (1913) as modified by Kulczynski (1927) to study the degree of similarity of different communities in Poland, reported that a plant community:

. . . cannot be conceived as a strict, limited systematic unit like a "species" in plant taxonomy. The distribution of plants and their groupings in communities is conformable to changes of intensity of ecological factor. If then, change is continual, we have also continual change in vegetation; with discontinuity [of environment] we observe the discontinued change of vegetation too. In each territory [there] may be distinguished some plant "associations," but they have only a strictly local meaning.

In the following year, Matuszkiewicz (1948) presented what was probably the first detailed account of a vegetational continuum, based on 220 stands in Poland. An objective treatment of the phytosociological data resulted in a two dimensional ordination of the stands, which was correlated with a multifactoral gradient of soil and climatic factors. The conclusions reached in this classical investigation should be read by all who are interested in the nature of plant communities.

Whittaker (1951) found that in the forests of the Great Smoky Mountains in Tennessee, "the species were not organized into association units" but rather, "Climax vegetation here is a complex continuum of plant populations." Motyka and Zawadzki (1953) reported, "the floral analysis does not show any existence of strictly segregated separate communities," in the meadow vegetation of Poland. Ehrendorfer (1954) arrived at the same view from his studies in Austria, where he reported that the plant populations (*Biosphäre*) form a continuum (*Kontinuum*) with no sharp boundaries in either time or space, but merely transition zones with corresponding transitional communities. Horikawa and Okutomi (1955) studied the *Pinus-Shiia* forest region of Japan and found, "The fact that the component species show a continuous change seems to demonstrate that those species segregating along an environmental gradient are forming a continuum of vegetation." Similar conclusions were expressed for the complex vegetation of the tropics by Hewetson (1956) who said, "I would describe the Tropical Forest as a continuum in which the parts are in unstable equilibrium. All the species can survive but some species are more closely adapted to the sum total of environmental factors . . . and are more likely to be successful."

Many other investigators have presented data showing a continuous change in species compositions for the communities they studied. Among them may be cited Hansen (1930) in Iceland, Sjörs (1950) in Sweden, Raabe (1952) in Germany, Zoller (1954) in Switzerland, Linteau (1955) in Quebec, and Dahl (1957) in Norway. Some of these workers have operated from a system of classification rather than a quantitative ordination of their data, but they have arranged the groups so that they are classified into an order which clearly shows the independent rise and fall of the component species.

Individual species, when plotted against an ordinational gradient, show Gaussian-type curves, with minima, optima, and maxima. They indicate specific individuality in amplitude, range, and position of the curves. A set of curves for the prominent members of a vegetation type, for example, those in Figure 11 or Figure 12, may be likened to a distant view of a complex range of mountains, such as the Front Range of the Rockies as seen from the High Plains. Each peak represents a different species and its location on a lateral gradient indicates its relationship to the species at either end of the set. It is essentially a one-dimensional view, however, with no indication of the possibility that one peak may be farther from the observer than another. A true understanding of the location of the mountain peaks is possible only with a two-dimensional contour map which shows the relative spatial position of each peak. Similarly a vegetational continuum is a one-dimensional image of a complex relationship. Species located near the middle of the compositional gradient are more or less equally related to the end species, but the possibility exists that they are not closely related to each other, i.e. that one is closer to the observer than another.

That this is the actual situation is shown clearly by the work of McIntosh (1957) on the York Woods in Green County. By analyzing the data from 60 quadrats, each 10 meters by 10 meters, through the use of an index of interspecific association (Cole, 1949), McIntosh was able to show that a definite gradient existed, ranging from quadrats dominated by white oak (*Quercus alba*) on pioneer xeric sites to those dominated by sugar maple (*Acer saccharum*) on the terminal mesic sites. Between these two extremes were two separate lines or pathways, whose members were significantly associated with each other but were disassociated with the species of the other line. Thus one line proceeded from white oak to red oak (*Q. borealis*), to ironwood (*Ostrya virginiana*), to basswood (*Tilia americana*); the other went from white oak, to black walnut (*Juglans nigra*), to slippery elm (*Ulmus rubra*), to sugar maple. A number of herbs showed comparable patterns. Thus red oak and slippery elm, which have similar adaptation values (Table V-1) and are located close together on the southern upland hardwood forest gradient (Figure 8), are actually not found together in the same quadrats in a stand in which they are both present. Becker (1954) found the same disassociation between red oak and slippery elm in Young's Woods near Poynette in Columbia County. As a matter of fact, examination of the P.E.L. data for all stands of this vegetation type revealed that the two species usually did not occur in the same stands.

Considerable other evidence of the same general pattern definitely indicates that the major vegetation types are multidimensional in nature and that a unidimensional continuum is an oversimplification of the interrelationships of the stands which constitute the type. Under these circumstances, a multidimensional ordination approach is indicated (Goodall, 1954b; Greig-Smith, 1957). Perhaps the most rigorous approach to this goal is provided by factor analysis, based on correlation coefficients between the quantitative occurrences of species. This method has been widely used in psychological research (Thurston, 1947). It received early attention by plant ecologists (Steward and Keller, 1936) but was fully developed only recently by Goodall (1953, 1954b). Factor analysis, according to Thurstone, is based upon the assumption that ". . . a variety of phenomena within a domain are related and that they are determined, at least in part, by a relatively small number of functional unities or factors." Although various measurable attributes of plant communities are no doubt related to environmental factors, there is no assurance that these factors are themselves independent; hence, there is no sound basis for concluding that the unities extracted by the analysis are real. Because of this doubt and because of the almost insuperable load of computation inherent in a factor analysis of communities which are as complex as those under consideration here, it became apparent that some less rigorous method was needed.

Such a treatment has been afforded the upland hardwood forest of southern Wisconsin by Bray and Curtis (1957). They used 59 stands, ranging from bur oak-black oak dry forests to sugar maple-basswood mesic forests. As the basis of the ordination, quantitative measurements were used for 26 selected species, which included the 12 most important trees, and 14 typical herbs and shrubs, rather than the total of more-than-300 species. The values for these species in each stand were compared with those in each other stand. The index of similarity was used as explained in Chapter 4; it was based on nonweighted measures of density and dominance for trees and frequency for groundlayer species, with all measures converted to comparable units as relative test scores. A total of 1711 indices was obtained. The assumption was made that the index was inversely proportional to the distance between stands when the latter were arranged in a spatial pattern. In such an arrangement, two very similar stands would have a large value for the similarity index, but would be located only a small distance apart. Using this assumption, it was easily possible to determine the two stands least similar to each other, place them at opposite ends of a linear axis, and then arrange all the other stands along the axis between them on the basis of their average relationship to the two terminal stands. The resulting one-dimensional arrangement of the 59 stands showed a very high correlation with the linear arrangement of the same stands as originally placed in the compositional continuum on the basis of adaptation values of the trees only (Chapter 5).

Examination of the group of stands in the center of this first axis revealed a number which had low indices of similarity with each other, that is to say, the central stands were equally different from the terminal stands but at the same time were unlike each other. Using the two least closely related stands of this central group as termini for another axis, the position of all remaining stands with respect to these new end points was determined by geometrical means as before. Each stand, therefore, received two values, one of which indicated its position on the first axis and the other its position on the second axis. Using these values as co-ordinates, each stand was located on a two-dimensional graph. Further examination showed that some of the new centrally located stands were dissimilar, with low indices, although they were alike with respect to the four terminal stands. Repetition of the process produced a third axis, which provided co-ordinate values for a three-dimensional placement of stands. With the stands used, no further division was necessary, since the central stands in the three-dimensional placements actually were very similar to each other. However, there is no theoretical reason why four or more dimensions might not be obtained in other, more complicated vegetation types.

With either the two or three dimensional placements of stands, it is possible to plot the behavior of individual species by using a figure or symbol for the measured quantity of that species in each stand at the locus of the stand on the ordination. The absolute basal area of red oak in square inches per acre is shown for three dimensions in Plate 65, and a two-dimensional side view of the same species is given in Figure 29 for comparison. A composite diagram on the latter basis for

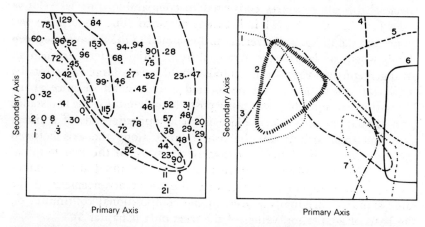

Figure 29 (left).—Behavior of red oak (*Quercus borealis*) on a two-dimensional presentation of the southern upland forest ordination. This represents a top view of the three-dimensional arrangement shown in Plate 65. The points represent individual stands; the numerals are absolute basal areas in 100's of square inches per acre of red oak in each stand.

Figure 30 (right).—Behavior of major trees of the southern upland forest on a two-dimensional ordination. The contour lines enclose all stands which possess the indicated species with an absolute basal area per acre at least half as great as the maximum attained by that species in any stand (after Bray and Curtis, 1957). 1, black oak (*Quercus velutina*); 2, shagbark hickory (*Carya ovata*); 3, white oak (*Quercus alba*); 4, red oak (*Quercus borealis*); 5, slippery elm (*Ulmus rubra*); 6, sugar maple (*Acer saccharum*); 7, white ash (*Fraxinus americana*).

the important tree species of the southern upland hardwoods is shown in Figure 30 by means of isorythms encompassing all the values in the two upper quartiles of absolute dominance for each species.

It is important to realize that the placement of the stands in either the two or three dimensional ordinations is in no way to be considered as an absolute portrayal of relationships in the sense of a physical constant. It does not even resemble an accurate topographic map where absolute distances between points on the map may be obtained directly by scalar measurement. The relative placement of stands is believed to accurately represent relative relationships, so that two stands which are spatially remote are less related than two stands which are located close together; yet the distances may not be directly equivalent to the relationship. The actual placement is influenced greatly by the particular index of similarity used, by the species used to compute the index, and by the particular pairs of stands that come to occupy the ends of the axes that are derived. There is some indication that the $2w/a+b$ formula does not result in values which are directly proportional to relationships. Motyka and Zawadzki (1953) adjusted the values by using the squares of the measurements for those species which exceeded a certain fixed level of cover; thus the more important species were accentuated. Other methods of weighting the index of similarity are possible, but no decision has been reached as to which is the most suitable technique.

More recently a study similar to that of Bray and Curtis has been completed for the upland conifer hardwood forests of northern Wisconsin and a similar portrayal of the important tree species is given in Figure 31. Maycock (1957) constructed a like ordination for the boreal-hardwood forests of the upper Great Lakes which included northern Wisconsin. The behavior of the major trees is shown in Figure 32.

It is believed that similar distribution patterns would be obtained for the lowland forest, the prairies, and the other communities, but the arduous calculations involved have prevented their production. It is clear that several of the major vegetation types (and probably all of them) are to be regarded as a more or less spherical constellation of stands, with intermediate or average assemblages in the center of the constellation and progressively more different groupings located at increasing distances from the center. Some of the peripheral stands may well be related as closely to another vegetation type as they were to the one in which they were first placed.

Similar two dimensional representations of the behavior of species in plant communities have been numerous in ecological studies of the past. Some of these have been produced by a direct alignment of stands on two or more environmental gradients, as soil moisture and soil nutrients, or soil moisture and elevation (Whittaker, 1956), or other measures. This is the general approach used by Ramensky (1926, 1930, 1953), Wiedemann (1929), and Pogrebnjak (1930), and is the basis

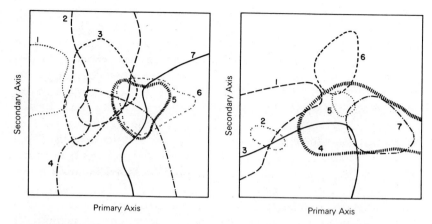

Figure 31 (left).—Behavior of major trees of the northern upland forest on a two-dimensional ordination. Contour lines as in Figure 30. 1, jack pine (*Pinus banksiana*); 2, white pine (*Pinus strobus*); 3, red pine (*Pinus resinosa*); 4, red oak (*Quercus borealis*); 5, hemlock (*Tsuga canadensis*); 6, yellow birch (*Betula lutea*); 7, sugar maple (*Acer saccharum*).

Figure 32 (right).—Behavior of major trees of the boreal forest of the Great Lakes region (after Maycock, 1957). 1, white pine (*Pinus strobus*); 2, jack pine (*Pinus banksiana*); 3, white spruce (*Picea glauca*); 4, balsam fir (*Abies balsamea*); 5, mountain-ash (*Sorbus americana*); 6, sugar maple (*Acer saccharum*); 7, black ash (*Fraxinus nigra*).

of the forest-site-regime classification in wide use in Russia today (Vorobiev, 1953). Other studies have used the degree of interspecific relationship to place species in a two dimensional framework (Tuomikoski, 1942; deVries, 1954; Guinochet, 1955; Agnew, 1957) while a number of investigators have used measures of stand similarity roughly comparable to that in the present study (Goodall, 1953; Clausen, 1957; Hopkins, 1957). The relative merits of these various methods are admirably discussed in Greig-Smith's recent (1957) book on *Quantitative Plant Ecology*.

THE INTERRELATIONS OF WISCONSIN COMMUNITIES

The interrelations of all of the plant communities of Wisconsin can be studied by the same methods used to compare the parts of a single vegetation type, namely by determining their degree of similarity and then placing them in a spatial arrangement with spacings inversely proportional to similarities. As test species for the determination of similarity, all of the prevalent groundlayer species from each of 28 native, terrestrial communities were used. These are the groundlayer species of highest presence, to the number of the average species density in each community. Altogether, there were 387 of these prevalent species, which was 44.6 per cent of the total of 867 species recorded in the entire P.E.L. study of these communities.

The sum of frequency of these prevalent species varied from 72 to 85 per cent of the total sum of frequency of all species for those communities for which frequency values were available; they, therefore, represent about three-quarters of all the plant individuals to be encountered within 50 centimeters of any point in any of the communities, since the frequencies were determined in quadrats of 1 square meter. The percentage presence of each of the test species was recorded for each of 28 communities on a special multipage score sheet and a complete matrix of 378 index of similarity values was calculated by the $2w/a+b$ formula directly from the presence figures. All of the terrestrial communities discussed in previous chapters were included except the two cliff types and the beach and dune communities. These were omitted because they were chosen originally on physiographic criteria and because they lacked sufficient data to permit ordering on an internal, sociological basis. All of the disturbance and weed communities were likewise omitted because of the preliminary and limited nature of the information available.

The indices of similarity were subtracted from unity to give inverse values which could be equated with spatial distances. Starting with the most central community (that which was most closely related to all others as shown by its highest sum of indices) and arbitrarily placing it in the middle of a hollow, cubical frame, all other communities were located by direct placement through triangulation procedures. The most central stand was found to be the dry-mesic northern forest and the two most closely related were the dry northern forest and the boreal forest. These formed a triangle which was located in space by means of suitable markers on movable rods in the frame. The other communities were located similarly by means of their distances from the first three or from later combinations. In general it was found easier to place a given community by its nearest neighbors, rather than by more remote groups, since very low indices of similarity with their inverse long distances were not very precise. This was due to the presence of a few species of very broad amplitude, such as woodbine (*Parthenocissus vitacea*) which occurred in 23 out of the 28 communities. Such ubiquitous species insure some degree of relationship between all communities. Actually, the two least related communities were the open bog and the dry prairie, with an index of similarity of only 0.4 per cent, caused by the presence of *P. vitacea* in both. The correction of Motyka and Zawadzki (1953) or a similar method for accentuating the contribution of the important species might alleviate these difficulties with the low indices.

This method results in a direct three-dimensional placement of the groups, without recourse to the calculation of independent axes as employed by Bray and Curtis (1957). The resulting spacings are directly related to the degree of similarity. Actually, the axis method, described above for the 59 stands of southern upland hardwoods, was also employed on the same data with generally similar results but with some deviations in the peripheral communities. Some difficulty was experienced in orienting the array within the framework, since no a priori information was available which could be used to govern the slope and direction of the original triangle. After repeated trials, the orientation shown in the figures was adopted since it seemed to offer fairly meaningful correlations with three major environmental factors along the three major axes. Thus, the vertical (A) axis appears to be related to temperature (Figure 33), with the coldest communities at the top and the warmest at the bottom. Similarly, the B axis is clearly correlated with soil moisture, from very dry at the left to very wet at the right. The third axis (C) is generally related to light intensity, with the dark forests (boreal, northern mesic, and southern mesic) at the top and the open prairie and meadow communities at the bottom (Figure 34). However, the phenomenon of compensating or interacting factors appears to enter here, since the bracken-grassland and the bog, both of which are fully open communities with low temperatures, are placed toward the middle of the C axis, rather than at the front where they would belong if they had been placed by actual measurement of light intensity. Other distortions of placement from the locations to be expected on the basis of direct measurement of temperature, water, or light indicate that this environmental interaction is quite general. Because of this, no direct connection between the three axes of the community ordination and any three simple environmental factors is either possible or to be expected. The immediate tendency of most ecologists to search for such causal factors is in part the result of long exposure to research papers on "the effect of pH on the growth of species X" or "the role of calcium in the development of species Y." Such information is beyond doubt of great value in itself but it should not obscure the fact that plants in nature are not subject to single factor causation but are responding continually to all of the factor of environment acting in concert.

The caution offered above with respect to the lack of absolute meaning to the placement of stands in the upland forest ordination applies here with equal force, since not only are the same variables which affect placement operative, but in addition, there is the problem of the angle from which the ordination is viewed. In the same way that an elephant looks very different if seen from a front view, a top view, or a side view, so any three-dimensional placement of stands in an ordination, whether absolutely or only relatively correct, will take on a different appearance. Under these circumstances, it is difficult, if not impossible, to decide which is the "correct" view. The choice must be made by criteria determined by the use to which we will put the information. If we were packing elephants into the hold of a ship, knowledge of their size and shape as gained from the top view might be sufficient, but a study of their means of locomotion would require at least a front and side view to avoid erroneous conclusions. In this work, the orientation finally chosen was the one thought to give maximum correlative information with the environment, but other views are equally valid and may be more useful for some purposes. It is believed that long-term experience with such three dimensional models may be very productive of valuable ecological ideas, but it is obvious that no one has yet achieved that experience.

The ordination of Wisconsin communities can be fully portrayed only in three dimensions, as in the model shown in Plate 65. However, the difficulties of preparation and illustration of these models prevent their use on a large scale. Accordingly, a two-dimensional presentation of the A and B axes (corresponding roughly to temperature and moisture) will be used to summarize certain pertinent factors of environment, life-form, and taxonomic relationships of the various communities and also to show the behavior of individual species. In one case, the B-C axes (moisture and light) will be used for the same purpose.

The A-B ordination of the communities is shown in Figure 33. The interrelations of the communities are fairly well indicated by the diagram, although it must be remembered that some communities which appear close together on the figure are actually separated by a considerable distance, with one being behind the other. Thus the southern wet forest (SW) seems relatively close to the wet prairie

Table 28

Key for Figure 33 and other ordination diagrams

AT	Alder Thicket	PB	Pine Barrens
BF	Boreal Forest	PD	Dry Prairie
BG	Bracken-Grassland	PDM	Dry-Mesic Prairie
BOG	Open Bog	PM	Mesic Prairie
CG	Cedar Glade	PWM	Wet-Mesic Prairie
FN	Fen	PW	Wet Prairie
ND	Northern Dry Forest	SB	Sand Barrens
NDM	Northern Dry-Mesic Forest	SC	Shrub-Carr
NM	Northern Mesic Forest	SD	Southern Dry Forest
NS	Northern Sedge Meadow	SDM	Southern Dry-Mesic Forest
NWM	Northern Wet-Mesic Forest	SM	Southern Mesic Forest
NW	Northern Wet Forest	SS	Southern Sedge Meadow
OB	Oak Barrens	SWM	Southern Wet-Mesic Forest
OO	Oak Opening	SW	Southern Wet Forest

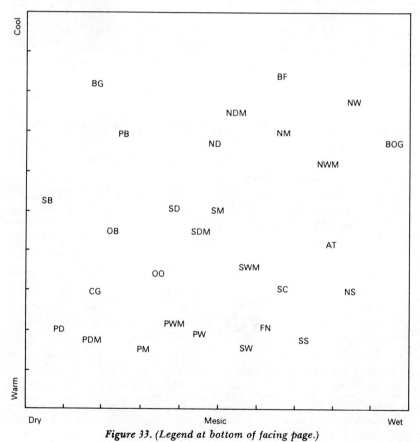

Figure 33. (Legend at bottom of facing page.)

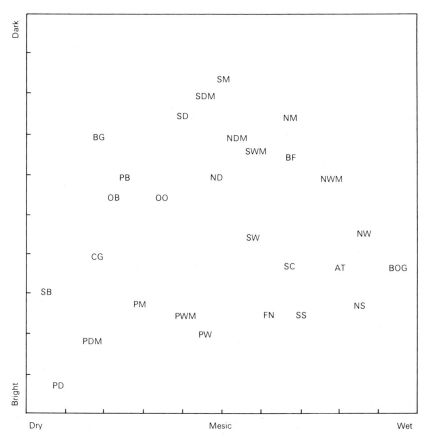

Figure 34.—Another two-dimensional view of the same ordination. In this case the vertical or C axis is related to the internal light conditions in the community, and the horizontal or B axis is the same as that in Figure 33.

(PW) but is really some distance back of it. This is shown in Figure 34 which gives the B-C ordination of moisture and light axes, where the wet forest and wet prairie are clearly separated. Communities which are relatively close on this second ordination may be separated in the third view, or the A-C axis combination. A key to the community

---

Figure 33.—Two-dimensional view of the ordination of 28 native terrestrial plant communities of Wisconsin. The vertical or A axis (ordinate), appears to be correlated with temperature and the horizontal or B axis (abscissa) is related to available moisture. The abbreviations on this and subsequent figures are keyed in Table 28.

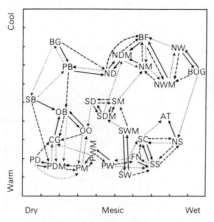

Figure 35.—A-B ordination, showing relationships of each community. The arrows indicate degree of relationship by the $2w / a + b$ test of similarity; the solid arrows point to the most closely related community, the dashed arrows to the second most closely related community, and the dotted arrows to the third most closely related community. Note that none of the communities of the prairie-forest province are the nearest relatives of any community in the northern pine-hardwood province.

abbreviations used on these and similar figures is given in Table 28.

Some of the community interrelationships are indicated in Figure 35 by means of arrows which point from each community to its three most closely related communities, as indicated by the index of similarity. The order of the relationship is given by the type of arrow, from solid lines for the closest, to dotted lines for the third closest.

The northern communities are separated from those south of the tension zone, as shown by the relatively wide gap between them. The northern sedge meadow and the alder thicket form a conspicuous exception to the last statement; their closest relationships are definitely with southern communities, in spite of their geographic position north of the tension zone.

ENVIRONMENTAL RELATIONS OF COMMUNITY ORDINATION

Certain features of the macroclimate, determined by Weather Bureau observers within the geographic area of each major community as described in Chapter 3, are plotted in Figures 36, 37, 38, and 39. In the case of the July temperatures for certain non-mapped communities, such as bracken-grassland, open bog, and fen, interpolations were made by actually measuring the average maximum and minimum temperatures in certain of these areas during July, 1957, and comparing their deviations from similar measurements in nearby weather stations in other communities during the same period. This method has obvious weaknesses but the results may be sufficiently close to the true values for present purposes. No such adjustment was attempted

INTERRELATIONS OF COMMUNITIES 493

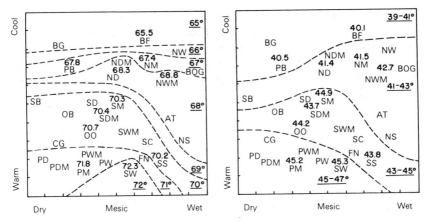

Figure 36 (left).—Average July temperature (in °F.) of all weather stations within the mapped area of each major community. A-B axes. Note that no communities lie in the band between 69°F. and 70°F., which clearly represents the tension zone.

Figure 37 (right).—Average annual temperature (in °F.) for the same stations. A-B axes.

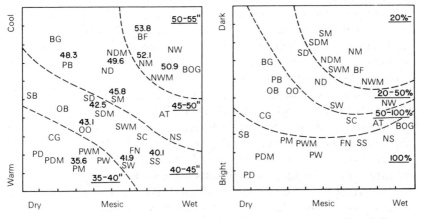

Figure 38 (left).—Average annual snowfall (in inches) for the same stations. A-B axes.

Figure 39 (right).—Estimated light intensity within the communities, expressed as a percentage of full sunlight, and plotted on the B-C axes.

for annual temperature values. The light intensities are estimates only, based on such direct measurements as were available plus the information on percentage canopy for forested communities, so no values are indicated for the individual communities.

The four climatic graphs show surprisingly definite correlations between the community ordination and the separate factor gradients, but the only one which is linearly related to an axis is that for July temperature. This shows a high correlation between temperature and ordination positions of the dry communities. In the wet group, however, the northern sedge meadow and the closely related alder thicket are displaced considerably downward from their expected position on the basis of actual temperatures. The same graph shows an excellent break at the tension zone, with the communities north of the zone having temperatures below 69°F., whereas those to the south are above 70°F.

The snowfall gradient is quite regular but runs at a diagonal to the two axes. Heavy snows are associated with the communities that are coldest and wettest, while minimum snows occur in the warmest and driest communities. The light intensity values are presented on the A-C ordination, rather than the A-B as used in the other graphs. The interaction of light and moisture is well illustrated, with communities of widely different light intensities located close to each other in wet sites, but with much greater spatial differentiation of communities of varying light intensities in dry places. Even in the dry communities, there is apparently some compensation, as shown by the displacement of the bracken-grassland to an intermediate position. In this case, it is probable that the bracken fern (*Pteridium aquilinum*) is providing shade in the same manner and amount as that which is provided by the shrubs in the tall shrub communities or the trees in the savannas and that the responses of the other community members is actually to partial light rather than to full sun.

Correlations between certain features of the soil environment and the ordination are shown in Figures 40 and 41. In each case, the values given are the average of all P.E.L. analyses in each community type. The most straightforward correlation of all is shown by the water-retaining capacities (Figure 41). Only one value is apparently misplaced, that for the southern wet forests of willow and cottonwood. The sandy nature of the overflow materials produces a low w.r.c. value, but in this case there is little relation between the value and the actual water supply, as the high water table from adjacent rivers or lakes provides a continuous source of water. Each isorythm on the w.rc. graph is double the previous one; a similar logarithmic relation was reported by Partch (1949) in his study of soil moistures. The pH gradient is also unusually regular. Many studies of soil acidity in the past have shown a wide variability; indeed, the basic figures in the current study were highly variable,

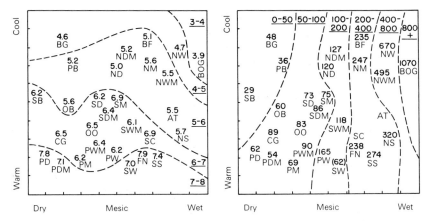

Figure 40 (left).—Average acidity of the A_1 layer of the soil (in pH units). A-B axes.

Figure 41 (right).—Average water-retaining capacity of the A_1 layer of the soil (in per cent dry weight). A-B axes. The southern wet forest (SW) is displaced from its proper contour interval.

also. The regularity of the averages presumably stems from the very large number of measurements involved. The similarity between the pH graph (Figure 40) and the July temperature graph (Figure 36) is worthy of special note. The vertical axis could as easily be termed an acidity axis as a temperature axis, since the two factors are themselves strongly correlated. The connection is not necessarily one of cause and effect, however, since the high temperature and high pH communities are in southwestern Wisconsin in a region of limestones, whereas the low temperature and low pH communities are in the north and northeast on granitic rocks or noncalcareous till. As demonstrated by Curtis and Dix (1956b), there is a relatively strong positive correlation between nature of bedrock, type and age of glacial covering, depth of loess, average July temperature, soil pH, and the alpha radioactivity of plant litter. All of these factors vary along a gradient from southwestern to northeastern Wisconsin.

MORPHOLOGICAL AND TAXONOMIC CORRELATIONS WITH COMMUNITY ORDINATION

When certain physiognomic aspects of vegetation are plotted against the ordination, clear-cut correlations are obtained, as had been demonstrated earlier for individual vegetation types by Randall

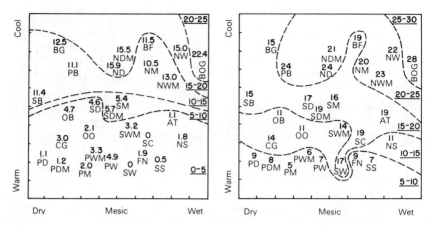

Figure 42 (left).—Evergreen species as a percentage of total understory species. A-B axes.

Figure 43 (right).—Shrubs and woody vines as a percentage of total understory species. A-B axes.

(1952) and Butler (1954). Thus, the number of evergreen species in the groundlayer, expressed as a percentage of the total number of species, is shown in Figure 42. There is a rather remarkable break between northern and southern communities, with 11 per cent or more evergreen species in the north and less than 6 per cent in the south. Again, the northern sedge meadow and the alder thicket show their anomalous nature by their very low content of evergreens. In the case of shrub species as a percentage of total groundlayer species (Figure 43), there is a definite though irregular trend, from less than 10 per cent in the prairies to nearly 30 per cent in the open bogs. The pattern bears some resemblance both to that for soil pH and annual snowfall although no causal relation is suspected.

Taxonomic groups above the level of species show definite correlations with the community ordination. To illustrate this, several families or higher categories have been plotted in Figures 44, 45, 46, and 47, with values indicating the number of species in a particular family as a percentage of the total number of species in the community. The ferns (*Filicales* and *Ophioglossales*) are essentially the reverse of the composites and grasses, and have optimum values in the sites which are both cold and wet, while the composites and grasses reach highest levels in either warm or dry communities. The pattern of the ericads, as might be expected, shows considerable resemblance to that of soil pH (Figure 40). Other families not illus-

INTERRELATIONS OF COMMUNITIES 497

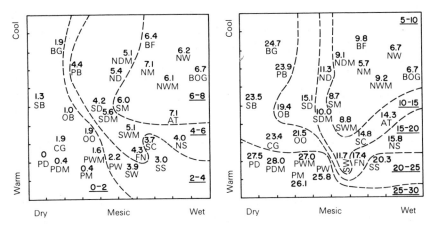

Figure 44 (left).—Number of fern species as a percentage of total groundlayer species. A-B axes.

Figure 45 (right).—Number of composite species as a percentage of total groundlayer species. A-B axes.

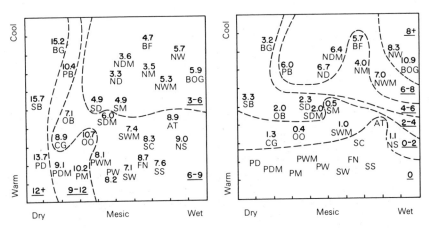

Figure 46 (left).—Number of grass species as a percentage of total groundlayer species. A-B axes.

Figure 47 (right).—Number of ericad species as a percentage of total groundlayer species. A-B axes.

trated here show different reactions to the environmental complex. Thus the legumes appear to be associated with the soil moisture gradient but the relation is not direct, since low values are attained in both very dry and very wet soils. Some families are more influenced by light than by temperature. The lilies, for example, reach an optimum in sites that are both wet and shaded, while the milkweeds are prominent only in light and dry communities.

BEHAVIOR OF INDIVIDUAL SPECIES

Trees.—When the measured quantities of individual species in each community are plotted at the proper locus for each community, the same kind of atmospheric distribution patterns are obtained as in the case of a stand ordination of a single community. Thus, in Figures

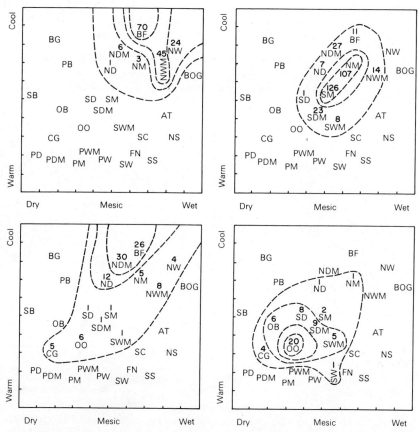

Figure 48.—Behavior of trees on the community ordination (A-B axes). (Upper left), balsam fir (*Abies balsamea*); (upper right), sugar maple (*Acer saccharum*); (lower left), white birch (*Betula papyrifera*); (lower right), shagbark hickory (*Carya ovata*). Figures are average importance values for all stands of each community. Contours are drawn as aids in interpretation.

INTERRELATIONS OF COMMUNITIES 499

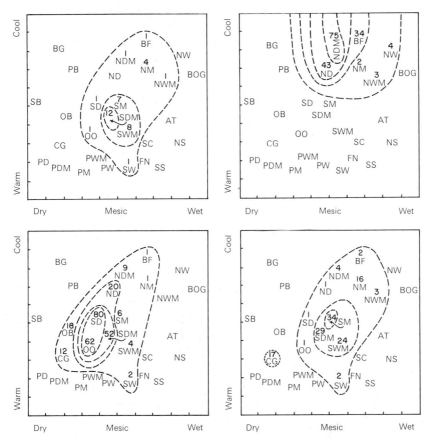

Figure 49.—Behavior of trees, as in Figure 48. (Upper left), white ash (*Fraxinus americana*); (upper right), white pine (*Pinus strobus*); (lower left), white oak (*Quercus alba*); (lower right), basswood (*Tilia americana*).

48 and 49, the average importance values for a number of tree species are shown on the A-B ordination. Each figure shows a limited number of communities in which the species reaches high importance, surrounded by a more or less concentric series of other communities, with successively lower importance values. In the figure for basswood, the high value for the cedar glades has been indicated as a disjunct island, since there is good indication that this is a different taxon (see Chapter 16). The contour lines in the figures were placed largely by subjective judgment, and they should not be accepted as rigid, quantitative demarcations, but simply as visual aids to the interpretation of the figures.

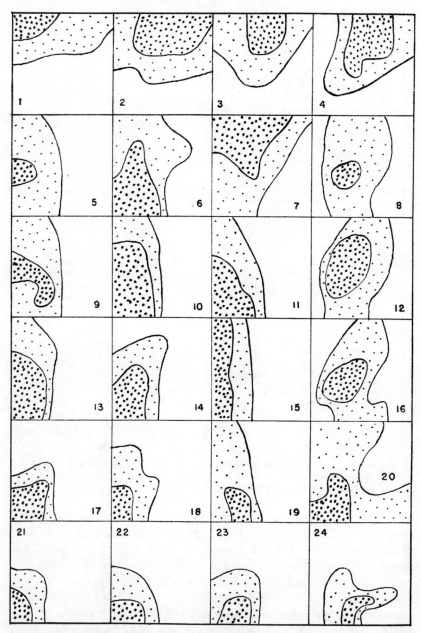

Figures 50 (above), *51* (facing page).—Behavior of the three most prevalent ground-layer species of the 28 communities included in the ordination, but excluding the dominants: The heavy shading includes the communities with presence values above 50 per cent; the light shading includes the communities with presence values from 1 to 50 per cent. The species are keyed by number in Table 29. In so far as possible,

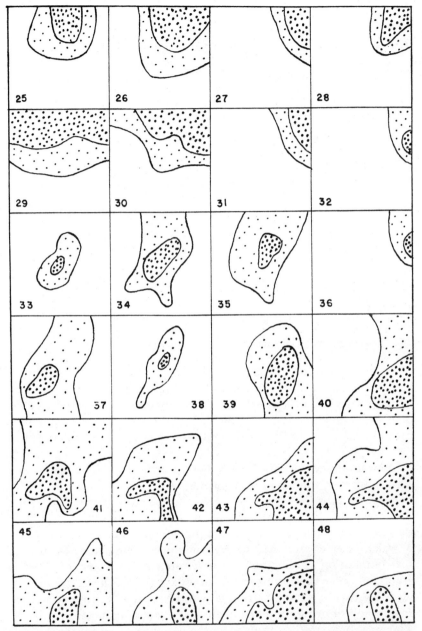

the ordinations have been arranged so as to have similar patterns close together. Note the gradual shifts in location of optimum areas along vertical, horizontal, or diagonal directions. These two figures offer strong support for Gleason's concept of the individualistic nature of plant communities.

Table 29

Key to herb ordination diagrams in Figures 50 and 51

1 Aster ciliolatus
2 Maianthemum canadense
3 Aster macrophyllus
4 Aralia nudicaulis
5 Helianthemum canadense
6 Monarda fistulosa
7 Pteridium aquilinum
8 Cornus racemosa
9 Lespedeza capitata
10 Euphorbia corollata
11 Tradescantia ohiensis
12 Smilacina racemosa
13 Andropogon scoparius
14 Amorpha canescens
15 Solidago nemoralis
16 Geranium maculatum
17 Petalostemum purpureum
18 Anemone patens
19 Aster laevis
20 Commandra richardsiana
21 Aster sericeus
22 Helianthus laetiflorus
23 Ratibida pinnata
24 Pycnanthemum virginianum
25 Streptopus roseus
26 Dryopteris austriaca
27 Ledum groenlandicum
28 Carex trisperma
29 Vaccinium angustifolium
30 Cornus canadensis
31 Smilacina trifolia
32 Sarracenia purpurea
33 Podophyllum peltatum
34 Osmorhiza claytoni
35 Polygonatum pubescens
36 Vaccinium macrocarpon
37 Amphicarpa bracteata
38 Allium tricoccum
39 Impatiens biflora
40 Eupatorium maculatum
41 Parthenocissus vitacea
42 Vitis riparia
43 Iris shrevei
44 Calamagrostis canadensis
45 Solidago gigantea
46 Lycopus americanus
47 Aster simplex
48 Asclepias incarnata

Little discussion of the patterns appears to be necessary, since the figures themselves are self-explanatory. The regularity or smoothness of the regression, as shown by the contour lines is rather remarkable. The other tree species not illustrated have similar patterns. In no case did any two species have identical patterns, either in amplitude or location.

Groundlayer species.—The behavior patterns of the herbs and shrubs are very similar to those of the trees. An objective selection of these patterns is shown in Figures 50 and 51. The species chosen are the three most prevalent species of the major communities, excluding the dominants. They are taken from the individual lists of the five most prevalent species tabulated in the community summaries. Because of space limitation, the individual species are shown in small-scaled diagrams, and no detailed numerical values are given. The diagrams were made by dividing the presence values of each species into two equal groups and drawing contours at the 1 per cent and 50 per cent levels. The areas between the contours are shaded, the darkest for the top half (50–100 per cent presence) and lightest for the bottom half (1–50 per cent presence). The areas on the ordination in which the species is lacking are not shaded. These figures are similar to the hypothetical diagrams given by Dansereau (1957) but differ because they are based on actual quantitative measurements.

Similar diagrams have been prepared for the entire 867 ground-layer species encountered in the P.E.L. studies of the native, terrestrial communities used in the ordination. They resemble those illustrated in Figures 50 and 51 and show no types of pattern not indicated there. There is every possible combination of amplitude and location of optimum community. Some species have patterns which are very similar to those of other species. This is most pronounced, but least meaningful, in the case of rare species restricted to a single community. In other cases, more widespread plants share a comparable behavior pattern. Thus, a number of species of the open bog are either restricted to the open bog or are found in the adjacent northern wet forest, the boreal forest, or the northern wet-mesic forest. Similarly, a group of species appears to center on the dry prairie in the opposite corner of the ordination, and shows a more or less uniform degree of overlap into the related prairie and savanna communities. Significantly, no such groups are apparent in the centrally located communities on the ordination. This result appears reasonable in view of the geographic position of Wisconsin midway between the prairie and the boreal forest. The major vegetation type is the deciduous forest which finds here a reasonable approximation of its optimum environment. The prairie and boreal forest are both marginal, and their groups of species of fairly high similarity no doubt would have lost their homogeneity if a full range of communities of these vegetations from surrounding areas had been included in the study.

The lack of groups of species with either identical patterns or with a series of concentric patterns has important consequences for any attempt to erect a hierarchic classification of the communities considered here. If such groups were present, then those species which shared a narrow range on the ordination might be used to characterize a unit centering there, while those species which had broader amplitudes centering on the first group might be considered as character species of the next higher unit in the hierarchy and so on. Since all possible patterns are actually present, it is not possible to erect such a system, as all the higher units would overlap to some degree with each other. A classification procedure can only pick out the areas with species of narrow amplitude and it thus devolves into a rather complicated method of describing the behavior of these rare species. Furthermore, there will be as many units as there are species of narrow tolerance, thus approaching the individualistic concept of Gleason (1926).

The lack of coincidence between the findings of classification and of ordination is well shown by the work of Dahl (1957), who studied the communities of a small region in southern Norway. The communities were described on the basis of single plots in "typical" examples of a few stands of each assemblage. The communities (or "associations") were arranged in alliances and orders according to the usual practice of phytosociologists of the region. Dahl then calculated an index of similarity between the communities, using the species presence basis as recommended by Sorensen (1948). In this way it was possible to compare the results of the two methods although a considerable element of bias in favor of the classification scheme was introduced by the subjective, non-random choice of the samples for initial study (Greig-Smith, 1957). In spite of this bias, it was found in many cases that the stands of a given "association" were more similar to stands in "associations" of other alliances and even other orders than they were to "associations" in the same alliance. Hanson (1955), using a similar comparative technique in the grasslands of Colorado, also found that some stands were more similar to other groups than they were to the other stands of the group in which they had been classified.

Thus there is widespread support for the idea that vegetation must be studied as a continuous variable. This situation means that it is not possible to erect a classification scheme which will place the plant communities of any large portion of the earth's surface into a series of discrete pigeonholes, each with recognizable, describable characteristics, and boundary limitations. The numerous attempts to do so in the past are attributed by Goodall (1954a) to the common learning patterns of childhood. He says:

Class-names are used in youth long before numbers are correctly recognized. . . . In performing measurements, the more mature mind passes beyond the habits of classification; but in objects less clearly suitable for measurement the habits of childhood are retained. It is to this that we must ascribe the wide-spread, wellnigh universal, attempts to fit vegetation to the Procrustean mould of a classification system.

THE QUESTION OF FIDELITY

In several European systems for the classification of plant communities, the practice has been to put great emphasis on characteristic species of high fidelity in the delimitation, description, and naming of communities. By fidelity is meant the degree to which a species is restricted to a particular community, a plant of high fidelity being completely or almost completely confined to a single community. It is obvious that the community initially must be determined by other criteria than fidelity, since the degree of fidelity can be determined only after all the communities of a region have been recognized and set apart from each other. The dangers of circular reasoning in this process have been well set forth by Katz (1933), Goodall (1954a), and Poore (1955).

The data from the P.E.L. studies, gathered in an area about the size of the combined areas of Switzerland and Austria, may be used to examine the question of fidelity. For this purpose, fidelity was determined by counting the number of communities in which each species occurred. Only groundlayer species in native terrestrial communities were used, since comparable presence values are available for all of them. There are 32 such communities, the same ones used in the three-dimensional ordination plus the shaded and exposed cliff communities, the beach community, and the lake dune community. One species, woodbine (*Parthenocissus vitacea*), occurred in 26 of the possible 32 communities. At the opposite extreme, 170 species were found in a single community only. There were 51 species which occurred in 50 per cent or more of the communities. These are listed in Table 30.

A complete frequency distribution diagram for the presence values of the entire 980 species is given in Figure 52. The number of species occurring in only a few communities is much greater than the number which occur in many. In fact, 50 per cent of the species occurred in less than four communities each. At first sight, this diagram seems to substantiate the use of fidelity as an ecological measure, since it indicates that there are many species of high fidelity in the flora. Further examination of the data, however, reveals a serious weakness of these possible diagnostic species. In Figure 53, the average presence attained by each species in the community of its optimum occurrence is determined according to the fidelity class of the species. It can be seen that species of high fidelity are also species of low presence, whereas the reverse is true for species of low fidelity. This means that a species

Table 30

Ubiquitous groundlayer species of terrestrial communities

All species are listed which occur in one-half or more of the total communities. Figures are communities of occurrence as per cent of total communities.

Species	%	Species	%
Achillea millefolium	53%	Monarda fistulosa	53%
Amphicarpa bracteata	59	Onoclea sensibilis	69
Andropogon gerardi	59	Parthenocissus vitacea	81
Anemone cylindrica	50	Pedicularis canadensis	53
A. quinquefolia	56	Phlox pilosa	56
Antennaria neglecta	53	Polygonatum canaliculatum	53
Apocynum androsaemifolium	56	P. pubescens	50
Aquilegia canadensis	59	Potentilla simplex	50
Aralia nudicaulis	53	Prenanthes alba	56
Aster umbellatus	50	Pteridium aquilinum	69
Calamagrostis canadensis	63	Rhus radicans	78
Carex pensylvanica	50	Ribes americanum	56
Celastrus scandens	56	Rubus pubescens	50
Cornus racemosa	66	Smilacina racemosa	59
Corylus americana	66	S. stellata	66
Diervilla lonicera	53	Solidago canadensis	63
Equisetum arvense	53	S. gigantea	50
Fragaria virginiana	78	Spiraea alba	50
Galium aparine	50	Steironema ciliatum	50
G. boreale	63	Thalictrum dasycarpum	59
G. triflorum	56	T. dioicum	50
Geranium maculatum	66	Vaccinium angustifolium	50
Heuchera richardsonii	56	Veronicastrum virginicum	56
Impatiens biflora	56	Viola cucullata	72
Lilium superbum	59	Vitis riparia	69
Maianthemum canadense	56		

of high fidelity, which might be used as a differentiating or characterizing species, is found in only a few stands of the community; hence it has limited value for the recognition of that community. The great number of such species means that each investigator is likely to choose a different species as being diagnostic. This will depend upon his personal predilections for the color, beauty, or the boyhood recollections it may bring to mind. The long lists of synonymy for various communities found in the publications of Tüxen (1950), Oberdorfer (1957), Dahl (1957), and other European writers perhaps are to be explained in this way.

Thus, on the average, species of high fidelity (those restricted to a single community) are rare plants in the true sense of the word. In common with other rare events, they are subject to very great sta-

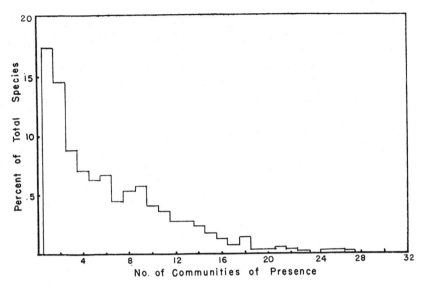

Figure 52.—Frequency distribution diagram of all terrestrial species in P.E.L. study by fidelity classes as indicated by number of communities of presence.

tistical errors (Cottam, Curtis, and Hale, 1955; Greig-Smith, 1957); hence they are subject to great possibilities for misinterpretation, simply because they occur so seldomly that we have little valid knowledge about them. To give a concrete example of this, consider two species of apparent high fidelity in the southern mesic forest, *Triphora trianthophora* and *Ellisia nyctelea*. Each of these occurred only once in the P.E.L. study, in two stands of maple forest several miles apart in Green County. On the basis of the quantitative data alone, both might be classed as diagnostic of this particular community, according to the fidelity scheme of community description. Actually, *Triphora* is truly confined to this forest type, but this can be determined only on the basis of extraneous information provided by extensive reports of orchidologists, who scour the countryside in search of their favorite rare plants, but not from the data themselves. On the other hand, *Ellisia* is a widespread weed of open fields and roadsides from central Illinois southward and in no sense can be considered a true mesic forest plant. It occurred as an accidental intruder in the Green County stand. The point is simply this, that rare species, because of their very rarity, cannot be assigned an important place in phytosociological investigations and definitely should not be used as characteristic hallmarks of a plant community, since inadequate sampling may lead to wholly erroneous conclusions and since truly adequate sampling is practicably impossible.

The above conclusions are based on the average results for 980 species in 908 stands of 32 communities in an area of about 52,000 square miles (134,680 square kilometers) and are believed valid for

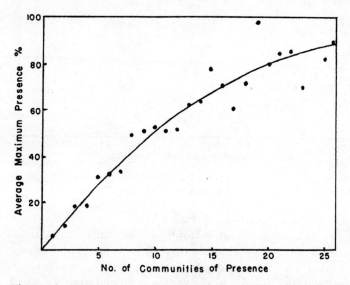

Figure 53.—Relation between presence and ubiquity. The curve shows the average maximum presence attained by species arranged according to the number of communities of their occurrence (out of a possible 32 communities).

the region and for the vegetations involved. As is the case with all averages of biological phenomena, there are many exceptions, or at least deviations from the average so great as to appear to be exceptions. These deviating species are themselves of great interest and deserve detailed study. For example, *Botrychium multifidum,* a leathery-leaved, evergreen grapefern, has a fidelity value of 10 per cent but a maximum presence of only 8 per cent instead of the 50 per cent called for by the average relation in Figure 53. It is widespread throughout the state and has been found in prairies, savannas, sedge meadows, sand barrens, bracken-grasslands, and a number of forests, which indicates a wide disregard for such environmental factors as moisture, temperature, and light. Yet it is always very rare, seldom occurring more often than one or two individuals per stand, or, as indicated by the presence value, in more then one stand in ten. The population mechanics of such a species are in marked contrast to another deviating type, exemplified by *Chaerophyllum procumbens*. This has a very high fidelity, and is found only in the southern wet-mesic forests. Its presence is also low, in accord with expectation, but its average frequency and density values are relatively high, due to its high population levels in the few stands where it does occur. Other similar examples are known, usually in species which are on the border of their range, as is the case with *Chaerophyllum*.

CONCLUSIONS CONCERNING THE NATURE OF PLANT COMMUNITIES

The outstanding characteristic of plant communities as described in this chapter and in the preceding basic accounts is their great variability. A single stand of any vegetation type shows place to place and time to time variation in the spatial distribution of its component species. These differences are most pronounced in small areas and become less important as larger areas of the stand are considered, because of the tendency for the patterns to be more or less repeated from place to place. This repetition is not very precise, since neighboring stands of the same community in the same local region resemble each other in quantitative composition only to the extent of 50 to 70 per cent. When stands in more remote regions are considered, the index of similarity falls rapidly.

On a broader scale, a single vegetation type, like the prairie or the hardwood forest, is similarly variable, with place to place and time to time differences in composition. These variations, like those within and between stands, are continuous, with no definite boundaries or clear-cut demarcations between subunits of the series.

The frequent sharp boundaries found locally at places of rapid environmental change do not invalidate this conclusion. The concept of continuous change in community composition does not deny that many stands are clearly and distinctly separate from their local neighbors which may be totally unrelated assemblages (as in the case of the conifer bog and the contiguous oak forest described in Chapter 3). It does deny that each of these sharply defined local groups is a representative subunit of a wide-ranging community which is meaningfully different from other wide-ranging communities. Rather, it says that there is a continuous variation from stand to stand of any particular community, such that those stands which are more remote are more different than those which are nearby, and that there are no inherent limits to the community *in toto*.

This difference between local and regional distinctness is perhaps best illustrated by Gleason's classic example of variation in the forests of the Mississippi Valley. At any point along this valley, stands of floodplain forest can be found which are clearly distinct from the community on the adjacent upland. At any such point, it might be said that two different communities are present and that the local differences are great, describable, and definite. However, when an attempt is made to define the floodplain forest at such a point as a biological entity, it is soon found that the local example is slightly different from other stands to be found either upriver or down-river. These differences increase in a gradual manner in both directions so that the riverside forest in Minnesota is almost totally different from the comparable forest in southern Louisiana, with no evident breaks or boundaries between subunits.

Ordinarily, the variation in a single vegetation type is associated with gradients in available moisture, temperature, or light, such that sites at one extreme of these conditions possess a very different assemblage from sites at the opposite extremes. The continuum nature of the stands in a particular vegetation type is also expressed between the communities of two or more vegetation types, such as the gradual change from hardwood forest through savanna to prairie. Actually, on a floristic basis, all native terrestrial communities in Wisconsin are related to each other to a certain extent, since all have some species in common with the others. In view of the continuous nature of these interrelations, the delimitation of separate communities becomes a pragmatic matter, to be decided on convenience for the purpose at hand. It is best accomplished following an ordination rather than prior to such a study.

A study of the behavior of particular species in single stands, communities, vegetation types, or the entire vegetational complex clearly reveals that each is behaving as an individual and not as a dependent member of a coherent biological group. The entire evidence of the P.E.L. study in Wisconsin can be taken as conclusive proof of Gleason's individualistic hypothesis of community organization, as summed up in his statement, ". . . an association is not an organism, scarcely even a vegetational unit, but merely a coincidence," (Gleason, 1926).

Each species grows where and when it does on the basis of its own environmental requirements and its own past history. No two species described have exactly similar requirements or backgrounds and no two species have been shown to grow together in, and only in, the same assemblages. The communities as we know them are chance gatherings of unrelated individuals, each concrete example of a community being composed of a momentarily distinct and limited set of species, arranged in a unique spatial pattern with a never-to-be-duplicated combination of numbers of individuals. The chance gatherings are not wholly random events, but rather they follow a broad pattern imprinted by the environment. In any given region, only a limited portion of the flora possesses the proper adaptations to grow together at a particular time and place; rarely, if ever, do all members of this limited portion actually occur together, since chance happenings of a historical nature have usually acted to prevent some of the potential members from reaching the community.

It is the task of phytosociology to describe the combinations of plants that do occur in each region of the world, to find how they came into being and how they maintain themselves, to relate them to their physical environment and to reach an understanding of the material and energy changes which occur within them. The classification of communities should be looked upon as a tool which might further these goals, rather than as an end in itself. Many of the findings of the present research suggest that a purely floristic basis may be inadequate for the task and that a behavioral approach would be more productive. More than three decades ago, Gleason suggested that much of the confusion in phytosociological investigations might be avoided "if those taxonomic units which have the same vegetational form and behavior could be considered as a single ecological unit." As usual, he was ahead of his time, but we would profit if we accepted his advice. Recent research along these lines has been encouraging and gives every promise of developing methods whereby communities which behave alike may be grouped together, regardless of floristic differences or similarities. The resulting system would be of obvious use in interregional comparisons now hampered by floristic differences which may have no ecological meaning. It would also be of great value in many phases of applied ecology such as forestry and range management where processes and behavioral responses are similar over a great range of floristic types.

Appendix

APPENDIX FOR CHAPTER 4
Table IV-1
Common and scientific names of trees mentioned in text

Apple
 Pyrus malus
Arborvitae
 Thuja occidentalis
Ash, black
 Fraxinus nigra
Ash, blue
 F. quadrangulata
Ash, green
 F. pennsylvanica subintegerrima
Ash, red
 F. pennsylvanica pennsylvanica
Ash, white
 F. americana
Aspen, large-toothed
 Populus grandidentata
Aspen, trembling
 P. tremuloides
Basswood
 Tilia americana
Beech
 Fagus grandifolia
Birch, river
 Betula nigra
Birch, sweet
 B. lenta
Birch, white
 B. papyrifera
Birch, yellow
 B. lutea
Boxelder
 Acer negundo
Buckeye
 Aesculus glabra
Butternut
 Juglans cinerea
Cedar, red
 Juniperus virginiana
Cedar, white
 Thuja occidentalis
Cherry, black
 Prunus serotina
Cherry, pin
 P. pensylvanica
Chestnut
 Castanea dentata

Coffeetree, Kentucky
 Gymnocladus dioica
Cottonwood
 Populus deltoides
Crab, Iowa
 Pyrus ioensis
Cypress
 Taxodium distichum
Dogwood, flowering
 Cornus florida
Elm, American
 Ulmus americana
Elm, rock
 U. thomasi
Elm, slippery
 U. rubra
Fir, balsam
 Abies balsamea
Gum, sour
 Nyssa sylvatica
Gum, sweet
 Liquidambar styraciflua
Hackberry
 Celtis occidentalis
Hemlock
 Tsuga canadensis
Hickory, shagbark
 Carya ovata
Hickory, yellowbud
 C. cordiformis
Ironwood
 Ostrya virginiana
Locust, black
 Robinia pseudoacacia
Locust, honey
 Gleditsia triacanthos
Magnolia
 Magnolia sp.
Maple, black
 Acer saccharum nigrum
Maple, mountain
 A. spicatum
Maple, Norway
 A. platanoides
Maple, red
 A. rubrum

Table IV-1 (cont.)

Maple, silver
   A. saccharinum
Maple, striped
   A. pensylvanicum
Maple, sugar
   A. saccharum
Mountain-ash
   Sorbus americana
Mulberry, red
   Morus rubra
Oak, black
   Quercus velutina
Oak, blackjack
   Q. marilandica
Oak, bur
   Q. macrocarpa
Oak, chinquapin
   Q. muhlenbergii Englm.
Oak, Hill's
   Q. ellipsoidalis
Oak, pin
   Q. palustris
Oak, post
   Q. stellata
Oak, red
   Q. borealis
Oak, swamp white
   Q. bicolor
Oak, white
   Q. alba
Pecan
   Carya pecan
Persimmon
   Diospyros virginiana

Pine, jack
   Pinus banksiana
Pine, red
   P. resinosa
Pine, short-leaf
   P. echinata
Pine, white
   P. strobus
Poplar, balsam
   Populus balsamifera
Poplar, black
   P. nigra
Poplar, white
   P. alba
Redbud
   Cercis canadensis
Silverbell
   Halesia monticola
Spruce, black
   Picea mariana
Spruce, red
   P. rubra
Spruce, white
   P. glauca
Sycamore
   Platanus occidentalis
Tamarack
   Larix laricina
Tuliptree
   Liriodendron tulipifera
Walnut, black
   Juglans nigra
Willow, black
   Salix nigra

Table IV-2
Statistical summary of native communities

Community	No. of stands	Species density	Total No. of species	No. of modal sp.	Index of homogeneity	Index of distinctness	Index of diversity
Southern Forest—Dry	30	61	289	25	59.6	24.1	67.0
Southern Forest—Dry Mesic	54	75	275	76	65.3	56.1	50.1
Southern Forest—Mesic	47	47	230	43	54.3	25.6	50.7
Southern Forest—Wet Mesic	64	46	333	34	44.5	10.2	69.0
Southern Forest—Wet	16	34	175	40	48.8	55.2	50.9
Northern Forest—Dry	38	46	264	19	54.8	23.1	59.9
Northern Forest—Dry Mesic	40	56	283	22	58.7	12.8	61.6
Northern Forest—Mesic	66	40	254	19	54.8	12.1	51.1
Northern Forest—Wet Mesic	26	56	253	34	53.8	35.4	60.5
Northern Forest—Wet	71	37	208	24	52.6	31.3	40.1
Boreal Forest (Wis. only)	39	70	328	65	57.7	43.1	70.4
Prairie—Dry	17	47	132	31	70.3	57.5	30.0
Prairie—Dry Mesic	66	55	245	26	61.3	14.6	45.4
Prairie—Mesic	45	55	264	40	59.5	34.6	54.9
Prairie—Wet Mesic	31	62	252	38	59.5	29.0	55.3
Prairie—Wet	22	44	186	23	56.3	22.7	45.9
Sand Barrens	20	33	159	34	52.7	48.5	42.1
Bracken-Grassland	27	32	158	36	57.1	56.3	38.2
Oak Opening	19	52	232	8	57.3	2.1	62.8
Oak Barrens	14	52	204	20	57.9	24.5	58.7
Pine Barrens	26	29	141	7	55.3	11.5	35.3
Cedar Glade	7	62	169	25	62.7	31.5	57.6
Shrub-Carr	10	28	116	15	51.8	14.3	38.2
Alder Thicket	12	31	115	25	60.5	45.2	33.8
Fen	6	41	115	37	61.3	51.2	41.3
Southern Sedge Meadow	44	28	197	17	51.1	17.9	44.7
Northern Sedge Meadow	35	29	177	24	51.9	27.6	41.6
Open Bog	17	27	130	30	53.7	51.8	36.3
Beach	23	14	131	32	34.1	28.6	37.3
Lake Dune	20	22	155	22	45.6	54.5	44.4
Exposed Cliff	13	23	157	27	35.2	39.1	52.3
Shaded Cliff	25	22	188	16	39.7	31.8	51.6
Emergent Aquatics	45	11	80	33	49.2	90.8	20.7
Submerged Aquatics	142	7	51	42	45.3	85.7	8.9

Table IV-3
Summary of climatic data averaged by communities
(Averages of all weather stations within mapped area of each type)

Community	Annual precipitation (cm)	Annual snowfall (cm)	Av. January temp. (°C.)	Av. July temp. (°C.)	Growing season (days)
Southern Forest—Xeric	78	108	−10.3°	21.3°	137
Southern Forest—Mesic	78	116	− 8.4	21.3	154
Southern Forest—Hydric	80	106	− 9.4	22.4	151
Northern Forest—Xeric	76	126	−11.8	20.2	123
Northern Forest—Mesic	78	133	−10.9	19.7	127
Northern Forest—Hydric	79	129	−10.4	20.4	136
Boreal Forest	68	136	−11.6	18.6	137
Prairie	79	93	− 8.9	22.1	152
Oak Opening	80	109	− 9.8	21.5	145
Pine Barrens	73	122	−12.7	19.9	118
Southern Sedge Meadow	76	102	−10.0	21.2	138

Table IV-4
Summary of soil analyses of A_1 layer averaged by communites

Community	W.R.C. %	pH	Available nutrients—p.p.m.			
			Ca	K	P	NH_4
Southern Forest—Dry	73	6.2	2210	84	25	12
Southern Forest—Dry Mesic	86	6.4	3025	86	28	15
Southern Forest—Mesic	75	6.9	3655	96	34	9
Southern Forest—Wet Mesic	118	6.1	3715	84	39	13
Southern Forest—Wet	62	7.0	3435	82	55	14
Northern Forest—Dry	120	4.9	1255	94	?	?
Northern Forest—Dry Mesic	127	5.2	1985	77	?	?
Northern Forest—Mesic	247	5.6	4215	93	?	?
Northern Forest—Wet Mesic	495	5.5	3780	84	9	18
Northern Forest—Wet	670	4.7	755	109	14	23
Boreal Forest (Wis. only)	223	5.1	2785	150	35	18
Prairie—Dry	62	7.8	5425	94	85	7
Prairie—Dry Mesic	54	7.1	3675	113	47	12
Prairie—Mesic	69	6.2	2700	129	38	29
Prairie—Wet Mesic	90	6.4	5125	123	50	25
Prairie—Wet	165	6.2	3900	89	32	21
Sand Barrens	29	6.2	50	18	30	5
Oak Opening	83	6.5	5220	124	22	12
Oak Barrens	60	6.5	1040	85	20	10
Pine Barrens	36	5.2	4025	69	22	?
Cedar Glade	89	6.5	4500	110	53	15
Southern Sedge Meadow	274	7.4	3750	90	14	14
Northern Sedge Meadow	320	5.7	1500	207	30	12
Open Bog	1070	3.9	750	100	10	17

APPENDIX FOR CHAPTER 5
Table V-1
Adaptation numbers for trees of southern forest

Species	Upland	Lowland	Species	Upland	Lowland
Acer negundo	1.0	3.0	Ostrya virginiana	8.5	9.0
A. rubrum	7.0	7.0	Platanus occidentalis	...	3.0
A. saccharinum	...	4.0	Populus deltoides	...	2.0
A. saccharum	10.0	10.0	P. grandidentata	4.5	...
Betula lutea	...	7.0	P. tremuloides	1.0	...
B. nigra	...	3.0	Prunus serotina	3.5	...
Carpinus caroliniana	8.0	9.0	Quercus alba	3.5	6.0
Carya cordiformis	8.5	7.0	Q. bicolor	...	4.0
C. ovata	4.5	6.0	Q. borealis	5.5	7.0
Celtis occidentalis	8.0	8.0	Q. ellipsoidalis	1.0	...
Fagus grandifolia	9.5	9.5	Q. macrocarpa	1.0	5.0
Fraxinus americana	6.5	8.0	Q. muhlenbergii Gray	1.0	...
F. nigra	...	6.0	Q. velutina	2.5	...
F. pennsylvanica	...	5.0	Salix nigra	...	1.0
Gymnocladus dioica	...	8.0	Tilia americana	7.5	7.0
Juglans cinerea	7.5	7.0	Ulmus americana	7.5	6.0
J. nigra	6.5	6.0	U. rubra	8.0	8.0

APPENDIX FOR CHAPTER 6

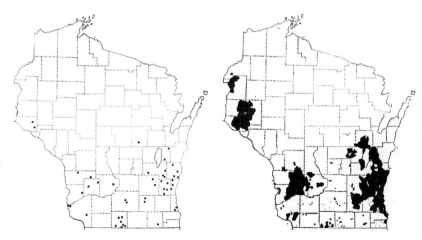

Figure VI-1.—Stands studied in southern mesic forest.

Figure VI-2.—Original mesic forest.

Table VI-1
Average tree composition of southern mesic forest

Species	Av. I.V.**	Constancy**	Species	Av. I.V.**	Constancy**
Acer saccharum*	126.0	100.0%	Carya ovata	1.8	19.1%
Tilia americana*	34.1	95.7	Celtis occidentalis	1.6	12.8
Fagus grandifolia	30.3	34.0	Betula papyrifera	1.2	6.4
Ulmus rubra*	25.5	72.3	Populus grandidentata	0.8	12.7
Quercus borealis	21.2	55.4	Juglans nigra	0.7	8.5
Ostrya virginiana*	15.8	78.8	Acer rubrum	0.6	4.2
Fraxinus americana	7.0	49.0	Fraxinus nigra	0.5	6.4
Quercus alba	5.7	38.3	Quercus bicolor	0.2	2.1
Carya cordiformis*	5.3	46.8	Fraxinus quadrangulata*	0.2	2.1
Juglans cinerea*	5.2	25.6	F. pennsylvanica	0.1	4.2
Ulmus americana	5.0	40.4	Quercus macrocarpa	0.1	2.1
Prunus serotina	1.9	31.9	Q. velutina	0.1	2.1
Gymnocladus dioica*	1.8	6.4	Ulmus thomasi	0.1	2.1

* Species which reach their optimum importance in Wisconsin in this forest.
** In this and similar tables, the quantitative figures for tree composition are given as average importance values (sum of relative density, relative frequency, and relative dominance) and as constancy percentages, which are based on the measured sample of 80 trees per stand rather than on a single sample of fixed area.

Table VI-2
Structure of a typical stand of southern mesic forest in Green County*

	Less than 1" d.b.h.		More than 1" d.b.h.				
Species	Less than 1' tall	More than 1' tall	1–4"	4–10"	10–20"	20–30"	30–40"
Acer saccharum	19,440	2,920	1,265.0	69.6	44.8	8.4	0.8
Carya cordiformis	114	6	11.0	6.8	0.2	0	0
Celtis occidentalis	8	2	0.2	0.4	0	0	0
Juglans cinerea	0	0	1.2	1.6	1.2	0	0
J. nigra	0	0	0	0	0	0.4	0
Ostrya virginiana	152	48	8.8	12.0	0	0	0
Prunus serotina	36	1	0.4	0.4	0	0	0
Quercus alba	0	0	0.2	0.4	0	0	0
Q. borealis	15	0	0.8	1.6	0.8	1.2	0.2
Tilia americana	110	8	19.0	12.4	6.0	1.2	0
Ulmus americana	0	0	4.0	5.2	2.0	0	0
U. rubra	224	7	2.0	19.6	12.0	1.2	0

* In this and other tables of forest structure, the figures represent numbers of plants per acre. (Multiply by 2.47 to get approximate numbers per hectare.)

Table VI-3
Prevalent groundlayer species of southern mesic forest

Species	Pres.	Av. freq.	Species	Pres.	Av. freq.
Actaea alba	38%	1.4%	Hepatica acutiloba*	62%	15.3%
Adiantum pedatum	43	4.0	Hydrophyllum virginianum*	60	12.3
Allium tricoccum*	78	9.3			
Amphicarpa bracteata	35	4.9	Laportea canadensis	43	7.7
Anemone quinquefolia	54	5.6	Osmorhiza claytoni	89	19.8
Arisaema triphyllum	57	12.9	Parthenocissus vitacea	65	12.8
Athyrium filix-femina	49	3.2	Phryma leptostachya	46	3.8
Botrychium virginianum	65	8.6	Podophyllum peltatum*	84	14.8
Brachyelytrum erectum	41	2.0	Polygonatum pubescens	70	19.2
Carex pensylvanica	60	8.3	Prenanthes alba	41	3.0
Caulophyllum thalictroides*	68	6.7	Ribes cynosbati	46	3.1
Celastrus scandens	38	1.9	Sanguinaria canadensis*	65	19.1
Circaea quadrisulcata	65	12.5	Sanicula gregaria	65	12.3
Claytonia virginica*	38	20.8	Smilacina racemosa	76	15.7
Cryptotaenia canadensis	35	6.3	Smilax ecirrhata	49	2.5
Galium aparine*	70	26.0	Thalictrum dioicum	57	7.6
G. concinnum	54	7.6	Trillium grandiflorum*	65	14.5
G. triflorum	41	2.9	Uvularia grandiflora	54	10.4
Geranium maculatum	78	15.7	Viola cucullata	65	15.4
Geum canadense	43	1.9	V. pubescens*	65	23.4

* Species are also modal, since their presence values are higher here than in any other Wisconsin community.

Additional modal species, with their presence (%) values: Aplectrum hyemale (8), Athyrium pycnocarpon (3), Camptosorus rhizophyllus (8), Cardamine douglassii (6), Carex convoluta (11), Carex davisii (3), Carex deweyana (3), Carex hirtifolia (3), Carex laxiflora (33), Carex tetanica (3), Dentaria laciniata (22), Ellisia nyctelea (3), Erythronium albidum (24), Floerkea proserpinacoides (14), Habenaria viridis (11), Hydrastis canadensis (5), Hydrophyllum appendiculatum (11), Impatiens pallida (11), Isopyrum biternatum (8), Phlox divaricata (19), Poa sylvestris (3), Solidago flexicaulis (22), Staphylea trifolia (16), Trillium recurvatum (16), Triphora trianthophora (3).

No. of modal species, 35.
No. of prevalent modal species as % of total prevalents, 25.6%.

Table VI-4
Community summary—southern mesic forest

Major dominants (I.V.): Acer saccharum (126), Tilia americana (34), Fagus grandifolia (30), Ulmus rubra (26), Quercus borealis (21).

Most prevalent groundlayer species (P%): Osmorhiza claytoni (89), Podophyllum peltatum (84), Allium tricoccum (78), Geranium maculatum (78), Smilacina racemosa (76).

Leading families (% of total species): Liliaceae (9.3), Compositae (8.7), Ranunculaceae (7.8), Cyperaceae (6.8), Caprifoliaceae (5.4).

Related communities (Index of similarity): Southern Dry-Mesic Forest (66), Southern Dry Forest (56), Northern Mesic Forest (49), Southern Wet-Mesic Forest (49), Northern Dry-Mesic Forest (44).

Species density: 39.

Index of Homogeneity: 54.4%.

No. of stands studied: 47.

Number of species: Trees 25, Shrubs 33, Herbs 172, Total 230.

Stability: Very stable—a terminal forest.

Climate: Total ppt. 30.6" (78 cm), Snowfall 45.8" (116 cm), Jan. Temp. 16.9°F. (−8.4°C.), July Temp. 70.3°F. (21.3°C.), Growing season 154 days. Typical weather stations: Milwaukee, Richland Center, West Bend.

Catena position: Middle. No gley layer.

Soil group: Brown Forest and Gray-Brown Podzolic.

Major soil series: Parr, Waupun, Warsaw, Elba, Fayette, and Downs.

Soil analyses: w.r.c. 75%, pH 6.9, ca 3655 p.p.m., K 95 p.p.m., P 33 p.p.m.

Approximate original area: 3,432,500 acres.

Typical examples: Wyalusing S.A., Wychwood S.A., Mauthe Lake S.A., Grant Park, Milwaukee, Petrifying Springs Park, Kenosha.

Major publications: Curtis and McIntosh (1951), Whitford (1951), Gilbert and Curtis (1953), Randall (1953), Ward (1956).

Geographical distribution: Southern Minnesota and eastern Iowa, south and east to optimum development in southern Appalachians.

APPENDIX FOR CHAPTER 7

Figure VII-1.—Stands studied in southern xeric forest.

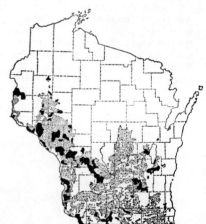

Figure VII-2. Original oak forests. The dotted areas were the closely related oak savannas.

Table VII-1
Average tree composition of southern xeric forest

Species	Dry Forest		Dry-mesic forest	
	Av. I. V.	Constancy	Av. I. V.	Constancy
Quercus alba	80.3*	88%	52.2	92%
Q. borealis	21.7	54	104.2*	97
Q. velutina	98.3	92	5.5	34
Tilia americana	0.8	10	28.6	77
Prunus serotina	23.2*	86	5.8	60
Quercus macrocarpa	25.6	64	2.3	19
Acer saccharum	0.2	2	22.9	47
Ulmus rubra	3.8	20	17.1	61
Carya ovata	8.2	53	8.5	49
Fraxinus americana	1.2	10	11.5*	44
Quercus ellipsoidalis	10.6	10	....	..
Ostrya virginiana	0.6	10	9.1	47
Populus grandidentata	1.3	18	6.6	32
Ulmus americana	3.7	24	4.0	30
Acer rubrum	1.5	14	5.1	25
Carya cordiformis	2.1	16	4.5	43
Juglans nigra	2.7	34	3.2*	25
J. cinerea	0.2	6	2.9	25
Quercus muhlenbergii	2.6*	2	....	..
Acer negundo	1.8	10	0.1	3
Populus tremuloides	1.8	16	0.1	3
Betula papyrifera	0.4	8	0.9	12
Fraxinus pennsylvanica	0.6	2	0.5	4
Fagus grandifolia	....	..	0.4	23
Ulmus thomasi	....	..	0.3	3
Celtis occidentalis	0.2	2	....	..
Fraxinus nigra	....	..	0.1	5
Quercus bicolor	....	..	0.1	1
Betula lutea	....	..	0.1	1

* Species which reach their optimum in Wisconsin in these communities.

Table VII-2
Structure of two typical stands of southern xeric forests

	Less than 1" d.b.h.		More than 1" d.b.h.				
Species	Less than 1' tall	More than 1' tall	1–4"	4–10"	10–20"	20–30"	30–40"
Dry oak forest in Dane County (in part after Cottam, 1949)							
Acer negundo	5	18	2	0	0	0	0
A. rubrum	0	0	1	1	0	0	0
Carya ovata	45	66	11	1	1	0	0
Ostrya virginiana	0	15	3	1	0	0	0
Populus grandidentata	0	0	2	1	0	0	0
P. tremuloides	0	13	1	0	0	0	0
Prunus serotina	996	235	50	5	1	0	0
Quercus alba	3000	810	29	38	48	4	1
Q. macrocarpa	0	2	0	3	1	1	1
Q. velutina	1580	198	11	9	23	5	0
Ulmus rubra	12	43	11	0	0	0	0
Structure of a dry-mesic oak forest in Columbia County							
Acer negundo	0	0	1	0	0	0	0
A. saccharum	1420	317	2	3	0	0	0
Carya cordiformis	110	22	13	3	0	0	0
C. ovata	0	0	6	2	0	0	0
Fraxinus americana	417	286	1	0	0	0	0
Juglans cinerea	0	0	4	1	0	0	0
Ostrya virginiana	76	654	62	12	0	0	0
Prunus serotina	233	100	1	2	0	0	0
Quercus alba	0	20	0	2	0	1	0
Q. borealis	41	87	2	4	12	7	1
Q. macrocarpa	0	0	0	2	1	0	0
Tilia americana	86	147	5	8	4	0	0
Ulmus americana	10	102	9	5	0	0	0
U. rubra	467	300	13	9	2	0	0

Table VII-3
Prevalent groundlayer species of southern dry forest

Species	Pres.	Av. freq.	Species	Pres.	Av. freq.
Adiantum pedatum	53%	1.6%	Lonicera prolifera	43%	3.1%
Agrimonia gryposepala*	60	1.5	Monarda fistulosa	47	6.2
Amphicarpa bracteata	77	16.7	Osmorhiza claytoni	77	15.0
Anemone cylindrica	40	1.9	Parthenocissus vitacea*	87	28.8
A. quinquefolia	47	8.1	Phryma leptostachya	70	11.0
Apocynum androsaemifolium	70	5.0	Podophyllum peltatum	43	6.7
Aralia nudicaulis	60	10.1	Polygonatum canaliculatum	50	4.6
Arisaema triphyllum	40	3.2	Potentilla simplex*	53	3.7
Aster sagittifolius	47	3.5	Prenanthes alba	70	3.9
Athyrium filix-femina	50	2.0	Pteridium aquilinum	67	9.3
Botrychium virginianum	67	4.9	Pyrola elliptica	50	3.5
Carex pensylvanica	53	18.9	Rhus radicans	70	10.2
Celastrus scandens	37	5.0	Ribes cynosbati	50	4.7
Circaea quadrisulcata	77	18.2	Rosa sp.	67	3.5
Cornus racemosa*	87	23.6	Rubus alleghenensis*	67	15.9
Corylus americana*	87	15.7	R. strigosus	53	10.6
Cypripedium pubescens*	40	0.8	Sambucus canadensis*	53	1.6
Desmodium glutinosum	80	17.1	Sanicula gregaria	50	7.5
Euphorbia corollata	37	4.8	Smilacina racemosa*	100	28.9
Fragaria virginiana	73	9.4	S. stellata*	63	7.4
Galium boreale	53	5.4	Smilax ecirrhata	43	2.2
G. concinnum	67	20.1	S. herbacea*	43	3.2
Geranium maculatum	90	41.2	Uvularia grandiflora	43	4.7
Geum canadense	50	5.1	Veronicastrum virginicum	40	0.9
Helianthus strumosus*	67	13.4	Viola cucullata	70	7.6
Hystrix patula	40	1.1	Vitis aestivalis	53	4.8
Lactuca biennis*	53	3.0	V. riparia	43	2.5

*Species are also modal, since their presence values are higher here than in any other Wisconsin community.

Additional modal species, with their presence (%) values: Arabis canadensis (10), Asclepias ovalifolia (7), Cynoglossum virginianum (3), Desmodium cuspidatum (10), Festuca obtusa (17), Liparis liliifolia (3), Lithospermum latifolium (7), Lonicera dioica (13), Polygonum scandens (7), Ribes missouriense (27).

No. of modal species 23.
No. of prevalent modal species as % of total prevalents 24.1%.

Table VII-4
Community summary—southern dry forest

Major dominants (I.V.): Quercus velutina (98), Q. alba (80), Q. macrocarpa (26), Prunus serotina (23), Q. borealis (22).

Most prevalent groundlayer species (P%): Smilacina racemosa (100), Geranium maculatum (90), Cornus racemosa (87), Corylus americana (87), Parthenocissus vitacea (87).

Leading families (% of total species): Compositae (15.1), Liliaceae (6.0), Leguminosae (5.3), Gramineae (4.9), Ranunculaceae (4.9).

Related communities (Index of similarity): Southern Dry-Mesic Forest (76), Southern Mesic Forest (56), Oak Opening (53), Northern Dry-Mesic Forest (45), Oak Barrens (43).

Species density: 54.

Index of Homogeneity: 59.6.

No. of stands studied: 30.

Number of species: Trees 24, Shrubs 44, Herbs 221, Total 289.

Stability: Unstable in absence of fire. Succeeded by more mesic types.

Catena position: Top. No gley layer.

Soil group: Gray-Brown Podzolic.

Major soil series: Fox, Rodman, Wyocena, Coloma, Hixton, Alvin, and Sogn.

Soil analyses: w.r.c. 73%, pH 6.2, Ca 2210 p.p.m., K 85 p.p.m., P 25 p.p.m.

Approximate original area: Uncertain. Probably 70% of total oak forest or 971,000 acres, but no accurate criteria are available for separating this community from the oak savannas.

Typical examples: U. W. Arboretum, Observatory S.A., Browntown S.A.

Major publications: Cottam (1949), Curtis and McIntosh (1951).

Geographical distribution: Upper Mississippi valley, south to Kentucky and the Ozark Mountains.

Table VII-5
Prevalent groundlayer species of southern dry-mesic forest

Species	Pres.	Av. freq.	Species	Pres.	Av. freq.
Adiantum pedatum*	81%	6.9%	C. racemosa	70%	11.5%
Agrimonia gryposepala	43	1.0	C. rugosa	43	2.2
Amphicarpa bracteata*	94	21.6	Corylus americana	82	10.7
Anemone quinquefolia	65	6.2	Cryptotaenia canadensis*	59	5.4
A. virginiana*	41	0.8	Desmodium glutinosum	93	7.9
Apocynum androsaemifolium	52	2.6	Dioscorea villosa*	57	2.7
Aralia nudicaulis	76	11.9	Fragaria virginiana	57	4.1
A. racemosa*	61	1.2	Galium aparine	50	10.1
Arisaema triphyllum*	81	17.2	G. concinnum*	93	26.0
Aster saggitfolius*	54	3.9	G. triflorum	50	4.3
A. shortii*	61	4.7	Geranium maculatum*	100	35.8
Athyrium filix-femina*	74	7.6	Geum canadense*	50	6.4
Botrychium virginianum*	83	6.3	Helianthus strumosus	63	6.7
Brachyelytrum erectum*	67	7.2	Hydrophyllum virginianum	44	5.5
Carex pensylvanica*	78	14.4	Hystrix patula*	67	2.6
Caulophyllum thalictroides	65	3.3	Lactuca spicata	52	2.0
Celastrus scandens*	67	7.8	Lonicera prolifera*	57	3.6
Circaea quadrisulcata*	89	24.6	Osmorhiza claytoni*	94	23.6
Cornus alternifolia*	48	1.6	Osmunda claytoniana*	46	3.9
			Parietaria pensylvanica*	43	4.6

Table VII-5 (cont.)

Species	Pres.	Av. freq.	Species	Pres.	Av. freq.
Parthenocissus vitacea	85%	23.9%	Sanicula gregaria*	83%	15.1%
Phryma leptostachya*	83	15.0	Smilacina racemosa	98	25.8
Podophyllum peltatum	70	6.9	Smilax ecirrhata*	72	2.6
Polygonatum pubescens	44	4.7	S. herbacea	61	2.1
Prenanthes alba*	80	4.0	Solidago ulmifolia*	59	6.8
Pteridium aquilinum	54	9.0	Thalictrum dioicum*	72	9.4
Ranunculus abortivus*	48	2.6	Triosteum perfoliatum*	52	1.0
Rhus radicans	72	9.8	Uvularia grandiflora*	93	16.6
Ribes cynosbati*	74	3.6	Veronicastrum virginicum	48	1.5
Rosa sp.	56	1.6	Viola cucullata*	70	11.1
Rubus allegheniensis	52	7.9	V. pubescens	59	12.9
R. strigosus	48	9.5	Vitis aestivalis*	69	4.5
Sambucus canadensis	44	0.8	Zanthoxylum americanum*	48	4.4
Sanguinaria canadensis	65	12.8			

* Species are also modal, since their presence values are higher here than in any other Wisconsin community.

Additional modal species, with their presence (%) values: Agastache scrophulariaefolia (7), Anemonella thalictroides (20), Asclepias exaltata (22), Aster prenanthoides (7), Bromus purgans (24), Campanula americana (22), Carex hitchcockiana (3), Carex jamesii (31), Cirsium altissimum (15), Conopholis americana (17), Corallorhiza odontorhiza (2), Desmodium nudiflorum (26), Elymus villosus (11), Erigeron pulchellus (39), Eupatorium purpureum (33), Eupatorium rugosum (28), Galium circaezans (37), Goodyera pubescens (20), Hackelia virginiana (20), Heracleum lanatum (7), Hieracium scabrum (9), Lathyrus ochroleucus (28), Orchis spectabilis (26), Oryzopsis racemosa (11), Osmorhiza longistylis (37), Panax quinquefolium (22), Panicum latifolium (24), Polemonium reptans (26), Polymnia canadensis (4), Ranunculus recurvatus (15), Rubus occidentalis (35), Scrophularia marilandica (4), Smilax hispida (41), Trillium gleasoni (28), Viburnum rafinesquianum (32), Viburnum lentago (15).

No. of modal species 73.

No. of prevalent modal species as % of total prevalents 56.1%.

Table VII-6
Community summary—southern dry-mesic forest

Major dominants (I.V.): Quercus borealis (104), Q. alba (52), Tilia americana (29), Acer saccharum (23), Ulmus rubra (17).

Most prevalent groundlayer species (P%): Geranium maculatum (100), Smilacina racemosa (98), Amphicarpa bracteata (94), Osmorhiza claytoni (94), Desmodium glutinosum (93).

Leading families (% of total species): Compositae (10.0), Ranunculaceae (7.2), Liliaceae (6.8), Gramineae (6.0), Rosaceae (5.2).

Related communities (Index of similarity): Southern Dry Forest (76), Southern Mesic Forest (66), Northern Dry-Mesic Forest (48), Southern Wet-Mesic Forest (43), Northern Mesic Forest (40).

Species density: 66.

Index of Homogeneity: 65.3.

No. of stands studied: 54.

Number of species: Trees 26, Shrubs 48, Herbs 201, Total 275.

Stability: One generation stands, succeeded by Southern Mesic Forest.

Climate: Total ppt. 30.6″ (78 cm), Snowfall 42.5″ (108 cm), Jan. Temp. 13.4°F. (−10.3°C.), July Temp. 70.4°F. (21.3°C.), Growing season—137 days. Typical weather stations: Blair, Osceola, Oshkosh.

Catena position: Middle. No gley layer.

Soil group: Gray-Brown Podzolic.

Major soil series: Fox, McHenry, Miami, Sogn, Dubuque, and Wyocena.

Soil analyses: w.r.c. 86%, pH 6.4, Ca 3015 p.p.m., K 85 p.p.m., P 28 p.p.m.

Approximate original area: Uncertain. Probably 30% of total oak forest, or 416,000 acres. Much more widespread in post-settlement times.

Typical examples: Devils Lake S.A., Wyalusing S.A., Perrot State Park, Wychwood S.A., South Woods Association Reserve at Ripon, Fond du Lac Co., New Glarus Roadside Park, U. W. Arboretum.

Major publications: Curtis and McIntosh (1951), Whitford (1951), Larsen (1953).

Geographical distribution: Largely confined to upper Mississippi valley.

APPENDIX FOR CHAPTER 8

Figure VIII-1.—Stands studied in southern lowland forest.

Figure VIII-2.—Original lowland hardwood forest.

Table VIII-1
Average tree composition of southern lowland forest

Species	Wet forest		Wet-mesic forest	
	Av. I. V.	Constancy	Av. I. V.	Constancy
Acer saccharinum	81.6*	81.5%	58.2	60.4%
Ulmus americana	26.5	66.7	73.7*	90.0
Salix nigra	64.0*	70.3	2.1	9.9
Populus deltoides	54.5*	70.4	0.7	8.6
Fraxinus pennsylvanica	8.2	51.9	26.9*	63.0
Betula nigra	24.4*	51.8	6.7	17.3
Quercus bicolor	15.2*	29.6	14.7	48.0
Tilia americana	1.6	11.1	24.4	64.1
Fraxinus nigra	2.9	18.5	16.2	50.6
Acer rubrum	....	....	10.2	19.6
Quercus borealis	0.3	3.7	9.7	32.1
Fraxinus americana	0.8	11.1	8.0	35.8
Acer saccharum	....	....	8.2	21.0
Quercus macrocarpa	5.8	3.7	2.4	12.3
Ulmus rubra	0.8	3.7	6.3	22.2
Carya ovata	0.2	3.7	5.1	14.8
Quercus alba	0.2	3.7	4.4	13.6
Q. velutina	3.6	3.7	1.0	11.1
Acer negundo	3.0*	22.2	1.2	12.3
Celtis occidentalis	....	....	3.9*	19.7
Carya cordiformis	0.4	7.4	2.8	30.8
Betula lutea	....	....	2.4	16.1
Prunus serotina	0.7	3.7	1.6	17.8
Ostrya virginiana	....	....	2.1	14.8
Fagus grandifolia	....	....	1.2	4.9
Juglans cinerea	....	....	0.9	9.9
Populus tremuloides	0.2	3.7	0.3	6.2
Ulmus thomasi	....	....	0.5*	2.5
Populus grandidentata	....	....	0.3	3.7
Gymnocladus dioica	....	....	0.3	2.5
Salix amygdaloides	0.2*	3.7	....	....
Betula papyrifera	....	....	0.2	3.7
Aesculus glabra	....	....	0.1*	1.2
Carpinus caroliniana	....	....	0.1	1.2
Gleditsia triacanthos	....	....	0.1*	1.2
Juglans nigra	....	....	0.1	2.5
Platanus occidentalis	....	....	0.1*	1.2

* Species which attain their optimum importance in these communities.

Table VIII-2
Structure of two typical stands of southern lowland forests

	Less than 1" d.b.h.		More than 1" d.b.h.					
Species	Less than 1' tall	More than 1' tall	1–4"	4–10"	10–20"	20–30"	30–40"	40+"
Wet forest in Dane County								
Acer negundo	154	0	4	13	0	0	0	0
A. saccharinum	422	158	0	0	0	0	0	0
Fraxinus pennsylvanica	154	0	3	0	0	0	0	0
Populus deltoides	0	0	2	78	66	2	0	0
Salix nigra	0	212	16	20	22	0	0	0
Ulmus americana	212	78	2	0	0	0	0	0
Wet-mesic forest in Rock County								
Acer saccharinum	428	0	6	26	29	4	0	2
Carya cordiformis	0	0	1	2	6	0	0	0
C. ovata	0	0	0	2	5	0	0	0
Fraxinus pennsylvanica	288	428	1	6	6	8	4	0
Quercus bicolor	0	0	0	4	4	1	4	2
Q. borealis	64	0	0	0	0	1	0	0
Platanus occidentalis	0	20	2	1	0	0	0	0
Tilia americana	112	0	0	4	0	0	0	0
Ulmus americana	1572	0	5	14	12	1	2	4
U. rubra	0	0	0	0	2	1	0	0

APPENDIX 531

Table VIII-3
Prevalent groundlayer species of southern wet forest

Species	Pres.	Av. freq.	Species	Pres.	Av. freq.
Amphicarpa bracteata	44%	14.5%	Onoclea sensibilis	44%	4.1%
Arisaema dracontium	31	0.6	Parthenocissus vitacea	56	14.9
Aster lateriflorus*	44	12.3	Pilea pumila*	47	15.2
Boehmeria cylindrica*	50	16.1	Rhus radicans*	81	6.5
Carex typhina*	31	6.6	Rudbeckia laciniata*	31	3.5
Cinna arundinacea*	31	3.5	Sambucus canadensis	50	4.8
Cryptotaenia canadensis	50	14.5	Scutellaria lateriflora*	56	10.1
Elymus virginicus*	50	12.0	Solidago gigantea	56	11.1
Geum canadense	44	9.6	Stachys hispida*	31	2.6
Impatiens biflora	50	24.7	Steironema ciliatum	31	7.1
Laportea canadensis*	81	19.5	Teucrium canadense*	56	14.6
Leersia virginica*	56	35.3	Urtica dioica	47	11.7
Lycopus uniflorus	50	4.1	Viola cucullata	50	3.4
Menispermum canadensis*	44	1.6	Vitis riparia*	81	6.3
Muhlenbergia frondosa*	50	8.6			

* Species are also modal, since their presence values are higher here than in any other Wisconsin community.

Additional modal species, with their presence (%) values: Apios americana (31), Carex grayii (19), Carex lupulina (27), Carex muskingumensis (7), Carex normalis (7), Carex retrorsa (7), Carex tribuloides (7), Carex vesicaria (7), Carex vulpinoidea (13). Cephalanthus occidentalis (25), Dracocephalum formosius (19), Euonymus atropurpureus (6), Lobelia cardinalis (19), Lysimachia nummularia (25), Rumex altissimus (7), Salix longifolia (7), Silphium perfoliatum (19).

No. of modal species 33.
No. of prevalent modal species as % of total prevalents 55.2%.

Table VIII-4
Community summary—southern wet forest

Major dominants (I. V.): Acer saccharinum (82), Salix nigra (64), Populus deltoides (55), Ulmus americana (27), Betula nigra (24).
Most prevalent groundlayer species (P%): Laportea canadensis (81), Rhus radicans (81), Vitis riparia (81), Leersia virginica (56), Parthenocissus vitacea (56).
Leading families (% of total species): Compositae (11.7), Cyperaceae (10.4), Gramineae (7.1), Rosaceae (5.8), Labiatae (5.8).
Related communities (Index of similarity): Southern Wet-Mesic Forest (63), Southern Sedge Meadow (30), Shrub-Carr (29), Southern Mesic Forest (27), Northern Sedge Meadow (25).
Species density: 29.
Index of Homogeneity: 48.8.
No. of stands studied: 16.
Number of species: Trees 21, Shrubs 26, Herbs 128, Total 175.
Stability: Subject to frequent catastrophe by flooding. Succeeded by more mesic forests.
Climate: Total ppt. 31.5" (80 cm), Snowfall 41.9" (106 cm), Jan. Temp. 15.0°F. (−9.4°C.), July Temp. 72.3°F. (22.4°C.), Growing season 151 days. Typical weather stations: Muscoda, Wisconsin Rapids, Necedah.

Table VIII-4 (cont.)

Catena position: Bottom. Gley layer near surface.
Soil groups: Alluvial and Humic-Gley.
Major soil series: Mostly undifferentiated colluvial soils.
Soil analyses: w.r.c. 62%, pH 7.0, Ca 3440 p.p.m., K 82 p.p.m., P 55 p.p.m.
Approximate original area: Very small, probably only 20% of total bottomland forest, or 84,000 acres.
Typical examples: Wyalusing S. A. Shot Tower State Park, Merrick State Park, U. W. Arboretum.
Major publications: Ware (1956 thesis).
Geographic distribution: Along rivers, Minnesota to New York and New Jersey, increasing in complexity rapidly toward the south in the Mississippi and Ohio valleys.

Table VIII-5
Prevalent groundlayer species of southern wet-mesic forest

Species	Pres.	Av. freq.	Species	Pres.	Av. freq.
Amphicarpa bracteata	34%	10.1%	Menispermum canadense	34%	4.8%
Arenaria lateriflora	34	9.0	Onoclea sensibilis	56	6.9
Arisaema triphyllum	66	17.2	Osmorhiza claytoni	33	6.6
A. dracontium*	44	2.5	Parthenocissus vitacea	80	23.0
Aster lateriflorus	41	12.5	Polygonatum pubescens	33	6.2
Athyrium filix-femina	39	3.8	Ranunculus abortivus	47	5.0
Boehmeria cylindrica	47	7.6	Rhus radicans	59	6.7
Circaea quadrisulcata	34	8.8	Ribes americanum	48	6.8
Cryptotaenia canadensis	45	12.9	Sambucus canadensis	42	3.2
Cuscuta gronovii*	31	3.3	Sanicula gregaria	36	13.1
Dioscorea villosa	31	3.3	Smilacina stellata	34	5.5
Elymus virginicus	39	9.1	Smilax ecirrhata	41	4.7
Galium triflorum	44	6.4	S. herbacea	41	2.0
Geum canadense	61	11.5	Solanum dulcamara*	39	4.2
Glyceria striata	41	6.7	Solidago gigantea	34	6.3
Impatiens biflora	67	21.4	Steironema ciliatum*	57	10.2
Laportea canadensis	77	39.7	Viola cucullata	63	16.3
Leersia virginica	36	11.8	V. pubescens	36	11.2
Lycopus uniflorus	36	5.7	Vitis riparia	58	3.4
			Zanthoxylum americanum	36	3.9

* Species are also modal, since their presence values are higher here than in any other Wisconsin community.

Additional modal species, with their presence (%) values: Bidens frondosa (20), Blephilia hirsuta (3), Botrychium dissectum (2), Bromus latiglumis (19), Carex alopecoidea (11), Carex amphibola (6), Carex bromoides (12), Carex cephalophora (2), Carex crinita (14), Carex cristatella (17), Carex debilis (2), Carex gracillima (11), Carex grisea (2), Carex molesta (2), Carex projecta (17), Carex rosea (23), Carex sparganioides (23), Carex sprengelii (3), Carex stipata (11), Carex tuckermani (6), Cassia marilandica (2), Chaerophyllum procumbens (2), Cornus purpusi (12), Echinocystis lobata (28), Hypericum punctatum (2), Osmunda regalis (19), Symplocarpus foetidus (20).

No. of modal species 31.
No. of prevalent modal species as % of total prevalents 10.2%.

Table VIII-6
Community summary—wet-mesic forest

Major dominants (I. V.): Ulmus americana (74), Acer saccharinum (58), Fraxinus pennsylvanica (27), Tilia americana (24), Fraxinus nigra (16).

Most prevalent groundlayer species (P%): Parthenocissus vitacea (80), Laportea canadensis (77), Impatiens biflora (67), Arisaema triphyllum (66), Viola cucullata (63).

Leading families (% of total species): Cyperaceae (12.5), Compositae (8.8), Gramineae (7.4), Ranunculaceae (6.7), Liliaceae (6.4).

Related communities (Index of similarity): Southern Wet Forest (63), Southern Mesic Forest (49), Southern Dry-Mesic Forest (43), Southern Dry Forest (42), Northern Wet Mesic (38).

Species density: 39.

Index of Homogeneity: 44.5.

No. of stands studied: 64.

Number of species: Trees 36, Shrubs 42, Herbs 255, Total 333.

Stability: Relatively stable. Succeeded by Southern Mesic Forest only by physiographic changes in water supply.

Catena position: Low. Gley layer present.

Soil group: Alluvial and Humic-Gley.

Major soil series: Genesee, Arenzville.

Soil analyses: w.r.c. 129%, pH 6.1, Ca 3715 p.p.m., K 85 p.p.m., P 38 p.p.m.

Approximate original area: Uncertain. Probably 80% of total bottomland or 336,000 acres.

Typical examples: Wyalusing S.A.; Shot Tower State Park.

Major publications: Ware (1956 thesis).

Geographical distribution: Along rivers and on flat lands, Minnesota to New Jersey, increasing in complexity rapidly toward the south in the Mississippi and Ohio valleys.

APPENDIX FOR CHAPTER 10

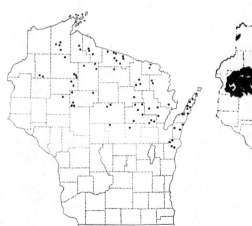

Figure X-1.—Stands studied in northern mesic forest.

Figure X-2.—Original northern mesic forest.

Table X-1
Average tree composition of northern mesic forest

Species	Av. I.V.	Constancy	Species	Av. I.V.	Constancy
Acer saccharum	106.6	96%	Pinus strobus	1.6	10%
Tsuga canadensis*	79.4	73	Fraxinus nigra	1.2	16
Fagus grandifolia*	40.2	41	Populus grandidentata	0.9	7
Betula lutea	29.4	77	Prunus serotina	0.9	7
Tilia americana	15.7	69	Carya cordiformis	0.7	6
Ostrya virginiana	7.2	64	Populus tremuloides	0.5	3
Quercus borealis	7.2	27	Juglans cinerea	0.4	9
Ulmus americana	6.6	37	Quercus alba	0.3	4
Acer rubrum	5.2	34	Acer spicatum	0.2	4
Betula papyrifera	5.0	33	Picea glauca	0.2	1
Fraxinus americana	4.0	37	Carya ovata	0.1	1
Ulmus rubra	3.3	19	Prunus pensylvanica	0.1	1
Abies balsamea	2.9	30	Quercus ellipsoidalis	0.1	1
Thuja occidentalis	1.7	17			

* Species which reach optimum values in this forest.

Table X-2
Structure of two typical stands of northern mesic forests

	Less than 1" d.b.h.		More than 1" d.b.h.				
Species	Less than 1' tall	More than 1' tall	1–4"	4–10"	10–20"	20–30"	30+"
Forest in Door County (Peninsula Park Scientific Area)							
Acer saccharum	17,280	960	11	17	41	0	0
Betula papyrifera	0	0	1	12	0	0	0
Fagus grandifolia	0	0	1	12	0	0	0
Fraxinus americana	2560	480	0	0	0	0	0
Ostrya virginiana	640	1440	3	0	0	0	0
Pinus strobus	0	0	0	0	2	1	1
Prunus serotina	160	0	0	1	0	0	0
Quercus borealis	160	0	0	0	2	0	0
Tsuga canadensis	0	0	6	0	4	1	0
Forest in Vilas County (Star Lake Scientific Area)							
Abies balsamea	40	15	23	1	0	0	0
Acer saccharum	20,160	5680	72	36	16	2	0
Betula lutea	7100	7	6	3	20	3	0
Ostrya virginiana	80	100	18	6	0	0	0
Picea glauca	0	0	0	0	1	0	0
Pinus strobus	0	0	0	0	2	2	1
Tilia americana	130	65	15	6	7	0	0
Tsuga canadensis	260	2	17	18	18	4	0
Ulmus americana	65	90	5	8	4	3	0

Table X-3
Prevalent groundlayer species of northern mesic forest

Species	Pres.	Av. freq.	Species	Pres.	Av. freq.
Actaea alba*	62%	2.7%	L. obscurum	47%	3.0%
Adiantum pedatum	41	3.6	Maianthemum canadense	85	31.0
Anemone quinquefolia	59	7.9	Mitchella repens	59	6.4
Aralia nudicaulis	71	11.9	Mitella nuda	67	1.1
A. racemosa	38	1.3	Oryzopsis asperifolia	52	4.3
Aster macrophyllus	59	11.5	Osmorhiza claytoni	61	12.9
Athyrium filix-femina	41	1.3	Polygonatum pubescens*	82	19.3
Botrychium virginianum	38	1.6	Sambucus pubens*	44	1.8
Clintonia borealis	58	6.8	Smilacina racemosa	48	8.3
Corylus cornuta	41	4.1	Streptopus roseus	73	12.6
Dirca palustris*	36	2.1	Trientalis borealis	62	8.4
Dryopteris austriaca	65	10.6	Trillium grandiflorum	52	10.4
Galium triflorum	67	9.4	Uvularia grandiflora	47	4.6
Gymnocarpium dryopteris	47	2.0	U. sessilifolia	39	4.8
Lonicera oblongifolia	52	1.7	Viola cucullata	50	5.2
Lycopodium annotinum	65	1.5	V. pubescens	64	1.1
L. lucidulum	52	6.6			

* Species are also modal, since their presence values are higher here than in any other Wisconsin community.

Additional modal species, with their presenec- (%) values: Athyrium thelypteroides (9), Botrychium lanceolatum (3), Carex nigro-marginata (2), Carex plantaginea (9), Claytonia caroliniana (14), Dentaria diphylla (5), Dicentra canadensis (14), Epifagus virginiana (6), Erythronium americanum (9), Galium lanceolatum (11), Milium effusum (5), Mitella diphylla (35), Viola canadensis (33).

No. of modal species 17.

No. of prevalent modal species as % of total prevalents 12.1%.

Table X-4
Community summary—northern mesic forest

Major dominants (I. V.): Acer saccharum (107), Tsuga canadensis (79), Fagus grandifolia (40), Betula lutea (29), Tilia americana (16).

Most prevalent groundlayer species (P%): Maianthemum canadense (85); Polygonatum pubescens (82), Streptopus roseus (73), Aralia nudicaulis (71), Mitella nuda (67).

Leading families (% of total species): Liliaceae (8.4), Cyperaceae (7.5), Rosaceae (5.7), Compositae (5.7), Ranunculaceae (4.8).

Related communities (Index of similarity): Northern Dry-Mesic Forest (65), Boreal Forest (56), Northern Wet-Mesic Forest (54), Southern Mesic Forest (49), Northern Dry Forest (48).

Species density: 33.

Index of Homogeneity: 54.8.

No. of stands studied: 66.

Number of species: Trees 27, Shrubs 46, Herbs 181, Total 254.

Stability: Very stable—a terminal forest.

Climate: Total ppt. 30.8″ (78 cm), Snowfall 52.1″ (132 cm), Jan. Temp. 12.4°F. (−10.9°C.), July Temp. 67.4°F. (19.7°C.), Growing season 127 days. Typical weather stations: Antigo, Butternut, Wausau.

Catena position: Middle. No gley layer.

Soil group: Gray-Brown and Brown Podzolic.

Major soil series: Kewaunee, Onaway, Kennan, Marenisco, Otterholt, and Spencer.

Soil analyses: w.r.c. 250%, pH 5.6, Ca 4215 p.p.m., K 95 p.p.m.

Approximate original area: 11,750,000 acres.

Typical examples: Plum Lake S. A., Flambeau Forest S. A., Peninsula Park Beech S. A., Menomonee Indian Reservation, Pike Lake Memorial Forest (in Price County).

Major publications: Potzger (1946), Stearns (1950), Brown and Curtis (1952).

Geographical distribution: From Wisconsin east through Michigan and S. Ontario to New York and Pennsylvania.

APPENDIX FOR CHAPTER 11

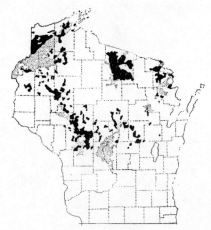

Figure XI-1.—Stands studied in northern xeric forest.

Figure XI-2.—Original pine forests. The dotted areas were pine barrens.

Table XI-1
Average tree composition of northern xeric forest

Species	Dry forest		Dry-mesic forest	
	Av. I. V.	Constancy	Av. I. V.	Constancy
Pinus strobus	43.0	50%	74.5*	71%
P. banksiana	64.7	57	0.1	4
P. resinosa	48.1*	68	15.6	42
Quercus borealis	11.7	48	36.2	67
Acer rubrum	10.1	50	36.6*	91
Quercus ellipsoidalis	37.3	66	6.3	27
Betula papyrifera	12.1	43	29.6*	71
Populus tremuloides	20.7	68	13.6	51
Quercus alba	19.9	36	9.2	27
Populus grandidentata	19.7*	50	8.3	47
Acer saccharum	6.8	27	27.0	69
Tsuga canadensis	0.1	2	15.0	31
Quercus macrocarpa	9.1	16	0.8	9
Abies balsamea	0.6	11	6.1	36
Betula lutea	....	..	4.3	40
Tilia americana	0.3	7	4.0	33
Thuja occidentalis	....	..	3.3	16
Prunus serotina	1.7	23	1.3	24
Fagus grandifolia	....	..	2.9	7
Ulmus americana	0.3	7	2.0	16
Fraxinus nigra	....	..	2.2	16
Picea glauca	0.7	9	0.8	11
Fraxinus americana	0.3	5	1.1	18
Ostrya virginiana	0.1	5	1.2	20
Populus balsamifera	1.1	5	0.2	4
Ulmus rubra	....	..	1.1	7
Prunus pensylvanica	0.7	2	....	..
Carya cordiformis	....	..	0.5	11
Fraxinus pennsylvanica	0.3	2	0.1	2
Juglans cinerea	....	..	0.4	9
Carya ovata	0.3	2	....	..
Betula nigra	0.2	2	....	..
Amelanchier sp.	....	..	0.2	7
Carpinus caroliniana	....	..	0.1	2

* Species which attain optimum importance in indicated segment.

Table XI-2
Structure of two typical stands of northern xeric forests

	Less than 1" d.b.h.		More than 1" d.b.h.			
Species	Less than 1' tall	More than 1' tall	1-4"	4-10"	10-20"	20-30"
Dry jack pine forest in Burnett County						
Acer rubrum	312	62	14	6	2	0
Betula papyrifera	0	0	0	9	0	0
Pinus banksiana	0	144	71	59	33	0
P. strobus	164	20	0	0	2	0
Populus tremuloides	0	0	0	7	0	0
Quercus ellipsoidalis	1568	865	68	43	11	0
Dry-mesic red pine forest in Oneida County (Finnerud Forest of the University of Wisconsin)						
Acer rubrum	3400	1000	80	25	0	0
Betula papyrifera	0	0	35	31	2	0
Pinus resinosa	0	0	5	7	67	0
P. strobus	2800	1400	27	5	3	4
Populus grandidentata	0	0	3	5	1	0
P. tremuloides	0	0	1	3	1	0
Quercus borealis	400	0	23	24	1	0

Table XI-3
Prevalent groundlayer species of northern dry forest

Species	Pres.	Av. freq.	Species	Pres.	Av. freq.
Anemone quinquefolia	74%	16.3%	Lysimachia quadrifolia	37%	4.3%
Apocynum androsaemifolium*	74	3.6	Maianthemum canadense	89	48.8
Aquilegia canadensis	34	1.9	Melampyrum lineare*	37	3.9
Aralia nudicaulis	76	14.1	Oryzopsis asperifolia	47	21.2
Aster ciliolatus	45	5.4	Polygala paucifolia*	47	3.8
A. macrophyllus	68	36.6	Pteridium aquilinum	87	52.4
Carex pensylvanica	55	26.8	Pyrola elliptica	39	3.4
Chimaphila umbellata*	68	8.4	P. rotundifolia	42	3.2
Clintonia borealis	47	5.2	P. secunda	55	5.1
Convolvulus spithamaeus	45	3.6	Rhus radicans	39	4.0
Cornus canadensis	47	7.6	Rosa sp.	42	4.9
Corylus americana	66	15.0	Rubus allegheniensis	50	7.9
C. cornuta	55	21.4	R. pubescens	82	16.8
Diervilla lonicera	76	21.9	R. strigosus	34	5.1
Epigaea repens*	47	4.4	Smilacina racemosa	79	8.9
Fragaria virginiana*	55	11.5	Streptopus roseus	34	3.5
Gaultheria procumbens*	66	19.8	Trientalis borealis	76	17.5
Linnaea borealis	34	5.7	Uvularia sessilifolia	53	10.3
Lycopodium obscurum	55	5.7	Vaccinium angustifolium*	92	37.1
			Waldsteinia fragarioides*	53	30.8

Table XI-4
Community summary—northern dry forest

Major dominants (*I. V.*): Pinus banksiana (65), P. resinosa (48), P. strobus (43), Quercus ellipsoidalis (37), Populus tremuloides (21).

Most prevalent groundlayer species (*P%*): Vaccinium angustifolium (92), Maianthemum canadense (89), Pteridium aquilinum (87), Rubus pubescens (82), Smilacina racemosa (79).

Leading families (% *of total species*): Compositae (11.3%), Liliaceae (6.3), Ranunculaceae (5.9), Rosaceae (5.9), Ericaceae (5.0).

Related communities (*Index of similarity*): Northern Dry Mesic Forest (70), Boreal Forest (56), Pine Barrens (49), Northern Mesic Forest (48), Southern Dry Forest (42).

Species density: 39.
Index of Homogeneity: 54.8.
No. of stands studied: 38.
Number of species: Trees 25, Shrubs 57, Herbs 182, Total 264.
Stability: Low—a one-generation forest in absence of fire. Succeeded by Dry-Mesic or Mesic Northern Forest.

Climate: Total ppt. 30.0" (76 cm), Snowfall 49.6" (126 cm), Jan. Temp. 10.8°F. ($-11.8°C$.), July Temp. 68.3°F. (20.2°C.), Growing season 123 days. Typical weather stations: Hatfield, Long Lake, St. Germain.

Catena position: Top, usually on sandy soils.
Soil group: Podzol and Gray-Brown Podzolic.
Major soil series: Vilas, Omega, Boone, Plainfield, and Coloma.
Soil analyses: w.r.c. 120%, pH 4.9, Ca 1255 p.p.m., K 95 p.p.m.

Approximate original area: Uncertain. Possibly 15% of Northern Xeric Forest or 340,000 acres, but not clearly differentiated from Pine Barren Savanna.

Typical examples: Castle Mound S. A., Cox Hollow S. A. (a southern relic), Necedah S. A.

Major publications: Kittredge (1938), Brown and Curtis (1952).

Geographical distribution: S. Manitoba, N. Minnesota, S. Ontario, N. New York and New England.

(Notes for Table XI-3)

* Species are also modal, since their presence values are higher here than in any other Wisconsin community.

Additional modal species, with their presence (%) *values:* Alnus crispa (3), Cynoglossum boreale (21), Gerardia gattingeri (3), Habenaria hookeri (3), Lycopodium complanatum (29), Lycopodium tristachyum (3), Monotropa hypopithys (13), Rubus hispidus (8).

No. of modal species 17.
No. of prevalent modal species as % *of total prevalents* 23.1%.

Table XI-5
Prevalent groundlayer species of northern dry-mesic forest

Species	Pres.	Av. freq.	Species	Pres.	Av. freq.
Actaea alba	48%	1.0%	Osmorhiza claytoni	45%	5.0%
Anemone quinquefolia*	78	18.1	Polygala paucifolia	45	5.8
Apocynum androsaemifolium	55	1.8	Polygonatum pubescens	75	12.3
			Prenanthes alba	50	1.7
Aquilegia canadensis	45	1.7	Pteridium aquilinum	88	28.4
Aralia nudicaulis*	98	26.1	Pyrola elliptica*	70	5.1
Aster macrophyllus	90	35.7	P. rotundifolia	35	1.7
Brachyelytrum erectum	38	1.5	P. secunda	65	4.0
Carex pensylvanica	45	9.5	Rhus radicans	40	3.2
Chimaphila umbellata	33	1.4	Ribes cynosbati	38	1.0
Clintonia borealis	78	12.5	Rubus alleghaniensis	58	8.3
Cornus alternifolia	38	1.4	R. pubescens	58	5.6
C. canadensis	58	5.6	R. strigosus	45	4.5
Corylus cornuta	68	17.4	Sanicula gregaria	35	2.7
Diervilla lonicera	68	11.2	Smilacina racemosa	78	8.7
Dryopteris austriaca	55	3.4	Streptopus roseus	63	10.1
Fragaria virginiana	53	6.0	Trientalis borealis	85	21.4
Galium triflorum	83	11.7	Trillium grandiflorum	43	8.0
Gaultheria procumbens	45	6.6	Uvularia sessilifolia*	58	14.3
Lonicera oblongifolia*	68	0.9	Vaccinium angustifolium	48	8.2
Lycopodium obscurum*	73	9.3	Vibernum acerifolium	60	5.4
Maianthemum canadense	100	50.1	Viola cucullata	38	4.5
Mitchella repens*	75	15.2	V. pubescens	40	3.3
Oryzopsis asperifolia	58	22.6	Waldsteinia fragarioides	35	0.3

* Species are also modal, since their presence values are higher here than in any other Wisconsin community.

Additional modal species, with their presence (%) values: Adenocaulon bicolor (3), Carex communis (3), Clematis verticillaris (5), Habenaria orbiculata (10), Hamamelis virginiana (15), Luzula campestris (10), Medeola virginiana (5), Monotropa uniflora (25), Panax trifolium (10), Pyrola minor (13), Schizachne purpurascens (25), Viburnum rafinesquianum (32).

No. of modal species 19.
No. of prevalent modal species as % of total prevalents 12.8%.

Table XI-6
Community summary—northern dry-mesic forest

Major dominants (I. V.): Pinus strobus (75), Acer rubrum (37), Quercus borealis (36), Betula papyrifera (30), Acer saccharum (27).
Most prevalent groundlayer species (P%): Maianthemum canadense (100), Aralia nudicaulis (98), Aster macrophyllus (90), Pteridium aquilinum (88), Trientalis borealis (85).
Leading families (% of total species): Compositae (9.1), Liliaceae (7.9), Ranunculaceae (6.4), Rosaceae (6.0), Ericaceae (4.4).
Related communities (Index of similarity): Northern Dry Forest (70), Boreal Forest (70), Northern Mesic Forest (65), Northern Wet-Mesic Forest (50), Southern Dry-Mesic Forest (48).
Species density: 47
Index of Homogeneity: 58.7.
No. of stands studied: 40.

Number of species: Trees 31, Shrubs 52, Herbs 200, Total 283.
Stability: Fairly stable. Succeeded by Mesic Northern Forest.
Catena position: Upper portion.
Soil group: Podzol and Gray-Brown Podzolic.
Major soil series: Cloquet, Omega, Hiawatha, Hixton, Gale, and Shawano.
Soil analyses: w.r.c. 130%, pH 5.2, Ca 1990 p.p.m., K 77 p.p.m.
Approximate original area: Uncertain. Possibly 85% of Northern Xeric Forest, or 1,930,000 acres.
Typical examples: Many young stands on Northern Highland State Forest, Council Grounds S.A., Potawatomi State Park.
Major publications: Stearns (1950), Brown and Curtis (1952).
Geographical distribution: N. Minnesota to New England and Pennsylvania.

APPENDIX FOR CHAPTER 12

Figure XII-1.—Stands studied in northern lowland forest.

Figure XII-2.—Original conifer swamps.

Table XII-1
Average tree composition of northern lowland forest

	Wet forest		Wet-mesic forest	
Species	Av. I. V.	Constancy	Av. I. V.	Constancy
Picea mariana	138.7*	86%	0.5	20%
Thuja occidentalis	45.0	37	91.1*	100
Abies balsamea	23.9	31	45.3	95
Larix laricina	55.6*	69	1.1	20
Tsuga canadensis	0.3	1	39.6	90
Betula lutea	1.1	9	33.9*	95
Fraxinus nigra	2.4	5	26.7*	70
Ulmus americana	....	..	16.3	40
Acer saccharum	....	..	13.8	10
Pinus banksiana	13.5	8	....	..
Acer rubrum	0.4	8	12.4	60
Betula papyrifera	3.5	28	7.7	65
Pinus strobus	3.7	28	2.5	15
Populus tremuloides	2.1	18	2.4	30
Picea glauca	0.6	6	2.9	20
Ulmus rubra	....	..	3.0	10
Tilia americana	....	..	2.7	20
Quercus borealis	....	..	2.5	5
Pinus resinosa	2.3	12	....	..
Fagus grandifolia	....	..	1.4	5
Ostrya virginiana	....	..	1.0	5
Populus balsamifera	....	..	1.0	5
Juglans cinerea	....	..	0.9	5
Fraxinus americana	....	..	0.5	5
Populus grandidentata	0.1	1	0.2	5
Prunus pensylvanica	....	..	0.1	5
Carya cordiformis	....	..	0.1	5

* Species which attain optimum presence in this segment.

Table XII-2
Structure of two typical stands of northern lowland forests

	Less than 1" d.b.h.		More than 1" d.b.h.		
Species	Less than 1' tall	More than 1' tall	1–4"	4–10"	10–20"
Wet swamp of spruce-tamarack in Vilas County					
Larix laricina	0	200	41	66	0
Picea mariana	2600	1000	187	290	0
Pinus strobus	200	0	3	15	0
P. resinosa	0	0	0	11	0
Wet-mesic swamp of cedar and fir in Vilas County (Trout Lake Scientific Area)					
Abies balsamea	3800	1000	56	200	0
Acer rubrum	3000	200	39	9	4
Betula lutea	1600	0	0	0	7
B. papyrifera	0	0	3	44	14
Fraxinus nigra	400	400	0	19	0
Picea mariana	0	0	0	3	0
Populus tremuloides	0	0	0	0	4
Sorbus americana	1000	0	9	6	0
Thuja occidentalis	10,400	0	76	31	14
Quercus borealis	400	0	0	0	0

Table XII-3
Prevalent groundlayer species of northern wet forest

Species	Pres.	Av. freq.	Species	Pres.	Av. freq.
Alnus rugosa	59%	7.5%	G. procumbens	45%	9.4%
Andromeda glaucophylla	49	10.1	Kalmia polifolia	56	....
Aralia nudicaulis	35	6.1	Ledum groenlandicum*	82	60.0
Carex disperma	37	8.5	Linnaea borealis	38	17.8
C. trisperma*	86	37.1	Maianthemum canadense	52	13.8
Chamaedaphne calyculata	58	34.6	Nemopanthus mucronatus*	42	3.5
Clintonia borealis	40	13.9	Osmunda cinnamomea	51	4.2
Coptis trifolia	47	26.7	Parthenocissus vitacea	41	0.6
Cornus canadensis	61	27.6	Rubus pubescens	46	24.5
Cypripedium acaule	45	3.5	Sarracenia purpurea	31	2.3
Dryopteris austriaca	63	11.5	Smilacina trifolia*	77	37.5
D. cristata	42	3.5	Trientalis borealis	58	14.6
Equisetum fluviatile*	29	3.8	Vaccinium angustifolium	55	14.6
Eriophorum spissum*	42	9.4	V. myrtilloides*	73	22.4
E. virginicum*	36	6.3	V. oxycoccus	28	28.5
Gaultheria hispidula*	76	24.5	Viola pallens*	49	18.1

*Species are also modal, since their presence values are higher here than in any other Wisconsin community.

Additional modal species, with their presence (%) values: Agrostis perennans (5), Arceuthobium pusillum (12), Carex canescens (6), Carex comosa (3), Carex interior (8), Carex pauciflora (19), Cinna latifolia (18), Equisetum scirpoides (4), Eriophorum tenellum (4), Habenaria clavellata (4), Lonicera villosa (7), Rhamnus alnifolius (17).

No. of modal species 22.
No. of prevalent modal species as % of total prevalents 31.3%.

Table XII-4
Community summary—northern wet forest

Major dominants (I. V.): Picea mariana (139), Larix laricina (56), Thuja occidentalis (45), Abies balsamea (24), Pinus banksiana (14).

Most prevalent groundlayer species (P%): Carex trisperma (94), Ledum groenlandicum (90), Smilacina trifolia (85), Gaultheria hispidula (83), Vaccinium myrtilloides (80).

Leading families (% of total species): Cyperaceae (13.5), Compositae (6.7), Ericaceae (6.7), Gramineae (5.7), Rosaceae (5.7).

Related communities (Index of similarity): Open Bog (60), Northern Wet-Mesic Forest (50), Boreal Forest (36), Northern Mesic Forest (27), Northern Dry-Mesic Forest (27).

Species density: 34.
Index of Homogeneity: 52.6.
No. of stands studied: 71.

Number of species: Trees 15, Shrubs 43' Herbs 150, Total 208.

Stability: Very stable. May be succeeded by Northern Wet-Mesic or Boreal Forest under some circumstances.

Catena position: Bottom.
Soil group: Azonal peat.
Major soil series: Saugatuck, Spalding.
Soil analyses: w.r.c. 670%, pH 4.7, Ca 755 p.p.m., K 110 p.p.m., P 14 p.p.m.

Approximate original area: Uncertain. Possibly 75% of Northern Lowland or 1,680,000 acres.

Typical examples: Cedarburg Swamp S.A., many stands in Northern Highlands State Forest.

Major publications: Clausen (1957).

Geographical distribution: Glaciated regions of NE. North America. Alberta to Minnesota, Wisconsin, NE. Illinois, N. Indiana, N. Ohio, Pennsylvania, Ontario, and Quebec.

Table XII-5
Prevalent groundlayer species of northern wet-mesic forest

Species	Pres.	Av. freq.	Species	Pres.	Av. freq.
Adiantum pedatum	42%	0.2%	Ilex verticillata*	27%	0.7%
Alnus rugosa	69	10.6	Impatiens biflora*	73	20.2
Anemone quinquefolia	31	1.8	Iris versicolor	31	0.4
Aralia nudicaulis	96	20.5	Laportea canadensis	31	2.5
Arisaema triphyllum	65	6.8	Linnaea borealis	62	21.8
Aster macrophyllus	46	3.0	Lonicera canadensis	62	5.5
Athyrium filix-femina	39	1.4	Maianthemum canadense	96	37.0
Botrychium virginianum	31	1.1	Mitella nuda*	77	30.8
Caltha palustris	42	3.0	Onoclea sensibilis	65	3.7
Carex disperma*	39	12.5	Osmunda cinnamomea*	62	6.9
C. leptalea*	35	9.8	Oxalis acetosella*	42	16.1
C. pedunculata	27	9.0	Parthenocissus vitacea	35	7.3
C. trisperma	62	31.2	Ribes americanum	39	3.6
Circaea alpina*	65	10.6	R. lacustre*	35	1.1
Coptis trifolia*	69	42.5	R. triste*	42	1.9
Cornus canadensis	73	39.1	Rubus pubescens*	92	32.0
C. stolonifera	27	0.8	Scutellaria lateriflora	35	6.6
Diervilla lonicera	35	0.9	Solanum nigrum*	39	3.9
Dryopteris austriaca*	92	29.6	Streptopus roseus	42	2.2
D. cristata*	58	4.8	Taxus canadensis*	27	3.7
Equisetum fluviatile	27	2.6	Thelypteris phegopteris*	42	3.9
Galium triflorum	89	20.9	Trientalis borealis	85	32.8
Gaultheria hispidula	50	10.8	Viola incognita	31	5.5
Gymnocarpium dryopteris*	69	14.1	V. pallens	42	26.5

* Species are also modal, since their presence values are higher here than in any other Wisconsin community.

Additional modal species, with their presence (%) values: Cardamine pensylvanica (4), Carex leptonervia (4), Corallorhiza trifida (12), Cypripedium reginae (4), Epilobium adenocaulon (19), Geum rivale (8), Habenaria hyperborea (8), Habenaria media (4), Habenaria obtusata (23), Ribes hirtellum (4), Rumex verticillatus (27), Salix pedicillaris (4), Stellaria calycantha (4).

No. of modal species 31.

No. of prevalent modal species as % of total prevalents 35.4%.

Table XII-6
Community summary—northern wet-mesic forest

Major dominants (I. V.): Thuja occidentalis (91), Abies balsamea (45), Tsuga canadensis (40), Betula lutea (34), Fraxinus nigra (27).

Most prevalent groundlayer species (P%): Aralia nudicaulis (90), Maianthemum canadense (96), Dryopteris austriaca (92), Rubus pubescens (92), Galium triflorum (89).

Leading families (% of total species): Compositae (9.2), Liliaceae (6.1), Cyperaceae (5.7), Ranunculaceae (5.7), Gramineae (5.3).

Related communities (Index of similarity): Boreal Forest (61), Northern Mesic Forest (54), Northern Wet Forest (50), Northern Dry-Mesic Forest (50), Northern Dry Forest (40).

Species density: 48.

Index of Homogeneity: 53.8.

No. of stands studied: 26.

Stability: Very stable. Succeeded by Northern Mesic Forest.

Climate: Total ppt. 31.0″ (79 cm), Snowfall 50.9″ (129 cm), Jan. Temp. 13.2°F. (−10.4°C.), July Temp. 68.8°F. (20.4°C.), Growing season 136 days. Typical weather stations: Mather, Marinette, Park Falls.

Catena position: Bottom.

Soil group: Azonal peat.

Major soil series: Spalding, Greenwood, and Dawson.

Soil analyses: w.r.c. 495%, pH 5.5, Ca 3780 p.p.m., K 85 p.p.m., P 9 p.p.m.

Approximate original area: Uncertain. Possibly 25% of Northern Lowland or 560,000 acres.

Typical examples: Trout Lake Cedar Swamp S.A., Point Beach State Forest, many stands in Northern Highlands State Forest.

Major publications: Christensen (1954 thesis).

Geographical distribution: Glaciated regions of NE. North America. Wisconsin, Michigan, Ontario, New York, New England, and Quebec.

APPENDIX FOR CHAPTER 13

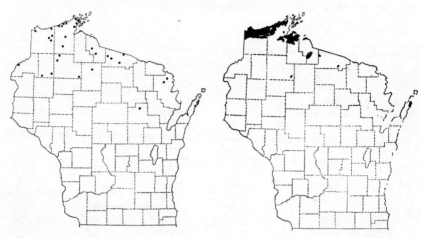

Figure XIII-1.—Stands studied in boreal forest.

Figure XIII-2.—Original boreal forest.

Table XIII-1
Average tree composition of boreal forest in Wisconsin

Species	Av. I. V.	Constancy	Species	Av. I. V.	Constancy
Abies balsamea*	69.5	100%	Populus grandidentata	3.0	31%
Picea glauca*	25.2	100	Ostrya virginiana	0.6	28
Betula papyrifera	25.9	87	Tilia americana	1.9	23
Acer spicatum*	1.1	82	Prunus serotina	0.3	23
Acer rubrum	9.2	77	Pinus banksiana	8.1	18
Populus tremuloides*	22.6	64	Ulmus americana	0.5	18
Pinus strobus	34.2	59	Populus balsamifera*	13.8	15
Thuja occidentalis	31.9	59	Fraxinus americana	0.7	15
Sorbus americana*	0.8	56	Quercus ellipsoidalis	1.6	10
Acer saccharum	10.7	54	Picea mariana	0.8	10
Fraxinus nigra	2.4	51	Quercus alba	0.3	8
Quercus borealis	5.1	46	Carpinus caroliniana*	0.1	8
Pinus resinosa	13.8	44	Larix laricina	0.4	5
Tsuga canadensis	14.8	41	Ulmus rubra	0.4	5
Betula lutea	13.4	41	Carya cordiformis	0.1	5
Prunus pensylvanica*	0.4	33	Fraxinus pennsylvanica	0.1	3

* Species which attain optimum importance values in this forest.

Table XIII-2
Prevalent groundlayer species of boreal forest in Wisconsin

Species	Pres.	Av. freq.	Species	Pres.	Av. freq.
Actaea rubra*	59%	1.8%	Lycopodium clavatum*	46%	2.8%
Alnus rugosa	46	2.2	L. obscurum	62	8.4
Anemone quinquefolia	62	18.8	Maianthemum canadense*	100	59.6
Apocynum androsaemifolium	49	1.5	Mitchella repens	39	5.0
Aralia nudicaulis	95	31.2	Mitella nuda	54	17.6
Aster macrophyllus*	90	56.7	Oryzopsis asperifolia	64	17.6
Athyrium filix-femina	59	4.1	Osmorhiza claytoni	39	2.6
Brachyelytrum erectum	44	3.5	Polygala paucifolia	36	5.6
Carex arctata*	39	7.9	Polygonatum pubescens	39	3.3
C. pedunculata*	41	9.3	Prenanthes alba	49	2.8
C. pensylvanica	39	5.3	Pteridium aquilinum	77	20.7
Circaea alpina	36	4.6	Pyrola elliptica	49	3.6
Clintonia borealis*	95	31.7	P. secunda*	67	5.8
Coptis trifolia	62	5.4	Ribes cynosbati	36	1.3
Cornus canadensis*	97	33.0	Rosa sp.	46	4.1
C. rugosa*	54	5.7	Rubus parviflorus*	41	2.3
Corylus cornuta*	80	13.9	R. pubescens	85	14.5
Diervilla lonicera*	80	10.8	R. strigosus*	59	3.1
Dryopteris austriaca	74	14.4	Sanicula marilandica*	49	6.9
Equisetum arvense	41	4.8	Smilacina racemosa	39	1.8
Fragaria virginiana	67	19.6	Streptopus roseus*	80	16.9
Galium triflorum*	100	26.9	Thelypteris phegopteris	36	2.1
Gaultheria procumbens	41	4.8	Trientalis borealis*	97	28.5
Gymnocarpium dryopteris	62	4.3	Trillium cernuum*	39	1.8
Hepatica americana*	49	7.1	Vaccinium angustifolium	44	6.1
Impatiens biflora	39	3.0	V. myrtilloides	39	3.4
Linnaea borealis*	74	19.7	Viola conspersa*	39	2.5
Lonicera canadensis*	97	6.6	V. incognita*	64	15.1
L. hirsuta*	51	2.3	V. pubescens	39	2.1

* Species are also modal, since their presence values are higher here than in any other Wisconsin community.

Additional modal species, with their presence (%) values: Botrychium matricariifolium (3), Carex brunnescens (10), Carex eburnea (5), Carex intumescens (33), Chrysosplenium americanum (2), Corallorhiza striata (13), Cynoglossum officinale (3), Equisetum sylvaticum (33), Geranium bicknellii (5), Goodyera decipiens (10), Goodyera repens (10), Goodyera tesselata (10), Halenia deflexa (10), Listera cordata (5), Luzula acuminata (33), Lycopodium annotinum (28), Mertensia paniculata (5), Moneses uniflora (31), Petasites frigidus (31), Prunella vulgaris (33), Pyrola asarifolia (27), P. virens (10), Ranunculus acris (19), Ribes glandulosum (3), Satureja vulgaris (18), Shepherdia canadensis (5), Streptopus amplexifolius (3), Tiarella cordifolia (3), Veronica scutellata (3), Viola renifolia (33), Viola selkirkii (3).

No. of modal species 57.

No. of prevalent modal species as % of total prevalents 43.1%.

Table XIII-3
Other prevalent species in P.E.L. study which attain an optimum in the boreal forest (but not in the Wisconsin stands)

Species	P%	Optimum Segment on Moisture Gradient	Species	P%	Optimum Segment on Moisture Gradient
Anaphalis margaritacea	73	D	L. tristachyum	36	D
Apocynum androsaemifolium	82	D	Melampyrum lineare	91	D
			Mitella nuda	86	W
Aralia nudicaulis	100	D	Moneses uniflora	50	W
Carex disperma	57	W	Oxalis acetosella	50	W
C. leptalea	57	W	Polygala paucifolia	82	D
Coptis trifolia	55	D	Prunella vulgaris	50	W
Epigaea repens	73	D	Pyrola asarifolia	35	WM
Gaultheria hispidula	86	W	Ribes triste	57	W
G. procumbens	91	D	Rubus pubescens	100	W
Gymnocarpium dryopteris	79	W	Thelypteris phegopteris	50	W
Habenaria obtusata	64	W	Vaccinium angustifolium	100	D
Lycopodium annotinum	73	D	V. myrtilloides	100	D
L. complanatum	64	D	Viola pallens	71	W
L. lucidulum	61	WM			

Table XIII-4
Community summary—boreal forest

Major dominants (I. V.): Abies balsamea (69), Pinus strobus (34), Thuja occidentalis (32), Betula papyrifera (26), Picea glauca (25).

Most prevalent groundlayer species (P%): Galium triflorum (100), Maianthemum canadense (100), Cornus canadensis (97), Lonicera canadensis (97), Trientalis borealis (97).

Leading families (% of total species): Compositae (9.8), Cyperaceae (5.7), Liliaceae (5.7), Ranunculaceae (5.7), Gramineae (4.7).

Related communities (Index of similarity): Northern Dry-Mesic Forest (70), Northern Wet-Mesic Forest (61), Northern Mesic Forest (56), Northern Dry Forest (56), Northern Wet Forest (36).

Species density: 58.

Index of Homogeneity: 57.7%.

No. of stands studied: 39.

Number of species: Trees 32, Shrubs 56, Herbs 240, Total 328.

Stability: Northernmost stands very stable. Southern stands tend to be replaced by Northern Mesic Forest.

Climate: Total ppt. 26.8" (68 cm), Snowfall 53.8" (137 cm), Jan. Temp. 11.1°F. (−11.6°C.), July Temp. 65.5°F. (18.6°C.), Growing season 137 days. Typical weather stations: Ashland, Cornucopia, Superior.

Catena position: Middle and low; poor internal drainage.

Soil groups: Grey Wooded and Podzol.

Major soil series: Ontanagon, Longrie, Gogebic, and Wakefield.

Soil analyses: w.r.c. 225%, pH 5.1, Ca 2780 p.p.m., K 150 p.p.m., P 35 p.p.m.

Approximate original area: 672,500 acres.

Typical examples: High Lake Spruce-Fir S.A. (second growth), Gillen Nature Reserve, Vilas Co., Red Cliff Indian Reservation, Bayfield Co.

Major publication: Maycock (1957 thesis).

Geographical distribution: Centers on Lake Superior, in region from central Ontario, northern Michigan to NW. Minnesota, north to limit of hardwoods in Manitoba and Ontario.

APPENDIX FOR CHAPTER 14

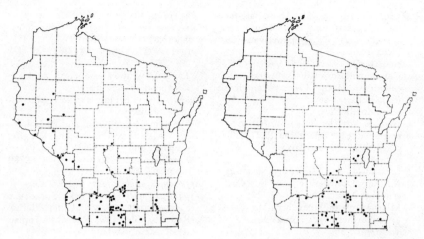

Figure XIV-1.—Stands studied in xeric prairies.

Figure XIV-3.—Stands studied in lowland prairies.

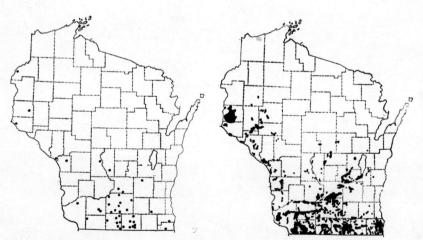

Figure XIV-2.—Stands studied in mesic prairies.

Figure XIV-4.—Original prairies.

Table XIV-1
Prevalent species of dry prairies

Species	Pres.	Av. freq.
Acerates viridiflora*	41%	1%
Ambrosia artemisiifolia*	77	15
Amorpha canescens*	100	20
Andropogon gerardi	94	23
A. scoparius*	100	45
Anemone cylindrica	59	2
A. patens*	94	3
Antennaria neglecta	59	2
Arabis lyrata	41	3
Arenaria stricta*	88	18
Artemisia caudata*	88	13
Asclepias verticillata*	88	19
Aster azureus	47	1
A. ericoides	77	14
A. laevis	41	2
A. oblongifolius*	59	3
A. ptarmicoides*	59	3
A. sericeus*	94	31
Bouteloua curtipendula*	100	36
Castilleja sessiliflora*	47	12
Comandra richardsiana	59	2
Coreopsis palmata*	82	11
Erigeron strigosus*	82	4
Euphorbia corollata	94	48

Species	Pres.	Av. freq.
Geum triflorum*	59%	3%
Hedeoma hispida*	65	15
Helianthus laetiflorus	59	8
Koeleria cristata	59	7
Kuhnia eupatorioides*	88	17
Liatris cylindracea*	71	20
Lithospermum incisum*	88	13
Monarda fistulosa	77	4
Oenothera biennis	59	3
Panicum perlongum*	77	32
Petalostemum purpureum*	100	23
Physalis longifolia*	65	7
P. virginiana	41	7
Potentilla arguta	53	6
Psoralea esculenta*	41	1
Rosa sp.	82	24
Scutellaria leonardi	53	9
Sisyrhincium campestre	47	10
Solidago nemoralis*	94	34
S. rigida	77	19
Sporobolus heterolepis*	82	25
Stipa spartea	53	7
Verbena stricta*	77	8

* Species are also modal, since their presence values are higher here than in any other Wisconsin community.

Additional modal species, with their presence (%) values: Acerates lanuginosa (29), Aronia melanocarpa (24), Physalis heterophylla (29), Sporobolus asper (12).

No. of modal species 31.

No. of prevalent modal species as % of total prevalents 57.5%.

Table XIV-2
Community summary—dry prairie

Major dominants (P%): Andropogon scoparius (100), Bouteloua curtipendula (100), Andropogon gerardi (94), Sporobolus heterolepis (82), Panicum perlongum (77).

Most prevalent groundlayer species (P%): Petalostemum purpureum (100), Anemone patens (94), Aster sericeus (94), Euphorbia corollata (94), Solidago nemoralis (94).

Leading families (% of total species): Compositae (27.5), Gramineae (13.7), Leguminosae (5.3), Rosaceae (4.6), Asclepiadaceae (4.6).

Related communities (Index of similarity): Dry-Mesic Prairie (79), Cedar Glade (59), Mesic Prairie (53), Oak Barrens (48), Oak Opening (46).

Species density: 47.

Index of Homogeneity: 70.3.

No. of stands studied: 17.

Number of species: Trees 3, Shrubs 12, Herbs 119, Total 134.

Stability: Very stable. Invaded by Southern Dry Forest very slowly, even in absence of fire.

Catena position: Top. Excessively good drainage.

Soil group: Brunizem and Melanized Rendzina.

Major soil series: Rodman, Fox, Hayden.

Soil analyses: w.r.c. 60%, pH 7.8, Ca 5425 p.p.m., K 95 p.p.m., P 85 p.p.m.

Approximate original area: Uncertain. Possibly 5% of total prairie, or 105,000 acres.

Typical examples: Brady's Bluff S.A., Nelson Dewey S.A., Interstate State Park, Kettle Moraine State Forest.

Major publications: Greene and Curtis (1949), Anderson (1954 thesis), Dix and Butler (1954), Curtis (1955).

Geographical distribution: S. Minnesota Iowa, W. Illinois, Missouri, Nebraska.

Table XIV-3
Prevalent species of dry-mesic prairies

Species	Pres.	Av. freq.
Ambrosia artemisiifolia	71%	10%
Amorpha canescens	90	41
Andropogon gerardi	97	61
A. scoparius	92	84
Anemone cylindrica	76	6
A. patens	50	9
Antennaria neglecta	70	14
Artemisia caudata	65	13
Asclepias syriaca	50	1
A. verticillata	59	16
Aster azureus*	68	14
A. ericoides*	80	21
A. laevis	44	6
A. oblongifolius	44	12
A. ptarmicoides	39	16
A. sericeus	76	51
Bouteloua curtipendula	77	69
Comandra richardsiana	58	11
Coreopsis palmata	76	31
Erigeron strigosus	59	3
Euphorbia corollata	92	40
Hedeoma hispida	61	9
Helianthus laetiflorus	65	25
H. occidentalis	42	12
Koeleria cristata	45	4
Kuhnia eupatorioides	61	10
Lespedeza capitata	47	2
Liatris aspera	80	20
L. cylindracea	47%	22%
Linum sulcatum	42	5
Lithospermum canescens	68	8
L. incisum	47	7
Monarda fistulosa	65	4
Oenothera biennis	64	2
Panicum leibergii	59	10
P. oligosanthes*	64	7
P. perlongum	59	31
Petalostemum candidum	42	4
P. purpureum	83	30
Physalis virginiana	45	3
Potentilla arguta	80	9
Ratibida pinnata	70	7
Rhus glabra*	61	3
Rosa sp.	88	13
Rudbeckia hirta	41	1
Scutellaria leonardi	45	7
Sisyrinchium campestre*	49	13
Solidago nemoralis	89	31
S. rigida	64	15
Sorghastrum nutans*	62	6
Sporobolus heterolepis	79	28
Stipa spartea*	91	19
Tradescantia ohiensis	50	1
Viola pedata	50	10
V. pedatifida*	45	7

* Species are also modal, since their presence values are higher here than in any other Wisconsin community.

Additional modal species, with their presence (%) values: Artemisia frigida (3), Artemisia ludoviciana (9), Asclepias tuberosa (24), Aster pilosus (8), Astragalus crassicarpus (2), Callirhoe triangulata (5), Cassia fasciculata (2), Cirsium hillii (15), Cirsium undulatum (2), Delphinium virescens (5), Gentiana quinquefolia (3), Lactuca ludoviciana (26), Liatris punctata (3), Microseris cuspidata (6), Muhlenbergia cuspidata (29), Orobanche fasciculata (2), Plantago rugelii (2), Prenanthes aspera (3).

No. of modal species 26.
No. of prevalent modal species as % of total prevalents 14.6%.

Table XIV-4
Community summary—dry-mesic prairie

Major dominants (P%): Andropogon gerardi (97), Andropogon scoparius (92), Stipa spartea (66), Sporobolus heterolepis (79), Bouteloua curtipendula (77).

Most prevalent groundlayer species (P%): Euphorbia corollata (92), Amorpha canescens (90), Solidago nemoralis (89), Rosa sp. (88), Petalostemum purpureum (83).

Leading families (% of total species): Compositae (28.0), Gramineae (9.1), Leguminosae (7.8), Rosaceae (3.9), Asclepiadaceae (3.5).

Related communities (Index of similarity): Dry Prairie (79), Mesic Prairie (70), Cedar Glade (62), Oak Barrens (58), Oak Opening (54).

Species density: 55.

Index of Homogeneity: 61.3.

No. of stands studied: 66.

Number of species: Trees 8, Shrubs 18, Herbs 214, Total 240.

Stability: Very stable in presence of fire, otherwise slowly invaded by Southern Dry Forest.

Catena position: Upper portion. No gley layer.

Soil group: Brunizem.

Major soil series: Warsaw, Waupun, Ostrander, Hayden, and Dakota.

Soil analyses: w.r.c. 55%, pH 7.1, Ca 3675 p.p.m., K 112 p.p.m., P 47 p.p.m.

Approximate original area: Uncertain. Possibly 30% of total prairie, or 630,500 acres.

Typical examples: None under public control. Frequent along railroads from St. Croix to Kenosha counties.

Geographical distribution: Mostly to the south and west from SW. Wisconsin.

Table XIV-5
Prevalent species of mesic prairies

Species	Pres.	Av. freq.	Species	Pres.	Av freq.
Achillea millefolium*	62%	3%	H. occidentalis*	44%	19%
Ambrosia artemisiifolia	51	23	Lactuca canadensis*	47	7
Amorpha canescens	73	32	Lathyrus venosus	51	2
Andropogon gerardi*	98	29	Lespedeza capitata	58	18
A. scoparius	69	28	Liatris aspera*	87	18
Anemone cylindrica	51	5	Lithospermum canescens	53	5
Antennaria neglecta	56	7	Monarda fistulosa	73	22
Apocynum androsaemifolium	42	6	Panicum leibergii*	62	46
Asclepias syriaca	76	13	Petalostemum purpureum	60	7
Aster azureus	56	16	Phlox pilosa	53	21
A. ericoides	76	44	Physalis virginiana	42	8
A. laevis*	89	35	Potentilla arguta	58	13
Baptisia leucophaea*	44	3	Quercus macrocarpa	40	2
Ceanothus americanus*	66	9	Ratibida pinnata	85	32
Cirsium discolor*	71	13	Rhus glabra	42	5
Comandra richardsiana	53	32	Rosa sp.*	91	36
Convolvulus sepium*	49	13	Rudbeckia hirta	44	3
Coreopsis palmata	76	34	Silphium integrifolium	40	3
Desmodium canadense	49	3	S. laciniatum*	78	8
D. illinoense*	64	20	Solidago missouriensis*	58	15
Dodecatheon meadia	53	3	S. rigida*	76	15
Elymus canadensis	42	5	S. speciosa*	62	11
Eryngium yuccifolium*	53	21	Sorghastrum nutans	58	13
Euphorbia corollata	86	75	Sporobolus heterolepis	64	35
Fragaria virginiana	56	16	Stipa spartea	69	58
Galium boreale	40	8	Tradescantia ohiensis	64	31
Helianthus grosseserratus	44	2	Viola pedatifida	42	13
H. laetiflorus*	87	40			

* Species are also modal, since their presence values are higher here than in any other Wisconsin community.

Additional modal species, with their presence (%) values: Allium cernuum (2), Apocynum cannabinum (36), Astragalus canadensis (11), Echinacea pallida (20), Eupatorium altissimum (4), Gaura biennis (4), Gentiana puberula (36), Gentiana saponaria (7), Heliopsis helianthoides (33), Hypericum canadense (2), Hypericum mutilum (2), Liatris ligulistylis (4), Lilium philadelphicum (20), Oxalis violacea (27), Parthenium integrifolium (9), Penstemon digitalis (4), Solidago juncea (29), Vicia angustifolia (9), Zizia aptera (20).

No. of modal species 39.
No. of prevalent modal species as % of total prevalents 34.6%.

Table XIV-6
Community summary—mesic prairie

Major dominants (*P*%): Andropogon gerardi (98), Stipa spartea (69), Andropogon scoparius (69), Sporobolus heterolepis (64), Panicum leibergii (62).

Most prevalent groundlayer species (*P*%): Aster laevis (89), Helianthus laetiflorus (87), Liatris aspera (87), Ratibida pinnata (85), Silphium laciniatum (78).

Leading families (% *of total species*): Compositae (26.1). Gramineae (10.2), Leguminosae (7.4), Labiatae (3.9), Liliaceae (3.5).

Related communities (*Index of similarity*): Wet-Mesic Prairie (70), Dry-Mesic Prairie (70), Oak Opening (58), Oak Barrens (57), Dry Prairie (53).

Species density: 55.

Index of Homogeneity: 59.5.

No. of stands studied: 45.

Number of species: Trees 5, Shrubs 13, Herbs 246, Total 264.

Stability: Very stable when burned frequently. Little information on nature of replacing forest.

Climate: Total ppt. 31.3" (80 cm), Snowfall 36.6" (93 cm), Jan. Temp. 15.9°F. (−8.8°C.), July Temp. 71.8°F. (22.1°C.), Growing season 152 days. Typical Weather Stations: Beloit, Dodgeville, New Richmond.

Catena position: Middle. Usually no gley layer.

Soil group: Brunizem.

Major soil series: Elliott, Parr, Corwin, Bristol, Waukegan, Tama, and Muscatine.

Soil analyses: w.r.c. 70%, *p*H 6.2, Ca 2700 p.p.m., K 130 p.p.m., P 37 p.p.m.

Approximate original area: Uncertain. Possibly 40% of total prairie, or 840,500 acres.

Typical examples: None under public control. Frequent along railroads in Rock County and elsewhere in south.

Major publications: Curtis and Greene (1949), Green (1950), Curtis (1955).

Geographical distribution: NW. Indiana to SW. Minnesota and E. South Dakota, south to Missouri and Kansas.

Table XIV-7
Prevalent species of wet-mesic prairies

Species	Pres.	Av. freq.	Species	Pres.	Av. freq.
Achillea millefolium	58%	16%	L. venosus*	55%	12%
Ambrosia artemisiifolia	48	5	Lespedeza capitata	52	7
Amorpha canescens	42	4	Liatris pycnostachya*	48	30
Andropogon gerardi	97	40	Lilium superbum	45	4
A. scoparius	48	29	Lithospermum canescens	52	9
Anemone canadensis	45	24	Monarda fistulosa	77	7
Asclepias syriaca*	81	19	Oxypolis rigidior	52	4
Aster azureus	61	43	Panicum leibergii	55	18
A. ericoides	45	5	Phlox pilosa*	65	22
A. laevis	58	5	Prenanthes racemosa*	42	4
A. novae-angliae*	71	10	Pycnanthemum virginianum*	84	24
Calamagrostis canadensis	74	14	Quercus macrocarpa	55	5
Cicuta maculata*	42	5	Ratibida pinnata*	90	48
Cirsium discolor	65	1	Rhus glabra	52	4
Comandra richardsiana	77	30	Rosa sp.	81	30
Corylus americana	52	4	Rudbeckia hirta*	74	24
Desmodium canadense*	65	5	Salix humilis	55	23
Dodecatheon meadia*	61	19	Silphium integrifolium*	52	15
Elymus canadensis	65	5	S. terebinthinaceum*	55	31
Equisetum arvense	48	25	Smilacina stellata	45	6
E. laevigatum*	45	4	Solidago gigantea	48	35
Euphorbia corollata	74	15	S. graminifolia	42	20
Fragaria virginiana	81	13	S. rigida	77	38
Galium boreale	68	15	Sorghastrum nutans	48	12
Gentiana andrewsii*	48	8	Spartina pectinata	74	28
Geranium maculatum	52	4	Spiraea alba	48	18
Helianthus grosseserratus*	81	41	Sporobolus heterolepis	55	8
H. laetiflorus	42	4	Thalictrum dasycarpum	87	16
Heuchera richardsonii	48	12	Tradescantia ohiensis	71	12
Lactuca canadensis	42	4	Veronicastrum virginicum*	74	17
Lathyrus palustris	42	3	Zizia aurea	65	8

* Species are also modal, since their presence values are higher here than in any other Wisconsin community.

Additional modal species, with their presence (%) values: Allium canadense (32), Artemisia serrata (3), Asclepias purpurascens (3), Baptisia leucantha (36), Blephilia ciliata (13), Cacalia tuberosa (23), Camassia scilloides (3), Carex bicknellii (19), Gentiana crinita (10), Habenaria flava (3), Habenaria leucophaea (23), Lythrum alatum (26), Napaea dioica (7), Oxybaphus nyctaginea (23), Polytaenia nuttallii (19), Prenanthes crepidinea (2), Rudbeckia subtomentosa (10), Spiranthes ochraleuca Rydb. (16), Vernonia fasciculata (19), Vicia americana (29).

No. of modal species 38.
No. of prevalent modal species as % of total prevalents 29.0%.

Table XIV-8
Community summary—wet-mesic prairie

Major dominants (P%): Andropogon gerardi (97), Calamagrostis canadensis (74), Spartina pectinata (74), Elymus canadensis (65), Panicum leibergii (55).

Most prevalent groundlayer species (P%): Ratibida pinnata (90), Pycnanthemum virginianum (84), Asclepias syriaca (81), Helianthus grosseserratus (81), Rosa sp. (81).

Leading families (% of total species): Compositae (27.0), Gramineae (8.1), Leguminosae (5.6), Umbelliferae (4.8), Labiatae (4.4).

Related communities (Index of similarity): Mesic Prairie (70), Wet Prairie (66), Cedar Glade (61), Dry-Mesic Prairie (54), Oak Barrens (50).

Species density: 62.

Index of Homogeneity: 59.5.

No. of stands studied: 31.

Number of species: Trees 4, Shrubs 15, Herbs 233, Total 252.

Stability: High in presence of fire, otherwise replaced by Southern Wet-Mesic Forest.

Catena position: Low. Gley layer within 4 feet.

Soil group: Brunizem.

Major soil series: Ashkum, Bryce, Brookston, Garwin, and Judson.

Soil analyses: w.r.c. 90%, pH 6.4, Ca 5125 p.p.m., K 122 p.p.m., P 50 p.p.m.

Approximate original area: Uncertain. Possibly 20% of total prairie or 420,000 acres.

Typical examples: Faville Prairie S.A., U. W. Arboretum.

Geographical distribution: South Dakota to S. Michigan, south to Oklahoma and Kansas.

APPENDIX 561

Table XIV-9
Prevalent species of wet prairies

Species	Pres.	Species	Pres.
Andropogon gerardi	68%	Liatris pycnostachya	32%
Anemone canadensis	36	Lilium superbum*	50
Apocynum cannabinum	32	Lobelia spicata	36
Asclepias syriaca	64	Monarda fistulosa	50
Aster novae-angliae	68	Muhlenbergia racemosa	36
A. simplex	59	Oxypolis rigidior*	73
Calamagrostis canadensis	91	Pedicularis lanceolata	32
Comandra richardsiana	36	Phlox pilosa	55
Cirsium muticum*	50	Pycnanthemum virginianum	82
Desmodium canadense	50	Ratibida pinnata	36
Dodecatheon meadia	41	Rudbeckia hirta	68
Equisetum arvense*	68	Salix humilis*	68
Erigeron strigosus	36	Saxifraga pensylvanica*	50
Eupatorium maculatum	41	Solidago gigantea*	73
Fragaria virginiana	77	Spartina pectinata*	91
Galium boreale	73	Spiraea alba	59
Gentiana andrewsii	36	Thalictrum dasycarpum	87
Helianthus grosseserratus	64	Thelypteris palustris	55
Heuchera richardsonii	41	Tradescantia ohiensis	46
Hypoxis hirsuta*	55	Veronicastrum virginicum	73
Iris shrevei	50	Viola cucullata	59
Lathyrus palustris	55	Zizia aurea*	73

* Species are also modal, since their presence values are higher here than in any other Wisconsin community.

Additional modal species, with their presence (%) values: Cacalia atriplicifolia (5), Cacalia suaveolens (5), Galium tinctorium (18), Gerardia aspera (5), Habenaria lacera (5), Helenium autumnale (14), Hierochloe odorata (27), Houstonia caerulea (5), Liparis loeselii (9), Oenthera perennis (14), Scleria triglomerata (5), Solidago ohioensis (18), Tofieldia glutinosa (5).

No. of modal species 23.
No. of prevalent modal species as % of total prevalents 22.7%.

Table XIV-10
Community summary—wet prairie

Major dominants (P%): Calamagrostis canadensis (91), Spartina pectinata (91), Andropogon gerardi (68), Muhlenbergia racemosa (36).

Most prevalent groundlayer species (P%): Thalictrum dasycarpum (87), Pycnanthemum virginianum (82), Fragaria virginiana (77), Oxypolis rigidior (73), Solidago gigantea (73).

Leading families (% of total species): Compositae (25.8), Gramineae (8.2), Leguminosae (6.0), Labiatae (4.4), Umbelliferae (3.8).

Related communities (Index of similarity): Wet-Mesic Prairie (66), Fen (54), Southern Sedge Meadow (46), Mesic Prairie (43), Oak Opening (41).

Species density: 44.
Index of Homogeneity: 56.3.
No. of stands studied: 22.

Number of species: Trees 1, Shrubs 12, Herbs 173, Total 186.

Stability: High in presence of frequent fire, otherwise quickly transformed to Southern Wet-Mesic Forest.

Catena position: Low. Gley layer within 24″.

Soil group: Brunizem and Azonal peat.

Major soil series: Carlisle, Kokomo, Peotone, and Abington.

Soil analyses: w.r.c. 165%, pH 6.2, Ca 3900 p.p.m., K 90 p.p.m., P 37 p.p.m.

Approximate original area: Uncertain. Possibly 5% of total prairie, or 105,000 acres.

Typical examples: Scuppernong S. A., Faville Prairie S. A., U.W. Arboretum.

Geographical distribution: SE. Wisconsin, south and east to Illinois, Indiana, S. Michigan, and Ohio.

APPENDIX FOR CHAPTER 15

Figure XV-1.—Stands studied in sand barrens.

Figure XV-2.—Stands studied in bracken-grasslands.

Table XV-1
Prevalent species of sand barrens

Species	Pres.	Species	Pres.
Ambrosia psilostachya*	50%	Leptoloma cognatum*	50%
Andropogon gerardi	40	Lespedeza capitata	65
A. scoparius	90	Liatris aspera	50
Arabis lyrata	55	Linaria canadensis*	40
Carex muhlenbergii*	50	Lithospermum caroliniense*	55
Cyperus schweinitzii*	45	Monarda punctata*	50
Danthonia spicata	40	Oenothera rhombipetala*	40
Erigeron strigosus	40	Panicum lanuginosum*	55
Euphorbia corollata	75	Physalis virginiana	45
Gnaphalium obtusifolium*	40	Polygala polygama*	45
Hedeoma hispida	40	Selaginella rupestris*	45
Helianthemum canadense	60	Silene antirrhina	35
Helianthus occidentalis	40	Solidago missouriensis	45
Hieracium longipilum*	45	S. nemoralis	40
Hudsonia tomentosa*	35	Tradescantia ohiensis	55
Koeleria cristata*	80	Viola pedata	45
Lechea tenuifolia*	40		

* Species are also modal, since their presence values are higher here than in any other Wisconsin community.

Additional modal species, with their presence (%) values: Antennaria plantaginifolia (15), Aristida basiramea (20), Cenchrus pauciflorus (5), Chrysopsis villosa (5), Croton glandulosus (5), Cyperus filiculmis (35), Eragrostis spectabilis (10), Euphorbia maculata (20), Froelichia floridana (5), Houstonia longifolia (15), Juncus greenii (5), Krigia virginica (4). Mollugo verticillata (20), Panicum virgatum (25), Plantago purshii (20), Polygonella articulata (30), Rhus aromatica (5), Specularia perfoliata (30).

No. of modal species 34.

No. of prevalent modal species as % of total prevalents 48.5%.

Table XV-2
Community summary—sand barrens

Major dominants (P%): Andropogon scoparius (90), Koeleria cristata (80), Panicum lanuginosum (55), Carex muhlenbergii (50), Leptoloma cognatum (50).

Most prevalent groundlayer species (P%): Euphorbia corollata (75), Lespedeza capitata (65), Helianthemum canadense (60), Arabis lyrata (55), Lithospermum caroliniense (55).

Leading families (% of total species): Compositae (23.5), Gramineae (15.7), Rosaceae (5.2), Leguminosae (3.3), Ericaceae (3.3).

Related communities (Index of similarity): Oak Barrens (42), Dry-Mesic Prairie (39), Pine Barrens (38), Cedar Glade (36), Dry Prairie (35).

Species density: 33.

Index of Homogeneity: 52.7.

No. of stands studied: 20.

Number of species: Trees 3, Shrubs 24, Herbs 135, Total 162.

Stability: Internally unstable, but invaded by Oak Barrens or Southern Dry Forest only very slowly.

Catena position: Top. No gley layer.

Soil group: Immature aeolian.

Major soil series: Some are derived from Boone and Coloma sands.

Soil analyses: w.r.c. 29%, pH 6.2, Ca 50 p.p.m., K 17 p.p.m., P 30 p.p.m.

Approximate original area: Very small. Largely the result of plowing Dry-Mesic prairies on sand.

Typical examples: None known under public control. Frequent on terraces of lower Wisconsin river, from Sauk City to Boscobel. Good examples at Gotham and Muscoda.

Geographical distribution: On suitable sites from Prairie Provinces of Canada to New England and southward. The Sand Hills of Nebraska, and the Illinois and Connecticut rivers have well-studied examples.

Table XV-3
Prevalent species of bracken-grassland

Species	Pres.	Species	Pres.
Achillea millefolium	44%	Myrica asplenifolia*	67%
Agropyron trachycaulum*	56	Oenothera biennis	48
Anaphalis margaritacea*	70	Oryzopsis asperifolia*	67
Antennaria neglecta	48	Pedicularis canadensis	41
Apocynum androsaemifolium	44	Phleum pratense*	44
Arabis glabra*	52	Poa compressa*	56
Aster ciliolatus*	93	P. pratensis	52
A. umbellatus*	41	Pteridium aquilinum*	100
Bromus kalmii*	85	Rubus pubescens	41
Convolvulus spithamaeus	44	Rumex acetosella*	44
Danthonia spictata*	78	Salix discolor	56
Erigeron annuus*	41	Solidago nemoralis	78
Fragaria virginiana	78	Vaccinium angustifolium	67
Hieracium aurantiacum*	63	V. myrtilloides	44
Lactuca scariola	37	Verbascum thapsus*	63
Muhlenbergia racemosa*	56	Viola adunca*	48

* Species are also modal, since their presence values are higher here than in any other Wisconsin community.

Additional modal species, with their presence (%) values: Agrostis hyemalis (33), Arctostaphylos uva-ursi (30), Berteroa incana (4), Chrysanthemum leucanthemum (4), Epilobium angustifolium (19), Erigeron glabellus (4), Festuca saximontana (22), Fragaria vesca (11), Hieracium canadense (26), Lechea intermedia (4), Malaxis unifolia (11), Oryzopsis pungens (15), Senecio pauperculus (15), Sonchus asper (15), Spiranthes gracilis (19), Stellaria graminea (4), Trifolium hybridum (22), Trifolium pratense (15).

No. of modal species 36.

No. of prevalent modal species as % of total prevalents 56.3%.

Table XV-4
Community summary—bracken-grassland

Major dominants (P%): Pteridium aquilinum (100), Bromus kalmii (85), Danthonia spicata (74), Agropyron trachycaulum (56), Poa compressa (56).

Most prevalent groundlayer species (P%): Aster ciliolatus (93), Solidago nemoralis (78), Fragaria virginiana (74), Anaphalis margaritacea (70), Vaccinium angustifolium (67).

Leading families (% of total species): Compositae (24.7), Gramineae (15.2), Rosaceae (7.0), Leguminosae (4.4), Caryophyllaceae (3.2).

Related communities (Index of similarity): Pine Barrens (38), Northern Dry Forest (32), Sand Barrens (25), Northern Dry Mesic Forest (23), Oak Barrens (23).

Species density: 32.

Index of Homogeneity: 57.1.

No. of stands studied: 27.

Number of species: Trees 4, Shrubs 24, Herbs 136, Total 164.

Stability: Fairly stable. Slowly invaded by balsam fir and aspens. Ultimate replacement uncertain.

Catena position: Middle to low.

Soil group: Melanized sandy Podzol.

Major soil series: Vilas, Plainfield, Omega, and Hiawatha.

Soil analyses: w.r.c. 48, pH 4.6.

Approximate original area: No information.

Typical examples: Many places in Northern Highland State Forest, Nicolet National Forest, and Chequamegon National Forest. None permanently dedicated for study purposes.

Major publications: None. Recognized only recently.

Geographical distribution: Apparently related to aspen parklands of N. Minnesota and the Prairie Provinces of Canada.

APPENDIX FOR CHAPTER 16

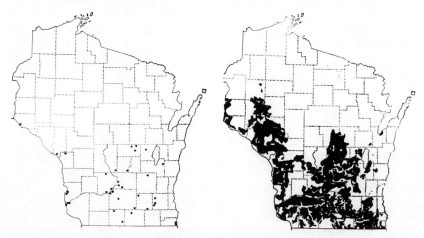

Figure XVI-1.—Stands studied in oak savanna and cedar glade.

Figure XVI-2.—Original oak savanna.

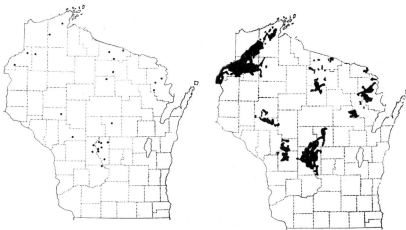

Figure XVI-3.—Stands studied in pine barrens.

Figure XVI-4.—Original pine barrens.

Table XVI-1
Average tree composition of oak openings

Species	Av. I. V.	Constancy	Species	Av. I. V.	Constancy
Quercus macrocarpa*	105.1	87.5%	Fraxinus pennsylvanica	2.0	6.2%
Q. velutina	71.5	87.5	Ulmus rubra	1.7	18.8
Q. alba	61.9	62.5	Juniperus virginiana	1.4	6.2
Carya ovata*	19.9	56.3	Acer negundo	0.8	12.5
Quercus ellipsoidalis	9.3	25.0	Populus grandidentata	0.7	6.2
Prunus serotina	6.8	18.8	Tilia americana	0.6	12.5
Betula papyrifera	6.3	25.0	Juglans nigra	0.3	6.2
Populus tremuloides	4.6	31.3	Fraxinus americana	0.2	6.2
Quercus borealis	3.9	12.5	Ulmus americana	0.2	6.2

* Species which reach their optimum in Wisconsin in this community.

Table XVI-2
Prevalent groundlayer species of the oak openings

Species	Pres.	Av. freq.	Species	Pres.	Av. freq.
Amorpha canescens	79%	33.8%	H. occidentalis	37%	5.3%
Amphicarpa bracteata	79	41.1	H. strumosus	47	13.3
Andropogon gerardi	68	19.4	Heliopsis helianthoides*	42	22.1
A. scoparius	58	12.2	Lespedeza capitata	47	7.9
Anemone cylindrica	58	7.5	Lithospermum canescens	68	14.6
Antennaria neglecta	53	7.8	Monarda fistulosa	84	23.1
Apocynum androsaemifolium	68	19.5	Panicum leibergii	58	10.9
Aquilegia canadensis	37	4.0	Parthenocissus vitacea	53	7.6
Aralia nudicaulis	42	8.0	Petalostemum purpureum	63	12.7
Asclepias syriaca	37	1.3	Phlox pilosa	42	3.0
Carex pensylvanica	32	8.4	Physalis virginiana	37	3.2
Ceanothus americanus	63	10.3	Poa pratensis	53	27.9
Coreopsis palmata	63	14.4	Polygonatum canaliculatum	53	3.8
Comandra richardsiana	84	11.0	Prenanthes alba	47	3.2
Cornus racemosa	79	21.8	Pteridium aquilinum	37	5.9
Corylus americana	68	20.8	Rhus glabra	53	8.4
Desmodium glutinosum	58	18.7	R. radicans	58	12.5
Euphorbia corollata	89	34.3	Rosa sp.	84	22.5
Fragaria virginiana	68	13.5	Smilacina racemosa	79	12.1
Galium boreale	79	33.2	Smilax herbacea	37	5.6
G. concinnum	42	12.1	Stipa spartea	58	8.5
Geranium maculatum	68	15.5	Viola cucullata	42	5.8
Helianthus laetiflorus	32	12.3	V. pedata	63	8.6
			Vitis riparia	79	11.6

* Species are also modal, since their presence values are higher here than in any other Wisconsin community.

Additional modal species, with their presence (%) values: Besseya bullii (16), Orobanche uniflora (5), Phlox glaberrima (5), Ranunculus fascicularis (11), Zygadenus elegans (21).

No. of modal species 6.

No. of prevalent modal species as % of total prevalents 2.1%.

Table XVI-3
Community summary—oak opening

Major dominants (I. V.): Quercus macrocarpa (105), Q. velutina (72), Q. alba (62), Carya ovata (20), Quercus ellipsoidalis (9).

Most prevalent groundlayer species (P%): Euphorbia corollata (89), Comandra richardsiana (84), Monarda fistulosa (84), Rosa sp. (84), Amphicarpa bracteata (79).

Leading families (% of total species): Compositae (21.5), Gramineae (10.7), Leguminosae (7.9), Liliaceae (6.1), Ranunculaceae (5.6).

Related communities (Index of similarity): Oak Barrens (67), Cedar Glade (61), Mesic Prairie (58), Dry Mesic Prairie (56), Southern Dry Forest (53).

Species density: 47.
Index of Homogeneity: 57.3.
No. of stand studied: 19.
Number of species: Trees 18, Shrubs 23, Herbs 191, Total 232.
Stability: High in presence of fire, otherwise quickly converted to Southern Dry Forest or Southern Dry-Mesic Forest.

Climate: Total ppt. 31.6" (80 cm), Snowfall 43.1" (109 cm), Jan. Temp. 14.3°F. (−9.8°C.), July Temp. 70.7°F. (21.5°C.), Growing season 145 days. Typical Weather Stations: Eau Claire, Madison, Wisconsin Dells.

Catena position: Low to high, with or without a gley layer.

Soil group: Gray-Brown Podzolic.

Major soil series: McHenry, Dane, Rodman, Fayette, Dubuque, Beecher, and Morley.

Soil analyses: w.r.c. 83%, pH 6.5, Ca 5220 p.p.m., K 125 p.p.m., P 22 p.p.m.

Approximate original area: Uncertain. Probably 75% of total oak savanna, or 5,500,000 acres.

Typical examples: Observatory Woods S. A., Scuppernong S.A., Kettle Moraine State Forest in Waukesha County.

Major publications: Stout (1944), Bray (1955 thesis).

Geographical distribution: South Dakota to S. Michigan, south to Ohio, Kentucky, and Oklahoma.

Table XVI-4
Average tree composition of oak barrens

Species	Av. I. V.	Constancy	Species	Av. I. V.	Constancy
Quercus velutina*	234.5	92.8%	Carya ovata	5.8	21.4%
Q. ellipsoidalis	23.8	14.3	Prunus serotina	4.4	28.6
Q. alba	18.3	57.1	Quercus borealis	1.4	14.3
Q. macrocarpa	12.4	28.6	Ulmus americana	0.8	7.1

* Black oaks, including Q. velutina and Q. ellipsoidalis, attain their optimum in this community.

Table XVI-5
Prevalent groundlayer species of the oak barrens

Species	Pres.	Av. freq.	Species	Pres.	Av. freq.
Achillea millefolium	43%	7.7%	Koeleria cristata	63%	3.9%
Amorpha canescens	93	24.8	Krigia biflora*	50	3.1
Amphicarpa bracteata	43	13.3	Lathyrus venosus	43	4.2
Andropogon gerardi	57	8.9	Lespedeza capitata*	93	8.6
A. scoparius	64	5.7	Liatris aspera	71	5.4
Anemone cylindrica	64	2.5	Lithospermum canescens*	71	3.9
Antennaria neglecta*	79	7.4	Lupinus perennis*	43	4.3
Apocynum androsaemifolium	57	6.9	Monarda fistulosa	50	10.4
Arabis lyrata	43	3.9	Oenothera biennis	50	2.6
Artemisia caudata	36	2.9	Physalis virginiana*	64	3.5
Asclepias syriaca	50	1.8	Polygonatum canaliculatum*	57	6.4
Aster linariifolius*	43	7.1	Potentilla arguta	43	2.0
A. sagittifolius	36	8.4	Pteridium aquilinum	57	15.4
Carex pensylvanica	50	11.9	Rhus glabra	57	1.8
Comandra richardsiana*	86	15.4	Rosa sp.	64	14.4
Coreopsis palmata	71	7.0	Rudbeckia hirta	71	3.9
Cornus racemosa	57	6.2	Salix humilis	50	5.4
Corylus americana	79	17.9	Smilacina racemosa	64	13.4
Euphorbia corollata*	100	36.0	S. stellata	71	12.9
Fragaria virginiana	71	10.6	Solidago nemoralis	57	9.4
Galium boreale	36	10.1	Tephrosia virginiana*	50	8.0
Geranium maculatum	57	10.0	Tradescantia ohiensis	64	11.6
Helianthemum canadense*	64	13.3	Viola pedata	43	9.3
Helianthus occidentalis	43	6.6	Vitis riparia	43	3.8
Heuchera richardsonii	43	2.1			

* Species are also modal, since their presence values are higher here than in any other Wisconsin community.

Additional modal species, with their presence (%) values: Asclepias amplexicaulis (29), Aureolaria pedicularia (4), Panicum praecocius (21), Penstemon digitalis (4), Penstemon grandiflorus (7), Rhus typhina (21), Viola sagittata (29).

No. of modal species 19.

No. of prevalent modal species as % of total prevalents 24.5%.

Table XVI-6
Community summary—oak barrens

Major dominants (I. V.): Quercus velutina (235), Q. ellipsoidalis (24), Q. alba (18), Q. macrocarpa (12), Carya ovata (6)

Most prevalent groundlayer species (P%): Euphorbia corollata (100), Amorpha canescens (93), Commandra richardsiana (86), Antennaria neglecta (79), Corylus americana (79).

Leading families (% of total species): Compositae (19.4), Gramineae (7.1), Leguminosae (7.1), Rosaceae (4.6), Liliaceae (3.6).

Related communities (Index of similarity): Oak Opening (67), Cedar Glade (64), Dry-Mesic Prairie (58), Mesic Prairie (57), Wet-Mesic Prairie (50).

Species density: 49.
Index of Homogeneity: 57.9.
No. of stands studied: 14.

Number of species: Trees 8, Shrubs 22, Herbs 174, Total 204.

Stability: Fairly high in presence of fire, otherwise quickly converted to Southern Dry Forest.

Catena position: Top, especially on sandy soils of outwash plains.

Soil group: Melanized or aeolian sands.

Major soil series: Wyocena, Plainfield, Coloma, Fox, and Alvin.

Soil analyses: w.r.c. 60%, pH 5.6, Ca 1025 p.p.m., K 85 p.p.m., P 20 p.p.m.

Approximate original area: Uncertain. Probably 25% of oak savanna or 1,800,000 acres.

Typical examples: Mazomanie H. & F., Black River State Forest.

Major publication: Bray (1955 thesis)

Geographical distribution: S. Minnesota, S. Wisconsin, N. Illinois.

Table XVI-7
Average tree composition of pine barrens

Species	Av. I.V.	Constancy	Species	Av. I.V.	Constancy
Pinus banksiana*	187.3	100%	Pinus resinosa	8.4	14%
Quercus ellipsoidalis*	57.7	86	Populus tremuloides	2.7	7
Quercus macrocarpa	28.0	21	Quercus borealis	1.2	7
Populus grandidentata	10.2	21			

* Species which reach their optimum in this community. (See black oaks in oak barrens.)

Table XVI-8
Prevalent groundlayer species of pine barrens

Species	Pres.	Species	Pres.
Apocynum androsaemifolium	69%	Lithospermum canescens	39%
Aster macrophyllus	46	Lupinus perennis	39
Ceanothus ovatus*	54	Lysimachia quadrifolia*	50
Chimaphila umbellata	50	Maianthemum canadense	62
Coreopsis palmata	34	Monarda fistulosa	50
Corylus americana	77	Myrica asplenifolia	54
Danthonia spicata	46	Pteridium aquilinum	58
Euphorbia corollata	77	Rosa sp.	65
Fragaria virginiana	73	Rubus pubescens	65
Gaultheria procumbens	50	Salix discolor	42
Gaylussacia baccata*	42	Smilacina racemosa	54
Helianthemum canadense	39	S. stellata	58
Leptoloma cognatum	35	Vaccinium angustifolium	85

* Species are also modal, since their presence values are higher here than in any other Wisconsin community.

Additional modal species, with their presence (%) values: Agastache foeniculum (6), Crepis tectorum (3).

No. of modal species 5.

No. of prevalent modal species as % of total prevalents 11.5%.

Table XVI-9
Community summary—pine barrens

Major dominants (I. V.): Pinus banksiana (187), Quercus ellipsoidalis (58), Q. macrocarpa (28), Populus grandidentata (10), P. resinosa (8).
Most prevalent groundlayer species (P%): Vaccinium angustifolium (84), Corylus americana (75), Euphorbia corollata (75), Fragaria virginiana (72), Apocynum androsaemifolium (69).
Leading families (% of total species): Compositae (23.9), Gramineae (10.4), Rosaceae (8.2), Liliaceae (6.7), Ericaceae (6.0).
Related communities (Index of similarity): Northern Dry Forest (49), Oak Barrens (44), Bracken-Grassland (38), Sand Barrens (38), Southern Dry Forest (35).
Species density: 26.
Index of Homogeneity: 55.3.
No. of stands studied: 26
Stability: Low, unless burned frequently. Replaced by Northern Dry Forest or by Dry-Mesic Prairie.

Climate: Total ppt. 28.6″ (73 cm), Snowfall 48.3″ (123 cm), Jan. Temp. 9.2°F. (−12.7°C.), July Temp. 67.9°F. (19.9°C.), Growing season 118 days. Typical Weather Stations: Danbury, Hayward, Tomahawk.
Catena position: Top. Largely on sand plains.
Soil group: Melanized Sandy Podzols.
Major soil series: Omega, Vilas, Boone, and Coloma.
Soil analyses: w.r.c. 36 %, pH 5.2, Ca 4025 p.p.m., K 70 p.p.m., P 22 p.p.m.
Approximate original area: 2,340,000 acres.
Typical examples: Marinette County Forest, Northern Highlands State Forest, Crex Meadows Conservation Area, Bayfield County Forest, Moquah S.A., Brule River State Forest.
Geographical distribution: Alberta to central Ontario and New York, south to central Michigan, central Wisconsin, and northern Minnesota.

Table XVI-10
Average tree composition of cedar glades

Species	Av. I. V.	Constancy	Species	Av. I. V.	Constancy
Juniperus virginiana*	182.3	100%	Quercus macrocarpa	10.9	43%
Quercus velutina	26.6	86	Betula papyrifera	4.9	43
Tilia americana (form)	17.4	43	Carya ovata	3.9	14
Quercus alba	12.0	43	Ostrya virginiana	3.4	29
Q. borealis	11.1	29	Populus tremuloides	2.0	29
Populus grandidentata	11.0	43			

* Species which attain optimum in this community.

Table XVI-11
Prevalent groundlayer species of the cedar glades

Species	Pres.	Av. freq.	Species	Pres.	Av. freq.
Ambrosia artemisiifolia	57%	5.8%	Heuchera richardsonii*	71%	5.4%
Amorpha canescens	86	20.8	Koeleria cristata	71	14.4
Amphicarpa bracteata	57	4.4	Lespedeza capitata	43	1.0
Andropogon gerardi	86	26.6	Linum sulcatum*	43	9.3
A. scoparius	100	38.0	Lithospermum incisum	43	1.6
Anemone cylindrica*	86	20.1	Lobelia spicata*	43	3.7
A. patens	71	5.1	Monarda fistulosa*	86	6.6
Antennaria neglecta	71	17.9	Oenothera biennis	57	4.4
Aquilegia canadensis*	100	16.6	Opuntia compressa*	43	15.0
Arabis lyrata*	57	3.1	Parthenocissus vitacea	71	3.7
Arenaria stricta	71	23.0	Pedicularis canadensis*	43	2.9
Artemisia caudata	57	8.6	Petalostemum candidum*	57	4.7
Asclepias verticillata	43	1.0	P. purpureum	86	16.0
Aster azureus	43	2.3	Physalis virginiana	57	3.1
A. laevis	57	3.0	Poa pratensis	43	5.7
A. sericeus	43	3.1	Potentilla arguta*	86	3.9
Bouteloua curtipendula	71	16.6	Rhus radicans	57	13.0
B. hirsuta*	86	19.4	Rosa sp.	71	7.3
Campanula rotundifolia*	57	5.0	Scutellaria leonardi*	86	7.4
Ceanothus americanus	57	5.4	Silene stellata*	57	6.7
Celastrus scandens	43	5.9	Smilacina racemosa	57	6.7
Comandra richardsiana	43	2.1	S. stellata	57	5.9
Coreopsis palmata	71	7.4	Solidago nemoralis	71	20.8
Dodecatheon meadia	43	3.0	Sporobolus cryptandrus*	57	7.4
Erigeron strigosus	57	3.1	S. heterolepis	43	6.4
Euphorbia corollata	71	27.1	Tradescantia ohiensis*	100	9.0
Geranium maculatum	57	12.3	Viola pedata*	86	13.1
Hedeoma hispida	43	3.0	Vitis riparia	71	8.6
Helianthus occidentalis	43	5.3			

* Species are also modal, since their presence values are higher here than in any other Wisconsin community.

Additional modal species, with their presence (%) values: Lespedeza violacea (14), Opuntia fragilis (14), Polygala sanguinea (14), Polygala verticillata (14) Scrophularia lanceolata (14), Sporobolus vaginiflorus (14).

No. of modal species 24.

No. of prevalent modal species as % of total prevalents 31.5%.

Table XVI-12
Community summary—cedar glade

Major dominants (I. V.): Juniperus virginiana (182), Quercus velutina (27), Tilia americana (17), Q. alba (12), Q. borealis (11).

Most prevalent groundlayer species (P%): Andropogon scoparius (100), Aquilegia canadensis (100), Tradescantia ohiensis (100), Bouteloua hirsuta (86), Scutellaria leonardi (86).

Leading families (% of total species): Compositae (23.4), Gramineae (8.9), Leguminosae (6.3), Rosaceae (5.1), Liliaceae (3.8).

Related communities (Index of similarity): Oak Barrens (64), Dry-Mesic Prairie (62), Wet-Mesic Prairie (61), Oak Opening (61), Dry Prairie (59).

Species density: 57.

Index of Homogeneity: 62.7.

No. of stands studied: 7.

Number of species: Trees 11, Shrubs 22, Herbs 136, Total 169.

Stability: High. Slowly invaded by oak forest from periphery.

Catena position: Top.

Soil group: Regosol (Skeletal Rendzina).

Major soil series: Bellefontaine, Rodman.

Soil analyses: w.r.c. 89%, pH 6.5, Ca 4500 p.p.m., K 110 p.p.m., P 52 p.p.m.

Approximate original area: Extremely small, perhaps several thousand acres only.

Typical examples: Wyalusing S.A., Brady's Bluff S.A., Nelson Dewey State Park, Wildcat Mountain State Park.

Major publication: Bray (1955 thesis).

Geographical distribution: Minnesota and Wisconsin, south to Tennessee and Arkansas.

APPENDIX FOR CHAPTER 17

Figure XVII-1.—Stands studied in shrub-carr.

Figure XVII-2.—Stands studied in alder thicket.

Table XVII-1
Prevalent species of shrub-carr

Species	Pres.	Species	Pres.
Asclepias incarnata	60%	L. uniflorus	40%
Aster lucidulus	50	Onoclea sensibilis	50
A. simplex	60	Phalaris arundinacea*	60
Calamagrostis canadensis	90	Poa palustris	50
Cicuta maculata	40	Rumex orbiculatus	40
Cirsium muticum	40	Salix bebbiana*	30
Cornus stolonifera*	70	S. discolor*	60
Dryopteris thelypteris	50	Solidago gigantea	70
Equisetum sylvaticum	30	S. canadensis	50
Eupatorium maculatum	60	Spiraea alba	50
E. perfoliatum	50	Stachys palustris	40
Glyceria nervata	60	Thalictrum dasycarpum	60
Impatiens biflora	60	Typha latifolia	50
Lycopus americanus	60	Viola cucullata	40

* Species are also modal, since their presence values are higher here than in any other Wisconsin community.

Additional modal species, with their presence (%) values: Carex rostrata (10), Cypripedium parviflorum Salisb. (25), Erigeron philadelphicus (30), Helianthus tuberosus (10), Leerzia oryzoides (10), Rhamnus catharticus (10), Scirpus lineatus (10), Silene nivea (10), Sium suave (20), Solidago patula (10), Steironema lanceolatum (10).

No. of modal species 15.
No. of prevalent modal species as % of total prevalents 14.3%.

Table XVII-2
Community summary—shrub-carr

Major dominants (P%): Cornus stolonifera (70), Salix discolor (60), Spiraea alba (50), Salix bebbiana (30), S. petiolaris (20).

Most prevalent groundlayer species (P%): Solidago gigantea (70), Asclepias incarnata (60), Aster simplex (60), Eupatorium maculatum (60), Thalictrum dasycarpum (60).

Leading families (% of total species): Compositae (14.8), Gramineae (8.3), Rosaceae (6.5), Cyperaceae (5.6), Labiatae (5.6).

Related communities (Index of similarity): Southern Sedge Meadow (66), Northern Sedge Meadow (61), Alder Thicket (60), Fen (55), Wet Prairie (39).

Species density: 28.

Index of Homogeneity: 51.8.

No. of stands studied: 10.

Number of species: Trees 2, Shrubs 21, Herbs 93, Total 116.

Stability: Fairly stable. Slowly invaded by Southern Wet Forest. May revert to Southern Sedge Meadow or Fen when burned.

Catena position: Bottom.

Soil group: Azonal peat and muck.

Soil analyses: pH 6.9.

Approximate original area: Unknown.

Typical examples: U.W. Arboretum, Mazomanie H. & F., Waterloo H. & F., Vernon H. & F.

Geographical distribution: Widespread, but limits uncertain.

Table XVII-3
Prevalent species of alder thicket

Species	Pres.	Species	Pres.
Alnus rugosa*	100%	Lycopus uniflorus	42%
Asclepias incarnata	42	Mentha arvensis*	50
Aster simplex	92	Onoclea sensibilis*	75
Aster puniceus	50	Poa palustris	58
Bromus ciliatus*	50	Polygonum sagittatum*	67
Calmagrostis canadensis	83	Potentilla palustris	42
Campanula aparinoides	67	Ribes americanum*	50
Chelone glabra*	50	Rumex orbiculatus*	50
Cornus stolonifera	50	Scirpus atrovirens*	75
Dryopteris cristata	50	Solidago canadensis*	58
Eupatorium maculatum*	92	S. gigantea	42
E. perfoliatum	58	Spiraea alba*	67
Galium asprellum*	67	Thalictrum dasycarpum	42
Glyceria grandis	58	Thelypteris palustris*	75
Impatiens biflora	67	Typha latifolia	50
Iris shrevei	50		

* Species are also modal, since their presence values are higher here than in any other Wisconsin community.

Additional modal species, with their presence (%) values: Botrychium multifidum (8), Clematis virginiana (17), Epilobium coloratum (25), Galium labradoricum (17), Geum aleppicum (33), Helianthus giganteus (25), Humulus americana (8), Hydrocotyle americana (8), Hypericum pyramidatum (8), Scirpus cyperinus (8), Viburnum opulus (17).

No. of modal species 25.

No. of prevalent modal species as % of total prevalents 45.2%.

Table XVII-4
Community summary—alder thicket

Major dominants (P%): Alnus rugosa (100), Spiraea alba (67), Cornus stolonifera (50), Ribes americanum (50).

Most prevalent groundlayer species (P%): Aster simplex (92), Eupatorium maculatum (92), Calamagrostis canadensis (83), Scirpus atrovirens (75), Thelypteris palustris (75).

Leading families (% of total species): Compositae (14.3), Gramineae (8.9), Rosaceae (8.1), Labiatae (6.3), Rubiaceae (4.5).

Related communities (Index of similarity): Northern Sedge Meadow (76), Shrub-Carr (60), Southern Sedge Meadow (54), Fen (49), Wet Prairie (29).

Species density: 31.

Index of Homogeneity: 60.5.

No. of stands studied: 12.

Number of species: Trees 3, Shrubs 21, Herbs 81, Total 115.

Stability: Fairly stable. May be replaced by Northern Wet-Mesic Forest under some circumstances.

Catena position: Bottom. Gley layer present.

Soil group: Azonal peat.

Major soil series: Uncertain.

Soil analyses: pH 5.5.

Approximate original area: Unknown.

Typical examples: Many places in northern State and Federal forests along lakes and streams.

Geographical distribution: Northeastern North America.

APPENDIX FOR CHAPTER 18

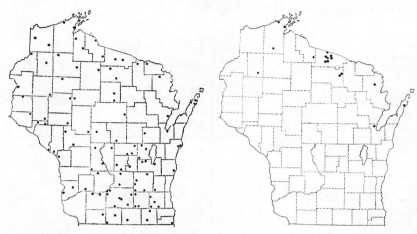

Figure XVIII-1.—Stands studied in sedge meadows.

Figure XVIII-3.—Stands studied of open bog.

Figure XVIII-2.—Original sedge meadows.

Table XVIII-1
Prevalent species of fen

Species	Pres.	Species	Pres.
Andropogon gerardi	33%	Lathyrus palustris	50%
Asclepias incarnata*	83	Lobelia kalmii*	33
Aster lucidulus*	83	Lycopus americanus*	100
A. simplex*	100	L. uniflorus	83
A. umbellatus	33	Mentha arvensis	50
Betula sandbergii*	33	Parnassia glauca*	67
Bromus ciliatus	50	Pedicularis lanceolata*	67
Calamagrostis canadensis*	100	Phlox pilosa	50
Caltha palustris*	50	Potentilla fruticosa*	33
Campanula aparinoides*	83	Pycnanthemum virginianum	67
Cirsium muticum	50	Solidago canadensis	50
Cornus stolonifera	67	S. gigantea	50
Cypripedium candidum*	33	S. riddellii*	33
Equisetum arvense	67	S. uliginosa	33
Eupatorium maculatum	83	Spartina pectinata	50
E. perfoliatum*	67	Thalictrum dasycarpum*	100
Galium boreale*	83	Thelypteris palustris	67
Gentiana procera*	67	Valeriana ciliata*	33
Gerardia paupercula	50	Viola cucullata	67
Glyceria striata*	67	Zizia aurea	50
Iris shrevei*	100		

* Species are also modal, since their presence values are higher here than in any other Wisconsin community.

Additional modal species, with their presence (%) values: Apocynum sibiricum (33), Asclepias sullivantii (17), Aster junciformis (17), Bidens connata (9), Carex sartwellii (17), Castilleja coccinea (17), Juncus dudleyi (17), Lobelia siphilitica (17), Ophioglossum vulgatum (17), Polygala senega (17), Scirpus cespitosus (17), Selaginella apoda (17), Sphenopholis intermedia (17), Spiranthes cernua (17), Steironema quadriflorum (17), Stellaria longifolia (33).

No. of modal species 37.

No. of prevalent modal species as % of total prevalents 51.2%.

Table XVIII-2
Community summary—fen

Major dominants (P%): Calamagrostis canadensis (100), Glyceria striata (67), Bromus ciliatus (50), Spartina pectinata (50), Andropogon gerardi (33).
Most prevalent groundlayer species (P%): Aster simplex (100), Iris shrevei (100), Lycopus americanus (100), Thalictrum dasycarpum (100), Aster luciduus (83).
Leading families (% of total species): Compositae (17.4), Gramineae (8.7), Labiatae (5.2), Cyperaceae (4.3), Umbelliferae (4.3).
Related communities (Index of similarity): Southern Sedge Meadow (59), Shrub-Carr (54), Wet Prairie (54), Northern Sedge Meadow (52), Alder Thicket (49).
Species density: 41.

Index of Homogeneity: 61.3.
No. of stands studied: 6.
Number of species: Trees 0, Shrubs 10, Herbs 105, Total 115.
Stability: Stable if burned, otherwise rapidly converted to Shrub-Carr.
Catena position: Bottom. Related to glacial lakes.
Soil group: Azonal peat
Major soil series: Carlisle
Approximate original area: Very small, probably only a few 100 acres.
Typical examples: U.W. Arboretum. No others under public control.
Geographical distribution: Little studied. Known from Iowa, Missouri, S. Michigan, Ohio, and Pennsylvania.

Table XVIII-3
Prevalent species (excluding Carex) of southern sedge meadows

Species	Pres.	Species	Pres.
Anemone canadensis*	55%	Iris shrevei	61%
Angelica atropurpurea*	41	Lathyrus palustris*	55
Asclepias incarnata	64	Lycopus americanus	64
Aster luciduus	43	L. uniflorus	52
A. simplex	91	Phalaris arundinacea	41
Calamagrostis canadensis	73	Poa palustris	36
Cicuta maculata	41	Solidago canadensis	57
Equisetum arvense	57	S. gigantea	48
Eupatorium maculatum	61	Spartina pectinata	52
E. perfoliatum	41	Stachys palustris*	45
Galium obtusum*	43	Thalictrum dasycarpum	50
Glyceria striata	41	Thelypteris palustris	39
Helianthus grosseserratus	39	Typha latifolia	57
Impatiens biflora	50	Viola cucullata	43

* Species are also modal, since their presence values are higher here than in any other Wisconsin community.

Additional modal species, with their presence (%) values: Amorpha fruticosa (2), Cardamine bulbosa (32), Cyperus erytherorhizos (2), Gratiola neglecta (2), Habenaria psycodes (5), Juncus effusus (5), Naumbergia thyrsiflora (30), Polygonum pensylvanicum (5), Ranunculus sceleratus (2), Rhexia virginica (2), Strophostyles helvola (2), Verbena urticifolia (14).

No. of modal species 17.
No. of prevalent modal species as % of total prevalents 17.9%.

Table XVIII-4
Average frequencies of southern sedge meadow species—21 stands in southeastern Wisconsin
(All species with average frequency above 4.0%)

Species	%	Species	%
Carex stricta	58%	Rumex verticillatus	6%
Calamagrostis canadensis	31	Helenium autumnale	6
Lycopus uniflorus	18	Verbena hastata	6
Carex incomperta	15	Carex sartwellii	6
Lycopus americanus	14	Solidago gigantea	5
Carex aquatilis	13	Epilobium adenocaulon	5
Glyceria striata	10	Anemone canadensis	5
Carex lasiocarpa	9	Equisetum arvense	4
Mentha arvensis	8	Eleocharis palustris	4
Thelypteris palustris	8	Scutellaria galericulata	4
Eupatorium perfoliatum	7	Agrostis stolonifera	4
E. maculatum	7	Bidens frondosa	4
Viola cucullata	7	Iris shrevei	4
Stachys palustris	7		

Table XVIII-5
Community summary—southern sedge meadow

Major dominants ($P\%$): Carex stricta and other sedges; plus Calamagrostis canadensis (73), Spartina pectinata (52), Glyceria striata (41), and Phalaris arundinacea (41).

Most prevalent groundlayer species ($P\%$): Aster simplex (91), Asclepias incarnata (64), Eupatorium maculatum (61), Iris shrevei (61), Equisetum arvense (57).

Leading families (% *of total species*): Compositae (20.3), Cyperaceae (8.1), Gramineae (7.6), Labiatae (5.1), Rosaceae (4.6).

Related communities (*Index of similarity*): Shrub-Carr (66), Northern Sedge Meadow (62), Fen (59), Alder Thicket (54), Wet Prairie (46).

Species density: 28.
Index of Homogeneity: 51.1.
No. of stands studied: 44.
Number of species: Trees 0, Shrubs 13, Herbs 184, Total 197.
Stability: Fairly stable. Invaded by Wet Prairie or Shrub-Carr.

Climate: Total ppt. 29.9" (76 cm), Snowfall 40.1" (102 cm), Jan. Temp. 14.0°F. (−10.0°C.), July Temp. 70.2°F. (21.2°C.), Growing season 138 days. Typical Weather Stations: Burnett, Grand River Locks, Meadow Valley.

Catena position: Bottom.
Soil group: Azonal peat and muck.
Major soil series: Badoura, Carlisle, Kokomo, and Arenzville.
Soil analyses: w.r.c. 275%, pH 7.4, Ca 3750 p.p.m., K 90 p.p.m., P 15 p.p.m.
Approximate original area: Uncertain. Perhaps 90% of total sedge meadow or *ca.* 1,000,000 acres.
Typical examples: Faville Prairie S.A., Scuppernong S.A., U.W. Arboretum.
Major publications: Stout (1914), Costello (1936), Frolik (1941).
Geographical distribution: Throughout southern United States, with most similar stands along eastern edge of prairie region.

Table XVIII-6
Prevalent species (excluding Carex) of northern sedge meadow

Species	Pres.	Species	Pres.
Asclepias incarnata	37%	Onoclea sensibilis	54%
Aster puniceus*	51	Poa palustris*	71
A. simplex	91	Polygonum natans*	37
Bromus ciliatus	37	P. sagittatum	54
Calamagrostis canadensis	94	Potentilla palustris	34
Campanula aparinoides	63	Scirpus atrovirens	60
Cicuta maculata	34	Solidago gigantea	43
Equisetum arvense	34	S. graminifolia*	49
Eupatorium maculatum	83	S. uliginosa*	34
E. perfoliatum	57	Spiraea alba	57
Glyceria canadensis*	40	Thalictrum dasycarpum	40
Impatiens biflora	40	Thelypteris palustris	57
Iris shrevei	66	Typha latifolia	60
Lycopus americanus	43	Verbena hastata*	34
L. uniflorus*	60		

* Species are also modal, since their presence values are higher here than in any other Wisconsin community.

Additional modal species, with their presence (%) values: Alopecurus pratensis (3), Bidens cernua (9), Cicuta bulbifera (31), Epilobium strictum (23), Galium trifidum (23), Gentiana rubricaulis (6), Gerardia tenuifolia (3), Lysimachia terrestris (20), Matteuccia struthiopteris (11), Mimulus ringens (17), Polygonum coccineum (23), Ranunculus flammula (6), Salix petiolaris (29), Scutellaria galericulata (29), Selaginella selaginoides (3), Spiraea tomentosa (26).

No. of modal species 24.

No. of prevalent modal species as % of total prevalents 27.6%.

Table XVIII-7
Community summary—northern sedge meadow

Major dominants (*P%*): Carex stricta and other sedges; plus Calamagrostis canadensis (94), Poa palustris (71), Scirpus atrovirens (60), Glyceria canadensis (40).
Most prevalent groundlayer species (*P%*): Aster simplex (91), Eupatorium maculatum (83), Iris shrevei (66), Campanula aparinoides (63), Lycopus uniflorus (60).
Leading families (*% of total species*): Compositae (15.8), Cyperaceae (9.0), Gramineae (9.0), Rosaceae (6.8), Labiatae (6.2).
Related communities (*Index of similarity*): Alder Thicket (76), Southern Sedge Meadow (62), Shrub-Carr (61), Fen (52), Wet Prairie (34).
Species density: 29.
Index of Homogeneity: 51.9.

No. of stands studied: 35.
Number of species: Trees 0, Shrubs 19, Herbs 158, Total 177.
Stability: Fairly stable. Slowly invaded by Alder Thicket.
Catena position: Bottom. Typically on stream sides.
Soil group: Azonal peat.
Major soil series: Greenwood, Carlisle, Spalding, Newton, Adolph, and Cable.
Approximate original area: Uncertain. Perhaps 10% of total sedge meadow or 115,000 acres.
Typical examples: Many places in northern State and Federal Forests.
Major publications: Jackson (1914), Catenhusen (1950).
Geographical distribution: Throughout northern United States and southern Canada.

Table XVIII-8
Prevalent species of open bog

Species	Pres.	Species	Pres.
Andromeda glaucophylla*	94%	Kalmia polifolia*	82%
Betula glandulosa*	29	Ledum groenlandicum	77
Calla palustris*	47	Menyanthes trifoliata*	41
Carex trisperma	53	Nemopanthus mucronatus	35
Chamaedaphne calyculata*	88	Potentilla palustris*	47
Cornus canadensis	53	Sarracenia purpurea*	77
Cypripedium acaule*	47	Smilacina trifolia	77
Drosera rotundifolia*	53	Thelypteris palustris	35
Dryopteris cristata	47	Trientalis borealis	35
Eriophorum angustifolium*	29	Vaccinium angustifolium	59
E. spissum	35	V. macrocarpon*	77
Gaultheria hispidula	47	V. myrtilloides	59
G. procumbens	41	V. oxycoccos*	65
Iris versicolor*	47		

* Species are also modal, since their presence values are higher here than in any other Wisconsin community.

Additional modal species, with their presence (%) values: Arethusa bulbosa (6), Bidens tripartita (6), Brasenia schreberi (6), Calopogon pulchellus (18), Carex lasiocarpa (29), Carex oligosperma (24), Cladium mariscoides (6), Drosera intermedia (24), Dryopteris bootii (6), Dulichium arundinaceum (12), Habenaria dilatata (6), Myrica gale (18), Pogonia ophioglossoides (18), Rhus vernix (12), Rhynchospora alba (24), Salix candida (6), Scirpus rubrotinctus (12), Triadenum virginicum (29), Triglochin maritima (12).

No. of modal species 30.
No. of prevalent modal species as % of total prevalents 51.8%.

Table XVIII-9
Community summary—open bog

Major dominants ($P\%$): Andromeda glaucophylla (94), Chamaedaphne calyculata (88), Kalmia polifolia (82), Ledum groenlandicum (77), Vaccinium angustifolium (59).

Most prevalent groundlayer species ($P\%$): Sarracenia purpurea (77), Smilacina trifolia (77), Vaccinium macrocarpon (77), Vaccinium oxycoccos (65), Carex trisperma (53).

Leading families (% of total species): Ericaceae (9.2), Cyperaceae (9.2), Compositae (6.7), Gramineae (5.9), Orchidaceae (5.0).

Related communities (*Index of similarity*): Northern Wet Forest (60), Northern Wet-Mesic Forest (25), Northern Sedge Meadow (20), Boreal Forest (19), Alder Thicket (18).

Species density: 27.
Index of Homogeneity: 53.7.

No. of stands studied: 17.
Number of species: Trees 11, Shrubs 32, Herbs 87, Total 130.
Stability: Very stable. Slowly invaded by Northern Wet Forest.
Catena position: Bottom.
Soil group: Azonal peat.
Soil analyses: w.r.c. 1070%, pH 3.9, Ca 750 p.p.m., K 100 p.p.m., P 10 p.p.m.
Approximate original area: No information. Probably less than 5% of conifer swamps, or *ca.* 110,000 acres.
Typical examples: Cedarburg Swamp S. A., many stands in northern State and Federal Forests, Hope Lake in Jefferson County.
Major publications: Jackson (1914), Rhodes (1933), Wilson (1939).
Geographical relations: Glaciated region of northeastern North America.

APPENDIX FOR CHAPTER 19

Figure XIX-1.—Stands studied of emergent aquatics.

Figure XIX-2.—Stands studied of submerged aquatics.

Table XIX-1
Prevalent species of emergent aquatic communities

Species	Pres.	Species	Pres.
Eleocharis acicularis*	38%	S. americanus*	42%
Iris shrevei	29	S. validus*	49
Phragmites communis*	38	Sparganium eurycarpum*	51
Pontederia cordata*	51	Typha latifolia*	71
Sagittaria latifolia*	62	Zizania aquatica*	53
Scirpus acutus*	73		

* Species are also modal, since their presence values are higher here than in any other Wisconsin community.

Additional modal species, with their presence (%) values: Alisma plantago-aquatica (11), Carex aquatilis (9), Carex trichocarpa (2), Cyperus strigosus (2), Decodon verticillatus (11), Echinochloa walteri (7), Eleocharis calva (4), Eleocharis palustris (18), Glyceria borealis (2), Juncus torreyi (11), Lemna minor (67), Lemna trisulca (22), Ludwigia palustris (2), Ranunculus flabellaris (2), Ranunculus reptans (2), Sagittaria rigida (29), Scirpus fluviatilis (31), Scirpus heterochaetus (16), Sparganium americanum (2), Spirodela polyrhiza (9), Typha angustifolia (29), Utricularia vulgaris (22), Veronica anagallis-aquatica (2).

No. of stands 45.
Species density 11.
Total no. of species 80.
Index of Homogeneity 49.2%.
No. of modal species 33.
No. of prevalent modal species as % of total prevalents 90.8%.

Table XIX-2
Average structure of southern cattail marshes

Species	Frequency	Density per sq. m.	Species	Frequency	Density per sq. m.
Lemna minor	100%	210.0	Typha latifolia	20	0.3
Typha angustifolia	95	27.7	Apocynum cannabinum	17	0.5
Utricularia vulgaris	93	130.2	Calamagrostis canadensis	17	3.1
Equisetum fluviatile	65	3.6	Carex hystericina	17	4.5
Eleocharis elliptica	53	70.9	Leersia oryzoides	17	2.0
Cicuta bulbifera	45	10.2	Salix interior	15	0.8
Scirpus americanus	40	33.1	Galium trifidum	13	1.7
Carex aquatilis	32	10.6	Carex stricta	13	5.5
Sagittaria latifolia	32	10.3	Rumex orbiculatus	11	0.2
Carex trichocarpa	23	7.9	Scirpus acutus	7	1.9
Lycopus uniflorus	23	5.8	Phragmites communis plus six other species	7	3.1
Carex lanuginosa	20	1.1			

Table XIX-3
Prevalent species of submerged aquatic communities

Species	Pres.	Species	Pres.
Anacharis canadensis*	42%	Potamogeton gramineus*	35%
Ceratophyllum demersum*	32	P. zosteriformis*	28
Eleocharis acicularis	26	Vallisneria americana*	39
Najas flexilis*	68		

Other species with presence over 10%

Bidens beckii*	11	P. illinoensis*	25
Isoetes macrospora*	14	P. natans*	11
Juncus pelocarpus*	12	P. pectinatus*	26
Myriophyllum exalbescens*	10	P. praelongus*	16
M. tenellum*	13	P. pusillus*	13
M. verticillatum*	16	P. richardsonii*	16
Potamogeton amplifolius*	19	Sagttaria graminea*	10
P. foliosus*	11	Zostierella dubia*	20
P. friesii*	16		

* Species are also modal, since their presence values are higher here than in any other Wisconsin community.

Additional modal species, with their presence (%) values: Anacharis nuttallii (1), Ceratophyllum echinatum (1), Elatine minima (4), Eriocaulon septangulare (6), Gratiola aurea (5), Lobelia dortmanna (8), Potamogeton diversifolius (1), Potamogeton epihydrus (6), Potamogeton filiformis (3), Potamogeton obtusifolius (3), Potamogeton pulcher (1), Potamogeton robbinsii (9), Potamogeton spirillus (5), Potamogeton strictifolius (7), Ranunculus aquatilis (9), Sagittaria cuneata (1), Sparganium angustifolium (4), Sparganium fluctuans (1), Utricularia intermedia (1), Utricularia purpurea (1).

No. of stands 142.
Species density 7.
Total no. of species 51.
Index of Homogeneity 45.3%.
No. of modal species 42.
No. of prevalent modal species as % of total prevalents 85.7%.

APPENDIX FOR CHAPTER 20

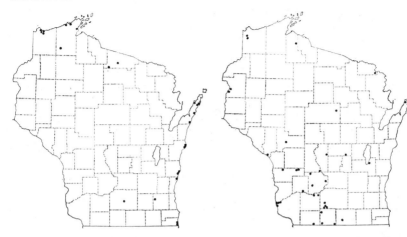

Figure XX-1.—Stands studied of beach and dune.

Figure XX-2.—Stands studied of rock cliffs.

Table XX-1
Prevalent species of lake beaches

Species	Pres.	Species	Pres.
Agropyron dasystachyum	22%	L. uniflorus	35%
Cakile edentula*	61	Mentha arvensis	35
Cerastium vulgatum*	22	Oenothera biennis	26
Elymus canadensis	39	Potentilla anserina	39
Equisetum arvense	22	Salix interior	30
Lathyrus maritimus	35	Salsola kali*	44
Lycopus americanus	22	Xanthium strumarium*	30

* Species are also modal, since their presence values are higher here than in any other Wisconsin community.

Additional modal species, with their presence (%) values: Acnida altissima (9), Avena fatua (4), Bidens connata (9), Bidens vulgata (4), Carex aurea (4), Carex hystericina (4), Carex viridula (13), Cirsium pitcheri (4), Cladium mariscoides (4), Corispermum hyssopifolium (22), Dianthus barbatus (13), Eleocharis compressa (9), Juncus alpinus (13), Juncus brevicaudatus (4), Juncus bufonius (9), Juncus nodosus (4), Muhlenbergia sylvatica (13), Oenothera serrulata (13), Penthorum sedoides (13), Polygonum hydropiper (13), Polygonum lapathifolium (9), Primula mistassinica (4), Ranunculus pensylvanicus (17), Rhynchospora capillacea (4), Satureja glabella (4), Sicyos angulatus (37), Thaspium barbinode (13), Triglochin palustris (9).

No. of stands 23.
Species density 14.
Total no. of species 131.
Index of Homogeneity 34.1%.
No. of modal species 32.
No. of prevalent modal species as % of total prevalents 28.6%.

Table XX-2
Prevalent species of lake dunes

Species	Pres.	Species	Pres.
Agropyron dasystachyum*	30%	J. horizontalis*	30%
Ammophila breviligulata*	70	Lathyrus maritimus*	55
Arabis lyrata	30	Oenothera biennis*	80
Artemisia caudata	80	Poa compressa	50
Asclepias syriaca	45	Potentilla anserina*	45
Cakile edentula	40	Rhus radicans	35
Calamovilfa longifolia*	30	Rosa sp.	55
Elymus canadensis*	75	Salix cordata*	35
Equisetum arvense	35	S. interior*	35
E. variegatum*	30	Smilacina stellata	45
Juniperus communis*	40	Solidago graminifolia	35

* Species are also modal, since their presence values are higher here than in any other Wisconsin community.

Additional modal species, with their presence (%) values: Bromus erectus (5), Coreopsis lanceolata (5), Cycloloma atriplicifolium (20), Equisetum hiemale (30), Fimbristylis autumnalis (5), Iris lacustris (10), Juncus balticus (15), Myosotis scorpioides (10), Panicum meridionale (5), Salix glaucophylloides (5), Salix lucida (10), Solidago spathulata (5), Veronica serpyllifolia (5).

No. of stands 20.
Species density 22.
Total no. of species 155.
Index of Homogeneity 45.6%.
No. of modal species 22.
No. of prevalent modal species as % of total prevalents 54.5%

Table XX-3
Structure of lake dune communities

Species	F. %	Density per 100 sq. meters	Species	F. %	Density per 100 sq. meters
Ammophila breviligulata	94	5780	Solidago juncea	16	43
Lathyrus maritimus	61	215	Galium boreale	14	11
Asclepias syriaca	34	21	Equisetum variegatum	14	21
Poa compressa	27	538	Rhus radicans	14	21
Artemisia caudata	27	248	Juniperus horizontalis	12	32
Smilacina stellata	26	54	Asparagus officinalis	10	75
Arabis lyrata	16	258	Agropyron dasystachyum	5	97

Table XX-4
Prevalent species of exposed rock cliffs

Species	Pres.	Species	Pres.
Andropogon scoparius	39%	Pellaea atropurpurea*	31%
Aquilegia canadensis	62	Polypodium vulgare	39
Arabis lyrata	39	Rhus radicans	46
Aralia hispida*	23	Selaginella rupestris	46
Campanula rotundifolia	46	Silene antirrhina*	54
Corydalis sempervirens*	23	Solidago nemoralis	46
Cryptogramma stelleri*	23	Talinum rugospermum*	23
Diervilla lonicera	31	Tradescantia ohiensis	46
Festuca octoflora*	31	Vaccinium angustifolium	31
Hedeoma hispida	39	Woodsia ilvensis*	23
Linaria canadensis	31	W. obtusa*	23
Parthenocissus vitacea	39		

* Species are also modal, since their presence values are higher here than in any other Wisconsin community.

Additional modal species, with their presence (%) values: Allium stellatum (8), Arenaria macrophylla (8), Cheilanthes feei (23), Commelina erecta (8), Galeopsis tetrahit (15), Gnaphalium uliginosum (8), Lespedeza violacea (8), Lobelia inflata (8), Myosotis virginica (8), Physocarpus opulifolius (15), Plantago aristata (15), Plantago virginica (8), Polygonum cilinode (15), Polygonum convolvulus (15), Polygonum tenue (15), Prunus pumila (23), Ranunculus rhomboideus (8), Stellaria aquatica (8), Symphoricarpos albus (8).

No. of stands 13.
Species density 23.
Total no. of species 157.
Index of Homogeneity 35.2%.
No. of modal species 27.
No. of prevalent modal species as % of total prevalents 39.1%.

Table XX-5
Prevalent species of shaded rock cliffs

Species	Pres.	Species	Pres.
Adiantum pedatum	24%	Dicentra cucullaria*	24%
Aquilegia canadensis	64	Diervilla lonicera	40
Arabis laevigata*	32	Dryopteris austriaca	40
A. lyrata	28	Hepatica americana	36
Aralia nudicaulis	64	Impatiens biflora	32
A. racemosa	36	Mitella diphylla	28
Arisaema triphyllum	40	Parthenocissus vitacea	36
Asarum canadense*	28	Polygonum virginianum*	64
Campanula rotundifolia	36	Polypodium vulgare*	64
Cystopteris bulbifera*	48	Prenanthes alba	56
C. fragilis*	60	Ribes americanum	36

* Species are also modal, since their presence values are higher here than in any other Wisconsin community.

Additional modal species, with their presence (%) values: Aconitum noveboracense (8), Aureolaria pedicularia (4), Dodecathon amethystinum (20), Dryopteris fragans (8), Dryopteris marginalis (8), Paronychia canadensis (43), Solidago hispida (16), Solidago sciaphila (12), Sullivantia renifolia (4).

No. of stands 25.
Species density 22.
Total no. of species 188.
Index of Homogeneity 39.7%.
No. of modal species 16.
No. of prevalent modal species as % of total prevalents 31.8%.

APPENDIX FOR CHAPTER 21

Figure XXI-1.—Stands studied of weeds.

Table XXI-1
Prevalent species of weed communities of southern moist, nitrogen-rich soils

Species	Pres.	Species	Pres.
Agropyron repens	80%	Melilotus albus	47%
Amaranthus retroflexus*	68	Nepeta cataria*	53
Ambrosia artemisiifolia	87	Panicum capillare*	60
A. trifida*	60	Phlcum pratense	67
Anthemis cotula*	47	Plantago major*	87
Arctium minus*	73	Poa pratensis	47
Asclepias syriaca	60	Polygonum aviculare	40
Bromus inermis*	40	P. convolvulus	47
Chenopodium album	87	P. pensylvanicum*	47
Cirsium arvense	60	Rumex crispus*	80
C. vulgare	47	Setaria lutescens	40
Digitaria sanguinalis*	40	Solanum dulcamara	40
Echinochloa crusgalli*	67	Taraxacum officinale	73
Lactuca scariola	73	Trifolium repens	40
Leonurus cardiaca*	60	Urtica dioica*	60
Lychnis alba	53	Verbascum thapsus	40
Malva neglecta*	53		

* Species are also modal, since their presence values are higher here than in any other Wisconsin community.

Additional modal species with their presence (%) values: Barbarea vulgaris (20), Brassica juncea (7), Bromus japonicus (7), Cannabis sativa (13), Descurainia sophia (7), Digitaria ischaemum (7), Eragrostis cilianensis (20), Galinsoga parviflora (7), Lamium amplexicaule (7), Matricaria matricarioides (27), Oxalis europaea (33), Pastinaca sativa (20), Portulaca oleracea (33), Setaria verticillata (20).

No. of stands 15.
Species density 34.
Total no. of species 132.
Index of Homogeneity 57.8%.
No. of modal species 27.
No. of prevalent modal species as % of total prevalents 41.2%.

Table XXI-2
Prevalent species of weed communities of southern heavy soils

Species	Pres.	Species	Pres.
Achillea millefolium	41%	M. sativa	36%
Agropyron repens	86	Melilotus albus	67
Amaranthus retroflexus	38	Oenothera biennis	52
Ambrosia artemisiifolia*	96	Oxalis stricta*	37
A. trifida	57	Panicum capillare	42
Asclepias syriaca	58	Phleum pratense	46
Capsella bursa-pastoris*	43	Plantago major	76
Chenopodium album*	71	Poa pratensis	63
Cirsium vulgare	41	Polygonum convolvulus	36
Conyza canadensis	52	Rumex crispus	68
Erigeron strigosus	37	Taraxacum officinale*	65
Lactuca scariola*	77	Trifolium hybridum	45
Lepidium virginicum*	37	T. pratense	66
Lychnis alba	35	T. repens	49
Medicago lupulina	52		

* Species are also modal, since their presence values are higher here than in any other Wisconsin community.

Additional modal species with their presence (%) values: Abutilon theophrasti (5), Ajuga genevensis (5), Brassica nigra (8), Chelidonium majus (5), Convolvulus arvensis (8), Coronilla varia (5), Dianthus prolifer (8), Dipsacus sylvestris (5), Euphorbia esula (5), Hibiscus trionum (14), Lolium multiflorum (5), Saponaria officinalis (21), Setaria viridis (27), Solanum triflorum (8), Solanum carolinense (5), Sorghum halepense (8), Thlaspi arvense (21), Trifolium procumbens (8), Veronica peregrina (21).

No. of stands 36.
Species density 29.
Total no. of species 163.
Index of Homogeneity 54.1%.
No. of modal species 27.
No. of prevalent modal species as % of total prevalents 27.6%.

Table XXI-3
Prevalent species of weed communities of southern sandy soils

Species	Pres.	Species	Pres.
Agropyron repens	57%	Melilotus albus	70%
Ambrosia artemisiifolia	83	Mollugo verticillata	48
Asclepias syriaca	44	Oenothera biennis	35
Cenchrus pauciflorus*	39	O. rhombipetala	35
Chenopodium album	70	Panicum capillare	44
Conyza canadensis	65	Physalis heterophylla	39
Euphorbia corollata	65	Plantago major	35
E. maculata	30	Polygonum convolvulus	44
Froelichia floridana	35	Rumex acetosella	48
Lactuca scariola	44	Setaria lutescens	44
Lepidium campestre*	52	Silene antirrhina	44
L. virginicum	35	Verbascum thapsus	35

Table XXI-4
Prevalent species of weed communities of northern heavy soils

Species	Pres.	Species	Pres.
Achillea millefolium	60%	Phleum pratense*	100%
Agropyron repens	80	Plantago major	80
Agrostis stolonifera	70	Poa pratensis	60
Ambrosia artemisiifolia	80	Polygonum convolvulus	50
Chenopodium album	50	Potentilla norvegica	40
Chrysanthemum leucanthemum	40	Prunella vulgaris*	40
Cirsium arvense*	70	Rumex crispus	50
C. vulgare*	60	Setaria glauca*	50
Conyza canadensis	40	Sonchus arvensis	50
Equisetum arvense	50	Taraxacum officinale	70
Erigeron strigosus	60	Trifolium hybridum*	80
Hieracium aurantiacum	50	T. pratense*	90
Lychnis alba	60	T. repens	50
Medicago lupulina	50	Verbascum thapsus	60
Oenothera biennis	60		

* Species are also modal, since their presence values are higher here than in any other Wisconsin community.

Additional modal species with their presence (%) values: Dactylis glomerata (30), Daucus carota (30), Erysimum cheiranthoides (10), Lolium perenne (10), Medicago sativa (40), Ranunculus acris (30), Silene cucubalus (10).

No. of stands 10.
Species density 29.
Total no. of species 89.
Index of Homogeneity 60.4%.
No. of modal species 14.
No. of prevalent modal species as % of total prevalents 24.1%

Notes for Table XXI-3

* Species are also modal, since their presence values are higher here than in any other Wisconsin community.

Additional modal species with their presence (%) values: Polanisia graveolens (4), Silene noctiflora (8), Trifolium arvense (8).

No. of stands 23.
Species density 24.
Total no. of species 145.
Index of Homogeneity 47.7%.
No. of modal species 5.
No. of prevalent modal species as % of total prevalents 8.3%.

Table XXI-5
Prevalent species of weed communities of northern sandy soils

Species	Pres.	Species	Pres.
Achillea millefolium	72%	Oenothera biennis	86%
Agropyron repens	86	Phleum pratense	62
Agrostis stolonifera*	72	Plantago major	67
Ambrosia artemisiifolia	62	Poa compressa	38
Anaphalis margaritacea	33	Polygonum convolvulus	43
Chenopodium album	67	Potentilla novegica	62
Chrysanthemum leucanthemum*	43	Rumex acetosella*	67
Cirsium vulgare	33	R. crispus	43
Conyza canadensis*	91	Setaria lutescens	48
Erigeron annuus*	62	Taraxacum officinale	62
E. strigosus	33	Trifolium hybridum	72
Hieracium aurantiacum	57	T. pratense	72
Lepidium virginicum	43	T. repens*	67
Lychnis alba*	76	Verbascum thapsus	57
Mollugo verticillata	57		

* Species are also modal since their presence values are higher than in any other Wisconsin community.

Additional modal species with their presence (%) values: Amaranthus graecizans (14), Carum carvi (10), Centaurea maculosa (10), Cerastium arvense (5), Euphorbia preslii (14), Potentilla argentea (10), Potentilla recta (10), Silene dichotoma (5).

No. of stands 21.
Species density 29.
Total no. of species 128.
Index of Homogeneity 54.8%.
No. of modal species 18.
No. of prevalent modal species as % of total prevalents 24.1%.

Table XXI-6
Prevalent species of weed communities of railroad yards

Species	Pres.	Species	Pres.
Achillea millefolium	38%	Medicago lupulina*	58%
Agropyron repens*	100	Melilotus albus*	83
Agrostis stolonifera	54	M. officinalis*	58
Ambrosia artemisiifolia	96	Oenothera biennis	54
Anthemis cotula	38	Oxybaphus nyctagineus	38
Asclepias syriaca	58	Panicum capillare	50
Bidens frondosa*	46	Phleum pratense	50
Chenopodium album	67	Plantago major	79
C. hybridum*	54	Poa compressa*	71
Cirsium arvense	50	P. pratensis*	71
C. vulgare	38	Polygonum aviculare*	46
Conyza canadensis	67	P. convolvulus*	58
Echinochloa crusgalli	50	P. persicaria*	46
Equisetum arvense	46	Potentilla novegica*	63
Euphorbia maculata	58	Rumex crispus	54
Hordeum jubatum*	54	Setaria lutescens	38
Lactuca scariola	50	Taraxacum officinale	67
Linaria vulgaris*	63	Trifolium hybridum	58
Lychnis alba	54	T. pratense	54

* Species are also modal, since their presence values are higher than in any other Wisconsin community.

Additional modal species with their presence (%) values: Agropyron smithii (4), Artemisia biennis (8), Artemisia vulgaris (4), Asparagus officinale (13), Atriplex patula (8), Bromus tectorum (21), Chenopodium capitatum (4), Chenopodium glaucum (4), Collomia linearis (13), Descurainia pinnata (4), Erucastrum gallicum (4), Grindelia squarrosa (8), Helianthus petiolaris (17), Hypericum perforatum (4), Iva xanthifolia (21), Kochia scoparia (21), Lepidium densiflorum (8), Lycium halimifolium (4), Lythrum salicaria (4), Plantago lanceolata (13), Polygonum erectum (25), Polygonum punctatum (4), Radicula armoracea (8), Ranunculus sceleratus (4), Ratibida columnifera (4), Sicyos angulatus (4), Sisymbrium altissimum (29), Sisymbrium officinale (4), Sonchus oleraceus (21), Spergula arvensis (8), Tanacetum vulgare (13), Triplasis purpurea (4), Verbena bracteata (8), Vicia cracca (4).

No. of stands 24.
Species density 38.
Total no. of species 183.
Index of Homogeneity 58.0%.
No. of modal species 48.
No. of prevalent modal species as % of total prevalents 36.8%.

Table XXI-7
Common species of cleared woodland pastures
(All species present in at least one-third of the stands)

Achillea millefolium	58%	Lepidium campestre	42%
Agropyron repens	58	Medicago lupulina	33
Agrostis stolonifera	50	Phleum pratense	50
Ambrosia artemisiifolia	50	Plantago major	58
Antennaria neglecta	50	Poa compressa	42
Asclepias syriaca	42	P. pratensis	100
Cerastium vulgatum	33	Potentilla argentea	42
Chenopodium album	33	P. norvegica	33
Chrysanthemum leucanthemum	33	Prunella vulgaris	33
Cirsium arvense	50	Rumex acetosella	42
C. vulgare	58	R. crispus	33
Conyza canadensis	50	Taraxacum officinale	75
Erigeron annuus	42	Trifolium pratense	33
Euphorbia maculata	42	T. repens	67
Fragaria virginiana	33	Verbascum thapsus	58

APPENDIX FOR CHAPTER 22

Figure XXII-1.—Map showing locations of fossil pollen studies. 1, Leopold (1946); 2, Truman (1937); 3, 4, 5, 6, Hansen (1937, 1939); 7, Voss (1934); 8, Natelson (1952); 9, 10, Potzger (1942); 11, Wilson and Webster (1942); 12, 13, Potzger (1943); 14, Voss (1934); 15, 16, 17, 18, Wilson (1932, 1938).

Glossary | Bibliography

GLOSSARY AND LIST OF ABBREVIATIONS

A_1 layer   That layer of the topsoil containing an incorporation of significant quantities of organic matter in the mineral matter.

A_2 layer   The layer of the soil immediately below the A_1 layer which is poor in available nutrient compounds and usually rich in silica. Most distinct in Podzol soils.

Adaptation value   An estimate of species behavior, indicated by an assigned number between 1 and 10. It is based on the observed joint occurrences of tree species in major forest types and may be interpreted as the probability of a species occurring in a forest dominated by sugar maple, with the chances proportional to the number. See Chapter 5.

Aggregation   A spatial dispersion of plants such that the individuals are clustered into dense clumps, separated by areas which are devoid of the individuals.

Amplitude   The width of the curve which represents the behavior of a species plotted against an environmental or compositional gradient. In general terms, a measure of the range of conditions which are tolerated by a species.

Association   The degree to which two or more species grow together at the same place, usually within the same quadrat, or other small sample.

"Association"   A term encountered in connection with literature quotations in this book which is used to describe a plant community that is presumably distinct and has discrete boundaries.

Barrens   A depauperate plant community with either a low total coverage or with stunted individuals of species which elsewhere reach considerable size as sand barrens or pine barrens. See Chapters 15 and 16.

Basal area   A measure of dominance in forests expressed as the area of the trunk of a tree at a height 4½ feet above the ground or as the total of such areas for all trees in a given space.

Behavior   Specialized use of the word to indicate the response of a plant to a factor or group of factors of the environment.

Canopy   The aerial branches of terrestrial plants together with their complement of leaves. Said to be a complete canopy when the ground is completely hidden by the leaves when viewed from above.

Catena   A series of related soils differing in intensity of a single factor. Here used exclusively for the soils which differ in degree of internal drainage. See Chapter 2.

Cline   A gradient or gradual change of an influencing factor or factor complex, of known or unknown makeup.

Community   A studiable grouping of organisms which grow together in the same general place and have mutual interactions.

Constancy   A specialized term to indicate the degree to which a species occurs in the separate stands of a given community. It is based upon a single sample of fixed size or fixed number of points in each stand and is expressed as number of occurrences as a percentage of number of stands examined.

Continuum   An adjectival noun referring to the situation where the stands of a community or larger vegetational unit are not segregated into discrete and objectively discernable sub-units but rather form a continuously varying series.

d.b.h. Diameter at breast height. The foresters' convention for a determination of the diameter of a tree at a height $4\frac{1}{2}$ feet above the ground.

Density A specialized term to indicate the number of plant individuals per unit area. May be expressed in absolute terms or as relative density, which is the number of individuals of a particular species as a percentage of the total number of individuals of all species on the same area.

Diagnostic species A plant of high fidelity to a particular community and one whose presence serves as a criterion of recognition of that community.

Differential species A plant which is distinctly more widespread or successful in one of a pair of plant communities than in the other. It may be still more successful in other communities not under discussion.

Dominance A measure of the total size, bulk, or weight of the individuals of a particular species on a particular area.

Dominant A species which is of great importance in a community through size, number, or other characters which enable it to receive the brunt of external environmental forces and modify them before they affect the lesser members of the community.

Driftless Area The region in southwestern Wisconsin and adjacent portions of neighboring states which was not covered by ice sheets of the Pleistocene epoch.

Ecotype A strain or race of a species which is differently adapted to the environment than other populations of the same species.

Fidelity A measure of the degree to which a species is confined to a particular community, a plant of high fidelity being wholly or largely so limited.

Flora The entire complement of plant species which grows spontaneously in a particular region. The size of a flora is determined by the number of such species and is uninfluenced by the number of individuals of each.

Floristic element A group of species in a particular flora which has a common origin or a uniform geographical center.

Floristic province A large area with a relatively uniform flora, delimited by a line or zone in which many species reach a common range boundary.

Forb A specialized term for any non-grassy herbaceous plant. Used particularly for the broad-leaved plants of prairies.

Frequency A measure of the commonness and widespread distribution of plant individuals in a single stand of a community. Usually determined from widely dispersed samples of fixed area and expressed as the ratio of occupied samples to total samples examined, on a percentage basis.

Frequency × Presence Index A measure of the commonness of a species in all of the stands of a community. The index ranges from 0 to 10,000, the latter value attained by a species which occurs in all samples in all stands. The highest value in the P.E.L. studies was 7900, given by the rough bedstraw (*Galium aparine*) in the mesic forests of Green County.

Gley layer A specialized term referring to a sticky, mottled layer at some depth in the soil, as produced by permanently water-logged and hence oxygen-deficient conditions.

Gradient A gradually changing factor.

GLOSSARY

Groundlayer The herbs, shrubs, and woody vines found beneath the trees in a forest. Excludes seedlings and saplings of the overhead trees.

Grub A colloquial term referring to the brushy growths of oaks and other trees when their tops are repeatedly burned but their root systems continue to increase in size.

H. and F. Hunting and Fishing Grounds, areas owned or leased by the Wisconsin Conservation Department for public use.

Importance Value (I. V.) A measure of the significance of a plant in a stand or a community, expressed as the total of its values for relative density, relative frequency, and relative dominance, with a possible range from 0 to 300.

Index of community distinctness The degree to which a community is visibly distinct from related communities as indicated by the number of prevalent groundlayer species which attain their highest measured values in that community expressed as a percentage of total prevalent species.

Index of diversity A measure of the floristic richness of a community, calculated by dividing the difference between the total number of species in all of the stands and the average number of species per stand by the Naperian logarithm of the number of stands in the community.

Index of homogeneity A measure of stand-to-stand variability in a particular community. Expressed as a ratio of the sum of presence of the prevalent species to the total sum of presence of all species in the community.

Index of similarity A measure of the degree to which two stands of a community (or of two communities of a particular vegetation) resemble each other. Expressed as the ratio of twice the sum of the measurements which are common in the two stands to the sum of the total measurements in both stands, ($2w/a$ plus b), with a possible range of 0 to 100 per cent.

Indicator A species whose occurrence in a stand can be used to tell something about the conditions present in that stand.

Introgression A result of interspecific hybridization in which certain characters, originally present only in one species, become transferred to a second species.

Isorythm A "contour" line on a map connecting or enclosing all stations with an indicated degree of a measured environmental or biological factor.

Loess A wind-borne soil material composed of silt particles.

Modal species A specialized term to indicate a species which achieves its highest presence percentage in an indicated community or other vegetational unit.

Mor humus A thick mat of partially decomposed and fibrous organic matter, comprising the remains of several years' leaf litter, as found on the surface of the soil especially in conifer forests.

Mull humus An intimate mixture of organic and mineral matter, as in the A_1 soil layer of mesic hardwood forests.

Muskeg An open bog with scattered and stunted conifer trees.

Ordination An arrangement of data of a continuous and non-discrete sort into an orderly spatial pattern.

P.E.L. Abbreviation of the Plant Ecology Laboratory of the University of Wisconsin.

pH Symbol for units in the measurement of acidity or alkalinity of soil or other media.

Pioneer species A plant found in early stages of a succession, usually able to grow in full sunlight under highly variable conditions of soil moisture and soil nutrients.

Plant assemblage A non-committal term for any studiable group of plants.

Presence A specialized term to indicate the degree to which a species occurs in the separate stands of a given community. It differs from constancy in that the entire stand is used as the sample, rather than a single fixed-area sample within the stand. Usually expressed as a percentage of the total stands examined.

Prevalent species When the total list of species in a plant community is arranged in decreasing order of their presence percentages, the prevalent species are the topmost species counted off to a number equal to the species density of that community. They are an objectively derived group of typical species which ordinarily includes about three-fourths of all the individuals to be found in the community.

Quadrat A sampling area, usually square, of relatively small but constant size.

Random dispersion A spatial distribution of individuals such that each item has equal chances of occurring at any locus in the area. The resulting pattern varies in density from place to place but shows no distinct clusters and no large empty spaces.

Relic A disjunct plant community, separated by other communities from its main geographical range.

Remnant A portion or fragment of an original plant community remaining after the destruction of the bulk of the community by the agricultural or exploitive actions of man.

S.A. Abbreviation for Scientific Area, a preserve maintained for scientific study purposes and so designated by the State Board for Preservation of Scientific Areas, an official body of the state of Wisconsin.

Serotinal cones A seed-bearing cone as in jack pine, which does not open to liberate the seeds but must receive additional heat, as by a forest fire. See Chapter 11.

Site A place or location. Not used here in the special sense employed by foresters.

Species density The number of species to be found in a stand of a plant community. Also the average number of species per stand in all stands of a community.

Spring ephemeral A groundlayer species which produces leaves and flowers before the development of a leaf canopy by the dominants and which dies down soon after that development. Most numerous in mesic hardwood forests.

Stand A particular example of a plant community.

Structure A specialized term referring to a quantitative analysis of a forest in which density figures are given separately for seedlings, saplings, and mature trees of different sizes.

Tension zone A band between two floristic provinces, marked by the intermingling of species from both. See Chapter 1.

Topocline A gradient along which biological change is evident but for which no environmental correlations are known except mere distance.

Vegetation The total of the plant communities of a region. Differs from the flora because quantitative aspects are considered; numerous or large species are given more attention than rare and inconspicuous species.

w.r.c. Abbreviation for water-retaining-capacity of the soil, a measure of the amount of water held by a finely divided layer of soil 1 cm thick against the pull of gravity.

BIBLIOGRAPHY

Agnew, A. D. Q. 1957. The ecology of British rushes with special reference to the invasion of re-seeded pastures. Ph.D. thesis, University of Wales. (in Greig-Smith, 1957).

Ahlgren, H. L., Wall, M. L., Muckenhirn, R. J., and Sund, J. M. 1946. Yields of forage from woodland pastures on sloping land in southern Wisconsin. *Jour. Forestry, 44:*709–11.

Aichinger, E. 1952. Rotföhrenwalder als Waldentwicklungstypen. *Angewandte Pflanzensoz, 6:*1–68.

Aikman, J. M., and Gilly, C. L. 1948. A comparison of the forest floras along the Des Moines and Missouri Rivers. *Proc. Iowa Acad. Sci., 55:*53–73.

———, and Smelser, A. W. 1938. The structure and environment of forest communities in central Iowa. *Ecology, 19:*141–50.

Alden, W. C. 1918. The quaternary geology of southeastern Wisconsin. *Prof. Paper U. S. Geol. Surv.* 106. Washington, D.C.

Aldrich, H. R., and Fassett, N. C. 1929. Botanical and geological evidence for an ancient lake. *Science, 70:*45–46.

Anderson, Edgar, 1949. *Introgressive hybridization.* John Wiley and Sons, Inc., New York.

———. 1952. *Plants, man and life.* Little, Brown, and Co., Boston.

———. 1948. Hybridization of the habitat. *Evolution, 2:*1–9.

Anderson, O. 1948. An ecological study of the climax vegetation of Taylor County, Wisconsin. M.S. thesis, University of Wisconsin.

———. 1954. The phytosociology of dry lime prairies of Wisconsin. Ph.D. thesis, University of Wisconsin.

Andrews, J. D. 1946. The macroscopic invertebrate populations of the larger aquatic plants in Lake Mendota. Ph.D. thesis, University of Wisconsin.

Anon. 1819. Journal of the march of the 5th Regiment, in June 1819, from Green Bay to Prairie du Chien, by an officer of that Regiment. *North Western Views,* Sept. 18, 1819.

Anthoney, R. B. 1937. Prairie plant distribution in Rock County. M.S. thesis, University of Wisconsin.

Ashby, E. 1948. Statistical ecology. II: A reassessment. *Bot. Rev., 14:*222–34.

Archbald, D. 1950. The effect of quadrat size and quadrat method on apparent plant dispersion. M.S. thesis, University of Wisconsin.

———. 1953. The effect of fire on woody plants in the prairie. *Bull. Ecol. Soc. Amer., 34:*69.

———. 1954. The effect of native legumes on the establishment of prairie grasses. Ph.D. thesis, University of Wisconsin.

Baas-Becking, L. G., and Nicolai, E. 1934. On the ecology of a sphagnum bog. *Blumea, 1:*10–45.

Baerreis, D. A. 1953. The airport village site, Dane County. *Wis. Archeol., 34:*149–64.

Baker, H. 1937. Alluvial meadows: a comparative study of grazed and mown meadows. *Jour. Ecol., 25:*408–20.

Barclay, H. G. 1924. The plant succession of the flood plain of the Mississippi River with special reference to the pioneer stage. M.A. thesis, University of Minnesota.

Bard, Lucia D. 1957. Relations between legumes and other prairie species on some relic prairie sites of Wisconsin. M.S. thesis, University of Wisconsin.
Barrett, S. A. 1933. Ancient Aztalan. *Bull. Publ. Mus. City Milwaukee, 13:* 1–602.
Bayley, C. C. 1954. Western trip. An address given at Manlius, N.Y. *Wis. Mag. Hist., 37:*237–39.
Beard, J. S. 1953. The savanna vegetation of northern Tropical America. *Ecol. Monogr., 23:*149–215.
Beavers, A. H. 1957. Source and deposition of clay minerals in Peorian loess. *Science, 126:*1285.
Becker, Virginia. 1954. A comparison of the vegetation in four sections of Young's Woods, with respect to the effect of microclimates. MSS Report of Bot. Dept., University of Wisconsin. Typescript 23 pp.
Bergseng, Margaret. 1955. The flora of Wisconsin as represented by specimens in the University of Wisconsin herbarium. MS Report of Bot. Dept., University of Wisconsin. Typescript 102 pp.
Bird, E. C. F. 1957. The use of the soil catena concept in the study of the ecology of the Wormley Woods, Hertfordshire. *Jour. Ecol., 45:*465–69.
Birge, E. A., and Juday, C. 1932. Solar radiation and inland lakes. Fourth report. *Trans. Wis. Acad. Sci. Arts Lett., 27:*523–62.
Blackman, G. E. 1935. A study by statistical methods of the distribution of species in grassland associations. *Ann. Bot., 49:*749–77.
Blatchley, W. S. 1920. *The Indiana weed book.* Nature Pub. Co., Indianapolis.
Bond, R. R. 1955. The birds of the upland forest communities of southern Wisconsin. Ph.D. thesis, University of Wisconsin.
———. 1957. Ecological distribution of breeding birds in the upland forests of southern Wisconsin. *Ecol. Monogr.* 27:351–84.
Borchert, J. R. 1950. The climate of the central North American grassland. *Annals Assoc. Amer. Geogr., 40:*1–29.
Bordner, J. S. 1942. Forests and land use. *Bull. Wis. Dept. Agric.,* 229. Madison. 56 pp.
Borisova, I. V., Isachenko, T. L., and Rachkovskaia, E. I. 1957. On the forest-steppe in northern Kazakhstan (in Russian). *Bot. Zhur.,* (Moscow), *42:* 677–90.
Bourdo, E. A. 1956. A review of the General Land Office survey and of its use in quantitative studies of former forests. *Ecology, 37:*754–68.
Braun, E. Lucy. 1950. *Deciduous forests of eastern North America.* Blakiston Co., Philadelphia.
———. 1951. Plant distribution in relation to the glacial boundary. *Ohio Jour. Sci., 51:*139–46.
Braun-Blanquet, J. 1951. *Pfllanzensoziologie: Grundzüge der Vegetationskunde.* Springer, Vienna.
Bray, J. R. 1955. The savanna vegetation of Wisconsin and an application of the concepts order and complexity to the field of ecology. Ph.D. thesis, University of Wisconsin.
———. 1956. Gap phase replacement in a maple-basswood forest. *Ecology, 37:*598–600.

———, and Curtis, J. T. 1957. An ordination of the upland forest communities of southern Wisconsin. *Ecol. Monogr., 27*:325–49.
Bray, W. L. 1921. History of forest development on an undrained sand plain in the Adirondacks. *Tech. Bull. N.Y. Coll. of Forestry*, No. 13.
Brendel, F. 1887. *Flora Peoriana.* J. W. Franks and Son, Peoria, Ill.
Britton, W. E. 1903. Vegetation of the North Haven sand plains. *Bull. Torrey Bot. Club, 30*:571–620.
Brown, R. T. 1950. Forests of the central Wisconsin sand plains. *Bull. Ecol. Soc. Amer., 31*:56.
———, and Curtis J. T. 1952. The upland conifer-hardwood forests of northern Wisconsin. *Ecol. Monogr., 22*:217–34.
Bruhin, T. A. 1876. Vergleichende Flora Wisconsins. *Verh. Zool.-bot. Gesell.,* (Vienna), *26*:229–86.
Bruncken, E. 1902. A tamarack swamp in Waukesha County. *Bull. Wis. Nat. Hist. Soc., 2*:164–67.
———. 1902. Studies in plant distributions. 1. On the succession of forest types in the vicinity of Milwaukee. *Ibid.*:17–28.
———. 1910. Studies in plant distribution. 9. The shore of Lake Michigan. *Ibid., 8*:145–57.
Buell, M. F., 1946. Jerome bog, a peat-filled "Carolina Bay." *Bull. Torrey Bot. Club, 73*:24–33.
Buell, M. F., and Niering, W. A. 1957. Fir-spruce-birch forest in northern Minnesota. *Ecology, 38*:602–10.
———, and Wistendahl, W. S. 1955. Flood plain forests of the Raritan River. *Bull. Torrey Bot. Club, 82*:463–72.
Bushnell, T. M. 1942. Some aspects of the soil catena concept. *Proc. Soil Sci. Soc. Amer., 1*:466–76.
Buss, I. O. 1956. Plant succession on a sand plain, northwest Wisconsin. *Trans. Wis. Sci. Arts Lett., 45*:11–20.
Butler, J. E. 1954. Interrelations of autecological characteristics of prairie herbs. Ph.D. thesis, University of Wisconsin.
Cabeza de Vaca. 1528. (In Stewart, O. C. Burning and natural vegetation in the United States. *Geogr. Rev., 41*:317–20, 1951.)
Cain, S. A. 1935. Studies on virgin hardwood forest: III. Warren's Woods, a beech-maple climax forest in Berrien Co., Michigan. *Ecology, 16*:500–513.
———, 1943. The Tertiary nature of the cove hardwood forests of the Great Smoky Mountains. *Torrey Bot. Club Bull., 70*:213–35.
———, 1944. *Foundations of plant geography.* Harper and Bros., New York.
———, 1947. Characteristics of natural areas and factors in their development. *Ecol. Monogr., 17*:185–200.
———, Castro, G. M., and Pires, J. M. 1956. Application of some phytosociological techniques to Brazilian rain forest. Part II. *Amer. Jour. Bot., 43*:915–28.
———, and Slater, J. V. 1948. The vegetation of Sodon Lake. *Amer. Midland Nat., 40*:741–62.
Carver, Jonathan. 1781. *Travels through the interior parts of North America in the years 1766, 1767, 1768.* London. 3rd ed.
Catenhusen, John. 1944. Some aquatic and sub-aquatic plants from the region

of Glacial Lake, Wisconsin. *Trans. Wis. Acad. Sci. Arts Lett., 36*:163–69.
Catlin, George, 1842. *Manners, customs, and conditions of the North American Indians.*
Caton, J. D. 1870. Origin of the prairies. *Fergus' Historical Series,* No. 7:35–53.
Chamberlin, T. C. 1877. Native vegetation of eastern Wisconsin. In *Geol. of Wis.,* Vol. 2:176–87.
Charlevoix, P. F. de. 1761. *Journal of Voyage to North America in 1721.* Vol. 2:199–200.
Chavannes, Elizabeth. 1940. The steep prairies of southwestern Wisconsin and their invasion by forest. Ph.D. thesis, University of Wisconsin.
Cheney, L. S. 1894. Is forest culture in Wisconsin desirable? *Trans. Wis. State Hist. Soc., 24*:163–70.
Cheyney, E. G. 1942. *American silvics and silviculture.* University of Minnesota Press, Minneapolis.
Christensen, E. M. 1954. A phytosociological study of the winter range of deer of northern Wisconsin. Ph.D. thesis, University of Wisconsin.
———, Clausen, Johanna Jones, and Curtis, J. T. 1959. Phytosociology of the lowland forests of northern Wisconsin. *Amer. Midland Nat.,* 60.
Christensen, Martha. 1956. Studies on the microfungi of the soil in relation to the conifer-hardwood forest continuum in northern Wisconsin. M.S. thesis, University of Wisconsin.
Clapham, A. R. 1936. Over-dispersion in grassland communities and the use of statistical methods in plant ecology. *Jour. Ecol., 24*:232–50.
Clark, P. J., and Evans, F. C. 1954. Distance to nearest neighbor as a measure of spatial relationships in populations. *Ecology., 35*:445–53.
Clausen, J. Jones. 1957. A phytosociological ordination of the conifer swamps in Wisconsin. *Ecology., 38*:638–48.
———. 1957. A comparison of some methods of establishing plant community patterns. *Botanisk Tidsskr., 53*:253–78.
Clausen, Jens, Keck, D. D., and Heisey, W. H. 1940. Experimental studies on the nature of species. I. *Publ. Carnegie Inst. Wash., 520*:1–452.
Clements, F. E. 1905. *Research methods in ecology.* University Publishing Co., Lincoln, Neb.
———. 1928. *Plant succession and indicators.* H. W. Wilson Co., New York.
Cole, L. C. 1946. A theory of analyzing contagiously distributed populations. *Ecology, 27*:329–41.
———. 1949. The measurement of interspecific association. *Ibid., 30*:411–24.
Collins, E. H. 1951. A study of the boreal forest formation in northern Cape Breton Island. M.A. Thesis, Acadia University.
Conard, H. S. 1952. The vegetation of Iowa. *Univ. Iowa Studies Nat. Hist., 19*:1–66.
Conway, Verona M. 1949. The bogs of central Minnesota. *Ecol. Monogr., 19*: 173–206.
Cook, D. B. 1941. Five seasons' growth of conifers. *Ecology, 22*:295–96.
Coombe, D. E. 1957. The spectral composition of shade light in communities. *Journ. Ecol., 45*:823–30.
Cooper, Fennimore. 1848. *The oak-openings.* New York.
Cooper, W. S. 1913. The climax forest of Isle Royale, Lake Superior and its development. *Bot. Gaz., 55*:1–44; 189–255.

———. 1936. The strand and dune flora of the Pacific Coast of North America. In Goodspeed's *Essays in geobotany in honor of W. A. Setchell.* University of California Press, Berkeley. pp. 141–89.

———, and Foot, H. 1932. Reconstruction of a late Pleistocene biotic community in Minneapolis, Minnesota. *Ecology, 13:*63–72.

Costello, D. F. 1936. Tussock meadows in southeastern Wisconsin. *Bot. Gaz., 97:*610–49.

Cottam, Grant. 1948. The phytosociology of an oak woods in southwestern Wisconsin. Ph.D. thesis, University of Wisconsin.

———. 1949. The phytosociology of an oak woods in southwestern Wisconsin. *Ecology, 30:*271–87.

———, and Curtis, J. T. 1948. The use of the punched card method in phytosociological research. *Ecology, 29:*516–19.

———, and ———. 1949. A method for making rapid surveys of woodlands by means of pairs of randomly selected trees. *Ibid., 30:*101–04.

———, and ———. 1955. Correction for various exclusion angles in the random pairs method. *Ibid., 36:*767.

———, and ———. 1956. The use of distance measures in phytosociological sampling. *Ibid., 37:*451–60.

———, and ———, and Catana, A. J. 1957. Some sampling characteristics of a series of aggregated populations. *Ecology, 38 (4):*610–22.

———, and ———, and Hale, B. W. 1955. Some sampling characteristics of a population of randomly dispersed individuals. *Ecology, 34:*741–57.

Coupland, R. T. 1950. Ecology of mixed prairie in Canada. *Ecol. Monogr., 20:*271–315.

———, and Brayshaw, T. C. 1953. The fescue grassland in Saskatchewan. *Ecology, 34:*386–405.

Cowles, H. C. 1899. The ecological relations of the vegetation on the sand dunes of Lake Michigan. *Bot. Gaz., 27:*95–117, 167–202, 281–308, 361–91.

Cox, H. J. 1910. Frost and temperature conditions in cranberry marshes in Wisconsin. *Bull. T., USDA Weather Bur.*

Cram, T. J. 1840. *Internal improvements in the territory of Wisconsin.* (Sen. Doc. *140;* 26th Congress, 1st Sess.), Washington, D.C.

Crocker, R. L., and Major, J. 1955. Soil development in relation to vegetation and surface age at Glacier Bay, Alaska. *Jour. Ecol., 43:*427–48.

———, and Wood, J. G. 1947. Some historical influences on the development of the South Australian vegetation communities and their bearings on concepts and classification in ecology. *Trans. Royal Soc. South Austral., 71:*81–136.

Culberson, W. L. 1955a. The corticolous communities of lichens and bryophytes in the upland forests of northern Wisconsin. *Ecol. Monogr., 25:* 215–31.

———. 1955b. Qualitative and quantitative studies on the distribution of corticolous lichens and bryophytes in Wisconsin. *Lloydia, 18:*25–36.

———. 1955c. The fossil mosses of the Two Creeks forest bed of Wisconsin. *Amer. Midland Naturalist, 54:*452–59.

Curtis, J. T. 1932. A new *Cypripedium* hybrid. *Rhodora, 34:*239–42.

———. 1936. The germination of some native orchid seeds. *Bull. Am. Orchid Soc.,* 5:42–47.

———. 1939. The relation of specificity of orchid mycorrhizal fungi to the problem of symbiosis. *Amer. Jour. Bot.,* 26:390–99.

———. 1941. Peloric flowers in *Cypripedium reginae* Walt. *Amer. Midl. Nat.,* 25:580–83.

———. 1943. Germination and seedling development in five species of *Cypripedium* L. *Amer. Jour. Bot.,* 30:199–206.

———. 1946. Use of mowing in management of white ladyslipper. *Jour. Wildlife Manage.,* 10:303–8.

———. 1947. Ecological observations on the orchids of Haiti. *Bull. Am. Orchid Soc.,* 16:262–68.

———. 1954a. Annual fluctuation in rate of flower production by native *Cypripediums* during two decades. *Bull. Torrey Bot. Club,* 81:340–52.

———. 1954b. Scientific areas in Wisconsin. *Wis. Cons. Bull.,* 19:13–16.

———. 1955. A prairie continuum in Wisconsin. *Ecology,* 36:558–66.

———. 1956. The modification of mid-latitude grasslands and forests by man. In Thomas, W. L. (ed.) *Man's role in changing the face of the earth.* University of Chicago Press. pp. 721–36.

———, and Cottam, Grant. 1950. Antibiotic and autotoxic effects in prairie sunflower. *Bull. Torrey Bot. Club,* 77:187–91.

———, and Dix, R. L. 1956a. Distribution of alpha-radioactivity in certain forest types. *Science,* 123:799–800.

———, and ———. 1956b. The distribution of alpha-radioactivity in native vegetation. *Bot. Gaz.,* 117:231–38.

———, and Greene, H. C. 1949. A study of relic Wisconsin prairies by the species-presence method. *Ecology,* 30:83–92.

———, and ———. 1953. Population changes in some native orchid of southern Wisconsin. *Orchid Jour.,* 2:152–55.

———, and Juday, Chancey. 1937. Photosynthesis of algae in Wisconsin lakes. III Observations of 1935. *Inter. Revue der gesamten Hydrobiol., und Hydrographie,* 35:122–33.

———, and McIntosh, R. P. 1950. The interrelations of certain analytic and synthetic phytosociological characters. *Ecology,* 31:434–55.

———, and ———. 1951. An upland forest continuum in the prairie-forest border region of Wisconsin. *Ibid.,* 32:476–96.

———, and Partch, M. L. 1948. Effect of fire on the competition between blue grass and certain prairie plants. *Amer. Midland Nat.,* 39:437–43.

———, and ———. 1950. Some factors affecting flower production in *Andropogon gerardi. Ecology,* 31:488–89.

Czekanowski, J. 1913. *Zarys metod statystycznych (Die Grundzuge der statischen Methoden).* (Warsaw.) 1913.

Dablon, C. 1670. In *Jesuit Relations,* 55:193–95, 1899.

Dachnowski, A. 1912. Peat deposits of Ohio. *Bull. Geol. Surv. Ohio,* No. 16. Columbus.

Dahl, E. 1957. Rondane; mountain vegetation in South Norway and its relation to the environment. *Norske Videnskaps-Akademi. Oslo, I. Mat. Naturv. Kl., 1956*:1–374.

Dansereau, P. 1943. L'erabliere laurentienne. *Canad. Jour. Res., 21*:66–93.
———. 1957. *Biogeography, an ecological perspective.* Ronald Press, New York.
———, and Segadas-Vianna, F. 1952. Ecological study of the peat bogs of eastern North America. *Canad. Jour. Bot., 30*:490–520.
Daubenmire, R. F. 1936. The "big woods" of Minnesota. *Ecol. Monogr., 6*: 233-268.
———. 1942. An ecological study of the vegetation of southeastern Washington and adjacent Idaho. *Ibid., 12*:53–79.
———. 1952. Forest vegetation of northern Idaho and adjacent Washington and its bearing on concepts of vegetation classification. *Ibid., 22*:301–30.
Davis, H. E. 1910. An ecological study of the southern shore of Lake Wingra. B.S. thesis, University of Wisconsin.
Dawkins, C. J. 1939. Tussock formation by *Schoenus nigricans*: the action of fire and water erosion. *Jour. Ecol., 27*:78–88.
Day, G. M. 1953. The Indian as an ecological factor in the northeastern forest. *Ecology, 34*:329–46.
Demianowiczowa, Zofia. 1952. Zbiorowiska chwastow zbozowych Lubelszczyzny i ich ekologie (weed communities and their ecology in cereal crops of the Province Lublin). *Ann. Univ. M. Curie-Sklodowska*, Sect. E., *7*:21–46.
Denisov, A. K. 1948. Origin of the oak groves of the southern taiga in river valleys. (In Russian.) *Akad. Nauk. S.S.S.R. Dok 61*:379–81.
Denniston, R. J. 1921. A survey of the larger aquatic plants of Lake Mendota. *Trans. Wis. Acad. Sci. Arts Lett., 20*:495–500.
De Selm, H. A. 1953. A comparison of certain prairie species of glaciated and unglaciated Ohio. *Bull. Ecol. Soc. Amer., 34*:70.
Desmarais, Y. 1952. Dynamics of leaf variations in the sugar maples. *Brittonia, 7*:347–88.
de Vries, D. M. 1954. Constellation of frequent herbage plants based on their correlation in occurrence. *Vegetatio, 5* and *6*:105–11.
de Vries, H., Barendsen, G. W., and Waterbolk, H. T. 1958. Groningen Radiocarbon dates. II. *Science, 127*:129–37.
Dietz, R. A. 1950. Bottomland hardwood forests of southern Wisconsin. MS report of Bot. Dept., University of Wisconsin. Typescript. 13 pp.
Dix, R. L. 1953. The effects of grazing on the species composition of dry prairies in Wisconsin. *Bull. Ecol. Soc. Amer., 34*:69.
———. 1955. Phytosociological changes on the thin-soil prairies of Wisconsin under the influence of grazing. Ph.D. thesis, University of Wisconsin.
———. 1957. Sugar maple in the climax forests at Washington, D.C. *Ecology, 38*:663–65.
———, and Butler, J. E. 1954. The effect of fire on a dry thin-soil prairie in Wisconsin. *Jour. Range Management, 7*:265–86.
Drew, W. B. 1942. The revegetation of abandoned cropland in the Cedar Creek area. Boone and Callaway Counties, Missouri. *Res. Bull. Missouri Agri. Exp. Sta.*, 344. 52 pp.
Duncan, D. P. 1954. A study of some of the factors affecting the natural regeneration of tamarack in Minnesota. *Ecology, 35*:498–521.

Dyksterhuis, E. J. 1957. The savannah concept and its use. *Ecology, 38*:435–42.
Ebling, W. H., Caparoon, C. D., Wilcox, E. C., and Estes, C. W. 1948. A century of Wisconsin agriculture. *Wis. Crop and Livestock Dept. Serv. Bull.* No. 290. 119 pp.
Eggler, W. A. 1938. The maple-basswood forest type in Washburn County, Wisconsin. *Ecology, 19*:243–63.
Ehrendorfer, F. 1954. Gedanken zur Frage der Struktur und Anordnung der Lebensgemeinschaften. *Festschrift für Erwin Aichinger.* Vol. I. Springer, Vienna. pp. 151–67.
Ellarson, R. S. 1949. The vegetation of Dane County, Wisconsin in 1835. *Trans. Wis. Acad. Sci. Arts Lett., 39*:21–45.
Ellen, Sister Mary. 1924. Some ferns of southwestern Wisconsin. *Trans. Wis. Acad. Sci. Arts Lett., 21*:249–50.
Ellenberg, H. 1952. *Landwirtschaftliche Pflanzensoziologie.* II. *Wiesen und Weiden und ihre standörtliche Bewertung.* Ulmer, Stuttgart.
Erickson, R. O., Brenner, L. G., and Wraight, J. 1942. Dolomitic glades of east-central Missouri. *Ann. Mo. Bot. Gard., 29*:89–101.
Etter, H. 1949. De l'analyse statistique des tableaux de vegetation. *Vegetatio, 1*:147–54.
Evans, R. I. 1945. Bottom deposits of the Brule River; Brule River Survey Report No. 9. *Trans. Wis. Acad. Sci. Arts Lett., 37*:305–23.
Evers, R. A. 1955. The hill prairies of Illinois. *Bull. Ill. Nat. Hist. Surv., 26*:368–446.
Eyre, F. H., and Le Barron, R. K. 1944. Management of jack pine stands in the Lake States. *Tech. Bull. U. S. Dept. Agri.* 863. 66 pp.
Ewing, M., and Donn, W. L. 1956. A theory of Ice Ages. *Science, 123*:1061–66.
Faegri, K., and Iversen, J. 1950. *Textbook of modern pollen analysis.* Copenhagen.
Fagerlind, F. 1952. The real signification of pollen diagrams. *Bot. Notiser., 1952*:185–224.
Fassett, N. C. 1929. Preliminary reports on the flora of Wisconsin. *Trans. Wis. Acad. Sci. Arts Lett., 24*:249–68.
———. 1930. The plants of some northeastern Wisconsin Lakes. *Ibid., 25*: 157–68.
———. 1931. Notes from the herbarium of the University of Wisconsin-VII. *Rhodora, 33*:224–28.
———. 1939. *The leguminous plants of Wisconsin.* University of Wisconsin Press, Madison. 157 pp.
———. 1943. Another Driftless Area endemic. *Bull. Torrey Bot. Club, 70*: 388–99.
———. 1944. Vegetation of the Brule Basin, past and present. *Trans. Wis. Acad. Sci. Arts Lett., 36*:33–56.
———. 1945. *Juniperus virginiana, J. horizontalis* and *J. scopulorum*, IV. Hybrid swarms of *J. virginiana* and *J. horizontalis. Bull. Torrey Bot. Club, 72*:379–84.
———. 1951. *Grasses of Wisconsin.* University of Wisconsin Press, Madison.
———, and Calhoun, Barbara. 1952. Introgression between *Typha latifolia* and *Typha angustifolia. Evolution, 6*:367–79.

Finley, R. W. 1951. The original vegetation cover of Wisconsin. Ph.D. thesis, University of Wisconsin.

Flint, R. F., and Deevey, E. S. 1951. Radiocarbon dating of late-Pleistocene events. *Amer. Jour. Sci., 249:*257–300.

———, and Rubin, M. 1955. Radiocarbon dates of pre-Mankato events in eastern and central North America. *Science, 121:*649–58.

Fogelberg, S. O. 1935. A pollen analysis of Hope Lake bog. B.S. thesis, University of Wisconsin.

Foot, Lyman. 1836. Remarks on Indian summers. *Amer. Jour. Sci. Arts, 30:* 8–13.

Fosberg, M. A. 1949. Soil and site condition typical of the maple basswood association in southern Wisconsin. M.S. thesis, University of Wisconsin.

Freeman, C. P. 1932. Ecology of the cedar glade vegetation near Nashville, Tennessee. *Jour. Tenn. Acad. Sci., 8:*143–228.

Frolick, A. L. 1941. Vegetation on the peat lands of Dane Co., Wisconsin. *Ecol. Monogr., 11:*117–40.

———, and Shepherd, W. O. 1940. Vegetative composition and grazing capacity of typical area of Nebraska sandhill range land. *Res. Bull. Neb. Agr. Exp. Sta., 117.* 39 pp.

Fuller, A. M. 1933. Studies on the flora of Wisconsin. Part I, the orchids; Orchidaceae. *Bull. Milw. Publ. Mus., 14:*1–284.

———. 1946. A plant "Eskimo" in Wisconsin. *Milw. Publ. Mus. Record., 2:* 23–24.

Gates, F. C. 1912. The vegetation of the beach area in northeastern Illinois and southeastern Wisconsin. *Bull. Ill. Lab. Nat. Hist., 9:*255–72.

———. 1942. The bogs of northern Lower Michigan. *Ecol. Monogr., 12:* 213–54.

———. 1950. The disappearing Sleeping Bear Dune. *Ecology, 31:*386–92.

Gilbert, M. L. 1953. The phytosociology of the understory vegetation of the upland forests of Wisconsin. Ph.D. thesis, University of Wisconsin.

Gilbert, M. L., and Curtis, J. T. 1953. Relation of the understory to the upland forest in the prairie-forest border region of Wisconsin. *Trans. Wis. Acad. Sci. Arts Lett., 42:*183–95.

Gleason, H. A. 1901. The flora of the prairies. B.S. thesis, University of Illinois.

———. 1907. On the biology of the sand areas of Illinois: II: A botanical survey of the Illinois River Valley sand region. *Bull. Ill. Lab Nat. Hist., 7:*149–94.

———. 1910. The vegetation of the inland sand deposits of Illinois. *Ibid., 9:*23–174.

———. 1912. An isolated prairie grove and its phytogeographical significance. *Bot. Gaz., 53:*38–49.

———. 1913. The relation of forest distribution and prairie fires in the Middle West. *Torreya, 13:*173–81.

———. 1917. A prairie near Ann Arbor, Michigan. *Rhodora, 19:*163–65.

———. 1918. On the development of two plant associations of northern Michigan. *Plant World, 21:*151–58.

———. 1920. Some application of the quadrat method. *Bull. Torrey Bot. Club, 47:21–33.*
———. 1923. Botanical observations in northern Michigan. *Jour. New York Bot. Gard., 24:276–83.*
———. 1926. The individualistic concept of the plant association. *Bull. Torrey Bot. Club, 53:7–26.*
———. 1939. The individualistic concept of the plant association. *Am. Midland Nat., 21:92–110.*
Goder, H. A. 1955. A phytosociological study of *Tsuga canadensis* near the termination of its range in Wisconsin. Ph.D. thesis, University of Wisconsin.
———. 1956. Pre-settlement vegetation of Racine County. *Trans. Wis. Acad. Sci. Arts Lett., 45:169–76.*
Goldthwait, J. W. 1907. The abandoned shore-lines of eastern Wisconsin. *Bull. Wis. Geol. Nat. Hist. Surv., 17:61–62.*
Goodall, D. W. 1953. Objective methods for the classification of vegetation. I. The use of positive interspecific correlation. *Australian Jour. Bot., 1: 39–63.*
———. 1953. Objective methods for the classification of vegetation. II. Fidelity and indicator value. *Ibid.* 434–56.
———. 1954a. Vegetational classification and vegetational continua. In *Festschrift für Erwin Aichinger.* Vol. 1. Springer, Vienna pp. 168–82.
———. 1954b. Objective methods for the classification of vegetation. III. An essay in the use of factor analysis. *Australian Jour. Bot., 2:304–24.*
Gorham, E. 1954. The ionic composition of some bog and fen waters in the English lake region. *Jour. Ecol., 44:142–52.*
———, and Pearsall, W. H. 1954. Acidity, specific conductivity and calcium content of some bog fen waters in northern Britain. *Ibid.,* 129–41.
Gould, F. W. 1937. The present status of Dane County prairie flora. M.A. thesis, University of Wisconsin.
———. 1941. Plant indicators of original Wisconsin prairies. *Ecology, 22:* 427–29.
Graham, E. H. 1944. *Natural principles of land use.* Oxford University Press, New York.
Graham, S. A. 1941. Climax forests of the upper penninsula of Michigan. *Ecology, 22:355–62.*
Grant, M. L. 1934. Climax forest community in Itasca County, Minnesota. *Ecology, 15:243–57.*
Green, Phoebe. 1950. Ecological composition of high prairie relics in Rock County, Wisconsin. *Trans. Wis Acad. Sci. Arts Lett., 40:159–72.*
Greene, C. W. 1935. *The distribution of Wisconsin fishes.* Wis. Cons. Comm., Madison.
Greene, H. C. 1932. Wisconsin Myxomycetes. *Trans. Wis. Acad. Sci. Arts Lett., 27:141–81.*
———. 1953. Preliminary reports on the flora of Wisconsin. XXXVII. Cyperaceae. Part I. *Proc. Wis. Acad. Sci. Arts Lett., 42:47–67.*
———, and Curtis, J. T. 1950. Germination studies of Wisconsin prairie plants. *Amer. Midl. Nat., 43:186–94.*

———, and ———. 1953. The re-establishment of prairie in the University of Wisconsin Arboretum. *Wild Flower, 29*:77–88.

———, and ———. 1955. A bibliography of Wisconsin vegetation. *Milw. Public Mus. Publ. in Bot.* 1.

Gregor, J. W. 1946. Ecotypic differentiation. *New Phytol., 45*:254–70.

Greig-Smith, P. 1957. *Quantitative plant ecology.* Academic Press, New York.

Griggs, R. F. 1914. Observations on the behavior of some species on the edges of their ranges. *Bull. Torrey Bot. Club, 41*:25–49.

Grosser, K. H. 1956. Waldvegetation und forstlicher Standort in der Oberlausitzer Heide. *Arch. Forstw., 5*:423–30.

Grümmer, G. 1955. *Die gegenseitige Beeinflussung höherer Pflanzen.* Fischer, Jena.

Guernsey, O., and Willard, J. F. 1856. *History of Rock County.* Janesville, Wisconsin.

Guinochet, M. 1955. *Logique et dynamique due peuplement vegetal.* Masson et Cie, Paris.

Habeck, J. R. 1956. Some aspects of the natural reproduction of northern white cedar in Wisconsin. MS progress report, Botany Dept., University of Wisconsin. Typescript. 18 pp.

Haight, Theron W. 1907. *Memoirs of Waukesha County.* West. Hist. Assoc., Chicago.

Hale, M. E. 1955. Phytosociology of corticolous cryptogams in the upland forests of southern Wisconsin. *Ecology, 36*:45–63.

———. 1955. A survey of upland forests in the Chautauqua Hills, Kansas. *Trans. Kans. Acad. Sci., 58*:165–68.

Hale, T. J. 1860. Additions to the flora of Wisconsin. *Trans. Wis. State Agr. Soc., 6*:258–63.

Hamel, A., and Dansereau, P. 1949. L'aspect ecologique du probleme des mauvaises herbes. *Bull. du Serv. de Biogeographie,* (Montreal). 5. 45 pp.

Hamerstrom, F., Hamerstrom, F., and Matson, D. E. 1952. Sharptails into the shadows. Wis. Cons. Dept., Madison. *Wis. Wildlife. 1.* 35 pp.

Hansen, H. H. 1930. *Studies on the vegetation of Iceland.* Frimodt, Copenhagen.

Hansen, H. P. 1933. Tamarack bogs of the driftless area of Wisconsin. *Bull. Milw. Publ. Mus., 7*:231–304.

———. 1937. Pollen analysis of two Wisconsin bogs of different ages. *Ecology, 18*:136–48.

———. 1939. Postglacial vegetation of the Driftless Area of Wisconsin. *Amer. Midland Nat., 21*:752–62.

Hanson, H. C. 1955. Characteristics of the *Stipa comata-Bouteloua gracilis-Bouteloua curtipendula* association of northern Colorado. *Ecology, 36*: 269–80.

———, and Whitman, W. 1938. Characteristics of major grassland types in western North Dakota. *Ecol. Monogr., 8*:57–114.

Harper, R. M. 1944. Preliminary report of the weeds of Alabama. *Geol. Surv. of Alab.*, Bull. 53. 275 pp.

Hart, C. A., and Gleason, H. A. 1907. On the biology of the sand areas of Illinois. *Bull. Ill. Lab. Nat. Hist., 7*:137–267.

Harrington, C. L. 1952. State Board for Preservation of Scientific Areas. *Wis. Conserv. Bull., 17*:14–15.

Hasler, A. D., and Jones, Elizabeth. 1949. Demonstration of the antagonistic action of large aquatic plants on algae and rotifers. *Ecology, 30*:359–64.

Hawkins, H. S. 1940. A wildlife history of Faville Grove, Wisconsin. *Trans. Wis. Acad. Sci Arts Lett., 32*:29–65.

Hayden, Ada. 1943. A botanical survey in the Iowa Lake Region of Clay and Palo Alto Counties. *Iowa State Coll. Jour. Sci., 17*:277–416.

Heddle, J. R. 1910. The plant geography of the University Bay region. B.S. thesis, University of Wisconsin.

Hedrick, U. P. 1933. *A history of agriculture in the state of New York.* Albany, N.Y.

Heimburger, C. C. 1934. Forest-type studies in the Adirondack region. *Mem. Cornell Agri. Exp. Sta.*, 165. 122 pp.

Hennepin, Louis. 1698. *A new discovery of a vast country in America.* London.

Henzel, T. 1938. Zagadnienia metodologiczne w okreslaniv rasowyn. *Przeglad Antropologiczny, 12,* No. 4.

Hewetson, C. E. 1956. A discussion on the "climax" concept in relation to the tropical rain and deciduous forest. *Empire Forestry Rev., 35*:274–91.

H'Doubler, Margaret. 1910. An ecological study of the flora of the southern bank of Lake Wingra. B.S. thesis, University of Wisconsin.

Hill, E. J. 1891. Notes on the flora of the St. Croix region. *Bot. Gaz., 16*:126–30.

Holdgate, M. W. 1955. The vegetation of some British upland fens. *Jour. Ecol., 43*:389–403.

Hole, F. D. 1943. Correlation of the glacial border drift of north central Wisconsin. *Am. Jour. Sci., 241*:498–516.

———, and Lee, G. B. 1955. *Introduction to the soils of Wisconsin,* Bull. Wis. Geol. and Nat. Hist. Surv., 79. 47 pp.

———, Peterson, F. F., and Robinson, G. H. 1952. The distribution of soils and slopes on the major terraces of southern Richland County, Wisconsin. *Trans. Wis. Acad. Sci. Arts Lett., 41*:73–81.

Hopkins, A. D. 1918. Periodical events and natural law as guides to agricultural research and practise. *U. S. Monthly Weather Rev. Suppl.* No. 9 (*Weather Bull.* No. 643.) Washington, D.C. 42 pp.

Hopkins, B. 1954. A new method for determining the type of distribution of plant individuals. *Ann. Bot., 18*:213–27.

———. 1957. Pattern in the plant community. *Jour. Ecol., 45*:451–63.

Horikawa, Y., and Okutomi, K. 1955. The continuum of the vegetation on the slopes of Mt. Shiroyama, Iwakuni City, Prov. Suwo. (Japanese, with English summary.) *The Seibutsugakkaishi, 6*:8–17.

Hough, F. A., and Forbes, R. D. 1943. The ecology and silvics of forests in the high plateaus of Pennsylvania. *Ecol. Monogr., 13*:299–320.

Hoyt, J. W. 1860. Natural resources of Wisconsin. *Trans. Wis. State Agri. Soc., 6*:46–49.

Huels, F. W. 1915. The peat resources of Wisconsin. *Bull. Wis. Geol. and Nat. Hist. Surv.,* 45. Econ. Ser. 20; Madison. 274 pp.

Hulten, E. 1937. *Outline of the history of arctic and boreal biota during the Quaternary period.* Stockholm.
Hutchinson, G. E. 1957. *A treatise on limnology.* Wiley, New York.
Huxley, J. 1938. Clines: an auxiliary taxonomic principle. *Nature, 142*:219–20.
Irving, R. D. 1880. Geology of the eastern Lake Superior district. In *Geology of Wisconsin,* Vol. *3*:89–91.
Jaccard, P. 1902. Lois de distribution florale dans le zone alpine. *Bull. Soc. Vaud. Sci. Natur., 38*:69–130.
———. 1928. Die statische-floristische Methode als Grundlage der pflanzensoziologie. *Handb. Biol. Arbeitsmeth. Abderhalden,* XI. *5*:165–202.
Jackson, H. H. T. 1914. The land vertebrates of Ridgeway Bog, Wisconsin: their ecological succession and sources of ingression. *Bull. Wis. Nat. Hist. Soc.,* NS *12*:4–54.
Jimbo, T. 1941. Ecological studies of peat bogs. VI. Studies of Suirennuma, a group of bog lakes. *Ecol. Rev., 7*:129–40.
Johnson, W. E., and Hasler, A. D. 1954. Rainbow trout production in dystrophic lakes. *Jour. Wildlife Management, 18*:113–34.
Jones, C. H. 1944. Studies in Ohio floristics. III. Vegetation of Ohio prairies. *Bull. Torrey Bot. Club, 71*:536–48.
Jones, J. Johanna. 1952. A survey of fifteen forest stands in the early Wisconsin drift plain in Indiana. *Butler Univ. Bot. Stud., 10*:182–204.
———. 1955. Conifer swamps of Wisconsin. Ph.D. thesis, University of Wisconsin.
———, and Zicker, Wilma A. 1955. A spruce-fir stand in the northern peninsula of Michigan. *Ecology, 36*:345.
Karlstrom, T. N. V. 1956. Radiocarbon-based Pleistocene correlations and world-wide climatic change. *Science, 124*:939.
Katz, N. J. 1926. Sphagnum bogs of central Russia: phytosociology, ecology, and succession. *Jour. Ecol., 14*:177–202.
———. 1933. Die Grundproblems und die neue Richtung der Phytosoziologie. *Beitr. Biol. der Pflanz.* (Breslau), *21*:133–66.
Keating, W. H. 1824. *Narrative of an expedition to the source of St. Peter's River, . . . under the command of Major Stephen H. Long.* Philadelphia.
Kellog, C. E. 1930. Preliminary study of the principal soil types of Wisconsin. *Bull. Wis. Geol. and Nat. Hist. Surv.,* 77A. (Soil Series 54.) 112 pp.
Kibbe, A. L. 1952. *A botanical study and survey of a typical midwestern county, Hancock County, Illinois, covering a period of 119 years, from 1883 to 1952.* Privately printed. Carthage, Ill.
King, F. L. 1882. Geology of the Upper Flambeau Valley; soils and vegetation. In *Geology of Wis.,* Vol. *4*:610–15.
Kittredge, Joseph. 1934. Evidence of the rate of forest succession on Star Island, Minnesota. *Ecology, 15*:24–35.
———. 1938. The interrelations of habitat, growth rate and associated vegetation in the aspen community of Minnesota and Wisconsin. *Ecol. Monogr., 8*:151–246.
Kitzke, E. D. 1949. Some ecological aspects of the Acrasiales in and near Madison, Wisconsin. *Papers Mich. Acad. Sci., 35*:25–32.

Kluender, W. A. 1934. A pollen analysis of the surface of Gibraltar Bog. B.A. thesis, University of Wisconsin.
Knapp, J. G. 1871a. The native vegetation of Wisconsin. *Trans. Wis. State Hort. Soc., 1*:119–25.
———. 1871b. The isothermal lines of Wisconsin. *Ibid. 1*:177–98.
Knapp, R. 1955. *Experimentelle Soziologie der höheren Pflanzen.* Ulmer, Stuttgart.
Krause, R. W., and Kent, G. N. 1944. Analysis and correlation of four New Hampshire bogs. *Ohio Jour. Sci., 44*:11–17.
Krylov, P. N. 1932. Die Abgrenzung von Steppen- und Waldsteppenzonen auf floristisch-statisticher Grundlage. *Handb. Biol. Arbeitsmeth. Abderhalden,* XI. *6*:129–36.
Kucera, C. L. 1952. An ecological study of a hardwood forest area in central Iowa. *Ecol. Monogr., 22*:283–99.
———, and Martin, S. C. 1957. Vegetation and soil relationships in the glade region of the Southwestern Missouri Ozarks. *Ecology, 38*:285–91.
———, and McDermott, R. E. 1955. Sugar maple-basswood studies in the forest-prairie transition of central Missouri. *Amer. Midland Nat., 54*:495–503.
Kulczynski, S. 1927. Zespoly roslin w Pieninach. (Die Pflanzen associationen der Pieninen.) *Bull. Intern. Polon. Acad. Sci. Lett., Cl. Sci. Math., et Nat.* Ser. B. Suppl. II, *1927*:57–203.
Laing, C. C. 1957. Mortality in a population of marram grass. *Bull. Ecol. Soc. Amer., 38*:79–80.
Laing, H. E. 1940. The composition of the internal atmosphere of *Nuphar advenum* and other water plants. *Amer. Jour. Bot., 27*:861–68.
Lane, G. H. 1941. Pollen analysis of interglacial peats of Iowa. *Iowa Geol. Survey, 37*:233–62.
Lapham, I. A. 1836. *A catalogue of plants and shells found in the vicinity of Milwaukee on the west side of Lake Michigan.* Milwaukee. 12 pp.
———. 1852. Plants of Wisconsin. *Trans. Wis. State Agr. Soc., 2*:375–419.
———. 1860. The Penokee Iron Range. *Ibid., 5*:391–400.
Larin, I. V., Matveeva, E. P., and Syrokomskaia, I. V. 1956. Dynamics of development of the meadow vegetation in the Kalingrad region. (In Russian.) *Akad. Nauk,* S.S.S.R., *Bot. Inst. Trudy,* Ser. 1, Geobot., *10*:31–101.
Larsen, J. A. 1953. A study of an invasion by red maple of an oak woods in southern Wisconsin. *Amer. Midland Nat., 49*:908–14.
Latrobe, C. J. 1833. *The rambler in North America.* London.
Lawson, P. V. 1902. Preliminary notice on the forest beds of the lower Fox. *Bull. Wis. Nat. Hist. Soc., 2*:170–73.
Le Barron, R. K. 1948. Silvicultural management of black spruce in Minnesota. *U. S. Dept. Agric. Circ.* 791. 60 pp.
Lee, M. B. 1945. An ecological study of the floodplain forest along the White River system in Indiana. *Butler Univ. Bot. Stud., 7*:155–65.
Leighton, M. M. 1957. Radiocarbon dates of Mankato drift in Minnesota. *Science, 125*:1037–38.
Leisman, G. A. 1953. The rate of organic matter accumulation on the sedge

mat zones of bogs in the Itasca State Park region of Minnesota. *Ecology.* *34:*81–101.

Leopold, Aldo. 1949. *A sand county almanac.* Oxford University Press, New York.

———, and Jones, Sara E. 1947. A phenological record for Sauk and Dane Counties, Wisconsin, 1935–1945. *Ecol. Monogr., 17:*81–122.

Leopold, Estella. 1946. Postglacial vegetation of Wingra Bog, Wisconsin. Manuscript Report of Bot. Dept., University of Wisconsin. 17 pp.

Lewis, F. J., Dowding, E. S., and Moss, E. H., 1928. The swamp, moor and bog forest vegetation of central Alberta. *Jour. Ecol., 16:*19–70.

Libby, W. F. 1952. *Radiocarbon dating.* University of Chicago Press, Chicago. 124 pp.

Lindsay, D. R. 1953. Climate as a factor influencing the mass ranges of weeds. *Ecology, 34:*308–21.

Linteau, A. 1955. Forest site classification of the northeastern coniferous section, boreal forest region, Quebec. *Canada Dept. North. Affairs and Nat. Res., Forestry Br., Bull.* 118. Ottawa. 85 pp.

Lippmaa, T. 1939. The unistratal concept of plant communities. *Am. Midland Nat., 21:*111–45.

Livingston, B. E. 1903. The distribution of the upland societies of Kent County, Michigan. *Bot. Gaz., 35:*36–55.

Loeffler, R. J. 1954. A new method of evaluating the distribution of planktonic algae in freshwater lakes. Ph.D. thesis, University of Wisconsin.

Lorenz, J. R. 1858. Allgemeine Resultate aus der pflanzengeographischen und genetischen Untersuchung der Moore in Präalpinen Hügellande Salzburg's. *Flora 41:*209 *et seq.*

Lothrop, J. 1856. Historical sketch of Kenosha County in 1855. *Coll. Wis. State Hist. Soc., 2:*450–79.

Ludi, W. 1935. Das Grosse Moos in west schweizerischen Seelande und die Geschiecte seiner Entstehung. *Veröff. Geobot. Inst. Rübel.,* (Zurich), *11:*1–344.

Lueders, H. F. 1895. Vegetation of the Town Prairie du Sac. *Trans. Wis. Acad. Sci. Arts Lett., 10:*510–24.

Lundegårdh, H. 1931. *Environment and plant development.* Arnold, London.

Marie-Victorin, F. 1943. Les hautes pinedes d'Haiti. *Contr. Inst. Bot., University of Montreal, 48:*47–60.

Marshall, Ruth. 1910. The vegetation of Twin Island. *Trans. Wis. Acad. Sci. Arts Lett., 16:*773–97.

———. 1910. *Ferns of the Dells of the Wisconsin River.* Appleton, Wis.

Marthaler, H. 1937. Die Stickstoffernahrung der Ruderalpflanzen. *Jarb. Wiss. Bot., 85:*76–106.

Martin, L. 1932. The physical geography of Wisconsin. *Bull. Wis. Geol. Nat. Hist. Surv.,* 36. Madison, 608 pp.

Matthews, J. T. 1955. The influence of weather on the frequency of beech mast years in England. *Forestry, 28:*107–16.

Matuszkiewicz, W. 1947. Zespoly lesne poludniowego Polesia (The forest associations of South Polessia). *Ann. Univ. M. Curie-Sklodowska.* Sect. E., *2:*69–138.

———. 1948. Roslinnosc lasow okolic lwowa (The vegetation of the forests of the environs of Lvov). *Ibid.*, Sect. C., *3*:119–93.
Maycock, P. F. 1956. Composition of an upland conifer community in Ontario. *Ecology, 37*:846–48.
———. 1957. The phytosociology of boreal conifer-hardwood forests of the Great Lakes region. Ph.D. thesis, University of Wisconsin.
McInteer, B. B. 1946. A change from grassland to forest vegetation in the "Big Barrens" of Kentucky. *Am. Midland Nat., 35*:276–82.
McIntosh, R. P. 1950a. Pine stands in southwestern Wisconsin. *Trans. Wis. Acad. Sci. Arts Lett., 40*:243–57.
———. 1950b. The phytosociology of the upland hardwood forest of southern Wisconsin. Ph.D. thesis, University of Wisconsin.
———. 1957. The York Woods, a case history of forest succession in southern Wisconsin. *Ecology, 38*:29–37.
McLaughlin, W. T. 1932. Atlantic coastal plain plants in the sand barrens of northwestern Wisconsin. *Ecol. Monogr., 2*:335–83.
McMillan, C. 1956. Nature of the plant community. I. Uniform garden and light period studies of five grass taxa in Nebraska. *Ecology, 7*:330–40.
———. 1957. Nature of the plant community. III. Flowering behavior within two grassland communities under reciprocal transplanting. *Am. Jour. Bot., 44*:144–53.
McVean, D. N. 1956. Ecology of *Alnus glutinosa* (L.) Gaertn. *Jour. Ecol., 44*:321–33.
Miller, Bonita J. 1954. Differential responses to cutting of six prairie grasses in Wisconsin. Ph.D. thesis, University of Wisconsin.
Milne, G. 1935. Some suggested units for classification and mapping, particularly for East African soils. *Soil Res., 4*:1–27.
Mizushima, M., and Yokouchi, I. 1953. A sketch of the plants of Nonomi Moor, Prov. Shinano. *Jour. Jap. Bot., 28*:348–52.
Moor, M. 1952. Die Fagion-gesellschaften in schweizer Jura. *Beitr. Z. Geobot. Landesaufnahme der Schweiz., 31*:1–201.
Morisita, M. 1954. Estimation of population density by spacing method. *Mem. Fac. Sci. Kyushu Univ.*, Ser E., *1*:187–97.
Moss, E. H. 1952. Grassland of the Peace River region, western Canada. *Canad. Jour. Bot., 30*:98–124.
———. 1953a. Marsh and bog vegetation in northwestern Alberta. *Ibid., 31*:448–70.
———. 1953b. Forest communities in northwestern Alberta. *Ibid.*:212–52.
———. 1955. The vegetation of Alberta. *Bot. Rev., 21*:493–567.
———, and Campbell, J. A. 1947. The fescue grassland of Alberta. *Canad. Jour. Res. C, 25*:209–27.
Motyka, J. 1947. O celachi metodach badan geobotanicznych. (Sur les buts et les methodes de recherches geobotaniques.) *Ann. Univ. M. Curie-Sklodowska. Sect. C. Suppl.* I:1–168.
———, Dobrzanski, B., and Zawadzki, S. 1950. Wstepne badania nad lakami poludniowo-wschodniej Lubelszczyzny (Preliminary studies on meadows in the southeast of the province Lublin). *Ann. Univ. M. Curie-Sklodowska. Sect. E., 5*:367–447.

―――, and Zawadzki, S. 1953. Badania nad lakami w dolinie Huczwy Kolo Werbkowic (Studies on meadows in the valley of the River Huczwa near Werbkowice). *Ann. Univ. M. Curie-Sklodowska. Sect. E., 8*:167–231.

Moyer, L. R. 1900. The prairie flora of southwestern Minnesota. *Bull. Minn. Acad. Sci., 4*:357–72.

Mueggler, W. F. 1953. The use of distance measures in the study of plant dispersion. M.S. thesis, University of Wisconsin.

Muir, John. 1913. *The story of my boyhood and youth.* New York.

Murphy, R. E. 1931. Geography of the northwest pine barrens of Wisconsin. *Trans. Wis. Acad. Sci. Arts Lett., 26*:69–120.

Natelson, Delle. 1952. A palynological study of a bog in Chippewa County, Wisconsin. MS report of Bot. Dept., University of Wisconsin. Typescript. 36 pp.

―――. 1954. The phytosociology of submerged aquatic macrophytes in Wisconsin lakes. Ph.D. thesis, University of Wisconsin.

Neiland, Bonita M., and Curtis, J. T. 1956. Differential responses to clipping of six prairie plants in Wisconsin. *Ecology, 37*:355–65.

Nelson, T. C. 1951. A reproduction study of northern white cedar. *Mich. Dept. of Conserv. Proj. Report,* 49-R.

Nielson, E. L. 1935. A study of a pre-Kansan peat deposit. *Torreya, 35*:53–56.

Niering, W. A. 1953. The past and present vegetation of High Point State Park, New Jersey. *Ecol. Monogr. 23*:127–48.

Norris, E. L. 1935. Ecological studies of the weed populations of eastern Nebraska. *Univ. Neb. Studies, 39*:29–91.

Norwood, J. G. 1852. Geological report of a survey of portions of Wisconsin and Minnesota made during the years 1847, '48, '49, and '50. In Owen's: *Report of a Geol. Surv. of Wis., Iowa, and Minn.,* pp. 213–424.

Oberdorfer, E. 1954. Über unkrautgesellschaften der Balkanhalbinsel. *Vegetatio, 4*:379–411.

―――. 1957. *Süddeutsche Pflanzengesellschaften.* Fischer, Jena.

Oehmcke, A. A. 1937. Ecology of the lake shore ravines and beaches of the Wisconsin coast of Lake Michigan. B.A. thesis, University of Wisconsin.

Olmsted, C. E. 1944. Growth and development in range grasses. IV. Photoperiodic responses in twelve geographic strains of side-oats grama. *Bot. Gaz., 106*:46–74.

Oosting, H. J., and Billings, W. D. 1951. A comparison of virgin spruce-fir forest in the northern and southern Appalachian system. *Ecology, 32*:84–103.

―――, and Bordeau, P. F. 1955. Virgin hemlock forest segregates in the Joyce Kilmer Memorial Forest of North Carolina. *Bot. Gaz., 116*:340–59.

―――, and Reed, J. F. 1952. Virgin spruce-fir forest in the Medicine Bow Mountains, Wyoming. *Ecol. Monogr., 22*:69–91.

Orpurt, P. A. 1955. Studies on the soil microfungi of Wisconsin prairies. Ph.D. thesis, University of Wisconsin.

―――, and Curtis, J. T. 1957. Soil microfungi in relation to the prairie continuum in Wisconsin. *Ecology, 38*:628–37.

Osborn, B. 1942. Prairie dogs in shinnery (oak scrub) savannah. *Ecology, 23*:110–15.

Osvald, H. 1925. Die Hochmoortypen Europas. *Veröff. Geobot. Inst. Rübel.*, *3*:707–23.
Owen, D. D. 1848. Report of a geological reconnaissance of the Chippewa Land District of Wisconsin. *Sen. Exec. Doc., 57; 30th Cong., 1st sess.*, Washington, D.C.: Government Printing Office. 57 pp.
———. 1852. *Report of a geological survey of Wisconsin, Iowa and Minnesota,* . . . Philadelphia.
Pammel, L. H. 1907. A comparative study of the vegetation of swamp, clay, and sandstone areas in western Wisconsin. *Proc. Davenport Acad. Sci., 10*:32–126.
Parmalee, G. W. 1953. The oak upland community in southern Michigan. *Bull. Ecol. Soc. Amer., 34*:84.
Partch, M. L. 1949. Habitat studies of soil moisture in relation to plants and plant communities. Ph.D. thesis, University of Wisconsin.
Passarge, H. 1956. Vegetationskundliche Untersuchungen in Waldern und Geholzen der Elbaue. *Arch. Forstw., 5*:339–58.
Pearsall, W. H. 1918. On the classification of aquatic plant communities. *Jour. Ecol., 6*:75–83.
Peattie, D. C. 1922. The Atlantic coastal plain element in the flora of the Great Lakes. *Rhodora, 24*:57–70, 80–88.
Perrot, N. 1667. On lakes and swamps in northern Wisconsin. *Repr. Coll. Wis. State Hist. Soc., 16*:17–19.
Pierce, R. S. 1951. Prairie-like mull humus, its physico-chemical and microbiological properties. *Proc. Soil Sci. Soc. Amer., 15*:362–64.
Place, I. C. M. 1955. The influence of seed-bed conditions on the regeneration of spruce and balsam fir. *Canada Forestry Research Bull.* 117. 87 pp.
Pogrebnjak, P. S. 1930. Über die Methodik der Standortuntersuchungen in Verbindung mit den Waldtypen. *Internatl. Congr. Forestry Expt. Stations, Stockholm Proc., 1929*:455–71.
Poliakov, P. P. 1950. On the flora of the Abies forests of Kazakhstan Altai. (In Russian.) *Bot. Zhur. S.S.S.R., 35*:301–03.
Pool, R. J. 1914. A study of the vegetation of the sandhills of Nebraska. *Minn. Bot. Studies, 4*:189–312.
Poore, M. E. D. 1955. The use of phytosociological methods in ecological investigations. I: The Braun-Blanquet System. *Jour. Ecol., 43*:226–44.
———. 1956. The ecology of Woodwalton fern. *Ibid., 44*:455–92.
Popov, M. G. 1953. The relation of forest to steppe in central Siberia. *Bull. Mosc. Soc. Invest. Nat., 58*:81–95.
Potzger, J. E. 1942. Pollen spectra from four bogs on the Gillen Nature Reserve along the Michigan-Wisconsin state line. *Amer. Midland Nat., 28*:501–11.
———. 1943. Pollen study of five bogs in Price and Sawyer Counties, Wisconsin. *Butler Univ. Bot. Stud., 6*:54–64.
———. 1946. Phytosociology of the primeval forest in central northern Wisconsin and upper Michigan, and a brief post-glacial history of the lake forest formation. *Ecol. Monogr., 16*:211–50.
———. 1948. A pollen study in the tension zone of lower Michigan. *Butler Univ. Bot. Stud., 8*:161–77.

———, and Friesner, R. C. 1940. What is climax in central Indiana? *Ibid.,* *4*:181–95.

———, and Van Engel, W. A. 1943. Study of the rooted aquatic vegetation of Weber Lake, Vilas County, Wis. *Trans. Wis. Acad. Sci. Arts Lett., 34*:149–66.

Quarterman, Elsie. 1950. Major plant communities of Tennessee Cedar Glades. *Ecology, 31*:234–54.

Quimby, G. I. 1954. Cultural and natural areas before Kroeber. *Amer. Antiquity, 19*:317–31.

Raabe, E. W. 1949. Der Zeigerwert der Ackerunkrauter in ostlichen Holstein. *Biol. Zentralbl., 68*:471–88.

———. 1952. Über den Affinitatswert in der Pflanzensoziologie. *Vegetatio, 4:53–68.*

———, and Raabe, H. 1949. Die Wiesen des Kossau-Tales. *Schrift. Naturwiss. Vereins. Schleswig-Holstein, 24*:16–29.

Ramensky, L. G. 1926. Die Grundgesetzmässigkeiten in Aufbau der Vegetationsdecke. *Bot. Zentralbl., 7*:453–55.

———. 1930. Zur Methodik der vergleichenden Bearbeitung und Ordnung von Pflanzenlisten und anderen Objeckten, die durch mehrere verschiedenartig wirkende Fakturen bestimmtwerden. *Beitr. Biol. Pflanz.* (Breslau), *18*:269–304.

———. 1953. Ecological study and systematization of groupings of vegetation. (In Russian.) *Mosk. Obshch. Isp. Priorody. B. Otd. Biol., 58*:35–54.

Randall, W. E. 1952 Interrelations of autecological characteristics of forest herbs. Ph.D. thesis, University of Wisconsin.

———. 1953. Water relations and chlorophyll content of forest herbs in southern Wisconsin. *Ecology, 34*:544–53.

Raunkiaer, C. 1934. *Life forms of plants and statistical plant geography.* Clarendon Press, Oxford.

Regel, C. 1947. The bogs and swamps of White Russia. *Jour. Ecol., 35*:96–104.

Rhodes, J. W. 1933. An ecological comparison of two Wisconsin peat bogs. *Milw. Publ. Mus. Bull., 7*:305–62.

Rickett, H. W. 1921. A quantitative study of the larger aquatic plants of Lake Mendota. *Trans. Wis. Acad. Sci. Arts Lett., 20*:501–27.

———. 1924. A quantitative study of the larger aquatic plants of Green Lake, Wisconsin. *Ibid., 21*:381–414.

Ritzenthaler, R. E. 1953. Prehistoric Indians of Wisconsin. *Milw. Publ. Mus. Pop. Sci. Handbook* No. 4. 43 pp.

Robinson, G. H. 1950. Soil carbonate and clay contents as criteria of rate and stage of soil genesis. Ph.D. thesis, University of Wisconsin.

Robocker, W. C. 1951. Certain factors affecting establishment and survival of several native grass species in Wisconsin. Ph.D. thesis, University of Wisconsin.

Rosendahl, C. O. 1948. A contribution to the knowledge of the Pleistocene flora of Minnesota. *Ecology, 29*:284–315.

Roth, Filibert. 1898. On the forestry conditions of northern Wisconsin. *Wis. Geol. and Nat. Hist. Surv. Bull., 1*: Econ. Series 1. Madison. 78 pp.

Rowe, J. S. 1956. Uses of undergrowth plant species in forestry. *Ecology, 37:* 461–73.
Rubner, K. 1954. Die Roterlengesellschaft der oberbayerischen Grundmoräne, *Forstarchiv.,* (Hannover), *25:*137–42.
Rübel, E. 1932. Die Buchenwalder Europas. *Veröff Geobot. Inst. Rübel., 8:*1–530.
Ruggles, D. 1835. Geological and miscellaneous notice of the region around Fort Winnebago, Michigan Territory. *Amer. Jour. Sci., 30:*No. 1.
Russel, H. 1914. Preliminary check-list of the flora of Wisconsin. *Bot. Committee of Wis. Nat. Hist. Soc.* In 9 parts (mimeo.). Milwaukee.
Russell, J. C. 1893. Second expedition to Mount St. Elias in 1891. *13th Ann. Rept. U. S. Geol. Surv.,* Part *2:*1–91.
Russell, N. H. 1953. Plant communities of the Apple River Canyon, Wisconsin. *Proc. Iowa Acad. Sci., 60:*228–42.
Ruttner, Franz. 1953. *Fundamentals of limnology.* University of Toronto Press, Toronto.
Sampson, H. C. 1921. An ecological survey of the prairie vegetation of Illinois. *Bull. Ill. Nat. Hist. Surv., 13:*519–77.
———. 1930. Succession in the swamp forest formation in northern Ohio. *Ohio Jour. Sci., 30:*340–57.
Sargent, C. S. 1884. Report on the forests of North America. *U. S. Dept. Interior, Census Off.,* 10th census.
Sauer, C. O. 1950. Grassland climax, fire, and man. *Jour. Range Management, 3:*16–21.
Sauer, J. 1950. The grain amaranths: a survey of their history and classification. *Ann. Mo. Bot. Gard., 37:*561–632.
———. 1952. A geography of pokeweed. *Ibid., 39:*113–25.
Schoolcraft, H. R. 1834. *Exploration of the St. Croix and Burntwood* [or Brule] *Rivers.* New York.
Schorger, A. W. 1949. Squirrels in early Wisconsin. *Trans. Wis. Acad. Sci. Arts Lett., 39:*195–247.
Schuette, H. A., and Alder, H. 1929. A note on the chemical composition of *Chara* from Green Lake, Wisconsin. *Trans. Wis. Acad. Sci. Arts Lett., 24:*141–45.
Scully, N. J. 1942. Root distribution and environment in a maple-oak forest. *Bot. Gaz., 103:*492–517.
Sears, P. B. 1942a. Postglacial migration of five forest genera. *Am. Jour. Bot., 29:*684–91.
———. 1942b. Xerothermic theory. *Bot. Rev., 8:*709–36.
Selleck, G. W., and Schuppert, K. 1957. Some aspects of microclimate in a pine forest and in an adjacent prairie. *Ecology, 38:*650–53.
Sernander, R. 1906. Entwurf einer Monographie der Europaischen Myrmechoren. *Kungl. Svenska Ventenskap, Handl., 41:*1–410.
Shaffer, P. R. 1956. Farmdale drift in northwestern Illinois. *Ill. State Geol. Surv. Rept. Invest.,* 198. 25 pp.
Shaffner, J. H. 1926. Observations on the grasslands of the central U. S. *Contr. Bot. Lab., Ohio State Univ., 178:*1–56.

Shanks, R. E. 1953. Forest composition and species association in the beech-maple forest region of western Ohio. *Ecology, 34*:455–66.

Shantz, H. L., and Zon, R. 1924. *Atlas of American agriculture; natural vegetation.* U. S. Dept. Agric. Washington, D.C.

Shea, J. G. 1853. *Discovery and exploration of the Mississippi valley with the original narratives of Marquette, Allouez, Membre, Hennepin and Anastase Douay.* New York.

———. 1861. *Early voyages up and down the Mississippi by Cavelier, St. Cosme, LeSueur, Gravier, and Guignas.* New York.

Sherff, E. E. 1912. The vegetation of Skokie Marsh, with special reference to the subterranean organs and their interrelationships. *Bot. Gaz., 53*: 415–35.

Shimek, B. 1911. The prairies. *Bull. Labs. Nat. Hist. Univ. Iowa, 6*:169–240.

———. 1924. The prairies of the Mississippi River bluffs. *Proc. Iowa Acad. Sci., 31*:205–12.

———. 1925. Papers on the prairie. *Univ. Iowa Studies Nat. Hist., 11.* 36 pp.

Shinners, L. H. 1940. Vegetation of the Milwaukee region. B.A. thesis, University of Wisconsin.

Shirley, H. S. 1932. Light intensity in relation to plant growth in a virgin norway pine forest. *Jour. Agric. Res., 44*:227–44.

Short, C. W. 1845. Observations on the botany of Illinois, more in reference to the autumnal flora of the prairies. *West. Jour. of Med. and Surg., 3*:185–98.

Shreve, F. 1910. The ecological plant geography of Maryland, mountain zone. In *The Plant Life of Maryland. Maryland Weather Service. Spec. Publ., 3*:275–91.

Shriner, F. A., and Copeland, E. B. 1904. Deforestation and creek flow about Monroe, Wisconsin. *Bot. Gaz., 37*:139–43.

Sjörs, H. 1950. Regional studies in North Swedish mire vegetation. *Bot. Notiser, 1950*:173–219.

Smith, W. R. 1837. *Incidents of a journey from Pennsylvania to the Wisconsin Territory, in 1837.* Wooster, Ohio.

Smith, C. C. 1940. The effect of overgrazing and erosion upon the biota of the mixed grass prairie of Oklahoma. *Ecology, 21*:381–98.

Smith, L. L., and Moyle, J. B. 1944. A biological survey and fishery management plan for the streams of the Lake Superior north shore watershed. *Tech. Bull. Minn. Dept. Conserv., 1.* 228 pp.

Sorensen, T. 1948. A method for establishing groups of equal magnitude in plant sociology based on similarity of species content. *Act. Kong. Danska Vidensk., Selsk. Biol. Skr. J., 5*:1–34.

———. 1954. Adaptation of small plants to deficient nutrition and a short growing season. *Bot. Tidsskr., 51*:339–61.

Stearns, F. W. 1949. Ninety years change in a northern hardwood forest in Wisconsin. *Ecology, 30*:350–58.

———. 1950. The composition of a remnant of white pine forest in the Lake States. *Ibid., 31*:209–92.

———. 1951. The composition of the sugar maple-hemlock-yellow birch association in northern Wisconsin. *Ibid., 32*:245–65.

Stebbins, G. L. 1950. *Variation and evolution in plants.* Columbia University Press, New York.
Steinbrenner, E. C. 1951. Effect of grazing on floristic composition and soil properties of farm woodlands in southern Wisconsin. *Jour. Forestry, 49:*906–10.
Steward, G., and Keller, W. 1936. A correlation method for ecology, as exemplified by studies of native desert vegetation. *Ecology, 17:*500–514.
Stewart, O. C. 1951. Burning and natural vegetation in the United States. *Geog. Rev., 41:*317–20.
Stokes, W. L. 1955. Another look at the Ice Age. *Science, 122:*815–21.
Stone, E. C. 1957. Dew as an ecological factor. I. A review of the literature. *Ecology, 38:*407–13.
Stone, Witmer. 1911. The plants of southern New Jersey. *Bull. New Jersey State Mus.* Trenton, N.J. 828 pp.
Stout, A. B. 1914. A biological and statistical analysis of the vegetation of a typical wild hay meadow. *Trans. Wis. Acad. Sci. Arts Lett., 17:*405–69.
———. 1946. The bur oak openings in southern Wisconsin. *Ibid., 36:*141–61.
Strong, Moses. 1877. The geology of the upper St. Croix district; soils and vegetation. In *Geology of Wisconsin,* Vol. *3:*375–81.
Struik, Gwendolyn J. 1957. The distribution and reproduction of some herbaceous plants in a maple forest. M.S. thesis, University of Wisconsin.
Sukachev, V. N. 1928. Principles of classification of the spruce communities of European Russia. *Jour. Ecol., 16:* 1–18.
Suzuki, T. 1953. The forest climaxes of East Asia. *Jap. Jour. Bot., 14:*1–12.
Sweet, E. T. 1880. Geology of the western Lake Superior district; climate, soils and timber. In *Geology of Wisconsin,* Vol. *3:*323–29.
Swezey, G. D. 1883. Catalogue of the phaenogamous and vascular cryptogamous plants of Wisconsin. In *Geology of Wisconsin,* Vol. *1:*376–95.
Swindale, Delle N., and Curtis, J. T. 1957. Phytosociology of the larger submerged plants in Wisconsin lakes. *Ecology, 38:*397–407.
Swindale, L. D., and Jackson, M. L. 1956. Genetic processes in some residual podzolised soils of New Zealand. *Rapp. VI Congress Internat. de la Sci. du Sol. V. 37:*233–39.
Tallos, P. 1954. A papakovacsi lapret norenytarsulasai es fasitasa. (Plant associations on the fen peatlands of Papakovacsi.) *Erdesz. Kutatas.* (Budapest), *4:*55–69.
Tansley, A. G. 1939. *The British Islands and their vegetation.* Cambridge University Press.
Taylor, Norman. 1923. The vegetation of Long Island. I. The vegetation of Montauk, a study of grassland and forest. *Mem. Brooklyn Bot. Gard., 2:*1–107.
Tchou, Yen-Tcheng. 1948. Etudes ecologiques et phytosociologiques sur les forets riveraines du Bas-Languedoc. *Vegetatio, 1:*2, 93, 217, 347.
Telford, C. J. 1926. Third report on a forest survey of Illinois. *Bull. Ill. Nat. Hist. Surv., 16.* 102 pp.
Thomson, J. W. 1940. Relic prairie areas in central Wisconsin. *Ecol. Monogr., 10:*685–717.

———. 1943 Plant succession on abandoned fields in the central Wisconsin sand plain area. *Bull. Torrey Bot. Club, 70:*34–41.
———. 1944. A survey of the larger aquatic plants and bank flora of the Brule River. *Trans. Wis. Acad. Sci. Arts Lett., 36:*57–76.
Thurstone, L. L. 1947. *Multiple factor analysis.* University of Chicago Press, Chicago.
Thwaites, F. 1953. *Outline of Glacial Geology.* Published by the author. Madison, Wisconsin.
Transeau, E. N. 1903. On the geographic distribution and ecological relations of the bog plant societies of North America. *Bot Gaz., 36:*401–20.
———. 1905–1906. The bogs and bog flora of the Huron River Valley. *Ibid., 40:*351–74, 418–48 (1905), *41:*17–42 (1906).
———. 1935. The prairie peninsula. *Ecology, 16:*423–37.
Tresner, H. E. 1952. Studies on the microfungal flora of the soil in relation to the hardwood forest continuum in southern Wisconsin. Ph.D. thesis, University of Wisconsin.
Tresner, H. D., Backus, M. P., and Curtis, J. T. 1954. Soil microfungi in relation to the hardwood forest continuum in southern Wisconsin. *Mycologia, 46:*314–33.
True, R. H. 1897. Botanizing in the Delles of the Wisconsin River. *Plant World, 1:*81–83.
Truman, H. V. 1937. Fossil evidence of two prairie invasions of Wisconsin. *Trans. Wis. Acad. Sci. Arts Lett., 30:*35–42.
Tryon, R. M., Fassett, N. C., Dunlop, D. W., and Diemer, M. E. 1940. *The ferns and fern allies of Wisconsin.* Madison.
Tüxen, R. 1950. Grundriss einer Systematik der nitrophilen Unkrautgesellschaften in der Eurosibirischen Region Europas. *Mitt. der Florist.-soziolog. Arbeitsgemeinshaft, 2:*93–175.
Tuomikoski, R. 1942. Untersuchungen uber die untervegetation der Bruchmoore in Ostfinnland. I. Zur Methodik der pflanzensoziologischen Systematik. *Ann. Soc. Zoo-Bot. Fenn., Vanamo., 17:*1–200.
Turesson, G. 1925. The plant species in relation to habitat and climate. *Hereditas, 6:*147–236.
Turrill, W. B. 1946. The ecotype concept. A consideration with appreciation and criticism, especially of modern trends. *New Phytol., 45:*34–43.
Van Arsdel, E. P. 1954. Climatic relations of the distribution of white pine blister rust in Wisconsin. *Univ. Wis. Forestry Res. Notes, 13.* 2 pp.
Vestal, A. G. 1913. An associational study of Illinois sand prairie. *Bull. Ill. State Lab. Nat. Hist., 10:*1–96.
———. 1914. A black-soil prairie station in northeastern Illinois. *Bull. Torrey Bot. Club, 41:*351–64.
Visher, S. S. 1954. *Climatic atlas of the United States.* Harvard University Press, Cambridge, Mass.
Vorobiev, D. V. 1953. *Tipi lesev europeiskei chastu USSR* (Forest types in the European part of the USSR). Moskow.
Voss, John. 1934. Postglacial migration of forests of central Illinois, Wisconsin, and Minnesota. *Bot. Gaz., 96:*3–43.

―――. 1937. Comparative study of bogs on Cary and Tazewell drift of Illinois. *Ecology, 18*:119–35.

―――. 1939. Forests of the Yarmouth and Sangamon interglacial period in Illinois. *Ibid., 20*:517–28.

Wagner, B. G. 1951. A study of prairie soils and vegetation in southern Wisconsin. M.S. thesis, University of Wisconsin.

Waite, P. J. 1958. Wisconsin's average winter snowfall. *Wis. Acad. Rev., 5*:1–2.

Ward, R. T. 1954. A phytosociological study of the beech forests of Wisconsin. Ph.D. thesis, University of Wisconsin.

―――. 1956. The beech forests of Wisconsin—changes in forest composition and the nature of the beech border. *Ecology, 37*:407–19.

Warder, J. A. 1881. Forests and forestry in Wisconsin. *Proc. Wis. State Hort. Soc., 11*:143–56.

Ware, G. H. 1955. A phytosociological study of lowland hardwood forests in southern Wisconsin. Ph.D. thesis, University of Wisconsin.

Waterman, W. G. 1923. Bogs of northern Illinois. *Trans. Ill. State Acad. Sci., 16*:214–25.

Watt, A. S. 1947. Pattern and process in the plant community. *Jour. Ecol., 35*:1–22.

Weaver, J. E. 1954. *North American Prairie*. Johnsen Publ. Co., Lincoln, Neb.

―――, and Clements, F. E. 1928. *Plant Ecology*. McGraw, New York.

―――, and Albertson, F. W. 1956. *Grasslands of the Great Plains*. Johnsen Publ. Co., Lincoln, Neb.

―――, and Fitzpatrick, T. J. 1934. The Prairie. *Ecol. Monogr., 4*:109–295.

Webb, D. A. 1954. Is the classification of plant communities either possible or desirable? *Bot. Tidsskrift, 51*:362–70.

Welty, C. 1957. State board for preservation of scientific areas. *Wis. Acad. Rev., 4*:105–7.

Wendelberger-Zelinka, E. 1952. Die Auwaldtypen von Oberösterreich. *Öst. Vjschr. Forstw., 93*:72–86.

Westveld, R. H. 1933. The relation of certain soil characteristics to forest growth and composition in the northern hardwood forest of northern Michigan. *Tech. Bull. Mich. Agr. Exp. Stat., 135*: 52 pp.

Whitford, P. W. 1948. Species distribution and age of hardwood stands in the prairie forest region of Wisconsin. Ph.D. thesis, University of Wisconsin.

―――. 1949. Distribution of woodland plants in relation to succession and clonal growth. *Ecology, 30*:199–208.

―――. 1951. Estimation of the age of forest stands in the prairie-forest border region. *Ibid., 32*:143–47.

―――, and Salamun, P. J. 1954. An upland forest survey of the Milwaukee area. *Ibid., 35*:533–40.

Whitman, W. C., Hanson, H. T., and Loder, G. 1943. Natural revegetation of abandoned fields in western North Dakota. *N. Dak. Agri. Exp. Sta. Bull., 321*. 18 pp.

Whitney, J. D. 1876. Plain, prairie, and forest. *Amer. Naturalist, 10*:577–588; 656–67.

Whitson, A. R., and Baker, O. E. 1928. The climate of Wisconsin and its relation to agriculture. *Wis. Agr. Exp. Sta. Bull., 223.* 46 pp.

Whittaker, R. H. 1951. A criticism of the plant association and climatic climax concepts. *Northwest Sci., 25:*17–31.

———. 1952. A study of summer foliage insect communities in the Great Smoky Mountains. *Ecol. Monogr., 22:*1–44.

———. 1954. Plant populations and the basis of plant indication. *Festschrift fur Erwin Aichinger.* Springer, Vienna. Vol. *1:*183–206.

———. 1956. Vegetation of the Great Smoky Mountains. *Ecol. Monogr., 26:*1–80.

———. 1957. Recent evolution of ecological concepts in relation to the eastern forests of North America. *Amer. Jour. Bot., 44:*197–206.

Whittlesey, C. 1852. Description of part of Wisconsin, south of Lake Superior. In Owen's *Report of a geological survey of Wisconsin.* pp. 425–71.

Wiedemann, E. 1929. Die ertragskundliche und waldbauliche Brauchbarkeit der Waldtypen nach Cajander in Sachsischen Erzgebirge. *Allgem. Forst und Jagd-Zeitung.* (Frankfurt), *105:*247–54.

Wight, O. W. 1877. Report of progress and results for the year 1875. *Geology of Wisconsin,* Vol. *2:*67–89.

Wilde, S. A. 1940. Classification of gley soils for the purpose of forest management and reforestation. *Ecology, 21:*34–44.

———. 1954. Mycorrhizal fungi: their distribution and effect on tree growth. *Soil Science, 78:*23–31.

———. 1958. *Forest soils, their properties and relation to silviculture.* Ronald, New York.

———, and Leaf, A. L. 1955. The relationship between the degree of soil podzolization and the composition of ground cover vegetation. *Ecology, 36:*19–22.

———, and Randall, G. W. 1951. Chemical characteristics of ground water in forest and marsh soils of Wisconsin. *Trans. Wis. Acad. Sci. Arts Lett., 40:*251–59.

———, Wilson, F. G., and White, D. P. 1949. Soils of Wisconsin in relation to silviculture. *Wis. Conservation Dept. Publ., 525.* 171 pp.

Williams, A. B. 1936. The composition and dynamics of a beech-maple climax community. *Ecol. Monogr., 6:*317–408.

Williams, C. B. 1944. Some applications of the logarithmic series and the index of diversity to ecological problems. *Jour. Ecol., 32:*1–44.

Wilson, L. R. 1932. The Two Creeks forest bed, Manitowoc County, Wisconsin. *Trans. Wis. Acad. Sci. Arts Lett., 27:*31–46.

———. 1935. Lake development and plant succession in Vilas County, Wisconsin. *Ecol. Monogr., 5:*207–47.

———. 1938. The postglacial history of vegetation in northwestern Wisconsin. *Rhodora, 40:*137–75.

———. 1939. A temperature study of a Wisconsin peat bog. *Ecology, 20:*432–33.

———. 1941. The larger aquatic vegetation of Trout Lake, Vilas Co., Wisconsin. *Trans. Wis. Acad. Sci. Arts Lett., 33:*135–46.

———, and Kosanke, R. M. 1940. The microfossils in a pre-Kansan peat deposit near Belle Plaine, Iowa. *Torreya, 40*:1–5.

———, and Webster, R. M. 1942. Microfossil studies of three north-central Wisconsin bogs. *Trans. Wis. Acad. Sci. Arts Lett., 34*:177–93.

Winter, A. G., and Bublitz, W. 1953. Uber die Keim und entwicklungshemmende Wirkung der Buchenstreu. *Naturwiss., 40*:416.

Winterringer, G. S., and Vestal, A. G. 1956. Rock-ledge vegetation in southern Illinois. *Ecol. Monogr., 26*:105–30.

Wittry, W. L., and Ritzenthaler, R. E. 1956. The Old Copper complex: an Archaic manifestation in Wisconsin. *Amer. Antiq., 21*:244–54.

Wright, H. E. 1957. Radiocarbon dates of Mankato drift in Minnesota. *Science, 125*:1038–39.

Wright, J. W. 1952. Pollen dispersion of some forest trees. *N.E. Forest Exp. Stat. Paper, 46*. 42 pp.

Wright, J. C., and Wright, E. A. 1948. Grassland types of south central Montana. *Ecology, 29*:449–60.

Wolfe, J. N. 1951. The possible role of microclimate (at the glacial border). *Ohio Jour. Sci., 51*:134–38.

———, Wareham, R. T., and Scofield, H. T. 1949. Microclimates and macroclimate of Neotoma, a small valley in central Ohio. *Bull. Ohio Biol. Surv., 41*. 267 pp.

Wooster, L. W. 1882. Geology of the St. Croix district; vegetation. In *Geology of Wisconsin*, Vol. *4*:146–54.

Yapp, R. H., Johns, D., and Jones, O. T. 1917. The salt marshes of the Dovey Estuary, II. *Jour. Ecol., 5*:65–103.

Yoshioka, K. 1949. Sociological studies of the pine forests in Japan, especially with regard to their structure and development. *Sci. Rpts. Tohuku University*, 4th Ser. *18*:229–42.

Youngberg, C. T. 1951. Evolution of prairie-forest soils under cover of invading northern hardwoods in the driftless area of southwestern Wisconsin. *Trans. Wis. Acad. Sci. Arts Lett., 40*:285–89.

Zicker, Wilma A. 1955. An analysis of Jefferson County vegetation using surveyor's records and present day data. M.S. thesis, University of Wisconsin.

Zimmerman, F. R. 1953. Waterfowl habitat surveys and food habit studies 1940–1943. *Wis. Conserv. Dept. Report R. R. Project* 6-R. Mimeo. 176 pp.

Zoller, H. 1954. Die Arten der Bromus erectus-wiessen des Schweizer Juras. *Veröff Geobot. Inst. Rübel, 28*:1–283.

Species list | Index

SPECIES LIST

This list includes all species which appear either in the tables of this book, have quantitative information of any sort, or which are discussed at length in the text. Page references for the last are given in the regular index. Species which are mentioned only incidentally are not included in the present list. The fidelity of each species is shown by the number following each name. This indicates the number of native communities in which it was found in the P.E.L. study. The symbol in capital letters following the fidelity numbers indicates the native community in which the species achieved maximum presence. These symbols are keyed below. If the species were confined to a weed community, pasture, or other disturbed area, this fact is indicated by an asterisk. No fidelity numbers are given for these weed species since the delimitation of disturbance communities was not equivalent to that employed for native communities and fidelity figures might therefore be misleading

AQE	Emergent Aquatic	NWM	Northern Wet-Mesic Forest
AQS	Submerged Aquatic	OB	Oak Barrens
AT	Alder Thicket	OO	Oak Opening
BEA	Lake Beach	PB	Pine Barrens
BF	Boreal Forest	PD	Dry Prairie
BG	Bracken-Grassland	PDM	Dry-Mesic Prairie
BOG	Open Bog	PM	Mesic Prairie
CG	Cedar Glade	PW	Wet Prairie
CLE	Exposed Cliff	PWM	Wet-Mesic Prairie
CLS	Shaded Cliff	SB	Sand Barrens
DUN	Lake Dune	SC	Shrub Carr
FN	Fen	SD	Southern Dry Forest
ND	Northern Dry Forest	SDM	Southern Dry-Mesic Forest
NDM	Northern Dry-Mesic Forest	SM	Southern Mesic Forest
NM	Northern Mesic Forest	SS	Southern Sedge Meadow
NS	Northern Sedge Meadow	SW	Southern Wet Forest
NW	Northern Wet Forest	SWM	Southern Wet-Mesic Forest

Abies balsamea 7 BF
Abutilon theophrasti*
Acalypha virginica 1 CLE
Acer negundo 8 SW
Acer rubrum 12 NDM
Acer saccharinum 2 SW
Acer saccharum 9 SM
Acer spicatum 4 BF
Acerates hirtella 5 PWM
Acerates lanuginosa 3 PD
Acerates viridiflora 5 PD
Achillea millefolium 17 PM
Acnida altissima 3 BEA
Aconitum noveboracense 1 CLS
Acorus calamus 4 SS
Actaea alba 7 NM

Actaea rubra 10 BF
Adenocaulon bicolor 2 NDM
Adiantum pedatum 11 SDM
Aesculus glabra 1 SWM
Agastache foeniculum 2 PB
Agastache scrophulariaefolia 2 SDM
Agrimonia gryposepala 9 SD
Agropyron dasystachyum 2 DUN
Agropyron repens 5 DUN
Agropyron smithii*
Agropyron trachycaulum 9 BG
Agrostis hyemalis 7 BG
Agrostis perennans 4 NW
Agrostis stolonifera 4 BG
Ajuga genevensis*
Alisma plantago-aquatica 4 AQE

Allium canadense 9 **PWM**
Allium cernuum 1 **PM**
Allium stellatum 1 **CLE**
Allium tricoccum 8 **SM**
Alnus crispa 1 **ND**
Alnus rugosa 9 **AT**
Alopecurus pratensis 1 **NS**
Amaranthus graecizans*
Amaranthus retroflexus*
Ambrosia artemisiifolia 12 **DP**
Ambrosia psilostachya 6 **SB**
Ambrosia trifida 6 **SW**
Ammophila breviligulata 2 **DUN**
Amorpha canescens 15 **PD**
Amorpha fruticosa 1 **SS**
Amphicarpa bracteata 19 **SDM**
Anacharis canadensis 1 **AQS**
Anacharis nuttallii 1 **AQS**
Anaphalis margaritacea 7 **BG**
Andromeda glaucophylla 2 **BOG**
Andropogon gerardi 19 **PM**
Andropogon scoparius 15 **PD**
Anemone canadensis 15 **SS**
Anemone cylindrica 16 **CG**
Anemone patens 10 **PD**
Anemone quinquefolia 18 **NDM**
Anemone virginiana 12 **SDM**
Anemonella thalictroides 7 **SDM**
Angelica atropurpurea 9 **SS**
Antennaria neglecta 17 **OB**
Antennaria plantaginifolia 5 **SB**
Anthemis cotula*
Apios americana 8 **SW**
Aplectrum hyemale 3 **SM**
Apocynum androsaemifolium 18 **ND**
Apocynum cannabinum 10 **PM**
Apocynum sibiricum 2 **FN**
Aquilegia canadensis 19 **CG**
Arabis canadensis 2 **SD**
Arabis glabra 2 **BG**
Arabis laevigata 2 **CLS**
Arabis lyrata 11 **CG**
Aralia hispida 10 **CLE**
Aralia nudicaulis 17 **NDM**
Aralia racemosa 11 **SDM**
Arceuthobium pusillum 2 **BOG**
Arctium minus 2 **SD**
Arctostaphylos uva-ursi 9 **BG**
Arenaria lateriflora 13 **SWM**
Arenaria macrophylla 1 **CLE**
Arenaria stricta 7 **PD**
Arethusa bulbosa 2 **BOG**
Arisaema dracontium 3 **SWM**
Arisaema triphyllum 13 **SDM**
Aristida basiramea 3 **SB**

Aronia melanocarpa 12 **PD**
Artemisia absinthium*
Artemisia biennis*
Artemisia caudata 12 **PD**
Artemisia frigida 1 **PDM**
Artemisia ludoviciana 4 **PDM**
Artemisia serrata 4 **PWM**
Artemisia vulgaris*
Asarum canadense 8 **CLS**
Asclepias amplexicaulis 6 **OB**
Asclepias exaltata 6 **SDM**
Asclepias incarnata 8 **FN**
Asclepias ovalifolia 3 **SD**
Asclepias purpurascens 2 **PWM**
Asclepias sullivantii 4 **FN**
Asclepias syriaca 15 **PWM**
Asclepias tuberosa 11 **PDM**
Asclepias verticillata 9 **PD**
Asparagus officinalis 1 **OO**
Aster azureus 14 **PDM**
Aster ciliolatus 9 **BG**
Aster cordifolius 1 **SWM**
Aster ericoides 9 **PDM**
Aster junciformis 2 **FN**
Aster laevis 13 **PM**
Aster lateriflorus 11 **SW**
Aster linariifolius 6 **OB**
Aster lucidulus 6 **FN**
Aster macrophyllus 14 **BF**
Aster novae-angliae 9 **PWM**
Aster oblongifolius 5 **PD**
Aster pilosus 4 **PDM**
Aster prenanthoides 6 **SDM**
Aster ptarmicoides 4 **PD**
Aster puniceus 9 **NS**
Aster sagittifolius 12 **SDM**
Aster sericeus 7 **PD**
Aster shortii 10 **SDM**
Aster simplex 14 **FN**
Aster umbellatus 16 **BG**
Astragalus canadensis 4 **PM**
Astragalus crassicarpus 1 **PDM**
Athyrium filix-femina 14 **SDM**
Athyrium pycnocarpon 2 **SM**
Athyrium thelypteroides 4 **NM**
Atriplex patula*
Aureolaria grandiflora 9 **CG**
Aureolaria pedicularia 1 **CLS**
Avena fatua 1 **BEA**

Baptisia leucantha 9 **PWM**
Baptisia leucophaea 7 **PM**
Barbarea vulgaris*
Berteroa incana 1 **BG**
Besseya bullii 4 **OO**

Betula glandulosa 5 BOG
Betula lutea 9 NWM
Betula nigra 5 SW
Betula papyrifera 15 NDM
Betula sandbergii 5 FN
Bidens beckii 2 AQS
Bidens cernua 6 NS
Bidens connata 1 BEA
Bidens coronata 8 FN
Bidens frondosa 8 SWM
Bidens tripartita 2 BOG
Bidens vulgata 2 BEA
Blephilia ciliata 4 PWM
Blephilia hirsuta 2 SWM
Boehmeria cylindrica 13 SW
Botrychium dissectum 1 SWM
Botrychium lanceolatum 1 NM
Botrychium matricariaefolium 2 BF
Botrychium multifidum 10 AT
Botrychium virginianum 12 SDM
Bouteloua curtipendula 5 PD
Bouteloua hirsuta 6 CG
Brachyelytrum erectum 10 SDM
Brasenia schreberi 1 BOG
Brassica juncea*
Brassica nigra*
Bromus ciliatus 11 AT
Bromus erectus 1 DUN
Bromus inermis*
Bromus japonicus*
Bromus kalmii 10 BG
Bromus latiglumis 2 SWM
Bromus purgans 8 SDM
Bromus tectorum*

Cacalia atriplicifolia 2 PW
Cacalia suaveolens 3 PW
Cacalia tuberosa 6 PWM
Cakile edentula 2 BEA
Calamagrostis canadensis 20 FN
Calamovilfa longifolia 3 DUN
Calla palustris 5 BOG
Callirhoe triangulata 1 PDM
Calopogon pulchellus 3 BOG
Caltha palustris 13 FN
Camassia scilloides 1 PWM
Campanula americana 6 SDM
Campanula aparinoides 8 FN
Campanula rotundifolia 14 CG
Camptosorus rhizophyllus 4 SM
Cannabis sativa*
Capsella bursa-pastoris*
Cardamine bulbosa 2 SS
Cardamine douglassii 2 SM
Cardamine pensylvanica 3 NWM

Carex alopecoidea 3 SWM
Carex amphibola 1 SWM
Carex aquatilis 3 AQE
Carex arctata 3 BF
Carex aurea 2 BEA
Carex bebbii 1 SS
Carex bicknellii 4 PWM
Carex bromoides 1 SWM
Carex brunnescens 2 NW
Carex canescens 2 NW
Carex cephalophora 1 SWM
Carex communis 1 NDM
Carex comosa 3 NW
Carex convoluta 5 SM
Carex crinita 5 SWM
Carex cristatella 3 SWM
Carex davisii 1 SM
Carex debilis 1 SWM
Carex deweyana 3 SM
Carex disperma 3 NWM
Carex eburnea 1 BF
Carex gracillima 1 SWM
Carex grayii 2 SW
Carex grisea 1 SWM
Carex haydenii 2 NS
Carex hirtifolia 2 SM
Carex hitchcockiana 1 SDM
Carex hystericina 3 BEA
Carex interior 3 NW
Carex intumescens 5 BF
Carex jamesii 1 SDM
Carex lacustris 3 NS
Carex lasiocarpa 3 BOG
Carex laxiflora 8 SM
Carex leptalea 3 NWM
Carex leptonervia 3 NWM
Carex lupulina 2 SW
Carex molesta 1 SWM
Carex muhlenbergii 6 SB
Carex muskingumensis 2 SW
Carex nigro-marginata 1 NM
Carex normalis 2 SW
Carex pauciflora 1 NW
Carex pedunculata 6 BF
Carex pensylvanica 16 SDM
Carex plantaginea 4 NM
Carex projecta 4 SWM
Carex retrorsa 1 NS
Carex rosea 2 SWM
Carex rostrata 3 SC
Carex sartwellii 1 FN
Carex siccata 1 ND
Carex sparganioides 1 SWM
Carex sprengelii 2 SWM
Carex stipata 4 SWM

Carex stricta 5 SS
Carex tetanica 4 SM
Carex tribuloides 2 SW
Carex trichocarpa 1 AQE
Carex trisperma 5 NW
Carex tuckermani 2 SWM
Carex typhina 2 SW
Carex vesicaria 1 SW
Carex viridula 1 BEA
Carex vulpinoidea 2 SW
Carpinus caroliniana 3 BF
Carum carvi*
Carya cordiformis 8 SM
Carya ovata 12 OO
Cassia fasciculata 2 PDM
Cassia marilandica 1 SWM
Castilleja coccinea 5 FN
Castilleja sessiliflora 5 PD
Caulophyllum thalictroides 10 SM
Ceanothus americanus 12 PM
Ceanothus ovatus 6 PB
Celastrus scandens 18 SDM
Celtis occidentalis 3 SWM
Cenchrus pauciflorus 2 SB
Centaurea maculosa*
Cephalanthus occidentalis 4 SW
Cerastium arvense*
Cerastium vulgatum 1 BEA
Ceratophyllum demersum 1 AQS
Ceratophyllum echinatum 1 AQS
Chaenorrhinum minus*
Chaerophyllum procumbens 1 SWM
Chamaedaphne calyculata 4 BOG
Cheilanthes feei 2 CLE
Chelidonium majus*
Chelone glabra 9 AT
Chenopodium album 2 SWM
Chenopodium capitatum*
Chenopodium glaucum*
Chenopodium hybridum*
Chimaphila umbellata 9 ND
Chrysanthemum leucanthemum 2 BG
Chrysopsis villosa 1 SB
Chrysosplenium americanum 1 BF
Cichorium intybus*
Cicuta bulbifera 7 NS
Cicuta maculata 14 PWM
Cinna arundinacea 3 SW
Cinna latifolia 3 NW
Circaea alpina 11 NWM
Circaea quadrisulcata 13 SDM
Cirsium altissimum 5 SDM
Cirsium arvense*
Cirsium discolor 10 PM

Cirsium hillii 2 PDM
Cirsium muticum 12 PW
Cirsium pitcheri 1 BEA
Cirsium undulatum 1 PDM
Cirsium vulgare*
Cladium mariscoides 1 BEA
Claytonia caroliniana 2 NM
Claytonia virginica 6 SM
Clematis verticillaris 2 NDM
Clematis virginiana 13 AT
Clintonia borealis 12 BF
Collomia linearis*
Comandra richardsiana 21 OB
Commelina erecta 1 CLE
Conopholis americana 4 SDM
Convolvulus arvensis*
Convolvulus sepium 10 PM
Convolvulus spithamaeus 11 ND
Conyza canadensis 5 SB
Coptis trifolia 8 NWM
Corallorhiza maculata 6 ND
Corallorhiza odontorhiza 1 SDM
Corallorhiza striata 4 BF
Corallorhiza trifida 4 NWM
Coreopsis lanceolata 2 DUN
Coreopsis palmata 13 PD
Corispermum hyssopifolium 2 BEA
Cornus alternifolia 13 SDM
Cornus canadensis 10 BF
Cornus purpusi 7 SWM
Cornus racemosa 21 SD
Cornus rugosa 12 BF
Cornus stolonifera 12 SC
Coronilla varia*
Corydalis sempervirens 7 CLE
Corylus americana 21 SD
Corylus cornuta 9 BF
Crepis tectorum 1 PB
Croton glandulosus 1 SB
Cryptogramma stelleri 2 CLE
Cryptotaenia canadensis 10 SDM
Cuscuta gronovii 2 SWM
Cycloloma atriplicifolium 2 DUN
Cynoglossum boreale 3 ND
Cynoglossum officinale 1 BF
Cynoglossum virginianum 2 SD
Cyperus erytherorhizos 1 SS
Cyperus filiculmis 4 SB
Cyperus schweintzii 2 SB
Cyperus strigosus 1 AQE
Cypripedium acaule 8 BOG
Cypripedium candidum 3 FN
Cypripedium parviflorum Salisb. 2 SC
Cypripedium pubescens Willd. 11 SD

Cypripedium reginae 1 NWM
Cystopteris bulbifera 5 CLS
Cystopteris fragilis 11 CLS

Dactylis glomerata*
Danthonia spicata 5 BG
Datura stramonium*
Daucus carota*
Decodon verticillatus 2 AQE
Delphinium virescens 1 PDM
Dentaria diphylla 1 NM
Dentaria laciniata 6 SM
Descurainia pinnata*
Descurainia sophia*
Desmodium canadense 9 PWM
Desmodium cuspidatum 2 SD
Desmodium glutinosum 11 SDM
Desmodium illinoense 8 PM
Desmodium nudiflorum 8 SDM
Dianthus barbatus 1 BEA
Dianthus prolifer*
Dicentra canadensis 3 CLS
Dicentra cucullaria 6 CLS
Diervilla lonicera 17 BF
Digitaria ischaemum*
Digitaria sanguinalis*
Dioscorea villosa 8 SDM
Dipsacus sylvestris*
Dirca palustris 8 NM
Dodecatheon amethystinum 2 CLS
Dodecatheon meadia 14 PWM
Draba reptans 3 CLE
Dracocephalum formosius 7 SW
Drosera intermedia 1 BOG
Drosera rotundifolia 3 BOG
Dryopteris austriaca 14 NWM
Dryopteris boottii 2 BOG
Dryopteris cristata 12 NWM
Dryopteris fragrans 1 CLS
Dryopteris goldiana 2 SM
Dryopteris marginalis 2 CLS
Dulichium arundinaceum 5 BOG

Echinacea pallida 3 PM
Echinochloa crusgalli*
Echinochloa walteri 1 AQE
Echinocystis lobata 9 SWM
Echium vulgare*
Elatine minima 1 AQS
Eleocharis acicularis 2 AQE
Eleocharis calva 2 AQE
Eleocharis compressa 1 BEA
Eleocharis palustris 2 AQE
Ellisia nyctelea 1 SM

Elymus canadensis 10 PWM
Elymus villosus 5 SDM
Elymus virginicus 5 SW
Epifagus virginiana 3 NM
Epigaea repens 9 ND
Epilobium adenocaulon 4 NWM
Epilobium angustifolium 10 BG
Epilobium coloratum 8 AT
Epilobium strictum 5 NS
Equisetum arvense 17 PW
Equisetum fluviatile 6 NW
Equisetum hiemale 9 DUN
Equisetum laevigatum 6 PWM
Equisetum scirpoides 3 NW
Equisetum sylvaticum 6 BF
Equisetum variegatum 2 DUN
Eragrostis cilianensis*
Eragrostis pectinacea*
Eragrostis spectabilis 2 SB
Erigeron annuus 5 BG
Erigeron glabellus 1 BG
Erigeron philadelphicus 12 SC
Erigeron pulchellus 11 SDM
Erigeron strigosus 15 PD
Ericaulon septangulare 1 AQS
Eriophorum angustifolium 3 BOG
Eriophorum spissum 2 NW
Eriophorum tenellum 1 NW
Eriophorum virginicum 5 NW
Erucastrum gallicum*
Eryngium yuccifolium 6 PM
Erysimum cheiranthoides*
Erythronium albidum 7 SM
Erythronium americanum 3 NM
Euonymus atropurpureus 3 SW
Eupatorium altissimum 1 PM
Eupatorium maculatum 11 AT
Eupatorium perfoliatum 12 FN
Eupatorium purpureum 5 SDM
Eupatorium rugosum 10 SDM
Euphorbia corollata 15 OB
Euphorbia esula*
Euphorbia maculata 1 SB
Euphorbia polygonifolia 2 DUN
Euphorbia preslii*

Fagopyrum esculentum 1 BG
Fagus grandifolia 6 NM
Festuca obtusa 10 SD
Festuca octoflora 2 CLE
Festuca saximontana 2 BG
Fimbristylis autumnalis 2 DUN
Floerkea proserpinacoides 1 SM
Fragaria vesca 6 NW

Fragaria virginiana 25 ND
Fraxinus americana 12 SDM
Fraxinus nigra 13 NWM
Fraxinus pennsylvanica 9 SWM
Fraxinus quadrangulata 2 SM
Froelichia floridana 2 SB

Galeopsis tetrahit 7 CLE
Galinsoga parviflora*
Galium aparine 16 SM
Galium asprellum 7 AT
Galium boreale 20 FN
Galium circaezans 5 SDM
Galium concinnum 12 SDM
Galium labradoricum 7 AT
Galium lanceolatum 9 NM
Galium obtusum 6 SS
Galium tinctorium 8 PW
Galium trifidum 7 NS
Galium triflorum 18 BF
Gaultheria hispidula 7 NW
Gaultheria procumbens 13 ND
Gaura biennis 3 PM
Gaylussacia baccata 9 PB
Gentiana andrewsii 8 PWM
Gentiana crinita 2 PWM
Gentiana flavida 2 PM
Gentiana procera 3 FW
Gentiana puberula 6 PM
Gentiana quinquefolia 1 PDM
Gentiana rubricaulis 2 NS
Gentiana saponaria 2 PM
Geranium bicknellii 2 BF
Geranium maculatum 21 SDM
Gerardia aspera 3 PW
Gerardia gattingeri 2 ND
Gerardia paupercula 2 FN
Gerardia tenuifolia 1 NS
Geum aleppicum 6 AT
Geum canadense 13 SDM
Geum rivale 5 NWM
Geum triflorum 8 PD
Glecoma hederacea*
Gleditsia triacanthos 1 SWM
Glyceria borealis 1 AQE
Glyceria canadensis 3 NS
Glyceria grandis 6 AT
Glyceria striata 13 FN
Gnaphalium obtusifolium 11 SB
Gnaphalium uliginosum 1 CLE
Goodyera decipiens 1 BF
Goodyera pubescens 5 SDM
Goodyera repens 5 BF
Goodyera tesselata 2 BF

Gratiola aurea 1 AQS
Gratiola neglecta 1 SS
Grindelia squarrosa 1 PDM
Gymnocarpium dryopteris 9 NWM
Gymnocladus dioica 2 SM

Habenaria clavellata 1 NW
Habenaria dilatata 1 BOG
Habenaria flava 2 PWM
Habenaria hookeri 2 ND
Habenaria hyperborea 5 NWM
Habenaria lacera 4 PW
Habenaria leucophaea 3 PWM
Habenaria media 2 NWM
Habenaria obtusata 3 NWM
Habenaria orbiculata 6 NDM
Habenaria psycodes 7 SS
Habenaria viridis 11 SM
Hackelia virginiana 7 SDM
Halenia deflexa 5 BF
Hamamelis virginiana 8 NDM
Hedeoma hispida 8 PD
Helenium autumnale 7 PW
Helianthemum canadense 15 OB
Helianthus annuus*
Helianthus giganteus 4 AT
Helianthus grosseserratus 11 PWM
Helianthus laetiflorus 8 PM
Helianthus occidentalis 13 PM
Helianthus petiolaris*
Helianthus strumosus 15 SD
Helianthus tuberosus 1 SC
Heliopsis helianthoides 12 PM
Hemerocallis fulva*
Hepatica acutiloba 9 SM
Hepatica americana 10 BF
Heracleum lanatum 4 SDM
Hesperis matronalis*
Heuchera richardsonii 18 CG
Hibiscus trionum*
Hieracium aurantiacum 5 BG
Hieracium canadense 14 BG
Hieracium longipilum 9 SB
Hieracium scabrum 5 SDM
Hierochloe odorata 4 PW
Hordeum jubatum*
Houstonia caerulea 1 PW
Houstonia longifolia 6 SB
Hudsonia tomentosa 2 SB
Humulus lupulus 3 AT
Hydrastis canadensis 1 SM
Hydrocotyle americana 4 AT
Hydrophyllum appendiculatum 4 SM
Hydrophyllum virginianum 10 SM

Hypericum boreale 1 CLS
Hypericum canadense 1 PM
Hypericum kalmianum 4 NS
Hypericum mutilum 1 PM
Hypericum perforatum*
Hypericum punctatum 1 SWM
Hypericum pyramidatum 3 AT
Hypoxis hirsuta 11 PW
Hystrix patula 9 SDM

Ilex verticillata 9 NWM
Impatiens biflora 18 NWM
Impatiens pallida 3 SM
Iris lacustris 3 DUN
Iris shrevei 13 FN
Iris versicolor 4 BOG
Isanthus brachiatus*
Isoetes macrospora 1 AQS
Isopyrum biternatum 6 SM
Iva xanthifolia*

Juglans cinerea 7 SM
Juglans nigra 5 SDM
Juncus alpinus 2 BEA
Juncus balticus 2 DUN
Juncus brevicaudatus 1 BEA
Juncus bufonius 2 BEA
Juncus dudleyi 3 FN
Juncus effusus 2 SS
Juncus greenii 3 SB
Juncus nodosus 1 BEA
Juncus pelocarpus 1 AQS
Juncus tenuis 3 SD
Juncus torreyi 2 AQE
Juniperus communis 8 DUN
Juniperus horizontalis 4 DUN
Juniperus virginiana 9 CG

Kalmia polifolia 3 BOG
Kochia scoparia*
Koeleria cristata 13 SB
Krigia biflora 13 PW
Krigia virginica 1 SB
Kuhnia eupatorioides 8 PD

Lactuca biennis 11 SD
Lactuca canadensis 14 PM
Lactuca ludoviciana 6 PDM
Lactuca scariola 1 BEA
Lamium amplexicaule*
Laportea canadensis 13 SW
Lappula echinata*
Larix laricina 5 NW
Lathyrus maritimus 2 DUN
Lathyrus ochroleucus 12 SDM

Lathyrus palustris 11 SS
Lathyrus venosus 13 PWM
Lechea intermedia 2 BG
Lechea tenuifolia 8 SB
Ledum groenlandicum 4 NW
Leerzia oryzoides 6 SC
Leerzia virginica 2 SW
Lemna minor 1 AQE
Lemna trisulca 1 AQE
Leonurus cardiaca*
Lepidium campestre*
Lepidium densiflorum*
Lepidium virginicum*
Leptoloma cognatum 5 SB
Lespedeza capitata 15 OB
Lespedeza violacea 1 CLE
Liatris aspera 13 PM
Liatris cylindracea 7 PD
Liatris ligulistylis 2 PM
Liatris punctata 1 PDM
Liatris pycnostachya 9 PWM
Liatris spicata 4 PM
Lilium philadelphicum 9 PM
Lilium superbum 19 PW
Linaria canadensis 4 SB
Linaria vulgaris*
Linnaea borealis 8 BF
Linum sulcatum 7 PDM
Liparis liliifolia 1 SD
Liparis loeselii 3 PW
Listera cordata 3 BF
Lithospermum canescens 15 OB
Lithospermum caroliniense 9 SB
Lithospermum incisum 6 PD
Lithospermum latifolium 4 SD
Lobelia cardinalis 4 SW
Lobelia dortmanna 1 AQS
Lobelia inflata 2 CLE
Lobelia kalmii 4 FN
Lobelia siphilitica 7 FN
Lobelia spicata 15 CG
Lolium multiflorum*
Lolium perenne*
Lonicera canadensis 10 BF
Lonicera dioica 8 SD
Lonicera hirsuta 1 BF
Lonicera oblongifolia 8 NDM
Lonicera prolifera 9 SDM
Lonicera villosa 3 NW
Ludwigia palustris 1 AQE
Lupinus perennis 7 OB
Luzula acuminata 6 BF
Luzula campestris 4 NDM
Lychnis alba 1 BG

Lycium halimifolium*
Lycopodium annotinum 7 BF
Lycopodium clavatum 6 BF
Lycopodium complanatum 9 ND
Lycopodium lucidulum 10 NM
Lycopodium obscurum 6 NDM
Lycopodium tristachyum 1 ND
Lycopus americanus 15 FN
Lycopus uniflorus 11 NS
Lysimachia nummularia 1 SW
Lysimachia quadrifolia 8 PB
Lysimachia terrestris 10 NS
Lythrum alatum 6 PWM
Lythrum salicaria*

Maianthemum canadense 18 BF
Malaxis unifolia 5 BG
Malva neglecta*
Matricaria matricarioides*
Matteuccia struthiopteris 9 NS
Medeola virginiana 3 NDM
Medicago lupulina*
Medicago sativa*
Melampyrum lineare 4 ND
Melilotus albus*
Melilotus officinalis*
Menispermum canadense 11 SW
Mentha arvensis 9 AT
Menyanthes trifoliata 3 BOG
Mertensia paniculata 1 BF
Microseris cuspidata 2 PDM
Milium effusum 5 NM
Mimulus ringens 5 NS
Mitchella repens 10 NDM
Mitella diphylla 9 NM
Mitella nuda 6 NWM
Mollugo verticillata 3 SB
Monarda fistulosa 17 CG
Monarda punctata 5 SB
Moneses uniflora 6 BF
Monotropa hypopithys 7 ND
Monotropa uniflora 11 NDM
Muhlenbergia cuspidata 4 PDM
Muhlenbergia frondosa 7 SW
Muhlenbergia mexicana*
Muhlenbergia racemosa 11 BG
Muhlenbergia schreberi*
Muhlenbergia sylvatica 1 BEA
Myosotis scorpioides 5 DUN
Myosotis virginica 1 CLE
Myrica asplenifolia 8 BG
Myrica gale 2 BOG
Myriophyllum verticillatum 1 AQS

Najas flexilis 1 AQS
Napaea dioica 2 PWM
Nasturtium officinale 1 NW
Naumbergia thyrsiflora 8 SS
Nemopanthus mucronatus 9 NW
Nepeta cataria*
Nuphar advena 1 BOG
Nymphaea odorata 7 BOG

Oenothera biennis 14 PDM
Oenothera perennis 4 PW
Oenothera rhombipetala 3 SB
Oenothera serrulata 1 BEA
Onoclea sensibilis 22 AT
Ophioglossum vulgatum 5 FN
Opuntia compressa 5 CG
Opuntia fragilis 1 CG
Orchis spectabilis 6 SDM
Orobanche fasciculata 1 PDM
Orobanche uniflora 2 OO
Oryzopsis asperifolia 9 BF
Oryzopsis pungens 1 PB
Oryzopsis racemosa 4 SDM
Osmorhiza claytoni 13 SDM
Osmorhiza longistylis 11 SDM
Osmunda cinnamomea 12 NWM
Osmunda claytoniana 14 SDM
Osmunda regalis 12 SWM
Ostrya virginiana 10 SM
Oxalis acetosella 5 NWM
Oxalis europaea*
Oxalis stricta 4 PDM
Oxalis violacea 11 PM
Oxybaphus nyctagineus 5 PWM
Oxypolis rigidior 8 PW

Panax quinquefolium 6 SDM
Panax trifolium 5 NDM
Panicum capillare*
Panicum dichotomiflorum*
Panicum lanuginosum 4 SB
Panicum latifolium 5 SDM
Panicum leibergii 10 PM
Panicum meridionale 1 DUN
Panicum oligosanthes 10 PDM
Panicum perlongum 7 PD
Panicum praecocius 5 OB
Panicum virgatum 10 SB
Parietaria pensylvanica 7 SDM
Parnassia glauca 4 FN
Paronychia canadensis 1 CLS
Parthenium integrifolium 3 PWM
Parthenocissus vitacea 26 SD
Pastinaca sativa*

SPECIES LIST 641

Pedicularis canadensis 17 CG
Pedicularis lanceolata 11 FN
Pellaea atropurpurea 2 CLE
Penstemon digitalis 1 PM
Penstemon gracilis 3 OB
Penstemon grandiflorus 1 OB
Penstemon hirsutus 2 CLE
Penthorum sedoides 2 BEA
Petalostemum candidum 8 CG
Petalostemum purpureum 8 PD
Petasites frigidus 1 BF
Phalaris arundinacea 11 SC
Phleum pratense 2 BG
Phlox divaricata 6 SM
Phlox glaberrima 3 OO
Phlox pilosa 18 PWM
Phragmites communis 4 AQE
Phryma leptostachya 12 SDM
Physalis heterophylla 9 PD
Physalis longifolia 2 PD
Physalis virginiana 14 OB
Physocarpus opulifolius 10 CLE
Picea glauca 8 BF
Picea mariana 4 NW
Pilea pumila 10 SW
Pinus banksiana 8 PB
Pinus resinosa 7 ND
Pinus strobus 12 NDM
Plantago aristata 1 CLE
Plantago lanceolata*
Plantago major 2 SW
Plantago purshii 1 SB
Plantago rugelii 1 PDM
Plantago virginica 1 CLE
Platanus occidentalis 1 SWM
Poa compressa 10 BG
Poa palustris 9 NS
Poa pratensis 14 OO
Poa sylvestris 1 SM
Podophyllum peltatum 8 SM
Pogonia ophioglossoides 1 BOG
Polanisia graveolens*
Polemonium reptans 10 SDM
Polygala paucifolia 6 ND
Polygala polygama 8 SB
Polygala sanguinea 6 CG
Polygala senega 8 FN
Polygala verticillata 3 CG
Polygonatum canaliculatum 17 OB
Polygonatum pubescens 16 NM
Polygonella articulata 4 SB
Polygonum arifolium 4 SC
Polygonum aviculare*
Polygonum cilinode 6 CLE

Polygonum coccineum 5 NS
Polygonum convolvulus 6 CLE
Polygonum erectum*
Polygonum hydropiper 4 BEA
Polygonum lapathifolium 1 BEA
Polygonum natans 7 NS
Polygonum pensylvanicum 2 SS
Polygonum persicaria 2 BEA
Polygonum punctatum 1 AQE
Polygonum sagittatum 9 AT
Polygonum scandens 1 SD
Polygonum tenue 3 CLE
Polygonum virginianum 6 CLS
Polymnia canadensis 3 SDM
Polypodium vulgare 5 CLS
Polytaenia nuttallii 3 PWM
Pontederia cordata 1 AQE
Populus alba 2 DUN
Populus balsamifera 4 BF
Populus deltoides 4 SW
Populus grandidentata 13 ND
Populus nigra 1 DUN
Populus tremuloides 21 BF
Portulaca oleracea*
Potamogeton amplifolius 1 AQS
Potamogeton diversifolius 1 AQS
Potamogeton epihydrus 1 AQS
Potamogeton filiformis 1 AQS
Potamogeton foliosus 1 AQS
Potamogeton friesii 1 AQS
Potamogeton gramineus 1 AQS
Potamogeton illinoensis 1 AQS
Potamogeton natans 1 AQS
Potamogeton obtusifolius 1 AQS
Potamogeton pectinatus 1 AQS
Potamogeton praelongus 1 AQS
Potamogeton pulcher 1 AQS
Potamogeton pusillus 1 AQS
Potamogeton richardsonii 1 AQS
Potamogeton robbinsii 1 AQS
Potamogeton spirillus 1 AQS
Potamogeton strictifolius 1 AQS
Potamogeton zosteriformis 1 AQS
Potentilla anserina 3 DUN
Potentilla argentea*
Potentilla arguta 14 CG
Potentilla fruticosa 4 FN
Potentilla norvegica 2 BEA
Potentilla palustris 6 BOG
Potentilla recta*
Potentilla simplex 16 SD
Prenanthes alba 18 SDM
Prenanthes aspera 2 PWM
Prenanthes crepidinea 1 PWM

Prenanthes racemosa 7 PWM
Primula mistassinica 1 BEA
Prunella vulgaris 14 BF
Prunus pensylvanica 7 BF
Prunus serotina 16 SD
Prunus virginiana 13 SD
Psoralea argophylla 1 PM
Psoralea esculenta 2 PD
Pteridium aquilinum 22 BG
Pycnanthemum virginianum 11 PWM
Pyrola asarifolia 1 BF
Pyrola elliptica 11 NDM
Pyrola minor 3 NDM
Pyrola rotundifolia 6 ND
Pyrola secunda 10 BF
Pyrola virens 1 BF

Quercus alba 12 SD
Quercus bicolor 4 SW
Quercus borealis 15 SDM
Quercus ellipsoidalis 12 PB
Quercus macrocarpa 17 OO
Quercus muhlenbergii 1 SD
Quercus velutina 8 OB

Radicula armoracea*
Ranunculus abortivus 13 SDM
Ranunculus acris 2 BF
Ranunculus aquatilis 1 AQS
Ranunculus fascicularis 8 OO
Ranunculus flabellaris 2 AQE
Ranunculus flammula 2 NS
Ranunculus pensylvanicus 1 BEA
Ranunculus recurvatus 7 SDM
Ranunculus reptans 1 AQE
Ranunculus rhomboideus 4 CLE
Ranunculus sceleratus 2 SS
Ranunculus septentrionalis 12 SWM
Ratibida columnifera*
Ratibida pinnata 11 PWM
Rhamnus alnifolius 3 NW
Rhamnus catharticus 2 SC
Rhexia virginica 1 SS
Rhus aromatica 1 SB
Rhus glabra 12 PDM
Rhus radicans 25 SW
Rhus typhina 10 OB
Rhus vernix 4 BOG
Rhynchospora alba 2 BOG
Rhynchospora capillacea 1 BEA
Ribes americanum 18 AT
Ribes cynosbati 14 SDM
Ribes glandulosum 1 BF
Ribes hirtellum 2 NWM

Ribes lacustre 4 NWM
Ribes missouriense 8 SD
Ribes triste 3 NWM
Rorippa austriaca*
Rorippa islandica 4 BEA
Rosa sp. 26 PM
Rubus alleghoniensis 14 SD
Rubus hispidus 5 ND
Rubus occidentalis 5 SDM
Rubus parviflorus 6 BF
Rubus pubescens 16 NWM
Rubus strigosus 13 BF
Rudbeckia hirta 13 PWM
Rudbeckia laciniata 13 SW
Rudbeckia subtomentosa 3 PWM
Rumex acetosella 7 BG
Rumex altissimus 2 SW
Rumex crispus*
Rumex orbiculatus 5 AT
Rumex verticillatus 6 NWM

Sagina procumbens*
Sagittaria cuneata 1 AQS
Sagittaria graminea 1 AQS
Sagittaria latifolia 8 AQE
Sagittaria rigida 1 AQE
Salix amygdaloides 1 SW
Salix bebbiana 2 SC
Salix candida 2 BOG
Salix cordata 2 DUN
Salix discolor 15 SC
Salix glaucophylloides 2 DUN
Salix humilis 10 PW
Salix interior 8 DUN
Salix longifolia 1 SW
Salix lucida 3 DUN
Salix nigra 2 SW
Salix pedicillaris 1 NWM
Salix petiolaris 5 NS
Salix sericea 1 FN
Salsola kali 2 BEA
Sambucus canadensis 12 SD
Sambucus pubens 9 NM
Sanguinaria canadensis 12 SM
Sanicula gregaria 11 SDM
Sanicula marilandica 15 BF
Saponaria officinalis 1 DUN
Sarracenia purpurea 2 BOG
Satureja glabella 1 BEA
Satureja vulgaris 1 BF
Saxifraga pensylvanica 13 PW
Schizachne purpurascens 8 BG
Scirpus acutus 2 AQE
Scirpus americanus 2 AQE

Scirpus atrovirens 6 AT
Scirpus cespitosus 1 FN
Scirpus cyperinus 8 AT
Scirpus fluviatilis 6 AQE
Scirpus heterochaetus 1 AQE
Scirpus lineatus 2 SC
Scirpus rubrotinctus 3 BOG
Scirpus validus 4 AQE
Scleria triglomerata 3 PW
Scrophularia lanceolata 6 CG
Scrophularia marilandica 1 SDM
Scutellaria galericulata 9 NS
Scutellaria lateriflora 10 SW
Scutellaria leonardi 9 CG
Selaginella apoda 1 FN
Selaginella rupestris 6 SB
Selaginella selaginoides 1 NS
Senecio aureus 11 PWM
Sencio pauperculus 6 BG
Setaria glauca*
Setaria verticillata*
Setaria viridis*
Shepherdia canadensis 2 BF
Sicyos angulatus 2 BEA
Silene antirrina 7 CLE
Silene cucubalus*
Silene dichotoma*
Silene nivea 1 SC
Silene noctiflora*
Silene stellata 6 CG
Silphium integrifolium 8 PWM
Silphium laciniatum 7 PM
Silphium perfoliatum 9 SW
Silphium terebinthinaceum 7 PWM
Sisymbrium altissimum*
Sisymbrium officinale*
Sisyrinchium campestre 9 PDM
Sium suave 10 SC
Smilacina racemosa 19 SD
Smilacina stellata 21 SD
Smilacina trifolia 4 NW
Smilax ecirrhata 13 SDM
Smilax herbacea 15 SD
Smilax hispida 12 SDM
Solanum carolinense*
Solanum dulcamara 4 SWM
Solanum nigrum 3 NWM
Solanum triflorum*
Solidago canadensis 20 AT
Solidago flexicaulis 9 SM
Solidago gigantea 16 PW
Solidago graminifolia 14 NS
Solidago hispida 6 CLS
Solidago juncea 12 PM

Solidago missouriensis 9 PM
Solidago nemoralis 14 PD
Solidago ohioensis 5 PW
Solidago patula 3 SC
Solidago riddellii 5 FN
Solidago rigida 10 PWM
Solidago sciaphila 1 CLS
Solidago spathulata 1 DUN
Solidago speciosa 9 PM
Solidago uliginosa 5 NS
Solidago ulmifolia 12 SDM
Sonchus arvensis 1 BEA
Sonchus asper 1 BG
Sonchus oleraceus*
Sorbus americana 4 BF
Sorghastrum nutans 10 PDM
Sorghum halepense*
Sparganium americanum 1 AQE
Sparganium angustifolium 1 AQS
Sparganium eurycarpum 5 AQE
Sparganium fluctuans 1 AQS
Spartina pectinata 9 PW
Specularia perfoliata 6 SB
Spergula arvensis*
Sphenopholis intermedia 1 FN
Spiraea alba 16 AT
Spiraea tomentosa 8 NS
Spiranthes cernua 4 FN
Spiranthes gracilis 1 BG
Spiranthes ochraleuca Rydb. 3 PWM
Spiranthes romanzoffiana 1 CLS
Spirodela polyrhiza 1 AQE
Sporobolus asper 3 PD
Sporobolus cryptandrus 7 CG
Sporobolus heterolepis 9 PD
Sporobolus vaginiflorus 3 CG
Stachys hispida 2 SW
Stachys palustris 11 SS
Staphylea trifolia 6 SM
Steironema ciliatum 16 SWM
Steironema lanceolatum 7 SC
Steironema quadriflorum 9 FN
Stellaria aquatica 1 CLE
Stellaria calycantha 2 NWM
Stellaria graminea 2 BG
Stellaria longifolia 9 FN
Stipa spartea 9 PDM
Streptopus amplexifolius 3 BF
Streptopus roseus 8 BF
Strophostyles helvola 1 SS
Sullivantia renifolia 1 CLS
Symphoricarpos albus 2 SM
Symphoricarpos occidentalis 4 OO
Symplocarpus foetidus 7 SWM

Taenidia integerrima 6 SD
Talinum rugospermum 3 CLE
Tanacetum vulgare*
Taraxacum officinale*
Taxus canadensis 8 NWM
Tephrosia virginiana 7 OB
Teucrium canadense 5 SW
Thalictrum dasycarpum 19 FN
Thalictrum dioicum 16 SDM
Thaspium barbinode 1 BEA
Thaspium trifoliatum 6 PM
Thelypteris palustris 13 AT
Thelypteris phegopteris 8 NWM
Thlaspi arvense*
Thuja occidentalis 10 NWM
Tiarella cordifolia 1 BF
Tilia americana 13 SM
Tofieldia glutinosa 2 PW
Tradescantia ohiensis 15 CG
Tragopogon pratensis*
Triadenum virginicum 4 BOG
Trientalis borealis 15 BF
Trifolium arvense*
Trifolium hybridum 2 BG
Trifolium pratense 3 BG
Trifolium procumbens*
Trifolium repens 2 BG
Triglochin maritima 2 BOG
Triglochin palustris 1 BEA
Trillium cernuum 5 BF
Trillium gleasoni 8 SDM
Trillium grandiflorum 10 SM
Trillium recurvatum 4 SM
Triosteum perfoliatum 7 SDM
Triplasis purpurea*
Tsuga canadensis 7 NM
Typha angustifolia 1 AQE
Typha latifolia 10 AQE

Ulmus americana 13 SWM
Ulmus rubra 10 SM
Ulmus thomasi 4 SWM
Urtica dioica 3 SWM
Utricularia intermedia 1 AQS
Utricularia purpurea 1 AQS
Utricularia vulgaris 3 AQE
Uvularia grandiflora 12 SDM
Uvularia sessilifolia 8 NDM

Vaccinium angustifolium 16 ND
Vaccinium macrocarpon 2 BOG
Vaccinium myrtilloides 11 NW

Vaccinium oxycoccos 4 BOG
Valeriana ciliata 4 FN
Vallisneria americana 1 AQS
Verbascum thapsus 1 BG
Verbena bracteata*
Verbena hastata 7 NS
Verbena stricta 6 PD
Verbena urticifolia 3 SS
Vernonia fasciculata 5 PWM
Veronica anagallis-aquatica 1 AQE
Veronica peregrina*
Veronica scutellata 1 BF
Veronica serpyllifolia 2 DUN
Veronicastrum virginicum 18 PWM
Viburnum acerifolium 10 NDM
Viburnum lentago 9 SDM
Viburnum opulus 3 AT
Viburnum rafinesquianum 8 SDM
Vicia americana 11 PWM
Vicia angustifolia 4 PM
Vicia cracca*
Vicia villosa*
Viola adunca 8 BG
Viola canadensis 9 NM
Viola conspersa 7 BF
Viola cucullata 23 SDM
Viola incognita 13 BF
Viola lanceolata 3 NS
Viola pallens 8 NW
Viola pedata 14 CG
Viola pedatifida 9 PM
Viola pubescens 11 SM
Viola renifolia 1 BF
Viola sagittata 9 OB
Viola selkirkii 1 BF
Vitis aestivalis 6 SDM
Vitis riparia 22 SW

Walsteinia fragarioides 8 ND
Woodsia ilvensis 3 CLE
Woodsia obtusa 2 CLE

Xanthium strumarium 3 DUN

Zanthoxylum americanum 14 SDM
Zizania aquatica 2 AQS
Zizia aptera 5 PM
Zizia aurea 11 PW
Zosterella dubia 1 AQS
Zygadenus elegans 5 OO

Index

Abies balsamea: description, 247; germination, 247; seed crop, 247; invader of bracken-grassland, 254; pollen as indicator of climate, 443; behavior on community ordination, 498
Abundance test for aggregation, 280, 281
Acer rubrum: invasion of oak forest, 147–48; hybridization with *Acer saccharinum,* 160; autumnal coloration, 207; bimodal curve for, 207
Acer saccharinum, 168
Acer saccharum: introgression with *Acer saccharum nigrum,* 22; reproduction rate, 104; flowers, 105; seed crop, 105; germination, 105; mortality rates, 106; shade tolerance, 107; suppression, 107; fire resistance, 108; on southern forest ordination, 100–101; reaction to selective logging, 130; as invader of oak forest, 148; response to deer, 189; dominance potential, 199; on community ordination, 498
Acer saccharum nigrum: introgression with *Acer saccharum,* 22
Adaptation numbers: defined, 94; for northern forest trees, 179; for savanna trees, 326; for southern forest trees, 519
Adirondack Mountains, N.Y.: pine forests, 217; boreal forests, 225
Aeolian sediments, 42, 45
Aftonian interglacial period, 446–47
Aggregation: in southern mesic forest, 117; in southern lowland forest, 162; in northern mesic forest, 194; in pine forest, 209; in conifer swamps, 224, 234; in xeric prairie, 274–75; frequency test for, 275; density test for, 280, 281; abundance test for, 280, 281; in mesic prairie, 280–82; in lowland prairie, 286–87; in oak opening, 331; in sedge meadow, 372; in cedar glade, 347; in plant communities, 477. *See also* Antibiotics
Air masses, 33, 34
Alder thicket: composition, 355–56; soils of, 356; succession, 356–57; geographical relations, 357
Algae, blue-green, 309
Algonquin Park, Ontario: 255

Alleghenian floristic element, 8–17 *passim,* 290
Alleröd glaciation, 29, 438
Alnus rugosa, 356
Ambrosia artemisiifolia, 422
Ambrosia trifida, 462
Ammophila breviligulata, 407
Amplitude of tolerance curves: described, 96; for *Quercus borealis,* 96; for southern trees, 97; for northern trees, 180–82; bimodal, 182, 183, 207, 248–49, 252; for boreal trees, 248; for prairie plants, 267; for savanna trees, 328; for submerged aquatics, 398; two-dimensional, 484, 486
Anacharis canadensis, 395
Anderson, Orlin, 178, 269, 271, 275
Annual plants: in mesic forest, 112–13; in xeric forest, 142; in sand barrens, 312; on beaches, 404; in weed communities, 420, 422
Antennaria neglecta, 282
Antibiotics: of *Aster macrophyllus,* 142; of *Fagus grandifolia,* 191–92; in mesic prairie, 281; of *Hieracium aurantiacum,* 316; in bracken-grassland, 318; in aquatic communities, 394; in relation to species association, 477
Ants, 115, 279
Apple River Canyon, 67
Aquatics, emergent: composition, 388–90; structure, 391–93
Aquatics, submerged: composition, 396–98; soils, 398–99; plankton, 399–401; river communities, 401
Arboretum, University of Wisconsin: acorn crop, 138; microclimates, 150; recovery of grazed forest, 155; mortality of *Salix* and *Populus,* 160; plantation of *Pinus banksiana,* 207; length of growing season, 283; prairie reestablishment, 294, 295, 475; controlled fires on, 300–301, 337; Wingra fen, 362; exclosure of rabbits, 364; dates of frost, 373; nettles on old peat burns, 375; cattail introgression, 391–92; Faville Prairie Preserve of, 426; description of area, 475
Arceuthobium pusillum, 229

646 INDEX

Archaeological methods, 456–57
Archaic Indian period, 458
Archbald, D., 115
Arctotertiary flora, 9
Arena, town of, 310
Ash, black, 230–31
Ash, white, 499
Aspen, large-toothed, 147
Aspen, trembling. See *Populus tremuloides*
Aspen park land, 320
Aster ciliolatus, 316
Aster macrophyllus, 142
Aster sericeus, 271
Atlantic Costal Plain floristic element, 10, 11, 362
Autogenic processes, 164, 211, 386
Autotoxic action, 281
Aztalan, 460

Baraboo Bluffs, 33, 42
Barberry, 425
Barron Hills, 33, 42
Basswood. See *Tilia americana*
Beaches: of Lake Michigan, 404; of Lake Superior, 404; of inland lakes, 405
Beachgrass, 407
Beaver, 222, 236
Beech. See *Fagus grandifolia*
Beer, 242
Beggar-ticks, 372
Belmont Mound, 264
Berberis vulgaris, 425
Berries, 233
Besseya bullii, 332
Betula lutea: germination, 174–75, 186–87; range, 186; ecotypes, 186; hybridization, 186; shade tolerance, 187
Betula nigra, 168
Betula papyrifera, 208, 498
Bidens coronata, 372
Big Woods of Minnesota, 125
Bihemispheric plant distribution, 9
Billings, W. D., 255
Bimodal amplitude curves: of northern forest plants, 182, 183; of *Acer rubrum*, 207; of boreal forest plants, 248–49, 252
Birch, river, 168
Birch, white, 208, 498
Birch, yellow. See *Betula lutea*
Birds: of southern mesic forest, 118; of southern xeric forest, 144–45; of southern lowland forest, 163–64; of northern xeric forest, 210; of boreal forest, 250; of xeric prairies, 273; of lowland prairies, 286; of sand barrens, 310; of pine barrens, 344; of bogs, 380; of marshes, 392
Bison, 263
Blooming season of herbs: southern mesic forest, 112; southern xeric forest, 141; southern lowland forest, 161; northern mesic forest, 191; northern xeric forest, 208; northern lowland forest, 231–32; boreal forest, 250; xeric prairie, 270; mesic prairie, 278; lowland prairie, 285; sand barrens, 309; bracken-grassland, 316; pine barrens, 341; fen, 363; sedge meadow, 367, 373; bog, 379; weed communities, 422. See also Phenology
Blowout, 310–11
Blueberries, 316, 340–41
Blue Mound State Park, 27, 33
Bog: composition, 378–80; soils of, 381; microclimate, 381–83; succession, 383–84; geographical relations, 384; utilization, 384
Bog mat, 235, 381
Bog shrubs, 239, 379
Bond, R. R. 118, 144–45
Boraginaceae, 430
Borchert, J. R., 34, 35, 42
Boreal forest: structure and composition, 246–47; life-histories of dominants, 247–48; groundlayer, 249–50; microclimate of, 250–51; soils of, 252–53; succession, 253–54; geographical relations, 255–57
Botrychium multifidum, 508–9
Bottomland forests. See Lowland forest, southern
Boulder Junction, town of, 38
Bracken fern, 494
Bracken-grasslands: composition, 314–17; origin of, 317–18; microclimate, 317; soils of, 318–19; geographical relations, 319–21; utilization, 321
Brady's Bluff Scientific Area, 554, 575
Braun, E. Lucy, 127
Bray, J. R., 110, 327, 332, 345, 482–85
Brodhead, town of, 38
Brown, R. T., 178, 341, 343
Brown Forest Soils, 44, 124
Browntown Scientific Area, 526
Brule river, 172, 251, 342
Bruncken, E., 66, 405
Brunizem Soils, 44

INDEX 647

Brush prairie, 299–300
Burrowing animals, 292–93
Butler, J. E., 270

Cadiz Township, Green County, 467
Cain, S. A., 357, 364
Cakile edentula, 405
Calcification, 44
Calypso bulbosa, 232
Cambrian rocks, 26, 215
Carex, 366–67
Carex stricta, 369–70, 371
Carrying capacity, 472
Carya cordiformis, 109
Carya ovata, 140, 147, 498
Cary ice: as substage of Wisconsin glaciation, 28–29; striations by, 41; relation to loess, 45; in relation to forest beds, 438; readvances of, 448; relation to Indians, 458
Castle Mound Scientific Area, 539
Catastrophe, 198, 253, 304
Catena. See Soil catena
Cattail, 391–92
Ceanothus americanus, 336
Cedar, red, 346, 347–48
Cedar, white. See *Thuja occidentalis*
Cedarburg bog, 229
Cedarburg Swamp Scientific Area, 545, 584
Cedar glades, composition, 344–47; structure, 347; origin, 347–48; geographical relations, 350; recent expansion, 351
Chaerophyllum procumbens, 509
Chamberlin, T. C., 4, 59, 88–90, 175–76
Chara, 395
Cheney, L. S., 15
Chequamegon Bay, 248
Cherry, black, 139–40, 147
Chippewa National Forest, Minn., 217
Chi-square test, 71
Christensen, E. M., 178, 223, 234
Christensen, Martha, 195, 210
Chrysanthemum leucanthemum, 431
Clambering plants, 355
Clausen, J. Jones, 178, 223, 237, 245
Climate of floristic provinces, 35–37
Climax concept, 111, 129, 215
Clisere succession, 305
Clones, age of, 116
Cochrane ice, 30
Coefficient of community, 83
Coffeetree, Kentucky, 463
Combined ordinations, 98–100, 179–80

Common names of plants, 5
Compensating factors, 488
Competition, 477. See also Antibiotics
Compositae, 269, 496, 497
Compositional index: for southern forests, 95; for northern forests, 179; for prairie, 266; for savanna, 327; for aquatic communities, 389
Condensation, 276
Conductivity, 393–94, 398–99
Coneflower, purple, 278
Conservation goals, 472
Conservatism of prairie plants, 293–94
Continuum. See Vegetational continuum
Controlled fires, 300–301, 304, 344. See also Fire
Cooper, Fennimore, 329
Copeland, E. B., 466–67
Copper Falls State Park, 229, 409
Copper Range, 244, 251. See also Penokee Range
Costello, D. F., 67, 371
Cottam, Grant, 143, 150, 281, 315, 331, 353
Cotton grass, 379
Council Grounds Scientific Area, 541
Court Oreille, Lake, 172
Cox Hollow Scientific Area, 215, 539
Crex Meadows, 304, 377
Criteria for selection of stands: general, 70; southern forests, 92; prairie, 262; savanna, 330–31; weed communities, 416
Crystal Lake, 387, 395, 439
Culgerson, W. L., 195, 209–10
Cultivated habitats, 413, 416
Cumberland Mountains, Ky., 128
Cutover area, 219
Cypripedium: nomenclature, 79; germination, 223, 233
Cypripedium acaule, 379
Cypripedium candidum, 364
Cypripedium parviflorum, 355
Cypripedium pubescens, 141, 142
Cypripedium reginae, 233, 355

Dahl, E., 504
Decreasers, 427–28
Deer, white-tailed: effect on *Acer saccharum*, 189; effect on *Taxus*, 191; utilization of *Fraxinus nigra*, 231; winter yards, 234; in bracken-grassland, 316
Deforestation, 466–67

Degraded habitats, 413
Density: annual fluctuations, 161–62; of shrubs, 191, 194; of forest herbs, 191, 194, 209; of prairie forbs, 280; of sedge meadow plants, 369
Density test for aggregation, 280, 281
Depodzolization, 44, 45
Desert pavement, 310, 311
Desmarais, Y., 22
Devil's Lake Scientific Area, 528
Devil's Track Lake, 233
Dietz, R. A., 98
Differential species, 368, 506
Dix, R. L., 197, 427–29
Dodecatheon amethystinum, 411
Dominance potential, 51
Dominants, 51, 54, 94
Door County peninsula, 251
Drainage, 236, 241, 375, 377
Driftless Area: bogs in, 11, 448; endemic plants in, 13, 410; distance from ice front, 13; as *refugium,* 14, 153; age, 27; loess deposits, 31; microclimate in valleys in, 40; as site of northern forest relics, 185, 215; ecotypes of *Tsuga* in, 216; invasion of prairies in, 299; relation to *Juniperus horizontalis,* 346–47; dripping cliffs in, 410; post-Cary vegetation of, 448
Dripping cliffs, 410
Drosera species, 379–80
Dune communities, 406–8

Eagle River, town of, 173
Early Woodland Indian period, 458
Echinacea pallida, 278
Ecotypes: definition, 20; origin, 21; role in community studies, 22; relation to phenology, 39, 40; of Ozark floristic element, 153; of *Pinus banksiana,* 206–7; of *Tsuga,* 216; of *Tilia,* 345; of *Phalaris,* 426; of *Chrysanthemum,* 431; in origin of weeds, 433
Edge effect, 321, 467
Effigy Mound Indian Culture, 459
Eggler, W. H., 67, 177
Electronic computers, 143
Elk, 263
Ellisia nyctelea, 507
Elm, slippery, 110, 482
Endemic plants: of Driftless Area, 13; on lake dunes, 407; on cliffs, 410, 411
Ephemerals, spring: description, 112–13; modified types, 113; origin, 113; in northern mesic forest, 192–94; relation to humus type, 197; in closed forest pastures, 432
Epibiotic flora, 10
Ericaceae, 496
Eriophorum sp., 379
Eryngium yuccifolium, 294
Erythronium albidum, 116–17
Escarpment: Magnesian, 26, 362, 403; Niagara, 26, 33, 403, 409; Prairie du Chien, 26, 362, 403
Evaporation: in oak woods, 40, 41; in southern mesic forests, 123; in mesic prairie, 284; in sedge meadow, 373; in beech forest, 373
Evergreen species, 496

Factor analysis, 482
Fagus grandifolia: fossil pollen in Wisconsin, 11; range of, 184; seed crops, 189; germination, 189–90; shade tolerance, 190; root sprouts, 190; hybridization, 190; antibiotics in, 191–92
Fairy rings, 109
Fameflower, 312
Family relations, 496–97
Farmdale ice, 28, 215
Fassett, N. C., 8, 13, 23, 394–95, 401
Faville Prairie Scientific Area, 286, 306, 426, 560, 562, 581
Fen: relation to lowland prairie, 286; definition, 361; composition, 362–63; succession, 363–64; geographical relations, 364–65
Fern, sweet, 340, 341
Ferns, 410, 496
Fidelity, 505
Finley, R. W., 64
Finnerud, C. W., 232
Fir, balsam. See *Abies balsamea*
Fire: effect on southern xeric forest, 151; effect on southern lowland forest, 159, 165; effect on northern xeric forest, 212–13; cause of retrogression cycles, 213; effective control of, 220; effect on northern lowland forest, 236; as cause of prairie, 296–301; relation to Indian summer, 297; effect on oak opening, 334; effect on oak grubs, 336–37; effect on blueberries, 341; effect on pine barrens, 342–43; effect on fens, 363; effect on sedge meadows, 374–75; effect on bogs, 383; as Indian tool, 461–62; set by lightning, 461; elimination of,

465; in logging slash, 469. *See also* Controlled fires
Fire protection, 469–71
Fire resistance, 108, 331
Flambeau Forest Scientific Area, 174, 201, 214, 536
Floerkea proserpinacoides, 116
Flooding, 161
Floodplain forests, 156
Flora, 7, 24, 49
Floristic-characteristic species combination, 80
Floristic composition, 54
Floristic element: Alleghenian, 8–17 *passim*, 290; Arctic-Alpine, 10; Boreal, 8, 15; Coastal Plain, 10–11; 362; Ozarkian, 9, 11; Prairie, 9–15 *passim*; Preglacial, 10; Western Mountain, 10
Floristic province, conifer-hardwood: description, 15; climate of, 35–37; forests of, 171, 184. *See also* Tension Zone
Floristic province, prairie-forest: description, 15; climate of, 35–37; forests of, 87, 90–91, 103; tamarack swamps in, 254; shift of boundary during Xerothermic, 452. *See also* Tension Zone
Flowage, 401
Flowing wells, 288
Forb, 262
Fort Winnebago, 264, 297, 298
Fraxinus americana, 499
Fraxinus nigra, 230–31
Frequency: meaning, 76; use in prairies, 273; relation to density and presence, 273; as probability value, 274; test for aggregation, 275, 280, 281
Frequency distribution, 505, 507
Frequency multiplication sign presence index, 81, 269, 274, 286
Frost pockets, 317–18
Frost resistance: of prairie plants, 283; of bracken, 316; of aspen, 318
Fuller, A. M., 233
Fungi, filamentous soil: in southern mesic forest, 118; in southern xeric forest, 144; in northern mesic forest, 195; in northern xeric forest; 210; in prairies, 272, 279, 286
Fungi, fleshy: in southern mesic forest, 118; in southern lowland forest, 163; in northern mesic forest, 309; in sand barrens, 309
Fungi, mycorrhizal, 233, 301

Gap phase concept: in southern forest, 110, 147; in northern forest, 198; in prairies, 293
Gap phase trees, 147
Gentiana sp., 283
Gentiana puberula, 271
Gentiana quinquefolia, 271
Geographical relations of communities: southern forests, 125–30, 140–45, 152–53, 167; northern forests, 198–200, 216–18, 239–41; boreal forest, 255–57; prairies, 288–92; sand barrens, 313; bracken-grassland, 319–21; savannas, 328–50; tall shrubs, 357; sedge meadows, 376; bog, 384; cliffs, 411
Geological formations, 25–26
Germination: *Acer saccharum*, 105; *Quercus* sp., 138–39; *Betula lutea*, 174–75; *Tsuga canadensis*, 187–88; *Fagus grandifolia*, 189–90; *Pinus strobus*, 205; *Cypripedium* sp., 223, 233; *Thuja occidentalis*, 229; *Abies balsamea*, 247
Gilbert, Margaret L., 92
Glacial drumlins, 375
Glacial forest, 448
Glacial Lake: Chippewa, 11, 30; Agassiz, 16; Oshkosh, 27, 29, 438, 450; Wisconsin, 27, 29, 222; Pecatonica, 28; Chicago, 29, 282 449; Algonquin, 30, 450, 452; Nippissing, 30; Barrens, 326, 340, 403
Gleason, H. A.: nomenclature of, 78–79; individualistic concept, 199, 504, 510; on stump prairies, 319; community variation with distance, 510; recommendation of behavioral approach, 511
Gley layer of soil, 166, 287, 350
Goder, H. A., 187–88
Goodall, D. W., 504
Goodyera pubescens, 141, 142
Gopher, pocket, 316
Grandfather Falls, 174
Grand Marsh, 387
Grant County, 16–17
Grasses, cool-season, 316
Grass pollen, 444
Grass rugs, 377
Gray-Brown Podzolic soils, 44, 124, 151, 196
Gray Wooded soils, 252
Grazing effects: on mesic forest, 131; on xeric forest, 154; on prairie, 307, 427; as disturbance factor, 425
Great Smoky Mountains, Tenn.: mesic

forest, 128; radioactivity of hemlock litter, 197; relation of forests to soil moisture, 253; boreal forests of, 255; vegetational continuum on, 480
Green, Phoebe, 280
Greene, H. C., 65, 195, 264, 272, 309, 353, 362
Green Lake, 394, 395
Grood soils, 151
Groundlayer species, 502-3
Ground water conductivity: in alder thicket, 356; in sedge meadows, 372, 374; in open bogs, 381
Growing season, 283
Grub: on brush prairie, 300; origin of word, 336; response to fire, 336-37; production by oaks, 337
Gymnocladus dioica, 463

Habeck, J. L., 229-30, 315
Hale, M. E., 118, 144
Hamamelis virginiana, 143-44
Hansen, H. P., 448
Hanson, H. C., 504
Hasler, A. D., 387
Hemlock. See *Tsuga canadensis*
Hickory, shagbark, 140, 147, 498
Hickory, yellowbud, 109
High Lake Spruce-Fir Scientific Area, 551
Homogeneity, 70, 71, 80
Honeysuckle, Tartarian, 417
Hope Lake bog, 441
Hopewell Indian culture, 459
Hopkins' Law, 38, 39, 40
Horicon Marsh, 387
Hoyt, J. W., 59
Hudson, town of, 38
Hudsonia tomentosa, 311, 407
Hummocks, 372
Humus: mor, 151, 196; mull, 151, 196, 197
Hybridization, See Introgressive hybridization

Illinoian glaciation, 28, 447
Impatiens pallida, 122
Importance value, 74
Increasers, 426-28
Index of distinctness, 82, 517
Index of diversity, 80, 517
Index of homogeneity, 80, 478, 517
Index of interspecific association, 397, 418, 481
Index of joint occurrence, 397, 418, 481

Index of similarity: defined, 83; in mesic forest, 119; between related stands, 478, 479; use in upland forest ordination, 483; adjustment of values, 485; use in community ordination, 487; in relation to classification schemes, 504
Indian, prehistoric culture periods: Paleo-Indian, 457-58; Archaic, 458; Early Woodland, 458; Middle Woodland, 458-59; Mississippi, 459-60
Indian, prehistoric cultures: Old Copper Culture, 458; Hopewell, 459; Effigy Mound, 459; Middle Mississippi, 459-60; Upper Mississippi, 459
Indians: early gardens, 172, 462; use of maple sugar, 172; use of fire, 341, 461-62, 463-64; signal fires, 348; agricultural influences, 462; use of ragweed, 462; use of coffeetree seeds, 463; plant introduction by, 463
Indian summer, 297
Indian tribes: Winnebago, 460; Menominee, 460; Santee Dakota, 460
Indicator plants: compared to modal species, 81; relation to breadth of amplitude, 183; of prairie soil catena, 256; of sandy prairies, 272; of sandy oak barrens, 339; of sandy pine barrens, 341; of submerged aquatic communities, 397; of weed communities, 423; of pastures, 430-31; of climate, 443
Individualistic concept, 51, 504, 510
Infiltration rates, 154
Insectivorous plants, 379
Introgressive hybridization: relation to ecotypes, 21; between *Acer* species, 22, 160; between *Quercus* species, 134, 331, 338; in linear forests along rivers, 160; in *Fagus*, 190
Invaders, 427-28, 430, 431-32
Iris lacustris, 408
Isle Royale, Mich., 255

Jackson, H. H. T., 32, 66
Jerome Bog, N. C., 384
Johnstown Moraine, 448
Joint occurrences, 479
Jones, J. J. *See* Clausen, J. J.
Jordan township, Green County, 466
Juglans nigra, 140
Juniperus horizontalis, 346, 407
Juniperus virginiana, 346, 347-48

Kettle Moraine, 29

INDEX 651

Key to Wisconsin communities, 56–59
King, F. L., 174
Kittredge, Joseph, 67
Knapp, Judge, 59

Lac du Flambeau, 173
Lacustrine forests, 156
Laing, H. E., 391
Lake Forest formation, 218
Lakes and Streams Committee (of U.W.), 387
Lapham, Increase, 23, 196
Larch, sawfly, 227
Larix laricina: life-history, 226–27; as upland tree, 245; possible ecotypes, 245; in southern swamps, 254; conversion of *Larix* forest to meadow, 354
Larsen, J. A., 147–48
Late Woodland Indian culture, 459
Layer societies, 194
Leading dominants, 93, 199
Leguminosae, 17
Leopold, Aldo, 474
Lianas: in southern mesic forest, 117–18; in lowland forest, 162–63; in shrub carr, 354
Lichen, communities of: southern mesic forest, 118; southern xeric forest, 144; northern mesic forest, 195; northern xeric forest, 209–10; *Pinus banksiana* relics, 216; northern lowland forest, 234; sand barrens, 309–10
Life-form classification, 394–95
Life-histories of trees of; southern mesic forest, 104–11; southern xeric forests, 134–40; southern lowland forests, 159–61; northern mesic forest, 186–90; northern xeric forests, 204–8; northern lowland forests, 225–31; boreal forest, 247–48
Light: in southern mesic forest, 122–23; amount needed for oak regeneration, 146; in prairies, 278; penetration in lake water, 395–96; average intensity on community ordination, 494
Limnology, 387
Lindsay, D. R., 419
Little Fox River, 263
Little John Lake, 394
Lizard, six-lined, 310
Lodi, town of, 351
Loeffler, R. J., 400
Loess, 45–46
London Marsh, 354

Lone Rock, town of, 166
Lonicera tatarica, 417
Lowland forest, northern: composition and structure, 223–25; life-histories of trees, 225–31; succession, 235–37; soils of, 237–38; microclimate of, 238–39; geographical relations, 239–41; utilization, 241–42
Lowland forest, southern: composition and structure, 157–61; groundlayer, 161–64; succession, 164–65; microclimate of, 165–66; soils of, 166–67; utilization, 168
Lumbering: influence on xeric forests, 155; Wisconsin history, 218–19, 468–69

Macrofossile, 437–38
Maianthemum canadense, 183
Maiden Rock, 296
Malaspina Glacier, 32
Mammals of: southern mesic forest, 118; southern xeric forest, 145; northern xeric forest, 210; boreal forest, 250; prairies, 273, 279–80; bracken-grassland, 316; bog, 380; marshes, 392. *See also* Beaver; Bison; Deer; Elk; Gopher; Muskrat; Rabbit
Mammoth Cave National Park, Ky., 128, 52
Manitou Falls, 409
Manitowish River, 173
Mankato ice, 29, 438
Maple, red. See *Acer rubrum*
Maple, silver, 168
Maple, sugar. See *Acer saccharum*
Maritime plants, 408
Marl, 395
Marshall, Ruth, 66
Matuszkiewicz, W., 480
Mauthe Lake Scientific Area, 522
Maycock, P. F., 245, 485
McIntosh, R. P., 148, 215, 481–82
McMillan, C., 39
Melanized sands, 318–19
Mendota, Lake, 396, 398, 400
Menominee Indian Reservation, 212, 214, 461
Menominee Indian tribe, 460
Mesic forest, northern: composition and structure, 185–86; life-histories, 186–90; groundlayer, 190–95; microclimate of, 195–96; soils of, 196–97; stability of, 197–98; geographical distribution, 198–200; utilization, 200–201

652 INDEX

Mesic forest, southern: composition and structure, 104–11; groundlayer, 111–18; regional variability, 118–21
Methods: point samples, 69; test for homogeneity, 70–71; quadrat, 72; random pairs, 72; quarter, 72–73; soil tests, 73; importance values, 74; electronic computers, 74; forest structure, 75; reliability, 75–76; summation of data, 79–82; use of weighted index values, 266; use of frequency, 273–74, 275; paired stand technique, 427; construction of ordination, 487–88
Miami soil, 46
Michigan, Lake, 404
Microclimate of: plant communities, 40–41; southern mesic forest, 122–23; southern xeric forest, 150; northern mesic forest, 195–96; northern xeric forest, 210–11; northern lowland forest, 238–39; boreal forest, 250–51; prairies, 276, 283–84, 287; sand barrens, 311–12; bracken-grasslands, 317; sedge meadow, 372; open bogs, 381–83
Microfossils, 438–39
Middle Mississippi Indian culture, 459, 460
Middle Woodland Indian period, 459
Milkwort, sand, 312
Miller, Bonita. See Neiland, B. M.
Milton Moraine, 448
Mississippi Indian period, 459–60
Mistletoe, dwarf, 229
Modal species, 81
Montreal river, 172, 177
Moquah Scientific Area, 573
Mortality rates: of *Acer saccharum*, 106–7; of *Salix nigra* and *Populus deltoides*, 160; of *Populus tremuloides*, 207
Mosses, 234, 386
Mountain-ash, 248
Mowing meadows: on prairies, 306; origin from tamarack swamps, 354; on fens, 363; composition, 432
Muhlenbergia cuspidata, 272
Muir, John, 335, 336, 345
Multiple factor causation, 382
Muscoda, town of, 310
Muskeg, 224
Muskrat, 392
Myrica asplenifolia, 340, 341

Najadaceae, 339
Namekagon River, 172

Natelson, Delle. See Swindale, D. N.
Naturalized species, 417, 467–68
Necedah Scientific Area, 539
Neiland, Bonita M., 307, 427
Nelson Dewey Scientific Area, 554, 575
Nettles, 375
Niagara escarpment, 26, 33, 403, 409
Nitrogen-fixation, 356
North Haven Sand Plains, Conn., 313
Norwood, J. G., 172–74
Noxious plants, 415
Nutrient pumping: by trees, 43, 108, 125; by grasses, 44

Oak, black, 135, 337
Oak, bur. See *Quercus macrocarpa*
Oak, chinquapin, 136
Oak, Hill's 135, 153
Oak, red. See *Quercus borealis*
Oak, swamp white, 137, 159
Oak, white. See *Quercus alba*
Oak barrens: description, 326; composition, 328–39, origin of, 339
Oak openings: description, 326; composition, 328–32; structure, 332–34; origin, 334–38; geographical relations, 348–49
Oak species. See *Quercus* sp.
Oak wilt, 146
Observation Hill, 42, 345
Observatory Woods Scientific Area, 526, 569
Oconto, town of, 458
Old Copper Indian culture, 458
Oosting, H. J., 255
Opuntia compressa, 312
Orchidaceae: species ranges, 17; in southern xeric forest, 141; in northern lowland forest, 232–33; germination, 233; in bogs, 379
Ordination: combined upland and lowland, 98–100; objective segmentation of, 102; multidimensional, 482
Ordovician rocks, 215
Orientation, 488, 489
Orpurt, P. A., 272, 279, 286
Osceola Landing, town of, 458
Owen, D. D., 172
Oxidation-reduction potential of ground water in: alder thicket, 356; sedge meadow, 373, 374; open bog, 381
Ozark Mountains: mesic forests, 128; xeric forests, 152–53; cedar glades, 350

Paired stand method, 427

Paleo-Indian period, 457–58
Partch, M. L., 268, 369
Parthenocissus vitacea, 488, 505
Pasture communities: of prairies, 425–29; of cleared forest land, 429–31; of closed forests, 431–32
Pattison State Park, 409
Peat, 238, 381
Pelican Lake, 39
Peninsula Park Beech Scientific Area, 536
Peninsula State Park, 408
Penokee Range, 33, 196, 244, 409
Pepin, Lake, 296
Perched bog, 362
Periglacial climates, 12, 346
Perrot, Nicolas, 221
Pestigo fire, 462
Phalaris arundinacea, 426
Phenology: definition, 37; relation to folklore, 38; of prairie plants, 270. *See also* Blooming season
Phragmites communis, 392
Physiognomy, 54
Physiographic changes, 164, 165
Phytoplankton, 399–401
Phytosociology, 511
Picea glauca, 247
Picea mariana, 227–29
Picea species, 443
Pine, jack. *See Pinus banksiana*
Pine, red, 205
Pine relics, 215–16
Pine, white. *See Pinus strobus*
Pine barrens: in pre-settlement times, 326; composition, 339–41; structure, 341; origin of, 342–44; geographical relations, 349–50
Pine Barrens of New Jersey, 349–50
Pinus banksiana: seed yield, 206; serotinal cones, 206; on pine relics, 215, 216; in bogs, 223–24
Pinus resinosa, 205
Pinus species, 443
Pinus strobus: dominance, 204; germination, 205; seed crop, 205; in swamp forests, 205; in pine relics, 215; on community ordination, 499
Pitcher-plant, 379
Plant communities: defined, 49, 50, 511; origin of, 50; nature of boundaries, 52
Plant introduction: by Indians, 463; by white man, 467–68
Platte Mounds, 33

Plum Lake Scientific Area, 536
Plum Lake township, Vilas County, 470, 471
Podzol, 43–44, 46, 196, 211, 252
Podzolization, 43, 44, 48
Point methods, 72, 73
Poison ivy, 163
Poison sumac, 233–34
Pollen analysis: effect of differential production and preservation, 440, 441, 442; surface accumulation, 441
Polygala paucifolia, 408
Polygala polygama, 312
Polytrichum piliferum, 309
Pomeranian glaciation, 28, 29
Pomme de prairie, 272, 463
Populus grandidentata, 147
Populus tremuloides: growth rate, 207; mortality rate, 207; shade tolerance, 207; vegetative reproduction, 207–8
Portage, town of, 263
Potentilla anserina, 407
Powell Marsh, 236, 383
Powers Bluff, 33
Prairie: soils of, 44, 124; rarity of remnants, 277; standing crop, 277; height of grasses, 279; expansion of, 290; on hills in Illinois, 291; stump, 317, 319; origin of, 296–305; pastures on, 425–29
Prairie chicken, 344
Prairie du Chien, town of, 263
Prairie du Chien escarpment, 26, 362, 403
Prairies, lowland: composition, 284–86; structure, 286–87; microclimate, 287; soils of, 287–88; geographical relations, 292; origin of, 305; utilization of, 306
Prairies, mesic: composition, 276–80; structure, 280–82; soils of, 282–83; microclimate, 283–84; geographical relations, 291; origin of, 304
Prairies, xeric: composition, 268–73; structure, 273–75; soils of, 275–76; microclimate of, 276; geographical relations, 291; secondary succession on, 294–95; origin of, 305; utilization of, 307
Preliminary Reports on the Flora of Wisconsin, 15, 23
Presence, 273–74, 505, 507
Prevalent species, 79–80, 81, 502
Prickly pear, 312
Primula mistassinica, 408
Pro-biotic substances, 142–43

Prunus serotina, 139–40, 147
Psoralea esculenta, 272, 463
Pteridium aquilinum, 494
Public Hunting and Fishing Grounds, 358

Quarter method, 72, 73
Quartzite, 41, 42
Quercus alba: description, 136–37; bark fungus on, 137; vivipary of acorns, 139; hybrids with *Quercus macrocarpa,* 331; on community ordination, 499
Quercus bicolor, 137, 159
Quercus borealis: amplitude of tolerance curve, 96; high importance values, 134; description, 135; origin in stands by retrogression, 149; seedlings in swamps, 224; disassociation with *Ulmus rubra,* 482
Quercus ellipsoidalis, 14, 135, 153
Quercus macrocarpa: races in, 136; description, 136; growth rate, 139; shape of crowns, 330; hybrids with *Quercus alba,* 331; grub production, 337; reaction to overtopping, 337
Quercus muhlenbergii Engelm., 136
Quercus species: variation in seed yield, 138; germination, 138–39; persistence of leaves, 139; light needed for regeneration, 146; nutrient content of leaves, 151; environmental tolerance, 442; value of fossil pollen as indicator, 443
Quercus velutina, 135, 337

Rabbit, 364
Radioactive carbon, 439–40
Radioactivity: in mesic forest litter, 125; of lowland soils, 166; in hemlock soils, 197; in northern lowland forest, 238; correlation with other environmental factors, 495
Ragweed, 422, 462
Railroad right-of-way, 306
Railroad yards, 421
Rainfall interception, 210
Randall, W. E., 114–15, 141, 194
Random pairs method, 72, 73
Rare plants, 506–9
Raritan River, N.J., 126, 167
Rattlesnake master, 294
Red cedar, 347
Red Cliff Indian Reservation, 245
Redroot, 336

Reference areas, 473–74
Refugium, 14
Reptiles, 310
Retrogression: in origin of red oak stands, 149; in northern xeric forest, 212–13; in conifer swamps, 236; of forests in postglacial time, 455
Rhododendron lapponicum, 411
Rhus radicans, 310
Rhus vernix, 233–34
Rib Mountain, 33, 42
Ridges Wild Flower Sanctuary, 408
Rock Lake, 345
Ruderal habitats, 413, 418

St. Croix, Lake, 173
St. Croix River, 251
Sand barrens: composition, 309; sand blows in, 310–11; microclimate, 311–12; soils, 312; geographical relations, 313; utilization, 314
Sand blows, 310–11
Sand Hills of Nebraska, 313, 376
Sand prairies, 272
Sangamon interglacial period, 28, 447
Santee Dakota Indian tribe, 460
Saprophytic angiosperms, 191
Sarracenia purpurea, 379
Savanna, 325
Savanna, tropical, 350
Scientific Areas: Wyalusing, 88–89, 522, 528, 532, 533, 575; Flambeau Forest, 174, 201, 214, 536; Cox Hollow, 215, 539; Faville Prairie, 286, 306, 426, 560, 562, 581; State Board for Preservation of, 474; Mauthe Lake, 522; Browntown, 526; Observatory Woods, 526; 569; Devil's Lake, 528; Peninsula Park Beech, 536; Plum Lake, 536; Castle Mound, 539; Necedah, 539; Council Grounds, 541; Cedarburg Swamp, 545, 584; Trout Lake Cedar Swamp, 547; High Lake Spruce-Fir, 551; Brady's Bluff, 554, 575; Nelson Dewey, 554, 575; Scuppernong, 562, 569, 581; Moquah, 573
Scirpus americanus, 392
Sclerophylls, 312
Scuppernong Marsh, 354
Scuppernong Scientific Area, 562, 569, 581
Sea rocket, 405
Sears, P. B., 450
Secondary succession: in prairies, 294–95; in weed communities, 413–14

Sedge bog, 236, 383
Sedge meadows, northern: composition, 373–74; soils of, 374; geographical relations, 376; utilization, 377
Sedge meadows, southern: composition and structure, 367–72; soils of, 372–73; microclimate, 373; succession, 374–76; geographical relations, 376; utilization, 376–77
Seed crops: *Acer saccharum*, 105; mesic forest herbs, 115; *Quercus* sp., 138; *Fagus*, 189; *Pinus strobus*, 205; *Pinus banksiana*, 206; *Thuja*, 229; *Abies*, 247
Selective logging, 200, 201, 424
Selleck, G. W., 284
Serotinal cones, 206
Shade plants, 113–14, 122
Shade tolerance: *Acer saccharum*, 107; *Prunus serotina*, 139–40; *Carya ovata*, 140; *Quercus* species, 146; *Betula lutea*, 187; *Tsuga canadensis*, 188–89; *Fagus grandifolia*, 190; *Populus tremuloides*, 207
Sharp-tailed grouse, 344
Shooting star, 270–71, 411
Shrub border, 349
Shrub-carr: composition, 353–54; succession, 354–55; geographical relations, 357; utilization; 357–58; origin, 375
Shrubs: of southern mesic forest, 117-18; density in southern xeric forest, 143; in southern lowland forest, 163; density in northern mesic forest, 191; in northern xeric forest, 209; in northern lowland forest, 233; in boreal forest, 250; in bracken-grasslands, 316; in savannas, 332, 339, 340, 346; on lake dunes, 407; on community ordination, 496. *See also* Bog shrubs; Shrub-carr; Alder thicket
Silphium laciniatum, 426
Silphium terebinthinaceum, 426
Silver Lake, 394
Single factor causation, 291
Sinsinawa Mound, 33
Skokie Marsh, Ill., 376
Snowfall, 474
Sodon Lake, Mich., 364
Soil: parent materials, 41–42; building processes, 42–44; major groups in Wisconsin, 46–48; of southern forests, 123–25, 150–51, 166–67; of northern forests, 196–97, 211, 237–38; of boreal forest, 252–53; of prairies, 275–76, 282–83, 287–88; of sand barrens, 312; of bracken-grasslands, 318–19; of alder thicket, 356; of sedge meadows, 372–73, 374; of open bogs, 381; of submerged aquatic communities, 398–99; weed indicators of, 423–24; analysis summaries, 518
Soil acidity, 494–95
Soil analyses, 73
Soil catena: definition, 46; relation to environmental gradients, 47–48; relation to prairies, 265; relation to sedge meadow, 365
Soil moisture, 252–53
Soil temperature, 123, 252–53, 311, 382
Spatial arrangement of plants. *See* Aggregation
Species association, 477
Species density: defined, 79; in northern forests, 209; of Wisconsin Communities, 478, 517
Species pairs, 192–93
Sphagnum, 239, 382, 384
Spruce, black, 227–29
Spruce, white, 247
Spruce budworm, 247
Standing crop: of mesic prairie, 277; of southern sedge meadow, 369; of open bog, 381; in lakes, 394, 396
Stand variability, 119
State Board for Preservation of Scientific Areas, 215, 474
State parks: Blue Mound, 27, 33; Wyalusing, 88, 89, 410; Governor Dodge, 215, 539; Copper Falls, 229, 409; Terry Andrae, 406; Penisula, 408; Pattison, 409; Devil's Lake, 528; Perrot, 528; Merrick, 532; Shot Tower, 532, 533; Potawatomi, 541; Interstate, 554; Wildcat Mountain, 575
Stearns, F. W., 178, 189, 191, 214–15
Stewart's woods, 134
Stonewort, 395
Stout, A. B., 67, 327, 368–71
Stratification, 457
Strong, Moses, 59
Struik, Gwendolyn, 115, 117
Sturgeon Bay, town of, 38
Succession in: southern xeric forest, 145–49; southern lowland forest, 164–65; northern xeric forest, 211–15; northern lowland forest, 235–37; boreal forest, 253–54; prairies, 292–95; shrub-carr, 354–55; alder thicket, 356–57; fens, 363–64; sedge meadows, 374–76; open bogs, 383–84

Succession, internal, 292
Succession, clisere, 305
Succulent plants, 312, 411
Sugar, maple: production in southern mesic forest, 131; from *Acer saccharinum*, 168; use by Indians, 172; production in northern mesic forest, 201
Sugar River bottoms, 158
Sundew, 379–80
Sun flecks, 122
Superior, Lake: climate, 244, 250–51; soils near shore of, 252; beaches of, 404; heath on dunes of, 407
Surveyors' records: use in vegetation mapping, 60, 63, 64; for study of vegetational change, 65; use for determining species association of trees, 90; relation to time of settlement, 176–77; density of savanna trees from, 331, 341; relation to Indian influences, 464–65
Swindale, Delle Natelson, 396, 596
Symbiotic fungi, 301

Talinum rugospermum, 312
Tamarack. See *Larix laricina*
Tamarack casebearer, 227
Taxus canadensis, 191
Tazewell-Cary interstadial period, 447
Tazewell ice, 28
Temperature, 150, 494. See also Microclimate; Soil temperature
Tension zone: location, 15; for animals, 16; climatic correlations, 35–37; relation to pine forest, 203; relation to weeds, 419; location during Xerothermic, 451, 452; in Michigan, 452; in New England, 452; postglacial shifts of, 455; on community ordination, 492, 494
Terry Andrae State Park, 406
Three-birds orchid, 114, 116, 507
Thuja occidentalis: seed crops, 229; germination, 229; vegetative reproduction, 230; in deer yards, 234; invasion of alder thicket, 236; utilization of, 242
Tilia americana: fairy rings in, 109; ecotypes, 345; on community ordination, 499
Tip-up mounds, 187–88
Touch-me-not, 122
Transect, 220
Tresner, H. D., 118, 144
Triphora trianthophora, 114, 116, 507
Trout Lake, 173, 395

Trout Lake Biological Station, 68, 69
Trout Lake Cedar Swamp Scientific Area, 547
Troutlily, 116–17
True prairie, 300
Tsuga canadensis: fossil pollen 11, 442; range in Wisconsin, 184–85; germination, 187–88; shade tolerance, 188–89; response to deer, 189; acidity of litter, 189; radioactivity of litter, 197; utilization, 200
Turions, 396
Tussock meadow, 371
Two Creeks forest bed, 437–38, 440, 449
Two Creeks interstadial period, 29, 440
Typha angustifolia, 391–92
Typha latifolia, 391–92
Type conversion, 155

Ulmus rubra, 110, 482
Undesirable plants, 415
Unions, 194
Upper Mississippi Indian culture, 459–60
Urtica dioica, 375
Utilization of communities: southern forests, 130–31, 153–55, 168; northern forests, 200–201, 218–20, 241–42; boreal forest, 257; prairies, 306–7; sand barrens, 314; savannas, 351; tall shrubs, 357–58; sedge meadows, 376–77; open bog, 384
Utricularia species, 380

Valders ice, 29, 439, 449
Van Arsdel, E. P., 40
Vegetation, 7, 49
Vegetational continuum: relation to soil catena, 47; definition, 52; segmentation, 55; in southern upland forest, 97; in entire southern forest, 101–2; in oak forests of Michigan, 152; in prairies, 268; in Austria, 480; in Great Smoky Mountains, 480; in Japan, 480–81; in tropical forest, 481; as one-dimensional image, 481; relation to local boundaries, 509–10
Vegetation maps, 59–62
Vegetative reproduction: in southern forest trees, 109, 110; in southern forest herbs, 116, 142; in *Fagus*, 190; in *Populus*, 208
Vitality, 148
Vivipary, 139
Voucher specimen, 77

INDEX 657

Wagner, B. G., 280
Walnut, black, 140
Ward, R. T., 64, 178
Ware, G. H., 158, 160–61, 164
Washburn Hills, 33
Water hardness, 389–90, 393–94
Water retaining capacity: determination of, 73; of conifer swamp soils, 238; of prairie soils, 283; of sedge meadow soils, 372; in relation to community ordination, 494
Weeds: in prairies, 293; in bracken-grasslands, 315–16, 320; on beaches, 405; definitions, 414–16; agricultural, 415, 417–18; obligate, 416; biological characteristics of, 416; naturalized, 417; of railroad yards, 421–22; from western United States, 422; nitrophilous species, 423; of forests, 424–25; source of native species, 433; in Indiana garden beds, 462
Weighted averages, 95
White Mountains, N.H., 255
Whitford, P. B., 116, 280
Whittaker, R. H., 253, 480
Wild flower display: in southern mesic forest, 131; in prairies, 270–71; 278–79, 285; in Door County, 408
Wind, 335
Wind throw, 198, 213, 214
Winnebago Indian tribe, 460
Wisconsin Conservation Department, 388

Wisconsin Dells, 411
Wisconsin Forest Inventory, 66
Wisconsin glaciation, 28–30
Wisconsin Land Economic Inventory, 65–66, 153, 220
Witchhazel, 143–44
Wolfe, J. N., 40
Wolf trees, 337
Woodbine, 488, 505
Wooster, L. C., 59, 174
Wyalusing Scientific Area, 88, 89, 522, 528, 532, 533, 575
Wyalusing State Park, 88, 89, 410, 411

Xeric forest, southern: structure, 133–34; life-histories of dominants, 134–40; groundlayer, 140–45; succession, 145–49
Xeric forests, northern: structure, 203–4; life histories of dominants, 204–8; groundlayer, 208–10; microclimate, 210–11; soils, 211; succession, 211–15; nature of pine relics, 215–16; geographical relations, 216–18; utilization, 218–20
Xerothermic period, 450–52

Yarmouth interglacial period, 447
Yew, Canadian, 191

Zicker, Wilma, 245, 375
Zimmerman, F. R., 388–89
Zonation, 235, 393